T0178661

Lecture Notes in Computer Science 14292

Advanced Research in Computing and Software Science
Subline of Lecture Notes in Computer Science

More information about this series at https://link.springer.com/bookseries/558

Henning Fernau · Klaus Jansen
Editors

Fundamentals
of Computation Theory

24th International Symposium, FCT 2023
Trier, Germany, September 18–21, 2023
Proceedings

 Springer

Editors
Henning Fernau (iD)
University of Trier
Trier, Germany

Klaus Jansen (iD)
University of Kiel
Kiel, Germany

ISSN 0302-9743   ISSN 1611-3349 (electronic)
Lecture Notes in Computer Science
ISBN 978-3-031-43586-7   ISBN 978-3-031-43587-4 (eBook)
https://doi.org/10.1007/978-3-031-43587-4

This Springer imprint is published by the registered company Springer Nature Switzerland AG
The registered company address is: Gewerbestrasse 11, 6330 Cham, Switzerland

Paper in this product is recyclable.

Preface

A Good Tradition

The 24th International Symposium on Fundamentals of Computation Theory (FCT 2023) was hosted by the University of Trier, Germany, during September 18th – 21st, 2023. The conference series "Fundamentals of Computation Theory" (FCT) was established in 1977 for researchers who are interested in all aspects of theoretical computer science and in particular algorithms, complexity, and formal and logical methods. FCT is a biennial conference. Previous symposia have been held in Poznań (Poland, 1977), Wendisch-Rietz (Germany, 1979), Szeged (Hungary, 1981), Borgholm (Sweden, 1983), Cottbus (Germany, 1985), Kazan (Russia, 1987), Szeged (Hungary, 1989), Gosen-Berlin (Germany, 1991), again Szeged (Hungary, 1993), Dresden (Germany, 1995), Kraków (Poland, 1997), Iaşi (Romania, 1999), Riga (Latvia, 2001), Malmö (Sweden, 2003), Lübeck (Germany, 2005), Budapest (Hungary, 2007), Wrocław (Poland, 2009), Oslo (Norway, 2011), Liverpool (UK, 2013), Gdańsk (Poland, 2015), Bordeaux (France, 2017), Copenhagen (Denmark, 2019), and Athens (Greece, 2021).

The Newest Edition: Trier 2023

The Program Committee of FCT 2023 (with two PC Chairs) included 33 scientists from 18 countries and was chaired by Henning Fernau (University of Trier, Germany) and by Klaus Jansen (University of Kiel, Germany).

This volume contains the accepted papers of FCT 2023. We received 79 abstract submissions in total. Each paper was reviewed by at least three PC members. As a result, the PC selected 30 papers for presentation to the conference and publication in these proceedings, evaluated based on quality, originality, and relevance to the symposium. The reviewing process was run using the EquinOCS conference system offered by Springer.

Highlights of the Conference

As the PC Chairs decided not to break ties of the voting, the Program Committee selected two papers to receive the Best Paper Award and one for the Best Student Paper Award, respectively. These awards were sponsored by Springer. The awardees are:

- Best Paper Award: Johanna Björklund. The Impact of State Merging on Predictive Accuracy in Probabilistic Tree Automata: Dietze's Conjecture Revisited.
- Best Paper Award: Michael Levet, Joshua A. Grochow. On the Parallel Complexity of Group Isomorphism via Weisfeiler–Leman.
- Best Student Paper Award: Zohair Raza Hassan, Edith Hemaspaandra, Stanisław Radziszowski. The Complexity of $((P_k, P_\ell)$-Arrowing.

The conference audience enjoyed four invited talks, given below in alphabetical order of the speakers:

- Karl Bringmann, MPI Saarbrücken, Germany:
  Fine-Grained Complexity of Knapsack Problems
- Stefan Glock, Universität Passau, Germany:
  Hamilton Cycles in Pseudorandom Graphs
- Lila Kari, University of Waterloo, Canada:
  Word Frequency and Machine Learning for Biodiversity Informatics
- Claire Mathieu, Univ. Sorbonne Paris, CNRS, France:
  Approximation Algorithms for Vehicle Routing

Finally, Big Thanks …

We would like to thank all invited speakers for accepting to give a talk at the conference, all Program Committee members who graciously gave their time and energy, and more than 100 external reviewers for their expertise, as well as the publishers Springer and MDPI to offer financial support for organizing this event. Also, we are grateful to Springer for publishing the proceedings of FCT 2023 in their ARCoSS subline of the LNCS series.

July 2023   Henning Fernau
                                                                  Klaus Jansen

Organization

Program Committee Chairs

Henning Fernau   University of Trier, Germany
Klaus Jansen   University of Kiel, Germany

Steering Committee

Bogdan Chlebus   University of Colorado, USA
Marek Karpinski (Chair)   University of Bonn, Germany
Andrzej Lingas   Lund University, Sweden
Miklos Santha   CNRS and Paris Diderot University, France
Eli Upfal   Brown University, USA

Program Committee

Akanksha Agrawal   IITM Chennai, India
Evripidis Bampis   Sorbonne University, France
Hans Bodlaender   University of Utrecht, The Netherlands
Ahmed Bouajjani   Paris Diderot University, France
Bogdan Chlebus   Augusta University, USA
Ugo Dal Lago   University of Bologna, Italy
Vida Dujmovic   University of Ottawa, Canada
Leah Epstein   University of Haifa, Israel
Piotr Faliszewski   AGH University of Science and Technology, Krakow,
                               Poland
Henning Fernau (Co-chair)   University of Trier, Germany
Robert Ganian   TU Vienna, Austria
Klaus Jansen (Co-chair)   University of Kiel, Germany
Artur Jeż   University of Wrocław, Poland
Mamadou Kanté   Clermont Auvergne University, France
Arindam Khan   IIS Bangalore, India
Ralf Klasing   CNRS, University of Bordeaux, France
Jan Kratochvíl   Charles University of Prague, Czech Republic
Sławomir Lasota   University of Warsaw, Poland
Florin Manea   University of Göttingen, Germany
Tomáš Masopust   Palacky University Olomouc, Czech Republic
Matthias Mnich   TU Hamburg, Germany
Nelma Moreira   University of Porto, Portugal
Norbert Müller   University of Trier, Germany
Yoshio Okamoto   University of Electro-Communication, Japan
Sang-il Oum   IBS/KAIST, South Korea

Kenta Ozeki	Yokohama National University, Japan
Markus L. Schmid	HU Berlin, Germany
Uéverton Souza	Fluminense Federal University, Brazil
Frank Stephan	NUS, Singapore
José Verschae	Pontifical Catholic University of Chile, Chile
Boting Yang	University of Regina, Canada
Guochuan Zhang	University of Zhejiang, China
Binhai Zhu	Montana State University, USA

Organizing Committee

Henning Fernau	University of Trier, Germany
Philipp Kindermann	University of Trier, Germany
Zhidan Feng	University of Trier, Germany
Kevin Mann	University of Trier, Germany

Additional Reviewers

Andreas Bärtschi
Laurent Beaudou
Thomas Bellitto
Ioana O. Bercea
Sebastian Berndt
Sebastian Bielfeldt
Davide Bilò
Niclas Boehmer
Nicolas Bousquet
Hauke Brinkop
Sabine Broda
Yixin Cao
Javier Cembrano
Dibyayan Chakraborty
Laura Codazzi
Simone Costa
Luis Felipe Cunha
Joel Day
Alexis de Colnet
Sanjana Dey
Konstantinos Dogeas
Gérard Duchamp
Guillaume Ducoffe
Foivos Fioravantes
David Fischer
Guilherme Fonseca
Mathew C. Francis
Eric Fusy

Esther Galby
Paweł Gawrychowski
Petr Golovach
Bruno Guillon
Michel Habib
Niklas Hahn
Yassine Hamoudi
Yo-Sub Han
John Haslegrave
Winfried Hochstättler
Piotr Hofman
Matheiu Hoyrup
Dmitry Itsykson
Sanjay Jain
Bart Jansen
Vincent Juge
Andrzej Kaczmarczyk
Kai Kahler
Naoyuki Kamiyama
Christos Kapoutsis
Debajyoti Kar
Jarkko Kari
Peter Kling
Florent Koechlin
Petr Kolman
Tore Koß
Pascal Kunz
Van Bang Le

Duksang Lee
Paloma Lima
Guohui Lin
Grzegorz Lisowski
António Machiavelo
Diptapriyo Majumdar
Andreas Maletti
Sebastian Maneth
Kevin Mann
Yaping Mao
Barnaby Martin
Simon Mauras
Damiano Mazza
Alexsander Melo
Lydia Mirabel Mendoza-Cadena
Martin Milanič
Neeldhara Mishra
Josefa Mula
Daniel Neuen
Panagiotis Patsilinakos
Daniel Paulusma
João Pedro Pedroso
Christophe Reutenauer

Cristobal Rojas
Suthee Ruangwises
Pavel Ruzicka
Philippe Schnoebelen
Jason Schoeters
Rafael Schouery
Pascal Schweitzer
Etsuo Segawa
Eklavya Sharma
Yongtang Shi
Dmitry Sokolov
Krzysztof Sornat
Alex Sprintson
Tobias Stamm
Aditya Subramanian
Prafullkumar Tale
Malte Tutas
Gabriele Vanoni
Ilya Vorobyev
Sebastian Wiederrecht
Michalis Xefteris
Lia Yeh
Thomas Zeume

Contents

Convergence of Distributions on Paths

Samy Abbes[⊠][iD]

Université Paris Cité, CNRS, IRIF, 75013 Paris, France
abbes@irif.fr

Abstract. We study the convergence of distributions on finite paths of weighted digraphs, namely the family of Boltzmann distributions and the sequence of uniform distributions. Targeting applications to the convergence of distributions on paths, we revisit some known results from *reducible* nonnegative matrix theory and obtain new ones, with a systematic use of tools from analytic combinatorics. In several fields of mathematics, computer science and system theory, including concurreny theory, one frequently faces non strongly connected weighted digraphs encoding the elements of combinatorial structures of interest; this motivates our study.

Keywords: Weighted digraph · Reducible nonnegative matrix · Uniform measure · Boltzmann measure

1   Introduction

Motivations. Given a weighted digraph, it is standard to consider for each vertex x and for each integer $k \geq 0$ the finite probability distribution μ_k which gives to each path of length k and starting from x the probability proportional to its multiplicative weight. If the underlying digraph is strongly connected and aperiodic, the classical results of Perron-Frobenius theory for primitive matrices show that the sequence $(\mu_k)_{k \geq 0}$ converges weakly toward a probability measure on the space of infinite paths starting from x. The sequence of random vertices visited by such a uniform infinite path is a Markov chain, which transition kernel is derived from the weights on the edges and from the Perron eigenvector of the adjacency matrix of the weighted digraph [10].

The same question of convergence is less standard in the case of a general, non strongly connected weighted digraph. It is however the actual framework for a variety of situations in mathematics, in computer science and in system theory. Indeed, the elements of many combinatorial or algebraic structures of interest can be represented by the paths of some finite digraph encoding the constraints of the "normal form" of elements, in a broad sense. Even when the initial structure is irreducible in some sense, it might very well be the case that the digraph itself being not strongly connected. A first example is the digraph of simple elements of a braid group or of a braid monoid, or more generally of a Garside group or monoid [2,7,8]; this digraph is never strongly connected. But also in automatic group theory: quoting [6], "a graph parameterizing a combing

H. Fernau and K. Jansen (Eds.): FCT 2023, LNCS 14292, pp. 1–15, 2023.
https://doi.org/10.1007/978-3-031-43587-4_1

of a hyperbolic group may typically fail to be recurrent". In system theory, the digraph of states-and-cliques introduced by the author to encode the trajectories of a concurrent system [1] is also not strongly connected in general, even for an irreducible concurrent system.

Understanding the asymptotic behavior of "typical elements of large length" is of great interest at some point, either for random generation purposes, or aiming for the probabilistic verification or the performance evaluation of systems, to name a few applications. This motivates the study of the weak limit of the uniform distributions on paths for a general weighted digraph, which is the main topic of this paper.

Framework and Contributions. Let $W = (V, \mathsf{w})$ be a weighted digraph, where $\mathsf{w} : V \times V \to \mathbb{R}_{\geq 0}$ is a non negative weight function, positive on edges. Let F be the adjacency matrix of W, that we assume to be of positive spectral radius ρ.

Let $\mathsf{H}(z)$ be the generating function with matrix coefficients defined by $\mathsf{H}(z) = \sum_{k \geq 0} F^k z^k$. The power series $\mathsf{H}(z)$, that we call the *growth matrix* of W, can also be seen as a matrix of generating functions with a well known combinatorial interpretation, which is recalled later in the paper. It is standard knowledge that all the generating functions $[\mathsf{H}(z)]_{x,y}$ are rational series. From this we can deduce the existence and uniqueness of a triple (λ, h, Θ) where λ is a positive real, h is a positive integer and Θ is a nonnegative and non zero square matrix indexed by $V \times V$, that we call the *residual matrix of* W—it can indeed be interpreted as the residue at its positive smallest singularity of the function $\mathsf{H}(z)$—, and such that:

$$\lim_{s \to \lambda} (1 - \lambda^{-1} s)^h \, \mathsf{H}(s) = \Theta \qquad s \in (0, \lambda) \qquad (1)$$

For instance if W is strongly connected and aperiodic, and so F is primitive, then it is well known that there is projector matrix Π of rank 1 and a square matrix R of spectral radius lower than 1 such that:

$$F = \rho(\Pi + R) \qquad\qquad \Pi \cdot R = 0 \qquad\qquad R \cdot \Pi = 0 \qquad (2)$$

In this case: $\lambda = \rho^{-1}$, $\Theta = \Pi$, $h = 1$, and the elements $[\Theta]_{x,y}$ are all positive.

For the less standard situation where W is not strongly connected, a contribution of this paper is to provide a recursive way of computing the residual matrix Θ and to characterize the pairs (x, y) such that $[\Theta]_{x,y} > 0$ (Theorem 4 and Sect. 4). For this we use a simple formula (Lemma 2) to compute the rectangular blocks of $\mathsf{H}(z)$ outside its diagonal blocks—the latter are occupied by the corresponding growth matrices of the access classes, or strongly connected components, of the digraph; despite its simplicity, this formula does not seem to have explicitly appeared in the literature.

We also show that the integer h occurring in (1) is the *height* of W, that is to say, the maximal length of chains of access equivalence classes of maximal spectral radius; in nonnegative matrix theory, access classes of maximal spectral radius are called *basic*. This is consistent with the known result saying that the dimension of the generalized eigenspace associated to ρ is the height of the digraph [11], a result that we also recover. This yields precise information on the growth of coefficients of the generating functions $[\mathsf{H}(z)]_{x,y}$.

We use these results to study in Sect. 5 the weak convergence of certain probability distributions on paths, in particular the sequence of *uniform distributions* and the family of *Boltzmann distributions* (see Definition 3). In the case of a general weighted digraph, we prove the convergence of the Boltzmann distributions toward a *complete cocycle measure* on a sub-digraph (see Definition 4 and Theorem 5). Under an aperiodicity assumption, the convergence also holds for the uniform distributions (Theorem 6).

The sub-digraph, support of the limit measure, has the property that its basic access classes coincide with its final access classes. We call such digraphs *umbrella digraphs*, and we devote Sect. 3 to their particular study. We give a decomposition of their adjacency matrix which generalizes the decomposition (2) for primitive matrices (Theorem 1 and 2), and reobtain in this way some results from nonnegative matrix theory, for instance that umbrella digraphs are exactly those for which there exists a positive Perron eigenvector (Corollary 2). Our point of view is influenced by our motivation toward probability measures on the space of infinite paths, and we characterize umbrella digraphs by the existence of a complete cocycle measure (see Definition 4 and Theorem 3).

2   Preliminaries

• *Weighted Digraphs, Adjacency Matrix, Paths.* A *weighted digraph*, or *digraph* for short, is a pair $W = (V, \mathsf{w})$ where V is a finite set of *vertices* and $\mathsf{w} : V \times V \to \mathbb{R}_{\geq 0}$ is a nonnegative real valued function. The set of pairs $E = \{(x,y) \in V \times V \; : \; \mathsf{w}(x,y) > 0\}$ is the set of *edges* of W. Given a bijection $\sigma : V \to \langle \nu \rangle$, where $\langle \nu \rangle = \{1, \ldots, \nu\}$, the function w identifies with the $\nu \times \nu$ nonnegative matrix F defined by $[F]_{\sigma(x),\sigma(y)} = \mathsf{w}(x,y)$. Changing the bijection σ results in a simultaneous permutation of the lines and of the columns of F.

A *finite path* of W is a sequence $u = (x_i)_{0 \leq i \leq k}$ of vertices such that $(x_i, x_{i+1}) \in E$ for all $i < k$. The *initial* and *final* vertices of u are denoted $\iota(u) = x_0$ and $\kappa(u) = x_k$, and its *length* is $\ell(u) = k$. Denoting by O the set of finite paths, we introduce the following notations, for $x, y \in V$ and for k any nonnegative integer:

$$O_x = \{u \in O \; : \; \iota(u) = x\} \qquad O_{x,y} = \{u \in O \; : \; \iota(u) = x \wedge \kappa(u) = y\}$$
$$O_x(k) = \{u \in O_x \; : \; \ell(u) = k\} \quad O_{x,y}(k) = \{u \in O_{x,y} \; : \; \ell(u) = k\}$$

The *infinite paths* are the sequences $\omega = (x_i)_{i \geq 0}$ such that $(x_i, x_{i+1}) \in E$ for all $i \geq 0$. The set $\overline{O}$ of paths, either finite or infinite, is equipped with the prefix ordering which we denote by $\leq$. For every $u \in O$, $x \in V$ and $k \geq 0$, we put:

$$\Omega_x = \{\xi \in \overline{O} \; : \; \iota(\xi) = x \wedge \ell(\xi) = \infty\} \qquad \Uparrow u = \{\xi \in \overline{O} \; : \; u \leq \xi\}$$
$$\uparrow u = \{\omega \in \overline{O} \setminus O \; : \; u \leq \omega\} \qquad \uparrow^k u = \{v \in O(k) \; : \; u \leq v\}$$

• **Access Relation, Access Equivalence Classes.** We write $x \Rightarrow y$ to denote that the vertex x has *access* to the vertex y, meaning that $O_{x,y} \neq \emptyset$. The relation $\Rightarrow$ is then the transitive and reflexive closure of E, seen as a binary relation on V; it is thus a preordering relation on V. We write $x \sim y$ to denote that x and y *communicate*, meaning that $x \Rightarrow y$ and $y \Rightarrow x$, collapse equivalence relation of $\Rightarrow$. The equivalence classes of $\sim$ are the *access classes* of W. In enumerative combinatorics context, access classes are also called *strongly connected components*. The set $\mathcal{D}$ of access classes is equipped with the partial ordering relation, still denoted $\Rightarrow$, induced by the preordering on vertices.

For each vertex x, we define a sub-digraph $V(x)$, identified with its set of vertices $V(x) = \{y \in V : x \Rightarrow y\}$.

By convention, the bijection $\sigma : V \to \langle \nu \rangle$ will always be chosen in such a way that the adjacency matrix F has the following block-triangular shape, corresponding to the so-called Frobenius normal form [12]:

$$F = \begin{pmatrix} F_1 & \mathsf{X} & \dots & \mathsf{X} \\ \vdots & \ddots & & \mathsf{X} \\ 0 & \dots & 0 & F_p \end{pmatrix} \tag{3}$$

where the diagonal blocks $F_1, \dots, F_p$ are the adjacency matrices of the access classes, and the Xs represent rectangular nonnegative matrices.

• **Spectral Radius, Basic and Final Classes.** By definition, the *spectral radius* $\rho(W)$ of W is the spectral radius $\rho(F)$ of its adjacency matrix F, i.e., the largest modulus of its complex eigenvalues. It is apparent on (3) that $\rho(W) = \max_{D \in \mathcal{D}} \rho(D)$. In (3), each matrix F_i is either the 1×1 block $[0]$ or an irreducible matrix. Hence it follows from Perron-Frobenius theory that $\rho(W)$ is itself an eigenvalue of F.

By definition, an access class D is *basic* if $\rho(D) = \rho(W)$; and *final* if D is maximal in $(\mathcal{D}, \Rightarrow)$.

For each vertex x, we set: $\gamma(x) = \rho(V(x))$.

• **Analytic Combinatorics, Growth Matrix, Residual Matrix.** We extend the function $\mathsf{w} : V \times V \to \mathbb{R}_{\geq 0}$ to finite paths by setting $\mathsf{w}(u) = \mathsf{w}(x_0, x_1) \times \dots \times \mathsf{w}(x_{k-1}, x_k)$ if $u = (x_0, \dots, x_k)$, with $\mathsf{w}(u) = 1$ if $\ell(u) = 0$, and we define for every integer $k \geq 0$ and for every pair $(x, y) \in V \times V$:

$$Z_{x,y}(k) = \sum_{u \in O_{x,y}(k)} \mathsf{w}(u), \qquad Z_x(k) = \sum_{u \in O_x(k)} \mathsf{w}(u) = \sum_{y \in V} Z_{x,y}(k)$$

The quantities $Z_{x,y}(k)$ are related to the powers of the adjacency matrix F through the well known formulas: $Z_{x,y}(k) = [F^k]_{x,y}$.

Definition 1. *The* growth matrix $\mathsf{H}(z)$ *of a weighted digraph with adjacency matrix F is the power series with matrix coefficients defined by:*

$$\mathsf{H}(z) = \sum_{k \geq 0} F^k z^k \tag{4}$$

Just as the adjacency matrix F, the growth matrix $\mathsf{H}(z)$ is defined up to a simultaneous permutation of its lines and columns. The element $[\mathsf{H}(z)]_{x,y}$ is itself a generating function:

$$[\mathsf{H}(z)]_{x,y} = \sum_{k\geq 0}[F^k]_{x,y}\, z^k = \sum_{k\geq 0}Z_{x,y}(k)z^k$$

Let $(x,y) \in V \times V$ and let $f(z) = [\mathsf{H}(z)]_{x,y} = \sum_k a_k z^k$, say of radius of convergence γ^{-1}. It is well known [9, Ch.V], [5] that $f(z)$ is a rational series; hence there exists polynomials A and $A_1, \ldots, A_p$ and complex numbers $\gamma_1, \ldots, \gamma_p$ distinct from γ and with $|\gamma_j| \leq |\gamma|$, such that $a_k = \gamma^k A(k) + \sum_j \gamma_j^k A_j(k)$ for k large enough. The degree of A, say d, is called the *subexponential degree* of $f(z)$, and is characterized by the property:

$$\lim_{s\to\gamma^{-1}} (1 - \gamma s)^{d+1}f(s) = t \in (0, +\infty) \tag{5}$$

where the limit is taken for $s \in (0, \gamma^{-1})$; and then the leading coefficient of A is $\frac{1}{d!}t$. When considering the whole matrix $\mathsf{H}(z)$, we obtain the following result.

Proposition 1. *Given a nonnegative matrix F of spectral radius $\rho > 0$ and the matrix power series $\mathsf{H}(z)$ defined as in* (4), *there exists a unique triple (λ, h, Θ) where λ is a positive real, h is a positive integer and Θ is a square nonnegative matrix, non zero and of the same size as F, such that:*

$$\lim_{s\to\lambda}(1 - \lambda^{-1}s)^h\,\mathsf{H}(s) = \Theta, \qquad s \in (0,\lambda) \tag{6}$$

Furthermore, $\lambda = \rho^{-1}$, and ρ^{-1} is the minimal radius of convergence of all the generating functions $[\mathsf{H}(z)]_{x,y}$. For each pair (x,y) such that $[\mathsf{H}(z)]_{x,y}$ has ρ^{-1} as radius of convergence, then the subexponential degree of $[\mathsf{H}(z)]_{x,y}$ is at most $h - 1$, and is equal to $h - 1$ if and only if $[\Theta]_{x,y} > 0$.

Definition 2. *Given a weighted digraph W of positive spectral radius and of growth matrix $\mathsf{H}(z)$, the square nonnegative and non zero matrix Θ as in* (6) *is the* residual matrix *of W. The nonnegative integer $h - 1$, where $h > 0$ is as in* (6), *is the* subexponential degree *of W.*

• *Boltzmann and Uniform Distributions, Weak Convergence.* This paragraph is needed for the reading of Sect. 5.

Definition 3. *Let $W = (V, \mathsf{w})$ be a weighted digraph. For each vertex x such that $\gamma(x) > 0$, we define the family of Boltzmann distributions $(\theta_{x,s})_{0<s<\gamma(x)^{-1}}$ and the sequence of uniform distributions $(\mu_{x,k})_{k\geq 0}$ as follows:*

$$\theta_{x,s} = \frac{1}{G_x(s)}\sum_{u\in O_x}\mathsf{w}(u)s^{\ell(u)}\delta_{\{u\}} \quad with \quad G_x(s) = \sum_{u\in O_x}\mathsf{w}(u)s^{\ell(u)}$$

$$\mu_{x,k} = \frac{1}{Z_x(k)}\sum_{u\in O_x(k)}\mathsf{w}(u)\delta_{\{u\}} \quad with \quad Z_x(k) = \sum_{u\in O_x(k)}\mathsf{w}(u)$$

where $\delta_{\{u\}}$ is the Dirac distribution concentrated on u.

The radius of convergence of $\sum_{u \in O_x} \mathsf{w}(u) z^{\ell(u)}$ is $\gamma(x)^{-1}$, hence both $G_x(s)$ and $\theta_{x,s}$ are well defined for $s < \gamma(x)^{-1}$. The assumption $\gamma(x) > 0$ is equivalent to the existence of paths in O_x of arbitrary length, hence $Z_x(k) > 0$ and $\mu_{x,k}$ is well defined for all $k \geq 0$.

All of these are discrete probability distributions on the set O_x of finite paths starting from x. We can also see them as probability distributions on the set $\overline{O}_x$, including infinite paths starting from x. The latter being compact in the product topology, it becomes relevant to look for their weak limits, either for $s \to \gamma(x)^{-1}$ for the Boltzmann distributions, or for $k \to \infty$ for the uniform distributions. Our basic analytical result in this regard will be the following elementary result (for the background on weak convergence, see for instance [4]).

Lemma 1. *Let W be a weighted digraph and let x be a vertex.*

1. *Assume that, for each integer $k \geq 0$, μ_k is a probability distribution on $O_x(k)$ and that, for every $u \in O_x$, the following limit exists: $t_u = \lim_{k \to \infty} \mu_k(\uparrow^k u)$. Then the sequence $(\mu_k)_{k \geq 0}$ converges weakly toward a probability measure μ on the space Ω_x of infinite paths starting from x, and μ is entirely characterized by: $\forall u \in O_x \quad \mu(\uparrow u) = t_u$.*
2. *Assume that $(\mu_s)_{0 < s < r}$ is a family of probability distributions on O_x such that, for each $u \in O_x$: 1: the following limit exists: $t_u = \lim_{s \to r} \mu_s(\Uparrow u)$; and 2: $\mu_s(\{u\}) \xrightarrow{s \to r} 0$. Then the family $(\mu_s)_{0 < s < r}$ converges weakly toward a probability measure μ on Ω_x as $s \longrightarrow r$, and μ is entirely characterized by: $\forall u \in O_x \quad \mu(\uparrow u) = t_u$.*

● *Cocycles and Cocycle Measures.* This paragraph is needed for reading Sect. 5 and the end of Sect. 3.

Let $W = (V, \mathsf{w})$ be a weighted digraph and let $[[\Rightarrow]] = \{(x,y) \in V \times V : x \Rightarrow y\}$, a subset of $V \times V$ which is of course distinct of E in general. A *cocycle* is a real valued and nonnegative function $\Gamma : [[\Rightarrow]] \to \mathbb{R}_{\geq 0}$ satisfying:

$$\forall (x, y, z) \in V \times V \times V \quad (x \Rightarrow y \Rightarrow z) \implies \Gamma(x, z) = \Gamma(x, y)\Gamma(y, z) \qquad (7)$$

Motivated by the form of the limit of the uniform distributions found in [10] for primitive matrices, we introduce below the notion of cocycle measure.

Definition 4. *A cocycle measure on a weighted digraph $W = (V, \mathsf{w})$ is a family $\mu = (\mu_x)_{x \in V}$ such that each μ_x is a probability measure on the space Ω_x, and such that for some real $\rho > 0$ and some cocycle Γ, one has:*

$$\forall x \in V \quad \forall u \in O_x \quad \mu_x(\uparrow u) = \rho^{-\ell(u)} \mathsf{w}(u) \Gamma(x, \kappa(u)) \qquad (8)$$

If Γ is positive on $[[\Rightarrow]]$, we say that μ is a complete cocycle measure.

Let $(X_i)_{i \geq 0}$ denote the sequence of canonical projections $X_i : \Omega_x \to V$. Assume that $\mu = (\mu_x)_{x \in V}$ is a cocycle measure as in (8), and let $x \in V$. Then, under μ_x, $(X_i)_{i \geq 0}$ is a Markov chain with initial distribution $\delta_{\{x\}}$ and with a transition kernel of the following special form, that we call a *cocycle kernel*:

$$q(x, y) = \rho^{-1} \mathsf{w}(x, y) \Gamma(x, y) \qquad (9)$$

This correspondence between cocycle measures and cocycle kernels is one-to-one.

3   Umbrella Digraphs

This section is devoted to the study of umbrella digraphs, for which several results from Perron-Frobenius theory for irreducible matrices can be transposed.

Definition 5. *A weighted digraph W of positive spectral radius is: **1:** an umbrella (weighted) digraph if the basic access classes of W coincide with its final access classes; **2:** an augmented umbrella (weighted) digraph if no two distinct basic classes of W have access to each other.*

Example 1. The digraph of simple elements, but the unit element, of a braid monoid on $n \geq 3$ strands is a digraph with two access classes [2,7]. The first class, say C_1, contains only the Δ element and has access to the second class, say C_2, which contains all the other simple elements. The spectral radii are $\rho(C_1) = 1$ and $\rho(C_2) > 1$, hence C_2 is the unique final and basic class: the digraph is an umbrella digraph.

Example 2. Let $W = (V, \mathsf{w})$ be a strongly connected digraph with adjacency matrix $F \neq [0]$. Then W is an umbrella digraph with a unique access class. For $p \geq 1$, let $W^{(p)} = (V, \mathsf{w}^{(p)})$ be the digraph with same vertices as W and with the weight function corresponding the pth power F^p. Then $W^{(p)}$ is irreducible for all $p \geq 1$ if and only if F is primitive, *i.e.*, if W is aperiodic. If W is periodic of period d, then $W^{(d)}$ is an umbrella digraph of spectral radius $\rho(W)^d$. The digraph $W^{(d)}$ has d basic classes which correspond to the periodic classes of the vertices (see [12]). All the d access classes of $W^{(d)}$ are both final and basic.

Under an aperiodicity assumption, the adjacency matrix of an augmented umbrella digraph has a decomposition that extends the well known decomposition of primitive matrices recalled in (2).

Theorem 1 (umbrella and augmented umbrella digraphs with aperiodicity). *Let F be the adjacency matrix of an augmented umbrella digraph of spectral radius $\rho > 0$ and with p basic classes. We assume that all the basic classes of W are aperiodic. Then:*

1. *There is a computable projector matrix of rank p and a matrix R satisfying:*

$$F = \rho(\Pi + R) \qquad \Pi \cdot R = R \cdot \Pi = 0 \qquad \rho(R) < 1 \qquad (10)$$

   and the following convergence holds: $\lim_{n\to\infty} (\rho^{-1}F)^n = \Pi$.
2. *There are two computable families of nonnegative line and column vectors $(\ell_i)_{1 \leq i \leq p}$ and $(r_i)_{1 \leq i \leq p}$ such that:*

$$\Pi = \sum_{i=1}^{p} r_i \cdot \ell_i \qquad \forall i,j \quad \ell_i \cdot r_j = \delta_i^j \qquad (11)$$

   The family $(r_i)_{1 \leq i \leq p}$ is a basis of the space of right ρ-eigenvectors of F, and $(\ell_i)_{1 \leq i \leq p}$ is a basis of the space of left ρ-eigenvectors of F.

3. *Only for umbrella digraphs: for every family $(\alpha_i)_{1 \le i \le p}$ of reals, the right ρ-eigenvector $r = \sum_i \alpha_i r_i$ is positive if and only if $\alpha_i > 0$ for all i.*

4. *The subexponential degree of W is 0. The residual matrix of W coincides with Π, i.e.: $\lim_{s \to \rho^{-1}}(1 - \rho s)\mathsf{H}(s) = \Pi$, where $\mathsf{H}(z)$ is the growth matrix of W. Furthermore, $[\Pi]_{x,y} > 0$ if and only if there is a basic class B such that $x \Rightarrow B \Rightarrow y$.*

Remark 1. If W is an umbrella digraph, then the condition for $[\Pi]_{x,y} > 0$ in point 4 above is equivalent to: $x \Rightarrow y$ and y belongs to some basic class—the same remark applies to next theorem.

In the following result, the aperiodicity assumption is dropped. We use the notation $W^{(d)}$ introduced in Example 2 above.

Theorem 2 (umbrella and augmented umbrella digraph). *Let F be the adjacency matrix of an umbrella (resp., augmented umbrella) digraph W of spectral radius $\rho > 0$. Let $C_1, \ldots, C_p$ be the basic access classes of W, say of periods $d_1, \ldots, d_p$. Let $C_{\nu,0}, \ldots, C_{\nu,d_\nu - 1}$ be the periodic classes of C_ν, and let $q = \sum_\nu d_\nu$. Let also d be a common multiple of $d_1, \ldots, d_p$.*

1. *$W^{(d)}$ is an umbrella (resp., augmented umbrella) weighted digraph of spectral radius ρ^d and with $\{C_{\nu,j} : 1 \le \nu \le p, \ 0 \le j < d_\nu\}$ as basic classes, which are all aperiodic. There is a computable projector Π_d of rank q and a matrix R_d such that:*

$$F^d = \rho^d(\Pi_d + R_d) \qquad \Pi_d \cdot R_d = R_d \cdot \Pi_d = 0 \qquad \rho(R_d) < 1 \qquad (12)$$

2. *There is a basis $(\ell_i)_{1 \le i \le p}$ of nonnegative left ρ-eigenvectors of F, and a basis $(r_j)_{1 \le j \le p}$ of nonnegative right ρ-eigenvectors of F satisfying $\ell_i \cdot r_j = \delta_i^j$ for all i, j.*

3. *Only for umbrella digraphs: for every family $(\alpha_i)_{1 \le i \le p}$ of reals, the ρ-eigenvector $r = \sum_i \alpha_i r_i$ is positive if and only if $\alpha_i > 0$ for all i.*

4. *The subexponential degree of W is 0. The residual matrix of W is given by*

$$\Theta = \frac{1}{d}\Big(\sum_{i=0}^{d-1} \rho^{-i} F^i\Big) \cdot \Pi_d \qquad (13)$$

and satisfies: $[\Theta]_{x,y} > 0$ if and only if there is a basic class B such that $x \Rightarrow B \Rightarrow y$.

In the particular case of a periodic and strongly connected digraph, the following result shows that some simplifications occur in (13); it is the basis of the convergence result for strongly connected digraphs, Theorem 7 in Sect. 5.

Corollary 1 (residual matrix of a strongly connected digraph). *Let F be the adjacency matrix of a strongly connected digraph of positive spectral radius. Then the subexponential degree of W is 0. Let (ℓ, r) be a pair of left and right Perron eigenvectors of F such that $\ell \cdot r = 1$. Then the residual matrix of W is $\Theta = \Pi$, the positive rank 1 projector given by $\Pi = r \cdot \ell$.*

Finally, umbrella weighted digraphs can be characterized by the existence of *complete* cocycle measures.

Theorem 3 (existence of a complete cocycle measure). *A digraph W is an umbrella digraph if and only if there exists a complete cocycle measure on W.*

If W is an umbrella digraph and is accessible from a vertex x, then the complete cocycle measures on W are parameterized by an open simplex of dimension $p - 1$, where p is the number of basic classes of W.

In particular there exists a unique complete cocycle measure if W is strongly connected. For F the adjacency matrix with $\rho = \rho(F)$, and for r a Perron right eigenvector of F, the cocycle kernel associated to this unique complete cocycle measure is given by:

$$q(x,y) = \rho^{-1}\mathsf{w}(x,y)\Gamma(x,y) \qquad\qquad \Gamma(x,y) = \frac{[r]_y}{[r]_x}$$

The following well known result [3,11] can be seen as a consequence of Theorem 3.

Corollary 2 (existence of a positive Perron eigenvector). *A weighted digraph W of spectral radius $\rho > 0$ is an umbrella weighted digraph if and only if its adjacency matrix has a positive ρ-eigenvector.*

4   Computing the Residual Matrix

This section is devoted to the recursive computation of the residual matrix of a general weighted digraph.

Recall that a subset A of a poset $(B, \leq)$ is: **1:]** *final* if: $\forall (a,b) \in A \times B \quad a \leq b \implies b \in A$; and **2:** *initial* if: $\forall (a,b) \in A \times B \quad b \leq a \implies b \in A$.

• *Theoretical Results.* The following elementary result will be instrumental.

Lemma 2 (recursive form of the growth matrix). *Let $W = (V, \mathsf{w})$ be a weighted digraph with adjacency matrix F. Assume that (S, T) is a partition of V and that T is final in V. Let $\mathsf{H}_S(z)$ and $\mathsf{H}_T(z)$ be the growth matrices of S and T, sub-weighted digraphs of W. Then:*

$$\mathsf{H}(z) = \begin{pmatrix} \mathsf{H}_S(z) & \mathsf{Y}(z) \\ 0 & \mathsf{H}_T(z) \end{pmatrix} \qquad \mathsf{Y}(z) = z\,\mathsf{H}_S(z) \cdot X \cdot \mathsf{H}_T(z) \qquad (14)$$

where X is the rectangular block submatrix of F corresponding to $S \times T$.

To determine the subexponentiel degree of a general weighted digraph, the notion of height that we introduce below and which is well known in nonnegative matrix theory [3,11], plays a key role.

Definition 6. *Let W be a weighted digraph, and let $(\mathcal{D}, \Rightarrow)$ be the poset of its access classes. A* dominant chain *is a chain of basic classes in $(\mathcal{D}, \Rightarrow)$ and of maximal lengTheorem The* height *of W is the length of the dominant chains.*

Remark 2. The augmented umbrella digraphs from Definition 5 are the weighted digraphs of positive spectral radius and of height 1.

Theorem 4 (subexponential degree of a digraph). *Let $W = (V, \mathsf{w})$ be a weighted digraph of positive spectral radius and of height h.*

1. *The subexponential degree of W is $h - 1$.*
2. *Let Θ be the residual matrix of W. Then $[\Theta]_{x,y} > 0$ if and only if there exists a dominant chain $(L_1, \ldots, L_p)$ such that $x \Rightarrow L_1$ and $L_p \Rightarrow y$.*

From the above result derives the following well knwown result originally proved by Rothblum [11].

Corollary 3. *Let F be the adjacency matrix of a weighted digraph of spectral radius ρ and of height h. Then the dimension of the generalized eigenspace of F associated to ρ is h.*

●**Recursive Computing.** We aim at recursively computing the residual matrix of a weighted digraph W of spectral radius $\rho > 0$. We first observe two facts.

FACT 1.   The growth matrix $\mathsf{H}(z)$ can be recursively computed by starting from its lower-right corner and extending the blocks already computed. In details, let $(D_1, \ldots, D_p)$ be an enumeration of the access classes of W. Put $V_i = D_1 \cup \ldots \cup D_i$ for $i = 0, \ldots, p$, and assume that the enumeration has been chosen such that V_i is final in V for each i. Then, with the obvious notations $\mathsf{H}_{D_i}(z)$ and $\mathsf{H}_{V_i}(z)$, one has for each $i > 0$:

$$\mathsf{H}_{V_i}(z) = \begin{pmatrix} \mathsf{H}_{D_i}(z) & \mathsf{Y}_i(z) \\ 0 & \mathsf{H}_{V_{i-1}}(z) \end{pmatrix} \qquad \mathsf{Y}_i(z) = z\,\mathsf{H}_{D_i}(z) \cdot X_i \cdot \mathsf{H}_{V_{i-1}}(z) \qquad (15)$$

where X_i is the rectangular block submatrix of F corresponding to $D_i \times V_{i-1}$. This results from Lemma 2 applied with the partition (D_i, V_{i-1}) of V_i. ∎

FACT 2.   If the height of W is 1, then the residual matrix of W is directly computable. This results from Theorem 2, *via* Remark 2 above. ∎

Let $r_i = \rho(D_i)$ for $0 < i \leq p$. To simplify the exposition, we assume that $r_i > 0$ for all i, and we consider Π_i the residual matrix of D_i, which is a computable rank 1 projector (see Corollary 1).

INITIALIZATION.   Denoting $\rho_i = \rho(V_i)$, we start the induction with the first i_0 such that $h_{i_0} \geq 2$ and $\rho_{i_0} = \rho$; in particular, $\rho(V_{i_0-1}) = \rho$. Indeed, as long as the height of V_i is 1, its residual matrix can be directly computed, as we observed in Fact 2 above.

Let Θ_i be the residual matrix of V_i, and let h_i denote the height of V_i. Since $\rho(V_i) = \rho$ for all $i \geq i_0$, we have:

$$\forall i \geq i_0 \quad \Theta_i = \lim_{s \to \rho^{-1}} (1 - \rho s)^{h_i} \mathsf{H}_{V_i}(s) \qquad (16)$$

INDUCTION STEP. We assume that the residual matrix Θ_{i-1} of V_{i-1} has been computed, and in certain cases, we also need to call for the computation of the residual matrix of some sub-digraph of V_{i-1}. We show how to compute Θ_i.

1. *Case where $h_i = h_{i-1}$.*
   (a) *If $\rho(D_i) < \rho$.* Referring to (15) and (16), and since $h_i \geq 2$, we have:

$$(1 - \rho s)^{h_i}\, \mathsf{Y}_i(s) = s\, \mathsf{H}_{D_i}(s) \cdot X_i \cdot \left((1 - \rho s)^{h_{i-1}}\, \mathsf{H}_{V_{i-1}}(s)\right)$$

$$\text{and thus} \quad \Theta_i = \begin{pmatrix} 0 & A \\ 0 & \Theta_{i-1} \end{pmatrix} \quad \text{with} \quad A = \rho^{-1}\, \mathsf{H}_{D_i}(\rho^{-1}) \cdot X_i \cdot \Theta_{i-1}$$

We have $\mathsf{H}_{D_i}(\rho^{-1}) = (\mathrm{Id} - \rho^{-1}F_i)^{-1}$, where F_i is the block sub-matrix of F corresponding to $D_i \times D_i$; so Θ_i can be computed.
   (b) *If $\rho(D_i) = \rho$.* Consider the partition $(\widetilde{T}, \overline{T})$ of V_{i-1} with

$$\widetilde{T} = \{y \in V_{i-1} \; : \; \exists x \in V_{i-1} \quad (y \Rightarrow x) \wedge (D_i \Rightarrow x)\} \qquad (17)$$

which is initial in V_{i-1}. Enumerating the vertices of V_i as those of D_i, then those of $\widetilde{T}$ and then those of $\overline{T}$ in this order, we have:

$$\mathsf{H}_{V_i}(z) = \begin{pmatrix} \mathsf{H}_{D_i}(z) & A(z) & 0 \\ 0 & \mathsf{H}_{\widetilde{T}}(z) & B(z) \\ 0 & 0 & \mathsf{H}_{\overline{T}}(z) \end{pmatrix} \quad A(z) = z\, \mathsf{H}_{D_i}(z) \cdot A \cdot \mathsf{H}_{\widetilde{T}}(z) \quad (18)$$

where A and B are the block submatrices of F corresponding to $D_i \times \widetilde{T}$ and to $\widetilde{T} \times \overline{T}$. Indeed, vertices of $\overline{T}$ are not accessible from D_i, whence the zero block in the upper right corner of $\mathsf{H}_{V_i}(z)$.

On the one hand, the down right corner of $\mathsf{H}_{V_i}(z)$ formed by the four blocks in (18) is nothing but $\mathsf{H}_{V_{i-1}}(z)$. On the other hand, we have:

$$(1 - \rho s)^{h_i}\, A(s) = s\left((1 - \rho s)\mathsf{H}_{D_i}(s)\right) \cdot A \cdot \left((1 - \rho s)^{h_i - 1}\, \mathsf{H}_{\widetilde{T}}(s)\right)$$

$$\xrightarrow{\; s \to \rho^{-1} \;} Y = \begin{cases} 0, & \text{if the height of } \widetilde{T} \text{ is } < h_i - 1 \\ \rho^{-1}\, \Pi_i \cdot A \cdot \widetilde{Y} & \text{if the height of } \widetilde{T} \text{ is } h_i - 1 \end{cases}$$

where $\widetilde{Y}$ is the residual matrix of $\widetilde{T}$. Hence $\Theta_i = \begin{pmatrix} 0 & Y & 0 \\ 0 & \Theta_{i-1} \end{pmatrix}$ is computable.

2. *Case where $h_i = h_{i-1} + 1$.* According to (15), we compute as follows:

$$(1 - \rho s)^{h_i}\, \mathsf{Y}(s) = s\left((1 - \rho s)\mathsf{H}_{D_i}(s)\right) \cdot X_i \cdot \left((1 - \rho s)^{h_i - 1}\, \mathsf{H}_{V_{i-1}}(s)\right)$$

$$\xrightarrow{\; s \to \rho^{-1} \;} Y = \rho^{-1}\, \Pi_i \cdot X_i \cdot \Theta_{i-1}$$

Hence $\Theta_i = \begin{pmatrix} 0 & Y \\ 0 & 0 \end{pmatrix}$ is computable.

5   Convergence of Distributions

This section is devoted to the convergence of distributions on paths of a weighted digraph, which was our main goal from the beginning. We focus on the family $(\theta_{x,s})_{0<s<\rho^{-1}}$ of Boltzmann distributions on the one hand, and on the sequence $(\mu_k)_{k\geq 0}$ of uniform distributions on the other hand (see Definition 3).

•**Boltzmann Distributions.** Let $W = (V, \mathsf{w})$ be a weighted digraph of spectral radius $\rho > 0$. In view of Lemma 1, point 2, we fix a pair (x, u) where $x \in V$ and $u \in O_x$, and we study the quantity $\theta_{x,s}(\Uparrow u)$, aiming at its convergence for $s \to \rho^{-1}$. For $s < \rho^{-1}$, we have:

$$\theta_{x,s}(\Uparrow x) = \frac{1}{G_x(s)} \sum_{v \in O_x\,:\,u \leq v} \mathsf{w}(v) s^{\ell(v)} \tag{19}$$

Every $v \in O_x$ with $u \leq v$ writes in a unique way as the concatenation $v = u \cdot v'$ for some $v' \in O_{\kappa(u)}$, and then $\mathsf{w}(v) = \mathsf{w}(u)\mathsf{w}(v')$. Hence (19) writes as:

$$\theta_{x,s}(\Uparrow x) = \frac{\mathsf{w}(u)s^{\ell(u)}}{G_x(s)} \sum_{v \in O_{\kappa(u)}} \mathsf{w}(v)s^{\ell(v)} = \mathsf{w}(u)s^{\ell(u)}\frac{G_{\kappa(u)}(s)}{G_x(s)} \tag{20}$$

Now for every vertex y and for every real $s \in (0, \rho^{-1})$, $G_y(s)$ is related to the growth matrix of W via:

$$G_y(s) = \sum_{t \in V}[\mathsf{H}(s)]_{y,t} = [\mathsf{H}(s) \cdot \mathbf{1}]_y \tag{21}$$

where $\mathbf{1}$ denotes the column vector filled with 1s.

Let $h(x)$ be the height of $V(x) = \{y \in V : x \Rightarrow y\}$, and let Θ be the residual matrix of $V(x)$. It follows from Theorem 4 that an equivalent of $[\mathsf{H}(s) \cdot \mathbf{1}]_x$ when $s \to \rho^{-1}$ is:

$$[\mathsf{H}(s) \cdot \mathbf{1}]_x \sim_{s\to\rho^{-1}} (h(x))!\,(1 - \rho s)^{-h(x)}[\Theta \cdot \mathbf{1}]_x \tag{22}$$

since $[\Theta \cdot \mathbf{1}]_x > 0$, as a sum of terms of which at least one is positive according to Theorem 4. Putting $y = \kappa(u)$, we note that $y \in V(x)$ and that two cases may occur. **1:** If there exists a dominant chain $(L_1, \ldots, L_{h(x)})$ of $V(x)$ such that $y \Rightarrow L_1$, then an equivalent for $[\mathsf{H}(s) \cdot \mathbf{1}]_y$ analogous to (22) and with the same exponent $h(x)$ holds. From (20), (21) and (22) we derive thus:

$$\lim_{s\to\rho^{-1}} \theta_{x,s}(\Uparrow u) = \rho^{-\ell(u)}\mathsf{w}(u)\frac{[\Theta \cdot \mathbf{1}]_{\kappa(u)}}{[\Theta \cdot \mathbf{1}]_x} \tag{23}$$

which is a positive number.

But, **2:** If not, it means that $[\mathsf{H}(s) \cdot \mathbf{1}]_y$ has either an exponential growth rate less that ρ, or an exponential growth rate equal to ρ but a subexponential degree less than h. In both situations, the ratio in (20) goes to zero as $s \to \rho^{-1}$.

The above discussion motivates the following definition, where we recall that $\gamma(x) = \rho(V(x))$.

Definition 7. *Let $W = (V, \mathsf{w})$ be a weighted digraph and let x be a vertex such that $\gamma(x) > 0$. The umbrella digraph spanned by x is the sub-digraph of $V(x)$, the vertices of which are the vertices $y \in V(x)$ such that $y \Rightarrow L_1$ for some dominant chain $(L_1, \ldots, L_{h(x)})$ of $V(x)$. We denote it by $\mathsf{U}(x)$.*

The digraph $\mathsf{U}(x)$ thus defined is indeed an umbrella digraph, of spectral radius $\gamma(x)$. Putting together the result of the above discussion and Lemma 1, we obtain the following convergence result.

Theorem 5 (convergence of Boltzmann distributions). *Let W be a weighted digraph of positive spectral radius ρ, with residual matrix Θ, and let x_0 be a vertex such that all vertices of W are accessible from x_0. Then the Boltzmann distributions $\theta_{x_0, s}$ converge weakly when $s \to \rho^{-1}$ toward the complete cocycle measure θ on $\mathsf{U}(x_0)$ which cocycle transition kernel is given on $\mathsf{U}(x_0)$ by:*

$$q(x, y) = \rho^{-1} \mathsf{w}(x, y) \Gamma(x, y) \qquad \Gamma(x, y) = \frac{[\Theta \cdot 1]_y}{[\Theta \cdot 1]_x} \qquad (24)$$

Remark 3. In the above result, the condition that V should be accessible from x_0 is not a severe restriction. In general, the theorem applies to the sub-digraph $V(x_0)$, which vertices are those accessible from x_0 and of spectral radius $\gamma(x_0)$. If $\gamma(x_0) = 0$, the theorem does not apply; and indeed, $\Omega_{x_0} = \emptyset$ in this case.

Remark 4. The limit distributions $(\theta_x)_{x \in V}$ are also obtained as the limits of the Boltzmann distributions relative to the sub-digraph $\mathsf{U}(x_0)$ itself. Hence if $\widetilde{\Theta}$ is the residual matrix of $\mathsf{U}(x_0)$, the cocycle kernel $q(x, y)$ in (24) can also be obtained as $q(x, y) = \rho^{-1} \mathsf{w}(x, y) \widetilde{\Gamma}(x, y)$ with $\widetilde{\Gamma}(x, y) = [\widetilde{\Theta} \cdot 1]_y / [\widetilde{\Theta} \cdot 1]_y$. And $\widetilde{\Theta}$ is directly computed according to Theorem 2 since $\mathsf{U}(x_0)$ is an umbrella digraph.

● *Uniform Distributions.* We now aim at using point 1 of Lemma 1 in order to derive the weak convergence of the sequence $(\mu_{x,k})_{k \geq 0}$ of uniform distributions on paths. We fix a vertex x and a path $u \in O_x$, and we consider the quantity $\mu_{x,k}(\uparrow^k u)$ for $k \geq \ell(u)$. The same change of variable that we used above yields the following expression:

$$\mu_x(\uparrow^k u) = \frac{1}{Z_x(k)} \left(\sum_{v \in O_x(k) \, : \, u \leq v} \mathsf{w}(v) \right) = \mathsf{w}(u) \frac{Z_{\kappa(u)}(k - \ell(u))}{Z_x(k)} \qquad (25)$$

We are thus brought to discuss the asymptotics of the kth coefficients of the generating functions $[\mathsf{H}(z) \cdot 1]_x$ and $[\mathsf{H}(z) \cdot 1]_{\kappa(u)}$, as $k \to \infty$. These are closely related to the values of these generating functions near their singularity ρ^{-1}, which connects to our previous discussion for the Boltzmann distributions. We have thus the following result, which is a sort of Tauberian theorem relatively to Theorem 5. Remarks 3 and 4 above apply to Theorem 6 as they did for Theorem 5.

Theorem 6 (convergence of uniform distributions with aperiodicity).
Let W be a weighted digraph of positive spectral radius ρ, and let x_0 be a vertex such that all vertices of W are accessible from x_0. We assume that all the basic classes of W are aperiodic. Then the sequence $(\mu_{x_0,k})_{k\geq 0}$ of uniform distributions converges weakly toward the complete cocycle measure on $\mathsf{U}(x_0)$ with the cocycle transition kernel q described in (24).

The aperiodicity condition in the above theorem is sufficient but not necessary, as shown by the following result, consequence of Corollary 1.

Theorem 7 (convergence of uniform distributions for a strongly connected digraph). *For every vertex x of a strongly connected digraph $W = (V, \mathsf{w})$ with positive spectral radius ρ, the sequence of uniform distributions converges weakly toward the unique complete cocycle measure on W, with transition kernel given by $q(x, y) = \rho^{-1}\mathsf{w}(x, y)[r]_y/[r]_x$ for any Perron eigenvector r.*

In general, without aperiodicity, the convergence of uniform distributions does not hold. This is for instance the case for the following umbrella digraph:

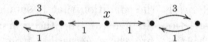

For each integer $n \geq 1$, there are exactly two paths of length n and starting from the initial vertex x, say c_n which goes to the left access class and d_n which goes to the right access class. For each $k \geq 1$, we have $\mathsf{w}(c_{2k}) = 3^{k-1}$, $\mathsf{w}(d_{2k}) = 3^k$ and $\mathsf{w}(c_{2k+1}) = \mathsf{w}(d_{2k+1}) = 3^k$. Henceforth $\mu_{x,2k} = \left[\frac{1}{4}\ \frac{3}{4}\right]$ and $\mu_{x,2k+1} = \left[\frac{1}{2}\ \frac{1}{2}\right]$. The two subsequences $(\mu_{x,2k})_k$ and $(\mu_{x,2k+1})_k$ do not converge toward the same limit, preventig $(\mu_{x,k})_k$ to be a convergent sequence of distributions. This example was communicated to the author by S. Gouëzel, I am happy to thank him here for it.

References

1. Abbes, S.: Introduction to probabilistic concurrent systems. Fundam. Inf. **187**(2–4), 71–102 (2022)
2. Abbes, S., Gouëzel, S., Jugé, V., Mairesse, J.: Asymptotic combinatorics of Artin-Tits monoids and of some other monoids. J. Algebra **525**, 497–561 (2019)
3. Berman, A., Plemmons, R.: Nonnegative Matrices in the Mathematical Sciences. SIAM (1994)
4. Billingsley, P.: Convergence of Probability Measures, 2nd edn. John Wiley, Hoboken (1999)
5. Bousquet-Mélou, M.: Rational and algebraic series in combinatorial enumeration. In: Proceedings of the ICM, pp. 789–826 (2006)
6. Calegari, D., Fujiwara, K.: Combable functions, quasimorphisms, and the central limit theorem. Ergodic Theory Dyn. Syst. **30**(5), 1343–1369 (2010)
7. Charney, R.: Geodesic automation and growth functions for Artin groups of finite type. Math. Ann. **301**, 307–324 (1995)

8. Dehornoy, P., Digne, F., Godelle, E., Krammer, D., Michel, J.: Foundations of Garside Theory. EMS Press (2014)
9. Flajolet, P., Sedgewick, R.: Analytic Combinatorics. Cambridge University Press, Cambridge (2009)
10. Parry, W.: Intrinsic Markov chains. Trans. Am. Math. Soc. **112**(1), 55–66 (1964)
11. Rothblum, U.: Nonnegative and stochastic matrices. In: Hogben, L. (ed.) Handbook of Linear Algebra, 2nd edn. Chapman & Hall (2014)
12. Seneta, E.: Non-Negative Matrices and Markov Chains, 2nd edn. Springer, Cham (1981). https://doi.org/10.1007/0-387-32792-4

Subhedge Projection for Stepwise Hedge Automata

Antonio Al Serhali[✉] and Joachim Niehren

Inria and University of Lille, Lille, France
antonio.al-serhali@inria.fr

Abstract. We show how to evaluate stepwise hedge automata (SHAs) with subhedge projection. Since this requires passing finite state information top-down, we introduce the notion of downward stepwise hedge automata. We use them to define an in-memory and a streaming evaluator with subhedge projection for SHAs. We then tune the streaming evaluator so that it can decide membership at the earliest time point. We apply our algorithms to the problem of answering regular XPath queries on XML streams. Our experiments show that subhedge projection of SHAs can indeed speed up earliest query answering on XML streams.

1 Introduction

Projection is necessary for running automata on words, trees, hedges or nested words efficiently without having to evaluate irrelevant parts of the input structure. Projection is most relevant for XML processing as already noticed by [7,13,14]. Saxon's in-memory evaluator, for instance, projects input XML document relative to an XSLT program, which contains a collection of XPath queries to be answered simultaneously [10]. When it comes to processing XML streams, quite some algorithms [5,9,12,15] are based on nested word automata (NWAs), for which an efficient projection algorithm exists [19].

More recently, it was noticed that stepwise hedge automata (SHA) [18] have important advantages over NWAs when it comes to determinization and earliest query answering [9]. SHAs are a recent variant of standard hedge automata that go back to the sixties [4,20]. They mix up bottom-up processing of standard tree automata with the left-to-right processing of finite word automata (NFAs), but do neither support top-down processing nor have an explicit stack in contrast to NWAs. In particular, it could be shown that earliest query answering for regular queries defined by deterministic SHAs [3] has a lower worst case complexity than for deterministic NWAs [9]. SHAs have the advantage that the set of states that are accessible over some hedge from a given set of start states can be computed in linear time, while for NWAs this requires cubic time.

Based on deterministic SHAs, earliest query answering for regular queries became feasible in practice [3], as shown for a collection of deterministic SHAs for real word regular XPath queries on XML documents [2]. On the other hand side, it is still experimentally slower than the best non-earliest approaches [5]. We

H. Fernau and K. Jansen (Eds.): FCT 2023, LNCS 14292, pp. 16–31, 2023.
https://doi.org/10.1007/978-3-031-43587-4_2

believe that this is due to the fact that projection algorithms for SHA evaluation are missing. Projecting in-memory evaluation assumes that the full graph of the input hedge is constructed at beforehand. Nevertheless, projection may still save time, if one has to run several queries on the same input hedge, or, if the graph got constructed for different reasons anyway. In the streaming case with subhedge projection, the situation is similar: the whole input hedge on the stream needs to be parsed. But only for the nodes that are not projected away, the automaton transitions need to be computed. Given that pure parsing is by two or three orders of magnitude faster, one can save considerable time as noticed in [19].

Consider the example of the XPath filter [self::list][child::item] that is satisfied by an XML document if its root is an list element that has some item child. When evaluating this filter on an XML document, it is sufficient to inspect its roots for having label list and then all its children until some item is found. The subhedges of these children can be projected away. However, one must memoize whether the level of the current node is 0, 1, or greater. This level information can be naturally updated in a top-down manner. The evaluators of SHAs, however, operate bottom-up and left-to-right exclusively. Therefore, projecting evaluators for SHAs need to be based on more general machines. It would not be sufficient to map SHAs to NWAs and use their projecting evaluators [19]. The NWAs obtained by compilation from SHAs do not push any information top-down, so no projection is enabled. Thus, the objective of the present paper is to develop evaluators with subhedge projection for SHAs.

As more general machines we propose *downward stepwise hedge automata* (SHA$^\downarrow$s), a variant of SHAs that support top-down processing in addition. They are basically Neumann and Seidl's pushdown forest automata [16], except that they apply to unlabeled hedges instead of labeled forests. NWAs are known to operate similarly on nested words [8], while allowing for more general visible pushdowns. We then distinguish subhedge projection states for SHA$^\downarrow$s, and show how to use them to evaluate SHAs with subhedge projection both in-memory and in streaming mode. Alternatively, subtree projecting evaluators for SHA$^\downarrow$s could be obtained by compiling them to NWAs, distinguishing irrelevant subtrees there [19], and using them for subtree projecting evaluation via projecting NWAs.

As a first and main contribution, we show how to compile SHAs to SHA$^\downarrow$s so that one can distinguish appropriate subhedge projection states. For instance, reconsider the XPath filter [self::list][child::item]. It can be defined by the deterministic SHA in Fig. 1, which our compiler maps to the SHA$^\downarrow$ in Fig. 2 (up to renaming of states). This compiler permits us to distinguish a projection state Π in which subhedges can be ignored. We note however that our compiler may in the worst case increase the size of the automata exponentially. Therefore, we avoid constructing the SHA$^\downarrow$s statically but rather construct only the needed part of the SHA$^\downarrow$s dynamically on the fly when using it to evaluate some hedge with subhedge projection. A sketch of the soundness proof of our compiler is provided.

Our second contribution is a refinement of the compiler from SHAs to SHA$^\downarrow$s for distinguishing safe states for rejection and selection. In this way, we obtain

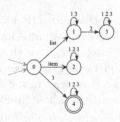

Fig. 1. A unique minimal deterministic SHA (with initial state equal tree initial state) for the XPath filter `[self::list][child::item]`.

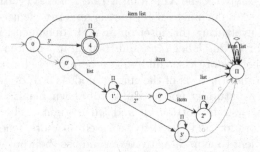

Fig. 2. The deterministic SHA$^\downarrow$ with subhedge projection state Π obtained by our compiler.

an earliest membership tester for deterministic SHAs in streaming mode which improves the recent earliest membership tester of [3] with subhedge projection. The property of being earliest carries over from there. We lifted this earliest membership tester to an earliest query answering algorithm with subhedge projection for monadic queries defined by deterministic SHAs but omit the details.

Our third contribution is an implementation and experimental evaluation of an earliest query answering algorithm for dSHAs with subhedge projection (not only of earliest membership testing), by introducing subhedge projection into the AStream tool [3]. For the evaluation, we consider the deterministic SHAs constructed with the compiler from [18] for the forward regular XPath queries of the XPathMark benchmark [6] and real-world XPath queries [2]. It turns out that we can reduce the running time for all regular XPath queries that contain only child axes considerably since large parts of the input hedges can be projected away. For such XPath queries, the earliest query answering algorithm of AStream with projection becomes competitive in efficiency with the best existing streaming algorithm from QuiXPath [5] (which is non-earliest on some queries though). The win is smaller for XPath queries with descendant axis, where only few subhedge projection is possible.

Outline. After some preliminaries in Sects. 2 and 3. In Sect. 4 we introduce SHA$^\downarrow$s and show that they enable in-memory evaluation with subhedge projection. In Sect. 5 we show how to compile SHAs to SHA$^\downarrow$s with subhedge projection states. Streaming evaluators for SHA$^\downarrow$s with subhedge projection follow in Sect. 6.

Section 7 improves the compiler from SHAs to SHA$^\downarrow$s for obtaining an earliest membership tester. Section 8 discusses our practical experiments. Supplementary material including the soundness proof of our compiler and further discussion on related work can be found in [1].

2 Preliminaries

Let A and B be sets and $r \subseteq A \times B$ a binary relation. The domain of r is $dom(r) = \{a \in A \mid \exists b \in B.\ (a, b) \in r\}$. We call r total if $dom(r) = A$. A partial function $f : A \hookrightarrow B$ is a relation $f \subseteq A \times B$ that is functional. A total function $f : A \to B$ is a partial function $f : A \hookrightarrow B$ that is total.

Words. Let $\mathbb{N}$ be the set of natural numbers including 0. Let the alphabet Σ be a set. The set of words over Σ is $\Sigma^* = \cup_{n \in \mathbb{N}} \Sigma^n$. A word $(a_1, \ldots, a_n) \in \Sigma^n$ where $n \in \mathbb{N}$ is written as $a_1 \ldots a_n$. We denote the empty word of length 0 by $\varepsilon \in \Sigma^0$ and by $v_1 \cdot v_2 \in \Sigma^*$ the concatenation of two words $v_1, v_2 \in \Sigma^*$.

Hedges. Hedges are sequences of letters and trees $\langle h \rangle$ with some hedge h. More formally, a hedge $h \in \mathcal{H}_\Sigma$ has the following abstract syntax:

$$h, h' \in \mathcal{H}_\Sigma ::= \varepsilon \quad | \quad a \quad | \quad \langle h \rangle \quad | \quad h \cdot h' \qquad \text{where } a \in \Sigma$$

We assume $\varepsilon \cdot h = h \cdot \varepsilon = h$ and $(h \cdot h_1) \cdot h_2 = h \cdot (h_1 \cdot h_2)$. Therefore, we consider any word in Σ^* as a hedge in $\mathcal{H}_\Sigma$, i.e., $\Sigma^* \ni aab = a \cdot a \cdot b \in \mathcal{H}_\Sigma$.

Nested Words. Hedges can be identified with nested words, i.e., words over the alphabet $\hat{\Sigma} = \Sigma \cup \{\langle, \rangle\}$ in which all parentheses are well-nested. This is done by the function $nw(h) : \mathcal{H}_\Sigma \to (\Sigma \cup \{\langle, \rangle\})^*$ such that: $nw(\varepsilon) = \varepsilon$, $nw(\langle h \rangle) = \langle \cdot nw(h) \cdot \rangle$, $nw(a) = a$, and $nw(h \cdot h') = nw(h) \cdot nw(h')$.

3 Stepwise Hedge Automata (SHAs)

Stepwise hedge automata (SHAs) are automata for hedges mixing up bottom-up tree automata and left-to-right word automata.

Definition 1. *A stepwise hedge automaton (SHA) is a tuple* $A = (\Sigma, \mathcal{Q}, \Delta, I, F)$ *where* Σ *and* $\mathcal{Q}$ *are finite sets,* $I, F \subseteq \mathcal{Q}$, *and* $\Delta = ((a^\Delta)_{a \in \Sigma}, \langle \rangle^\Delta, @^\Delta)$ *where:* $a^\Delta \subseteq \mathcal{Q} \times \mathcal{Q}$, $\langle \rangle^\Delta \subseteq \mathcal{Q}$, *and* $@^\Delta : (\mathcal{Q} \times \mathcal{Q}) \times \mathcal{Q}$. *A SHA is deterministic or equivalently a dSHA if* I *and* $\langle \rangle^\Delta$ *contain at most one element, and all relations* $(a^\Delta)_{a \in \Sigma}$ *and* $@^\Delta$ *are partial functions.*

The set of state $q \in \mathcal{Q}$, subsumes a subset I of initial state, a subset F of final states, and a subset $\langle \rangle^\Delta$ of tree initial states. The transition rules in Δ have three forms: If $(q, q') \in a^\Delta$ then we have a letter rule that we write as $q \xrightarrow{a} q'$ in Δ. If $(q, p, q') \in @^\Delta$ then we have an apply rule that we write as: $q@p \to q'$ in Δ. And if $q \in \langle \rangle^\Delta \in \mathcal{Q}$ then we have a tree initial rule that we denote as $\xrightarrow{\langle \rangle} q$ in Δ.

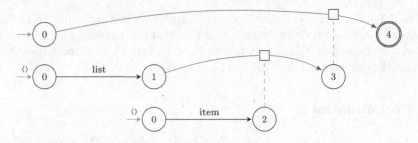

Fig. 3. A successful run of the SHA in Fig. 1 on $\langle list \cdot \langle item \rangle \rangle$.

For any hedge $h \in \mathcal{H}_\Sigma$ we define the transition relation $\xrightarrow{h}$ wrt Δ such that for all $q, q' \in \mathcal{Q}$, $a \in \Sigma$, and $h, h' \in \mathcal{H}_\Sigma$:

$$\frac{true}{q \xrightarrow{\varepsilon} q \text{ wrt } \Delta} \qquad \frac{q \xrightarrow{a} q' \text{ in } \Delta}{q \xrightarrow{a} q' \text{ wrt } \Delta} \qquad \frac{q \xrightarrow{h} q' \text{ wrt } \Delta \quad q' \xrightarrow{h'} q'' \text{ wrt } \Delta}{q \xrightarrow{h \cdot h'} q'' \text{ wrt } \Delta}$$

$$\frac{\xrightarrow{()} q' \text{ in } \Delta \quad q' \xrightarrow{h} p \quad q@p \rightarrow q'' \text{ in } \Delta}{q \xrightarrow{\langle h \rangle} q'' \text{ wrt } \Delta}$$

A run of the dSHA in Fig. 1 on the tree $\langle h \rangle$ with subhedge $h = list \cdot \langle item \rangle$ is illustrated graphically in Fig. 2. It justifies the transition $0 \xrightarrow{\langle h \rangle} 4$ wrt Δ. The run starts on the top-most level of $\langle h \rangle$ in the initial state 0 of the automaton. The run on the topmost level is suspended immediately. Instead, a run on the tree's subhedge h on the level below is started in the tree initial state, which is 0 since $\xrightarrow{()} 0$ in Δ. This run eventually ends up in state 3 justifying the transition $0 \xrightarrow{h} 3$ wrt Δ. The run of the upper level is then resumed from state 0. Given that $0@3 \rightarrow 4$ in Δ it continues in state 4. In the graph, this instance of the suspension/resumption mechanism is illustrated by the box in the edge $0 \; \text{—}\square\text{—}\!\!\rightarrow$ 4. The box stands for a future value. Eventually, the box is filled by state 3, as illustrated by $3 \; \text{-}\text{-}\text{-}\square$ so that the computation can continue. But in state 4, the upper hedge ends. Since state 4 is final the run ends successfully. The run on the subhedge justifying $0 \xrightarrow{h} 3$ wrt Δ works in analogy (Fig. 3).

A hedge is accepted if its transition started in some initial states reaches some final state. The language $\mathcal{L}(A)$ is the set of all accepted hedges:

$$\mathcal{L}(A) = \{ h \in \mathcal{H}_\Sigma \mid q \xrightarrow{h} q' \text{ wrt } \Delta, \; q \in I, \; q' \in F \}$$

For any subset $Q \subseteq \mathcal{Q}$ and hedge $h \in \mathcal{H}_\Sigma$ we define the in-memory evaluation: $[\![h]\!](Q) = \{ q' \mid q \xrightarrow{h} q' \text{ wrt } \Delta, \; q \in Q \}$. An in-memory membership tester for $h \in \mathcal{L}(A)$ can be obtained by computing $[\![h]\!](I)$ by applying the transition relation to all elements recursively and testing whether it contains some final state in F.

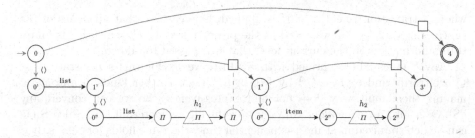

Fig. 4. A run of the $\text{SHA}^{\downarrow}$ in Fig. 2 on $\langle list \cdot \langle list \cdot h_1 \rangle \cdot \langle item \cdot h_2 \rangle\rangle$.

4 Downward Stepwise Hedge Automata (SHA$^{\downarrow}$s)

SHAs process information bottom-up and left-to-right exclusively. We next propose an extension to downward stepwise hedge automata with the ability to pass finite state information top-down. These can also be seen as an extension of Neumann and Seidl's pushdown forest automata [17] from (labeled) forests to (unlabeled) hedges.

Definition 2. *A downward stepwise hedge automaton (SHA$^{\downarrow}$) is a tuple $A = (\Sigma, Q, \Delta, I, F)$ where Σ and Q are finite sets, $I, F \subseteq Q$, and $\Delta = ((a^{\Delta})_{a \in \Sigma}, \langle\rangle^{\Delta}, @^{\Delta})$. Furthermore, $a^{\Delta} \subseteq Q \times Q$, $\langle\rangle^{\Delta} \subseteq Q \times Q$, and $@^{\Delta} : (Q \times Q) \times Q$. A SHA$^{\downarrow}$ is deterministic or equivalently a dSHA$^{\downarrow}$ if I contains at most one element, and all relations $\langle\rangle^{\Delta}$, a^{Δ}, and $@^{\Delta}$ are partial functions.*

The only difference to SHAs is the form of the tree opening rules. If $(q, q') \in \langle\rangle^{\Delta} \in Q$ then we have a tree initial rule that we denote as: $q \xrightarrow{\langle\rangle} q'$ in Δ. So here the state q' where the evaluation of a subhedge starts depends on the state q of the parent. The definition of the transition relation and thus the evaluator of a SHA$^{\downarrow}$ differs from that of a SHA by the following equation:

$$\frac{q \xrightarrow{\langle\rangle} q' \text{ in } \Delta \quad q' \xrightarrow{h} p \text{ wrt } \Delta \quad q@p \to q'' \text{ in } \Delta}{q \xrightarrow{\langle h \rangle} q'' \text{ wrt } \Delta}$$

This means that the evaluation of the subhedge h starts in some state of q' such that $q \xrightarrow{\langle\rangle} q'$ in Δ. So the restart state q' now depends on the state q above. This is how finite state information is passed top-down by SHA$^{\downarrow}$s. SHAs in contrast operate purely bottom-up and left-to-right.

An example of an in-memory evaluation on the dSHA$^{\downarrow}$ in Fig. 2 for the filter [self::list][child::item] is shown in Fig. 4. The run of SHA$^{\downarrow}$s works quite similarly to the runs of SHAs, just that when restarting a computation in the subhedge of some tree in state q, then it will start in some state q' such that $q \xrightarrow{\langle\rangle} q'$ (rather than in some tree initial state that is independent of q). This can be noticed for example when opening the first subtree labeled with item where a transition rule $1' \xrightarrow{\langle\rangle} 0''$ is applied. One can see that all nodes of the subtrees h_1 and h_2 are evaluated to the projection state Π, which holds finite-state information on the current level that was passed top-down.

Any SHA can be identified with a SHA$^\downarrow$: we fix for this some state $q_0 \in \mathcal{Q}$ arbitrarily and replace $\langle\rangle^\Delta$ by $\{q_0\} \times \langle\rangle^\Delta$. As for other kinds of automata, making them multi-way does not add expressiveness. So we can convert any dSHA$^\downarrow$ A into an equivalent SHA by introducing nondeterminism. Since SHAs can be determinized in at most exponential time, the same holds for SHA$^\downarrow$s. It is sufficient to convert it to a SHA, determinize it, and identify the resulting dSHA with a dSHA$^\downarrow$.

We next show how to get subhedge projection for SHA$^\downarrow$s. Two notions will be relevant here, automata completeness and subhedge projection states.

So let $\mathcal{A} = (\Sigma, \mathcal{Q}, \Delta, I, F)$ be a SHA$^\downarrow$. We call Δ complete if all its relations $(a^\Delta)_{a\in\Sigma}$, $\langle\rangle^\Delta$ and $@^\Delta$ are total. We call A complete if Δ is complete and $I \neq \emptyset$.

Definition 3. *We call a state $q \in \mathcal{Q}$ a subhedge projection state of Δ if there exists $q' \in \mathcal{Q}$ called the witness of q such that the set of transition rules of Δ containing q' or with q on the leftmost position is included in:*

$$\{q \xrightarrow{\langle\rangle} q',\ q@q' \to q,\ q' \xrightarrow{\langle\rangle} q',\ q'@q' \to q'\}$$
$$\cup\{q' \xrightarrow{a} q',\ q \xrightarrow{a} q \mid a \in \Sigma\}$$

In the example SHA$^\downarrow$ in Fig. 2 Π is a subhedge projection state with witness Π, but also the states $3'$, 4, and $2''$ are subhedge projection states with witness Π. Note that only inclusion holds for the latter but not equality since this automaton is not complete.

For complete SHA$^\downarrow$s A, the above set must be equal to the set of transition rules of Δ with q or q' on the leftmost position. In the soundness expressed in Proposition 4, completeness will be assumed and the proof relies on it. In the examples, however, we will consider automata that are not complete. Still they are "sufficiently complete" to illustrate the constructions.

Note that a subhedge projection state q may be equal to its witness q'. Therefore the witness q' of any subhedge projection state is itself a subhedge projection state with witness q'.

Let $\mathcal{P} \subseteq \mathcal{Q}$ be a subset of subhedge projection states of Δ. We define the transition relation with projection $\xrightarrow{h}_\mathcal{P} \subseteq \mathcal{Q} \times \mathcal{Q}$ with respect to Δ such that for all hedges $h, h' \in \mathcal{H}_\Sigma$ and letters $a \in \Sigma$:

$$\frac{q \in \mathcal{P}}{q \xrightarrow{h}_\mathcal{P} q \text{ wrt } \Delta} \qquad \frac{q \notin \mathcal{P} \quad q \xrightarrow{a} q' \text{ in } \Delta}{q \xrightarrow{a}_\mathcal{P} q' \text{ wrt } \Delta} \qquad \frac{q \notin \mathcal{P}}{q \xrightarrow{\varepsilon}_\mathcal{P} q \text{ wrt } \Delta}$$

$$\frac{q \notin \mathcal{P} \quad q \xrightarrow{h}_\mathcal{P} q' \text{ wrt } \Delta \quad q' \xrightarrow{h'}_\mathcal{P} q'' \text{ wrt } \Delta}{q \xrightarrow{h \cdot h'}_\mathcal{P} q'' \text{ wrt } \Delta}$$

$$\frac{q \notin \mathcal{P} \quad q \xrightarrow{\langle\rangle} q' \text{ in } \Delta \quad q' \xrightarrow{h}_{\mathcal{P}} p \quad q@p \to q'' \text{ in } \Delta}{q \xrightarrow{\langle h \rangle}_{\mathcal{P}} q'' \text{ wrt } \Delta}$$

Transitions with respect to $\mathcal{P}$ stay in states $q \in \mathcal{P}$ until the end of the current subhedge is reached. This is correct if p is a subhedge projection state since transitions without subhedge projection don't change state p nor if the run is not blocking.

Proposition 4. *Let* $\mathcal{A} = (\Sigma, \mathcal{Q}, \Delta, I, F)$ *be a complete* SHA$^{\downarrow}$ *and* $\mathcal{P}$ *a subset of subhedge projection states for* Δ. *Then for all hedges* $h \in \mathcal{H}_{\Sigma}$ *and states* $q, q' \in \mathcal{Q}$: $q \xrightarrow{h} q'$ *wrt* Δ *iff* $q \xrightarrow{h}_{\mathcal{P}} q'$ *wrt* Δ.

For any subset $Q \subseteq \mathcal{Q}$ and hedge $h \in \mathcal{H}_{\Sigma}$, we define the in-memory evaluation with subhedge projection: $[\![h]\!]_{\mathcal{P}}(Q) = \{q' \in \mathcal{Q} \mid q \xrightarrow{h}_{\mathcal{P}} q' \text{ wrt } \Delta, \ q \in Q\}$. An in-memory membership tester for $h \in \mathcal{L}(A)$ with subtree projection can be obtained by computing $[\![h]\!]_{\mathcal{P}}(I)$ and testing whether it contains some state in F.

5   Compiling SHAs to SHA$^{\downarrow}$s with Projection States

We show how to compile any SHA to some SHA$^{\downarrow}$ with subhedge projection states, yielding an evaluator with appropriate subhedge projection for the SHA via the SHA$^{\downarrow}$. This compiler is the most original contribution of the paper.

Let $A = (\Sigma, \mathcal{Q}, \Delta, I, F)$ be a SHA. For any set $Q \subseteq \mathcal{Q}$ we define the set $acc^{\Delta}(Q) = \{q' \in \mathcal{Q} \mid \exists q \in Q, \ h \in \mathcal{H}_{\Sigma}. \ q \xrightarrow{h} q' \text{ wrt } \Delta\}$. We note that $acc^{\Delta}(Q)$ can be computed in linear time in the size of Δ. We define:

$$safe^{\Delta}(Q) = \{q \in \mathcal{Q} \mid acc^{\Delta}(\{q\}) \subseteq Q\}$$

If A is complete and deterministic then safety can be used to characterize universal states, since for all $q \in \mathcal{Q}$: $L(A[I/\{q\}]) = \mathcal{H}_{\Sigma}$ if and only if $q \in safe^{\Delta}(F)$. See Lemma 5 of [3]. Note that $safe^{\Delta}(Q)$ can be computed in linear time in the size of Δ. We consider pairs (q, Q) consisting of a current state q and a set of forbidden states Q that must not be reached at the end of the hedge. We define:

$$sdown^{\Delta}(q, Q) = safe^{\Delta}(\{p \in \mathcal{Q} \mid q@^{\Delta}p \subseteq Q\})$$
$$no\text{-}change^{\Delta}(q, Q) = sdown^{\Delta}(q, Q \cup \{q\})$$

The states in $sdown^{\Delta}(q, Q)$ can only access states $p \in \mathcal{Q}$ such that $q@^{\Delta}p$ is included in Q. They are safe for $\{p \in \mathcal{Q} \mid q@^{\Delta}p \subseteq Q\}$. The states in $no\text{-}change^{\Delta}(q, Q)$ safely either go to states p whose application doesn't change q or lead to Q. For instance in Fig. 1, $no\text{-}change^{\Delta}(1, \{2, 4\}) = sdown^{\Delta}(1, \{1, 2, 4\}) = \{1, 3, 4\}$. Note that, not only $1@^{\Delta}1 = 1 \subseteq \{1, 2, 4\}$, but also $1@^{\Delta}0 = 1@^{\Delta}4 = \emptyset \subseteq \{1, 2, 4\}$. Therefore, $sdown^{\Delta}(1, \{1, 2, 4\}) = safe^{\Delta}(\{0, 1, 3, 4\}) = \{1, 3, 4\}$.

We next compile the SHA A to a SHA$^{\downarrow}$ $A^{\pi} = (\Sigma, \mathcal{Q}^{\pi}, \Delta^{\pi}, I^{\pi}, F^{\pi})$. For this let Π be a fresh symbol and consider the state set: $\mathcal{Q}^{\pi} = \{\Pi\} \uplus (\mathcal{Q} \times 2^{\mathcal{Q}})$. A

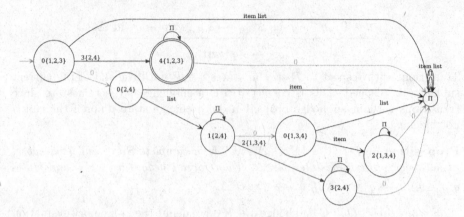

Fig. 5. The dSHA$^{\downarrow}$ A^{π} constructed from the dSHA A in Fig. 1 except for useless state transitions leading out of the schema of our application.

pair (q, Q) means that the evaluator is in state q but must not reach any state in Q. We next define projection of such pairs with respect to Δ:

$$\pi(q, Q) = if \ q \in Q \ then \ \Pi \ else \ (q, Q)$$

The sets of initial and final states are defined as follows:

$$I^{\pi} = \{\pi(q, safe^{\Delta}(\mathcal{Q} \setminus F)) \mid q \in I\} \qquad F^{\pi} = \{(q, safe^{\Delta}(\mathcal{Q} \setminus F)) \mid q \in F\}$$

So at the beginning, the set of forbidden states are those in $safe^{\Delta}(\mathcal{Q} \setminus F)$. The transition rules in Δ^{π} are given by the following inference rules where $p, q \in \mathcal{Q}$, $P, Q \subseteq \mathcal{Q}$ and $a \in \Sigma$.

$$\frac{q \xrightarrow{a} q' \ in \ \Delta}{\pi(q, Q) \xrightarrow{a} \pi(q', Q) \ in \ \Delta^{\pi}} \qquad \frac{\xrightarrow{\langle\rangle} q' \ in \ \Delta \qquad P = no\text{-}change^{\Delta}(q, Q)}{\pi(q, Q) \xrightarrow{\langle\rangle} \pi(q', P) \ in \ \Delta^{\pi}}$$

$$\frac{q@p \to q' \ in \ \Delta \qquad P = no\text{-}change^{\Delta}(q, Q)}{\pi(q, Q)@\pi(p, P) \to \pi(q', Q) \ in \ \Delta^{\pi}}$$

When going down to a subhedge from state q with forbidden states Q, the next set of forbidden states is $no\text{-}change^{\Delta}(q, Q)$. This is where finite state information is passed down. States in $no\text{-}change^{\Delta}(q, Q)$ cannot lead to any change of state q, so that subhedge projection can be applied.

When applied to the SHA in Fig. 1 for [self::list][child::item], the construction yields the SHA$^{\downarrow}$ in Fig. 5 which is indeed equal to the SHA$^{\downarrow}$ from Fig. 2 up to state renaming. When run on the hedge $\langle list \cdot \langle list \cdot h_1 \rangle \cdot \langle item \cdot h_2 \rangle \rangle$ as shown in Fig. 4, it does not have to visit the subhedges h_1 nor h_2, since all of them will be reached starting from the projection state Π.

Proposition 5 (Soundness). $\mathcal{L}(A^\pi) = \mathcal{L}(A)$ *for any complete* SHA *A*.

Proof. For the inclusion $\mathcal{L}(A) \subseteq \mathcal{L}(A^\pi)$ we can show for all hedge $h \in \mathcal{H}_\Sigma$ and states $q, q' \in \mathcal{Q}$ that $q \xrightarrow{h} q'$ wrt Δ implies $q \xrightarrow{h} q'$ wrt Δ^π. This is straightforward by induction on the structure of h.

The inverse inclusion $\mathcal{L}(A^\pi) \subseteq \mathcal{L}(A)$ is less obvious. We have to show that not inspecting projected subhedges does not change the language. Intuitively, this is since the projected subhedges are irrelevant for detecting acceptance. They either don't change the state or lead to rejection.

Claim. For all $h \in \mathcal{H}_\Sigma$, $q \in \mathcal{Q}$, $Q' \subseteq \mathcal{Q}$, $Q = safe^\Delta(Q')$, and $\mu \in \mathcal{Q}^\pi$ the hypothesis $(q, Q) \xrightarrow{h} \mu$ wrt Δ^π implies:

1. if exists $q' \in \mathcal{Q}$ such that $\mu = (q', Q)$ then $q \xrightarrow{h} q'$ wrt Δ.
2. if $\mu = \Pi$ then there exists $q' \in Q$ such that $q \xrightarrow{h} q'$ wrt Δ.

The lengthy proof is by induction on the structure of h. It remains to show that the Claim implies $\mathcal{L}(A^\pi) \subseteq \mathcal{L}(A)$. So let $h \in \mathcal{L}(A^\pi)$. Then there exist $q \in I$ and $q' \in F$ such that for $Q = safe^\Delta(\mathcal{Q} \setminus F)$: $(q, Q) \xrightarrow{h} (q', Q)$ wrt Δ^π Part 1. of the Claim implies that $q \xrightarrow{h} q'$ wrt Δ, so that $h \in \mathcal{L}(A)$. (Part 2 intervenes only for proving Part 1 by induction.) □

The projecting in-memory evaluator of A^π will be more efficient than that the nonprojecting evaluator of A. Note, however, that the size of A^π may be exponentially bigger than that of A. Therefore, for evaluating a dSHA A with subhedge projection on a given hedge h, we create only the needed part of A^π on the fly. This part has size $O(|h|)$ and can be computed in time $O(|A|\,|h|)$, so the exponential preprocessing time is avoid.

Example 6. *In order to see how the exponential worst case may happen, we consider a family of regular languages, for which the minimal left-to-right* DFA *is exponentially bigger than the minimal right-to-left* DFA. *The classical example languages with this property are* $L_n = \Sigma^*.a.\Sigma^n$ *where* $n \in \mathbb{N}$ *and* $\Sigma = \{a, b\}$. *Intuitively, a word in* Σ^* *belongs to* L_n *if and only its* $n+1$-*th letter from the end is equal to "a". The minimal left-to-right* DFA *for* L_n *has* 2^{n+1} *many states, since needs to memoize a window of* $n+1$-*letters. In contrast, its minimal right-to-left* DFA *has only* $n+1$ *states; in this direction, it is sufficient to memoize the distance from the end modulo* $n+1$.

We next consider the family of hedge languages $H_n \in \mathcal{H}_\Sigma$ *such that each node of* $h \in H_n$ *is labeled by one symbol in* Σ *and so that the sequence of labels of some root-to-leave path of* h_n *belongs to* L_n. *Note that* H_n *can be recognized in a bottom-up manner by the* dSHA A_n *with* $O(n+1)$ *states, which simulates the minimal deterministic* DFA *of* L_n *on all paths of the input hedge. For an evaluator with subhedge projection the situation is different. When moving top-down, it needs to memoize the sequence of labels of the* $n+1$-*last ancestors, possibly filled with* b's, *and there a* 2^{n+1} *such sequences. If for some leaf, its sequence starts with an "a" then the following subhedges with the following leaves can be projected away. As a consequence, there cannot be any* $\mathrm{SHA}^{\downarrow}$ *recognizing* H_n *that projects away all irrelevant subhedges with less than* 2^{n+1} *states. In particular, the size of* A_n^{π} *must be exponential in the size of* A_n.

6   Streaming Evaluators for SHA↓s

Any $\mathrm{SHA}^{\downarrow}$ yields a visibly pushdown machine [11] that evaluates nested words in a streaming manner. The same property was already noticed for Neumann and Seidl's pushdown forest automata [8].

Let $\mathcal{A} = (\Sigma, \mathcal{Q}, \Delta, I, F)$ be a $\mathrm{SHA}^{\downarrow}$. A configuration of the corresponding visibly pushdown machine is a pair in $\mathcal{K} = \mathcal{Q} \times \mathcal{Q}^*$ containing a state and a stack of states. For any word $v \in \hat{\Sigma}^*$ we define the transition relation of the visibly pushdown machine $\overset{v \ \mathrm{str}}{\to} \subseteq \mathcal{K} \times \mathcal{K}$ such that for all $q, q' \in \mathcal{Q}$ and $\sigma \in \mathcal{Q}^*$:

$$\frac{true}{(q, \sigma) \overset{\varepsilon \ \mathrm{str}}{\to} (q, \sigma) \ \mathrm{wrt} \ \Delta} \qquad \frac{(q, \sigma) \overset{v \ \mathrm{str}}{\to} (q', \sigma) \quad (q', \sigma) \overset{v' \ \mathrm{str}}{\to} (q'', \sigma) \ \mathrm{wrt} \ \Delta}{(q, \sigma) \overset{v \cdot v' \ \mathrm{str}}{\to} (q'', \sigma) \ \mathrm{wrt} \ \Delta}$$

$$\frac{q \overset{a}{\to} q' \ \mathrm{in} \ \Delta}{q \overset{a \ \mathrm{str}}{\to} q' \ \mathrm{wrt} \ \Delta} \qquad \frac{q \overset{\langle\rangle}{\to} q' \ \mathrm{in} \ \Delta}{(q, \sigma) \overset{\langle \ \mathrm{str}}{\to} (q', \sigma \cdot q) \ \mathrm{wrt} \ \Delta} \qquad \frac{q@p \to q' \ \mathrm{in} \ \Delta}{(p, \sigma \cdot q) \overset{\rangle \ \mathrm{str}}{\to} (q', \sigma) \ \mathrm{wrt} \ \Delta}$$

The same visibly pushdown machine can be obtained by compiling the SHA to an NWA. In analogy to Theorem 4 of [8], we can show for any hedge h that the streaming transition relation $\overset{nw(h) \ \mathrm{str}}{\longrightarrow}$ wrt Δ is correct for its in-memory transition relation $\overset{h}{\to}$ wrt Δ:

Proposition 7. $L(A) = \{h \in \mathcal{H}_{\Sigma} \mid (q, \varepsilon) \overset{nw(h) \ \mathrm{str}}{\longrightarrow} (q', \varepsilon) \ \mathrm{wrt} \ \Delta, \ q \in I, \ q' \in F\}$.

Any nested word $v \in \hat{\Sigma}^*$ can be evaluated in streaming mode on any subset of configurations $K \subseteq \mathcal{K}$: $[\![v]\!]_{str}(K) = \{(q', \sigma') \mid (q, \sigma) \overset{v \ \mathrm{str}}{\to} (q', \sigma') \ \mathrm{wrt} \ \Delta, \ (q, \sigma) \in K\}$. So any hedge can be evaluated in streaming mode by computing $[\![nw(h)]\!]_{str}(I \times \{\varepsilon\})$. The hedge is accepted if it can reach some final configuration in $F \times \{\varepsilon\}$.

Going one step further, we show how to enhance the streaming evaluator of an $\mathrm{SHA}^{\downarrow}$ with subhedge projection, in analogy to the in-memory evaluator. This approach yields a similar result in a more direct manner, as obtained by mapping $\mathrm{SHA}^{\downarrow}$s to NWAs, identifying subtree projection states there, and mapping NWAs with subtree projection states to projecting NWAs [19].

Let $\mathcal{P} \subseteq \mathcal{Q}$ be the subset of subhedge projection states of Δ. We define a transition relation with subhedge projection $\overset{h \ \mathrm{str}}{\to}_{\mathcal{P}} \subseteq \mathcal{K} \times \mathcal{K}$ with respect to Δ such that for all nested words $v, v' \in \mathcal{N}_{\Sigma}$, letters $a \in \Sigma$, states $p, q, q', q'' \in \mathcal{Q}$ and stacks $\sigma, \sigma', \sigma'' \in \mathcal{Q}^*$:

$$\frac{q \in \mathcal{P}}{(q, \sigma) \overset{v \ \mathrm{str}}{\to}_{\mathcal{P}} (q, \sigma) \ \mathrm{wrt} \ \Delta} \qquad \frac{q \notin \mathcal{P} \quad q \overset{a}{\to} q' \ \mathrm{in} \ \Delta}{(q, \sigma) \overset{a \ \mathrm{str}}{\to}_{\mathcal{P}} (q', \sigma) \ \mathrm{wrt} \ \Delta} \qquad \frac{q \notin \mathcal{P}}{q \overset{\varepsilon \ \mathrm{str}}{\to}_{\mathcal{P}} q \ \mathrm{wrt} \ \Delta}$$

$$\frac{q \notin \mathcal{P} \qquad (q,\sigma) \xrightarrow{v}{}^{str}_{\mathcal{P}} (q',\sigma') \text{ wrt } \Delta \qquad (q',\sigma') \xrightarrow{v'}{}^{str}_{\mathcal{P}} (q'',\sigma'') \text{ wrt } \Delta}{(q,\sigma) \xrightarrow{v \cdot v'}{}^{str}_{\mathcal{P}} (q'',\sigma'') \text{ wrt } \Delta}$$

$$\frac{q \notin \mathcal{P} \qquad q \xrightarrow{\langle\rangle} q' \text{ in } \Delta}{(q,\sigma) \xrightarrow{\langle}{}^{str}_{\mathcal{P}} (q',\sigma \cdot q) \text{ wrt } \Delta} \qquad\qquad \frac{p \notin \mathcal{P} \qquad q@p \to q' \text{ in } \Delta}{(p,\sigma \cdot q) \xrightarrow{\rangle}{}^{str}_{\mathcal{P}} (q',\sigma) \text{ wrt } \Delta}$$

The projecting transition relation stays in a configuration with a projection state until the end of the current subhedge is reached. This is correct since the state of the non-projecting transition relation would not change the state either, while the visible stack comes back to its original value after the evaluation of a nested word (that by definition is well-nested).

Proposition 8. *Let v be a word in $\hat{\Sigma}^*$, Δ a set of transition rules of a complete* SHA$^{\downarrow}$ *with state set $\mathcal{Q}$, $q \in \mathcal{Q}$ a state and $\sigma \in \mathcal{Q}^*$ a stack. For any subset $\mathcal{P} \subseteq \mathcal{Q}$ of subhedge projection states of Δ: $(q,\sigma) \xrightarrow{v}{}^{str} (q',\sigma')$ wrt Δ iff $(q,\sigma) \xrightarrow{v}{}^{str}_{\mathcal{P}} (q',\sigma')$ wrt Δ.*

For any subset $K \subseteq \mathcal{K}$ and nested word $v \in \mathcal{N}_\Sigma$ we define the streaming evaluation with subhedge projection:

$$[\![v]\!]^{str}_{\mathcal{P}}(K) = \{(q',\sigma') \mid (q,\sigma) \xrightarrow{v}{}^{str}_{\mathcal{P}} (q',\sigma') \text{ wrt } \Delta, \ (q,\sigma) \in K\}$$

A streaming membership tester for $h \in \mathcal{L}(A)$ with subtree projection can be obtained by computing $[\![nw(h)]\!]^{str}_{\mathcal{P}}(I \times \{\varepsilon\})$ and testing whether it contains some state in $F \times \{\varepsilon\}$.

7   Earliest Membership with Subhedge Projection

We next enhance our compiler from SHAs to SHA$^{\downarrow}$s for introducing subtree projection such that it can take safe rejection and safe selection into account. The streaming version for deterministic SHAs leads us to an earliest membership tester, which enhances the previous earliest membership tester for dSHAs from [3] with subtree projection.

The idea is as follows: A state is called safe for rejection if whenever the evaluator reaches this state on some subhedge then it can safely reject the hedge independently of the parts that were not yet evaluated. In analogy, a state is safe for selection if whenever the evaluator reaches this state for some subhedge, the full hedge will be accepted.

Given an $\mathcal{A} = (\Sigma, \mathcal{Q}, \Delta, I, F)$, at the beginning all states in $S_0 = safe^{\Delta}(F)$ are safe for selection and all states in $R_0 = safe^{\Delta}(\mathcal{Q} \setminus F)$ are safe for rejection. We will have to update these sets when moving down a tree. The states of our SHA$^{\downarrow}$ will contain tuples (q, Q, R, S) stating that the evaluator is in state q, that the states in Q are safe no-changes, the states in R are safe for rejection, and the states in S safe for selection. We define:

$$no\text{-}change^{\Delta}_e(q, Q, R, S) = no\text{-}change^{\Delta}(q, Q) \setminus \ sdown^{\Delta}(q, R) \setminus sdown^{\Delta}(q, S)$$

That is, state changes are now relevant only if they don't move to states that are safe for rejection or selection. Let *sel* and *rej* be two fresh symbols beside of Π. We adapt tuple projection as follows:

$$\pi_e(q, Q, R, S) = \begin{cases} \textit{if } q \in S \textit{ then sel else if } q \in R \textit{ then rej} \\ \textit{else if } q \in S \textit{ then } \Pi \textit{ else } (p, Q, R, S) \end{cases}$$

We next compile the given SHA A to a SHA$^\downarrow$ $A_e^\pi = (\Sigma, \mathcal{Q}_e^\pi, \Delta_e^\pi, I_e^\pi, F_e^\pi)$. The state sets of A_e^π are:

$$\mathcal{Q}_e^\pi = \pi_e(Q \times 2^{\mathcal{Q}} \times 2^{\mathcal{Q}} \times 2^{\mathcal{Q}}) \quad I_e^\pi = \{\pi_e(q, R_0, R_0, S_0) \mid q \in I\}$$
$$F_e^\pi = \{\pi_e(q, R_0, R_0, S_0) \mid q \in F\}$$

The transition rules in Δ_e^π are given by the following inference rules where $p, q \in \mathcal{Q}$, $P, Q, R, S \subseteq \mathcal{Q}$ and $a \in \Sigma$.

$$\frac{q \xrightarrow{a} q' \text{ in } \Delta}{\pi_e(q, Q, R, S) \xrightarrow{a} \pi_e(q', Q, R, S) \text{ in } \Delta_e^\pi}$$

$$\frac{\xrightarrow{\langle\rangle} q' \quad P = \textit{no-change}_e^\Delta(q, Q, R, S)}{\pi_e(q, Q, R, S) \xrightarrow{\langle\rangle} \pi_e(q', P, \textit{sdown}^\Delta(q, R), \textit{sdown}^\Delta(q, S)) \text{ in } \Delta_e^\pi}$$

$$\frac{q@p \to q' \text{ in } \Delta \quad P = \textit{no-change}_e^\Delta(q, Q, R, S)}{\pi_e(q, Q, R, S)@\pi_e(p, P, \textit{sdown}^\Delta(q, R), \textit{sdown}^\Delta(q, S)) \to \pi_e(q', Q, R, S) \text{ in } \Delta_e^\pi}$$

The dSHA from Fig. 1 is not sufficiently complete to obtain the expected results. The problem is that $\textit{sdown}^\Delta(0, \{4\}) = \mathcal{Q}$ there, but only state 3 is really safe for selection. We therefore add a sink state to it.

Running the SHA$^\downarrow$s A_e^π in streaming mode with subtree projection yields an earliest membership tester for *d*SHAs with subtree projection. Two adaptations are in order. Whenever the safe rejection state *rej* is reached, the computation can stop and the hedge on the input stream is rejected. And whenever the safe selection state *sel* is reached, the evaluation can be stopped and the input hedge on the stream is accepted.

Theorem 1. *For any d*SHA $\mathcal{A} = (\Sigma, \mathcal{Q}, \Delta, I, F)$ *and hedge* $h \in \mathcal{H}_\Sigma$ *with* $[\![h]\!](I) \neq \emptyset$ *wrt* Δ *the streaming evaluator* $[\![h]\!]_{\{\Pi, sel, rej\}}^{str}(I_e^\pi)$ *with respect to* Δ_e^π *can check membership* $h \in \mathcal{L}(A)$ *at the earliest event when streaming* $nw(h)$.

The hedge h is accepted once the evaluator reaches state *sel* and rejected once the evaluator reaches state *rej*. If neither happens the truth value of $q \in F$ is returned where q is the state in the final tuple.

Proof (sketch). The streaming membership tester with safe selection and rejection by computing $[\![h]\!]_{\{sel, rej\}}^{str}(I_e^\pi)$ wrt. Δ_e^π can be shown to be similar to the earliest membership tester from Proposition 6 of [3] enhanced with safe rejection.

So we can rely on the definitions of earliest membership testing and the result given there. When adding subtree projection by computing $[\![h]\!]^{str}_{\{\Pi,sel,rej\}}(I^{\pi}_e)$ wrt. Δ^{π}_e, the only difference is that the evaluator ignores some subtrees in which the state does not change. Clearly, this does not affect earliest selection and rejection, so we still have an earliest membership tester, but now with subtree projection.

8   Experimental Evaluation

We integrated subhedge projection into the earliest query answering tool AStream [3]. It is implemented in Scala while computing safety with ABC Datalog.

In order to benchmark AStream 2.01 with subhedge projection for efficiency, and to compare it to AStream 1.01 without projection, we considered the regular XPath queries from the XPathMark [6] $A1-A8$. We used the deterministic SHAs for all these XPath queries constructed by the compiler from [18]. These were evaluated on XML documents of variable size created by the XPathMark generator. We did further experiments on a sub-corpus of 79 regular XPath queries extracted by Lick and Schmitz from real-world XSLT and XQUERY programs, for which dSHAs are available [2]. These experiments confirm the results presented here, so we don't describe them in detail.

The XPath queries of the XPathMark without descendant axis are A1,A4 and A6-A8. The evaluation time on these queries a 1.2 GB document are reduced between $88-97\%$. In average, it is 92.5%, so the overall time is divided by 12. While the parsing time remains unchanged the gain on the automaton evaluation time is proportional to the percentage of subhedge projection for the respective query. This remains true for the other queries with the descendant axis, just that the projection percentage is much lower.

Finally, we compared AStream with for QuiXPath [5], the best previous streaming tool that can answer A1-A8 in an earliest manner. QuiXPath compiles regular XPath queries to possibly nondeterministic early NWAs, and evaluates them with subtree and descendant projection [19]. QuiXPath is not generally earliest though. On the queries without descendant axis, AStream 1.01 without projection is by a factor of 60 slower than QuiXPath [3]. With subhedge projection in version 2.01, the overhead goes down to a factor of $5 = 60/12$. So our current implementation is close to becoming competitive with the best existing streaming tool while guaranteeing earliest query answering in addition.

9 Conclusion and Future Work

We developed evaluators with subhedge projection for SHAs in in-memory mode and in streaming mode. One difficulty was how to push the needed finite state information for subtree projection top-down given that SHAs operate bottom-up. We solved it based on a compiler from SHAs to downward SHAs. This compiler propagates safety information about non-changing states, similar to the propagation of safety information proposed for earliest query answering for dSHA queries on nested word streams. We confirmed the usefulness of our novel subhedge projection algorithm for SHAs experimentally. We showed that it can indeed speed up the best previously existing earliest query answering algorithm for dSHA queries on nested word streams, as needed for answering regular XPath queries on XML streams. In future work, we plan to improve on subhedge projection for SHAs with descendant projection for SHAs and to use it for efficient stream processing. Another question is whether and how to obtain completeness results for subhedge projection.

References

1. Al Serhali, A., Niehren, J.: Subhedge projection for stepwise hedge automata
2. Al Serhali, A., Niehren, J.: A benchmark collection of deterministic automata for XPath queries. In: XML Prague 2022, Prague, Czech Republic (2022)
3. Al Serhali, A., Niehren, J.: Earliest query answering for deterministic stepwise hedge automata (2023)
4. Comon, H., et al.: Tree automata techniques and applications (1997). http://tata.gforge.inria.fr (2007)
5. Debarbieux, D., Gauwin, O., Niehren, J., Sebastian, T., Zergaoui, M.: Early nested word automata for XPath query answering on XML streams. Theor. Comput. Sci. **578**, 100–125 (2015)
6. Franceschet, M.: XPathmark performance test. https://users.dimi.uniud.it/~massimo.franceschet/xpathmark/PTbench.html. Accessed 25 Oct 2020
7. Frisch, A.: Regular tree language recognition with static information. In: Exploring New Frontiers of Theoretical Informatics, IFIP 18th World Computer Congress, TCS 3rd International Conference on Theoretical Computer Science, pp. 661–674 (2004)
8. Gauwin, O., Niehren, J., Roos, Y.: Streaming tree automata. Inf. Process. Lett. **109**(1), 13–17 (2008)
9. Gauwin, O., Niehren, J., Tison, S.: Earliest query answering for deterministic nested word automata. In: Kutyłowski, M., Charatonik, W., Gębala, M. (eds.) FCT 2009. LNCS, vol. 5699, pp. 121–132. Springer, Heidelberg (2009). https://doi.org/10.1007/978-3-642-03409-1_12
10. Kay, M.: The Saxon XSLT and XQuery processor (2004). https://www.saxonica.com
11. Kumar, V., Madhusudan, P., Viswanathan, M.: Visibly pushdown automata for streaming XML. In: 16th International Conference on World Wide Web, pp. 1053–1062. ACM-Press (2007)
12. Madhusudan, P., Viswanathan, M.: Query automata for nested words. In: Královič, R., Niwiński, D. (eds.) MFCS 2009. LNCS, vol. 5734, pp. 561–573. Springer, Heidelberg (2009). https://doi.org/10.1007/978-3-642-03816-7_48

13. Maneth, S., Nguyen, K.: XPath whole query optimization. VLPB J. **3**(1), 882–893 (2010)
14. Marian, A., Siméon, J.: Projecting XML documents. In: VLDB, pp. 213–224 (2003)
15. Mozafari, B., Zeng, K., Zaniolo, C.: High-performance complex event processing over XML streams. In: Candan, K.S., et al. (eds.) SIGMOD Conference, pp. 253–264. ACM (2012)
16. Neumann, A., Seidl, H.: Locating matches of tree patterns in forests. In: Arvind, V., Ramanujam, S. (eds.) FSTTCS 1998. LNCS, vol. 1530, pp. 134–145. Springer, Heidelberg (1998). https://doi.org/10.1007/978-3-540-49382-2_12
17. Neumann, A., Seidl, H.: Locating matches of tree patterns in forests. In: Arvind, V., Ramanujam, S. (eds.) FSTTCS 1998. LNCS, vol. 1530, pp. 134–145. Springer, Heidelberg (1998). https://doi.org/10.1007/978-3-540-49382-2_12
18. Niehren, J., Sakho, M.: Determinization and minimization of automata for nested words revisited. Algorithms **14**(3), 68 (2021)
19. Sebastian, T., Niehren, J.: Projection for nested word automata speeds up XPath evaluation on XML streams. In: Freivalds, R.M., Engels, G., Catania, B. (eds.) SOFSEM 2016. LNCS, vol. 9587, pp. 602–614. Springer, Heidelberg (2016). https://doi.org/10.1007/978-3-662-49192-8_49
20. Thatcher, J.W.: Characterizing derivation trees of context-free grammars through a generalization of automata theory. J. Comput. Syst. Sci. **1**, 317–322 (1967)

The Rectilinear Convex Hull of Line Segments

Carlos Alegría[1], Justin Dallant[2], Pablo Pérez-Lantero[3], and Carlos Seara[4(✉)]

[1] Dipartimento di Ingegneria, Università Roma Tre, Rome, Italy
carlos.alegria@uniroma3.it
[2] Computer Science Department, Faculté des Sciences, Université libre de Bruxelles, Bruxelles, Belgium
justin.dallant@ulb.be
[3] Departamento de Matemática y Computación, Universidad de Santiago de Chile, Santiago, Chile
pablo.perez.l@usach.cl
[4] Departament de Matemàtiques, Universitat Politècnica de Catalunya, Barcelona, Spain
carlos.seara@upc.edu

Abstract. We explore an extension to rectilinear convexity of the classic problem of computing the convex hull of a collection of line segments. Namely, we solve the problem of computing and maintaining the rectilinear convex hull of a set of n line segments, while we simultaneously rotate the coordinate axes by an angle that goes from 0 to 2π.

We describe an algorithm that runs in optimal $\Theta(n \log n)$ time and $\Theta(n\alpha(n))$ space for segments that are non-necessarily disjoint, where $\alpha(n)$ is the inverse of the Ackermann's function. If instead the line segments form a simple polygonal chain, the algorithm can be adapted so as to improve the time and space complexities to $\Theta(n)$.

Keywords: rectilinear convex hull · line segments · polygonal lines

1 Introduction

The problem of computing the convex hull of a finite set of points is one of the foundational problems of Computational Geometry. Since it was first proposed in the late seventies, this problem has been extensively studied and remains a central topic in the field, as demonstrated by the rich body of variations that span across different types of point sets, algorithmic approaches, and high dimensional spaces; see for example [3,4,22]. In this paper we extend the previous work on the convex hull problem, by introducing a variation that combines two classic research directions: the problem of computing the convex hull, but of a collection of line segments instead of a finite set of points; and the problem of computing the convex hull, but using a non-traditional notion of convexity called *Orthogonal Convexity*[1] instead of the standard notion of convexity.

---

[1] In the literature, orthogonal convexity is also known as *ortho-convexity* [19] or *x-y convexity* [14].

H. Fernau and K. Jansen (Eds.): FCT 2023, LNCS 14292, pp. 32–45, 2023.
https://doi.org/10.1007/978-3-031-43587-4_3

Throughout this paper, let P be a set of n line segments in the plane. The convex hull of P, $\mathcal{CH}(P)$, is the closed region obtained by removing from the plane all the open half planes whose intersection with P is empty. The ortho-convex analog of the (standard) convex hull is called the *Rectilinear Convex Hull* [15]. The rectilinear convex hull of P, $\mathcal{RCH}(P)$, is the closed region obtained by removing from the plane all the open and axis-aligned wedges of aperture angle $\frac{\pi}{2}$, whose intersection with P is empty (a formal definition is given in Sect. 2). See Fig. 1.

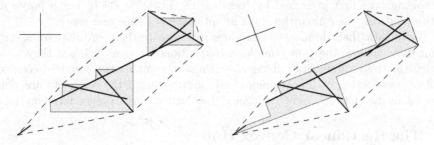

Fig. 1. A collection P of segments, and the rectilinear convex hull of P for two different orientations of the coordinate axes, which are shown in the top-left of each figure. The interior and the boundary of $\mathcal{RCH}(P)$ are shown respectively, in light and dark brown. The dotted line is the boundary of the standard convex hull of P. On the left, $\mathcal{RCH}(P)$ has two connected components, one of which is a single line segment. On the right, $\mathcal{RCH}(P)$ is formed by a single connected component.

The rectilinear convex hull introduces two important differences with respect to the standard convex hull. On one hand, observe that $\mathcal{RCH}(P)$ might be a simply connected set, yielding an intuitive and appealing structure. However, if the union of the segments in P is disconnected, then $\mathcal{RCH}(P)$ can have several simply connected components, some of which may be single line segments of P. On the other hand, note that $\mathcal{RCH}(P)$ is *orientation-dependent*, meaning that the orientation of the empty wedges changes along with the orientation of the coordinate axes, hence the shape of $\mathcal{RCH}(P)$ changes as well.

The problem we study in this paper consists in computing and maintaining the rectilinear convex hull of a collection of line segments, while we simultaneously rotate the coordinate axes by an angle that goes from 0 to 2π. We describe algorithms to solve this problem for the cases in which the segments in P are not necessarily disjoint, or form a simple polygonal chain. The algorithm for the first case takes optimal $\Theta(n \log n)$ time and $\Theta(n\alpha(n))$ space, where $\alpha(n)$ is the extremely slowly-growing inverse of Ackermann's function. We adapt this algorithm to solve the second case in optimal $\Theta(n)$ time and space.

Background and Related Work. The problem of computing the rectilinear convex hull has been predominantly explored on scenarios where the input is a finite set of points. Several results can be found for both the cases in which the coordinate axes are kept fixed or simultaneously rotating; see for example [1,2,6,16]. As far as we are aware, there are no previous results related to the computation of the rectilinear convex hull of a collection of line segments while the coordinate axes are rotating. Nonetheless, there are related results for the case where the coordinate axes are kept fixed. The problem of computing the rectilinear convex hull of a polygon was first studied by Nicholl et al. [14]. The authors considered a collection of orthogonal polygons, and presented an algorithm that runs in $O(n \log n)$ time and $O(n)$ space, where n is the number of edges of all the polygons in the collection. An algorithm for polygons that are non-necessarily orthogonal was first presented by Rawlins [18, Theorem 7.4.1]. For a polygon with n vertices, their algorithm runs in optimal $\Theta(n)$ time and space.

We remark that, despite we introduce rotations to the coordinate axes, our algorithms achieve the time complexities mentioned above. Notably, they also match the time complexities of the algorithms to compute the standard convex hull of a collection of line segments: $\Theta(n \log n)$ time if the segments are not necessarily disjoint [17], and $\Theta(n)$ time if they form a simple polygonal chain [12].

2   The Rectilinear Convex Hull

We start with a formal definition of the rectilinear convex hull[2]. For the sake of completeness, we also briefly describe the properties of the rectilinear convex hull that are relevant to our work. More details on these and other properties can be found in [7,15].

The *orientation* of a line is the smallest of the two possible angles it makes with the X^+ positive semiaxis. A *set of orientations* is a set of lines with different orientations passing through some fixed point. Throughout this paper, we consider a set of orientations that is formed by two orthogonal lines. For the sake of simplicity, we assume that both lines are passing through the origin and are parallel to the coordinate axes. We denote such an orientation set with $\mathcal{O}$. We say that a region is $\mathcal{O}$-*convex*, if its intersection with a line parallel to a line of $\mathcal{O}$ is either empty, a point, or a line segment[3,4].

Let ρ_1 and ρ_2 be two rays leaving a point $x \in \mathbb{R}^2$ such that, after rotating ρ_1 around x by an angle of $\theta \in [0, 2\pi)$, we obtain ρ_2. We refer to the two open regions in the set $\mathbb{R}^2 \setminus (\rho_1 \cup \rho_2)$ as *wedges*. We say that both wedges have vertex x and sizes θ and $2\pi - \theta$, respectively. For an angle ω, we say that an ω-*wedge* is a wedge of size ω. A *quadrant* is a $\frac{\pi}{2}$-wedge whose rays are parallel to the lines of $\mathcal{O}$. We say a region is *free of points of P*, or P-*free* for short, if its intersection with P is empty. The rectilinear convex hull of P is the closed and $\mathcal{O}$-convex set

$$\mathcal{RCH}(P) = \mathbb{R}^2 \setminus \bigcup_{q \in \mathcal{Q}} q,$$

---

[2] See the definition of the *maximal r-convex hull* in [15].

[3] In the literature, $\mathcal{O}$-convexity is also known as *D-convexity* [20], *Directional convexity* [8], *Set-theoretical D-convexity* [9], or *Partial convexity* [13].

[4] We remark that, since the set $\mathcal{O}$ is formed by two orthogonal lines parallel to the coordinate axis, in this paper $\mathcal{O}$-convexity is equivalent to Orthogonal Convexity.

where $\mathcal{Q}$ denotes the set of all (open) P-free quadrants of the plane.

We now describe three important properties of the rectilinear convex hull. Please refer again to Fig. 1 for an illustration. First, it is known that $\mathcal{RCH}(P) \subseteq \mathcal{CH}(P)$ [17, Theorem 4.7]. In particular, if $\mathcal{CH}(P)$ is a rectangle whose sides are parallel to the lines of $\mathcal{O}$, then $\mathcal{RCH}(P) = \mathcal{CH}(P)$. If this is not the case then $\mathcal{RCH}(P)$ is not convex. Second, observe that, if the segments in P form a polygonal chain, then $\mathcal{RCH}(P)$ is connected. If instead $\mathcal{RCH}(P)$ is disconnected, then a connected component is either a line segment of P or a closed polygon that is $\mathcal{O}$-convex. Note the polygon is not necessarily orthogonal and may even not be simple, since some edges may not be incident to the interior of the polygon. Finally, we have the property we call *orientation dependency*: except for some particular cases, like simultaneously rotating the lines of $\mathcal{O}$ by $\frac{\pi}{2}$ in the counter-clockwise direction, two rectilinear convex hulls of the same set at different orientations of the lines of $\mathcal{O}$ are non-congruent to each other.

Hereafter, we denote with $\mathcal{O}_\theta$ the set resulting after simultaneously rotating the lines of $\mathcal{O}$ in the counterclockwise direction by an angle of θ. We denote with $\mathcal{RCH}_\theta(P)$ the rectilinear convex hull of P computed with respect to $\mathcal{O}_\theta$.

3   Rectilinear Convex Hull of Line Segments

Let $P = \{s_1, s_2, \ldots, s_n\}$ be a set of n line segments (or segments for short) in the plane, that are not necessarily disjoint. In this section we describe an optimal $\Theta(n \log n)$ time and $\Theta(n\alpha(n))$ space algorithm to compute and maintain $\mathcal{RCH}_\theta(P)$ while θ is increased from 0 to 2π.

We denote the four orientations: North, South, East, and West, by N-orientation, S-orientation, E-orientation, and W-orientation, respectively.

Definition 1. *A point $x \in \mathbb{R}^2$ is ω-wedge P-free with respect to N-orientation, if there exists an ω-wedge with apex at x such that (i) it contains the N-orientation through x, and (ii) it contains no point of a segment of P in its interior.*

The same definition can be given analogously for each of the other orientations: S-orientation, E-orientation, and W-orientation, respectively. To compute $\mathcal{RCH}_\theta(P)$ we are interested in P-free $\frac{\pi}{2}$-wedges with apex at points $x \in \mathbb{R}^2$ or at points x of the segments of P which contain at least one of the orientations: N-orientation, S-orientation, E-orientation, or W-orientation in the $\frac{\pi}{2}$-wedge.

In Fig. 2 we consider an example for the N-orientation, where we can see some points in the segment s which are apices of P-free $\frac{\pi}{2}$-wedges. From Fig. 2 it is easy to see that a necessary condition for a point x in a segment s to be P-free ω-wedge with respect to the N-orientation, $\omega > 0$, is that x is *visible from the infinity north*, i.e., there is a P-free ω-wedge, $\omega > 0$, with apex at x containing the N-orientation. In other words, the point $x \in s$ has to belong to the upper envelope of P, which is equivalent to the fact that $\omega > 0$, and thus, if x belongs to the upper envelope of P, we only have to check that $\omega \geq \frac{\pi}{2}$.

Clearly, for the other three orientations: S-orientation, E-orientation, and W-orientation, we can do a similar analysis, i.e., for the S-orientation we can

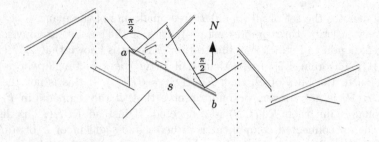

Fig. 2. Points in the segment $s \in P$ which are apices of P-free $\frac{\pi}{2}$-wedges for the N-orientation. In blue the upper envelope of P. Notice that only the part of a segment s that belongs to the upper envelope of P can see the N-orientation.

consider the lower envelope or equivalent seeing the segments of P from the infinity south. And analogously, for the E-orientation, and W-orientation by either rotating the coordinate system by $\frac{\pi}{2}$ and computing the new upper and lower envelopes, or defining the east envelope and the west envelope as seeing the set of segments of P from the infinity east and the infinity west, respectively.

The four envelopes: north envelope, south envelope, east envelope, and west envelope of the set of segments P can be computed in $O(n \log n)$ time and $O(n\alpha(n))$ space. These envelopes are formed of segments of P or parts of them, with total complexity $O(n\alpha(n))$, see Hersberger [10].

First, we need to determine when a given segment of P (or part of it) can belong to the boundary of $\mathcal{RCH}_\theta(P)$. To do this, we select one of the four orientations N, S, E, or W for the corresponding $\frac{\pi}{2}$-wedges of the points of the boundary of $\mathcal{RCH}_\theta(P)$ such that those $\frac{\pi}{2}$-wedges will contain at least one of these orientations N, S, E, or W.

Assume that we are computing which segments of P (or parts of them) belong to the boundary of $\mathcal{RCH}_\theta(P)$; concretely, to the N-orientation.

Lemma 1. *Let s be a segment in P. If a non-empty subset of s belongs to the boundary of $\mathcal{RCH}_\theta(P)$ for some angle $\theta \in [0, 2\pi)$, then a non-empty subset of s belongs to at least one of the four envelopes of P described above.*

Proof. If a point x in the segment s belongs to the boundary of $\mathcal{RCH}_\theta(P)$, then x is the apex of a P-free $\frac{\pi}{2}$-wedge containing at least one of the N-orientation, S-orientation, E-orientation, or W-orientation. See Fig. 3.

In order to know the segments of P (or the parts of them) that are $\frac{\pi}{2}$-wedge P-free, as a first step we have to compute the upper envelope of P, which we will denote by $\mathcal{N}(P)$. It is known that the complexity of $\mathcal{N}(P)$ is $O(n\alpha(n))$, where $\alpha(n)$ is the inverse of Ackermann's function, and that $\mathcal{N}(P)$ can be computed in optimal $O(n \log n)$ time and $O(n\alpha(n))$ space [10].

Analogously, the lower envelope of P corresponding to the S-orientation, denoted by $\mathcal{S}(P)$, has $O(n\alpha(n))$ complexity and can be computed in optimal $O(n \log n)$ time and $O(n\alpha(n))$ space. In a similar way, we define and compute the righter and lefter envelopes, denoted by $\mathcal{E}(P)$ and $\mathcal{W}(P)$, respectively, corresponding to the East and West envelopes of P, respectively; and defined by the E-orientation and W-orientation, respectively, again with the same complexities. These four envelopes are composed of segments or parts of segments of P with total complexity $O(n\alpha(n))$. See Fig. 3.

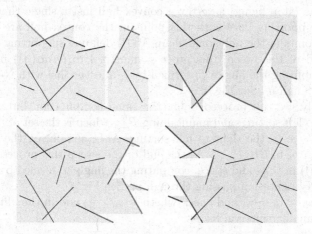

Fig. 3. The envelopes of P for the N-orientation, the S-orientation, the E-orientation, and the W-orientation. The horizontal and vertical segments are not part of the envelopes, and are added for the sake of clarity.

Proposition 1. $\mathcal{N}(P)$, $\mathcal{S}(P)$, $\mathcal{E}(P)$, and $\mathcal{W}(P)$ can be computed in optimal $O(n \log n)$ time and $O(n\alpha(n))$ space, and have $O(n\alpha(n))$ total complexity.

Now, we compute the ω-wedge of all the points of the four envelopes, i.e., compute which of these points have $\omega \geq \frac{\pi}{2}$. We proceed as follows: first, in Subsect. 3.1, we compute the ω-wedges of the endpoints in the four envelopes; and then, in Subsect. 3.2, we compute the ω-wedges of the (non-vertical) segments in the four envelopes.

3.1 Computing the ω-Wedges of Endpoints of Segments

We consider first the north envelope formed by the polygonal chain $\mathcal{N}(P)$. We do computations for the non-vertical segments of $\mathcal{N}(P)$. Clearly, the number of endpoints of segments in $\mathcal{N}(P)$ is $O(n\alpha(n))$. For each of these endpoints, say p_i, we will compute the ω_i-wedge with apex at p_i. The containing of the N-orientation is guaranteed because the point p belongs to $\mathcal{N}(P)$. See Fig. 4.

The ω_i-wedge with apex at an endpoint p_i can be computed as follows. Sort the endpoints in $\mathcal{N}(P)$ according to their x-coordinates, both from left to right and from right to left, in $O(n\alpha(n) \log n)$ time. Well, in fact, the computation of the upper envelope outputs the endpoints and the segments in a sorted way (from left to right). Then, we do the following.

1. For each endpoint p_i maintain (update) in $O(\log(n\alpha(n))) = O(\log n)$ time the convex hull of the point set P_{i-1}, $\mathcal{CH}(P_{i-1})$, where P_{i-1} is the set of the endpoints $p_j < p_i$ in the sorting above.

2. Compute the supporting line l_i from p_i to $\mathcal{CH}(P_{i-1})$: instead of using a $O(\log n)$ time binary search, we traverse the boundary of $\mathcal{CH}(P_{i-1})$ until we find the tangent vertex of $\mathcal{CH}(P_{i-1})$. Since every endpoint p_j in $\mathcal{N}(P)$ can become (stop being) a vertex a convex hull just a single time, then the different supporting lines with this point in the convex hull are charged to different points in the polygonal chain $\mathcal{N}(P)$ defining supporting lines.

   At the end of the sweep, this process amortizes to $O(n\alpha(n))$ in time and space, computing the angles β_i formed by l_i and the line with N-orientation passing through p_i. See Fig. 4.

3. Analogously, we can proceed doing the same computation but considering the right to left sorting, and maintaining P'_{i-1} which is the set of the extreme points $p_j > p_i$ in the right to left sorting. Also, computing the supporting line r_i from p_i to $\mathcal{CH}(P'_{i-1})$. At the end of the sweep this process amortizes to $O(n\alpha(n))$ in time and space, computing the angle γ_i formed by r_i and the line with N-orientation passing through p_i. See Fig. 4.

4. Compute $\omega_i = \beta_i + \gamma_i$, and check whether $\omega_i \geq \frac{\pi}{2}$, and in the affirmative let ω_i be the angular interval for p_i.

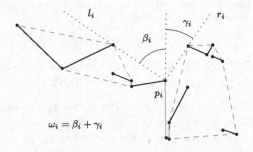

Fig. 4. The angles β_i and γ_i for an extreme point p_i in $\mathcal{N}(P)$, and $\omega_i = \beta_i + \gamma_i$.

Since there are $O(n\alpha(n))$ endpoints, the complexity of these steps is $O(n\alpha(n))$ time and space. Clearly, we proceed analogously with the other envelopes $\mathcal{S}(P)$, $\mathcal{E}(P)$, and $\mathcal{W}(P)$. Therefore, the total complexity of this process for all the four envelopes is $O(n\alpha(n))$ time and space. Notice that we are considering the at most $O(n\alpha(n))$ endpoints p_i of the four envelopes and computing which of these p_i have $\omega_i \geq \frac{\pi}{2}$ for some of the four orientations. A point p_i can not have the four corresponding $\omega_i \geq \frac{\pi}{2}$ for the four orientations, since otherwise P is formed by a unique point. We translate the angles ω_i into angular intervals in $[0, 2\pi]$ or into angular intervals in the unit circle also in $[0, 2\pi]$.

A question that arises from the process above is whether the number of endpoints in the envelope $\mathcal{N}(P)$ (resp. $\mathcal{S}(P)$, $\mathcal{E}(P)$, and $\mathcal{W}(P)$) having an angle $\omega_i \geq \frac{\pi}{2}$ can be at most $O(n)$. We solve this question with the following result.

Proposition 2. *The number of endpoints of the upper envelope of $\mathcal{N}(P)$ having an angle $\omega_i \geq \frac{\pi}{2}$ is $\Omega(n\alpha(n))$.*

Proof. Take a set of n segments which upper envelope has $\Omega(n\alpha(n))$ complexity. To do that follow the construction of the collection of segments with this $\Omega(n\alpha(n))$ complexity that appears in [23] (see also Theorem 4.11 in [21]).

Every endpoint of the envelope is either: (i) the endpoint of a segment, (ii) right below the endpoint of a segment, or (iii) the intersection of two segments. The number of endpoints of type (i) and (ii) is easily seen to be $O(n)$. This means there are $\Omega(n\alpha(n))$ endpoints of type (iii). Each endpoint of type (iii) has angles $\beta > 0$ and $\gamma > 0$ (as defined above). Call ϵ the minimum over all β and γ. Now stretch the whole set horizontally by a factor of $\frac{1}{\sin \epsilon}$. This results in a set where every endpoint of type (iii) has $\beta \geq \frac{\pi}{4}$ and $\gamma \geq \frac{\pi}{4}$, thus is a $\frac{\pi}{2}$-wedge. Therefore, we get $\Omega(n\alpha(n))$ endpoints which are $\frac{\pi}{2}$-wedge. In fact, one can stretch even further to get larger angles up to π.

Observation 1. *Notice that if the segments of P are non-intersecting, then the complexity of their upper envelope is $O(n)$, and the envelope can be constructed in optimal $O(n \log n)$ time [10]*

Clearly, the same $\Omega(n\alpha(n))$ lower bound applies for the number of endpoints of the envelopes $\mathcal{S}(P)$, $\mathcal{E}(P)$ and $\mathcal{W}(P)$, having an angle $\omega_i \geq \frac{\pi}{2}$. From the discussion above we have the following result.

Theorem 1. *The number of endpoints p_i of the envelopes $\mathcal{N}(P)$, $\mathcal{S}(P)$, $\mathcal{E}(P)$, and $\mathcal{W}(P)$ having an angle $\omega_i \geq \frac{\pi}{2}$ is $\Theta(n\alpha(n))$. These angles and their angular intervals can be computed in $O(n \log n)$ time and $O(n\alpha(n))$ space.*

3.2 Computing the ω-Wedges of the Points Inside the Segments

Now we show how to compute which segments of P or part of them belong to $\mathcal{RCH}_\theta(P)$. First, we illustrate this for the segments in $\mathcal{N}(P)$. Given a non-vertical segment s_i of $\mathcal{N}(P)$ with endpoints p_{i-1} and p_i, we show how to compute the angles β and γ for the interior points of s_i, see Fig. 5(a). And then we compute the parts of s_i (if any) such that $\beta + \gamma = \omega \geq \frac{\pi}{2}$.

Let $s_i = p_{i-1}p_i$ be a segment in $\mathcal{N}(P)$, and consider the sequence of the endpoints in $\mathcal{N}(P)$ from left to right $< \ldots, p_{i-2}, p_{i-1}, p_i, p_{i+1}, \cdots >$. Let P_{i-1} be the set of those endpoints in $\mathcal{N}(P)$ till p_{i-1}, i.e., $P_{i-1} = \{\ldots, p_{i-4}, p_{i-3}, p_{i-2}, p_{i-1}\}$, and let $\mathcal{CH}(P_{i-1})$ be the convex hull of these points. Without loss of generality, assume that $< \ldots, p_{i-4}, p_{i-3}, p_{i-2}, p_{i-1} >$ is exactly the sequence of the points of $\mathcal{CH}(P_{i-1})$. See Fig. 5(a) and (b).

Assume that the lines containing the segments $p_{i-3}p_{i-2}$, $p_{i-4}p_{i-3}$, $\ldots$ intersect the segment s_i, splitting the segment s_i into parts such that to each of

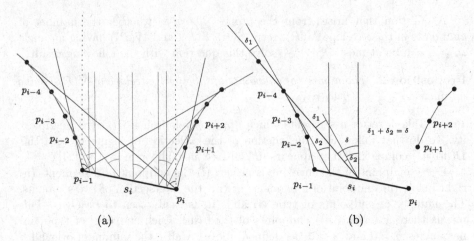

Fig. 5. (a) Computing the angles β and γ of the (non-vertical) segments s_i of $\mathcal{N}(P)$. (b) Computing the different angles β and γ in s_i of $\mathcal{N}(P)$ from the previous angles using the angles between consecutive edges of $CH(\mathcal{N}(P))$, i.e., using that $\delta_1 + \delta_2 = \delta$.

these parts correspond β angular intervals. By the convexity of the $\mathcal{CH}(P_{i-1})$ the endpoints of these β angular intervals can be computed in constant time (see Fig. 5(b)), since we only have to add the value of the before angular interval. Of course, each of these operations can be assigned to the new supporting point in the convex hull $\mathcal{CH}(P_{i-1})$.

Clearly, just doing the reverse process, i.e., from right to left, we can compute the γ angular intervals, and merge the two sets of intervals into one set of intervals. For each one of these new intervals we can compute whether the sum $\beta + \gamma \geq \frac{\pi}{2}$ at points inside the interval in constant time as follows.

For simplicity, assume that the segment s_i is horizontal, moreover, the segment s_i is on the X-axis, as in Fig. 6. Consider an interval inside s_i in the same figure. In order to know whether a point $(t, 0) \in s_i$ is the apex of P-free $\frac{\pi}{2}$ wedge we have to check when its corresponding angle $w \geq \frac{\pi}{2}$.

$$\boldsymbol{u} = (x_1 - t, y_1 - 0), \quad \boldsymbol{v} = (x_2 - t, y_2 - 0),$$

$$\cos w = \frac{|\boldsymbol{u} \cdot \boldsymbol{v}|}{|\boldsymbol{u}||\boldsymbol{v}|} = \frac{|(x_1 - t)(x_2 - t) + y_1 y_2|}{\sqrt{(x_1 - t)^2 + y_1^2}\sqrt{(x_2 - t)^2 + y_2^2}},$$

$$\cos^2 w = \frac{(t^2 - (x_1 + x_2)t + x_1 x_2 + y_1 y_2)^2}{[(x_1 - t)^2 + y_1^2][(x_2 - t)^2 + y_2^2]}$$

$$\cos w = 0 \iff \cos^2 w = 0 \iff t^2 - (x_1 + x_2)t + x_1 x_2 + y_1 y_2 = 0$$

Since the function is a quadratic function, either there is no solution, or one solution, or two solutions, and in any case, any solution t has to verify that $t_1 \leq t \leq t_2$. Then, in constant time we can check whether the values of w inside the computed intervals verify that $w \geq \frac{\pi}{2}$, and determine the constant number of intervals where $w \geq \frac{\pi}{2}$. See Fig. 6.

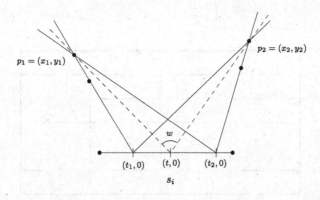

Fig. 6. The function to compute if and where the angle $w \geq \frac{\pi}{2}$.

Theorem 2. *The total number of intervals inside the segments of the envelope $\mathcal{N}(P)$ is $\Omega(n\alpha(n))$. These intervals and the ones corresponding to values of $w \geq \frac{\pi}{2}$ can be computed in $O(n \log n)$ time and $O(n\alpha(n))$ space.*

Proof. By the construction illustrated in Fig. 5 we see that the points of $\mathcal{N}(P)$ causing a split in a segment (the points p_{i-1}, p_{i-2}, p_{i-3}, p_{i-4}) are captured by the next *left* convex hull $\mathcal{CH}(P_i)$; and similarly for the points to the right by the next *right* convex hull. Only the last point p_{i-5} which belongs to the boundary of $\mathcal{CH}(P_{i-1})$ will be considered in the next step for the next segment to the right, and so the point p_{i-5} can be used at most two times.

Because the equation above has degree two, an interval defined by supporting lines can be split into at most 3 sub-intervals thus, the total number of sub-intervals is $\Omega(n\alpha(n))$, and in each sub-interval we have spent constant time. The total number of steps of the process above is upper bounded by $\Omega(n\alpha(n))$ in time and space since we follow the order of the elements of $\mathcal{N}(P)$, we don't need to do binary search, and the computations in each of the steps takes constant time.

Theorem 3. *The same statement of Theorem 2 above is also true for the envelopes $\mathcal{S}(P)$, $\mathcal{E}(P)$, and $\mathcal{W}(P)$.*

3.3 Computing $\mathcal{RCH}_\theta(P)$

Now we show how to compute and maintain $\mathcal{RCH}_\theta(P)$ as $\theta \in [0, 2\pi]$ from the information computed in the Subsects. 3.1 and 3.2. From Theorems 2 and 3, once we have computed the angular intervals of angles $w \geq \frac{\pi}{2}$, then we can translate all these intervals for all the segments s_i (or part of it) in $\mathcal{N}(P)$, $\mathcal{S}(P)$, $\mathcal{E}(P)$, and $\mathcal{W}(P)$, to angular intervals inside $[0, 2\pi]$ in the real line, see Fig. 7.

Thus, doing a line sweep with vertical lines corresponding to angles θ, $\theta + \frac{\pi}{2}$, $\theta + \pi$, and $\theta + \frac{3\pi}{2}$ (in a circular way, i.e., completing a $[0, 2\pi]$ round with each line), and then inserting and deleting the changes of segments that belong to each of the four staircases of $\mathcal{RCH}_\theta(P)$ as θ changes in $[0, 2\pi]$.

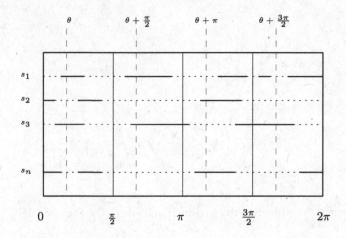

Fig. 7. Angular intervals inside $[0, 2\pi)$ of the angles ω of the points of the segments s_i (or part of it) in $\mathcal{N}(P)$, $\mathcal{S}(P)$, $\mathcal{E}(P)$, and $\mathcal{W}(P)$, such that have $w \geq \frac{\pi}{2}$.

Now, considering that an endpoint or an interior point of a segment of P (or part of it) in any of the envelopes $\mathcal{N}(P)$, $\mathcal{S}(P)$, $\mathcal{E}(P)$, and $\mathcal{W}(P)$ can be apex of a P-free $\frac{\pi}{2}$-wedge, and because Theorems 2 and 3, we can compute the endpoints, segments, or part of segments of P that belong to $\mathcal{RCH}_\theta(P)$ as $\theta \in [0, 2\pi]$ in $O(n \log n)$ time and $O(n\alpha(n))$ space. Therefore, we obtain the following result.

Theorem 4. *Computing and maintaining $\mathcal{RCH}_\theta(P)$ as $\theta \in [0, 2\pi]$ can be done in $O(n \log n)$ time and $O(n\alpha(n))$ space.*

3.4   Rectilinear Convex Hull of a Set of Simple Polygons

Let P be a collection of simple polygons with n total complexity, i.e., the number of edges of all the polygons is n. Again, the problem is to compute and maintain $\mathcal{RCH}_\theta(P)$ as $\theta \in [0, 2\pi]$. In fact, it is exactly the same problem as a collection of segments because we take the set, say S, of segments corresponding to all the sides of the simple polygons and apply the Theorem 4 to the set S. Thus, by the discussion above, we have the same result but for a set of simple polygons.

Theorem 5. *Computing and maintaining $\mathcal{RCH}_\theta(P)$ as $\theta \in [0, 2\pi]$ can be done in $O(n \log n)$ time and $O(n\alpha(n))$ space.*

4   Rectilinear Convex Hull of a Simple Polygonal Chain

Let P be a simple polygonal chain formed by a sequence of n (non-intersecting) line segments. The goal in this subsection is to compute the $\mathcal{RCH}_\theta(P)$ using the techniques applied in the section above. A first question to determine is how to compute the upper envelope (resp. lower envelope) of P. Obviously, it can be computed as the upper envelope (resp. lower envelope) of a set of line segments in $O(n \log n)$ time and $O(n\alpha(n))$ space, but the question is to determine whether these envelopes have $O(n)$ complexity or $O(n\alpha(n))$ complexity. This fact will fix the time and space complexity of our algorithms. Notice that now the segments are not intersecting and that we have the order of the segments forming the polygonal chain. See Fig. 8.

Proposition 3. *The complexity of the upper envelope of P is $O(n)$.*

Proof. Assume that we follow the simple polygonal chain P from its origin on the left to its end on the right. Assume that we are computing the upper envelope of P, $\mathcal{U}(P)$. If a segment s_i or part of it appears in $\mathcal{U}(P)$ for first time, then either it belongs to $\mathcal{U}(P)$ till the endpoint of s_i or another segment of $\mathcal{U}(P)$ appears before the endpoint of s_i. In the second case the segment s_i does not appear in $\mathcal{U}(P)$ to the right in $\mathcal{U}(P)$ by the continuity of P, i.e., the right part of P after s_i is like a ray with origin at the right endpoint of s_i. See Fig. 8.

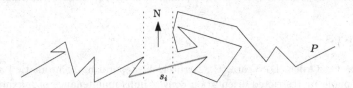

Fig. 8. Illustration of the proof of Proposition 3.

Observation 2. *Notice that by Observation 1 the complexity of the upper enve-lope of n non-intersecting segments is $O(n)$ and can be computed in $O(n \log n)$ time and $O(n)$ space. But now, for the simple polygonal chain P of n segments the complexity of the upper envelope is $O(n)$ and can be computed in $O(n)$ time and space because the set of segments are sorted in the polygonal chain P.*

Theorem 6. *Computing and maintaining $\mathcal{RCH}_\theta(P)$ as $\theta \in [0, 2\pi]$ can be done in $O(n)$ time and space.*

Proof. By Proposition 3 the complexity of each envelope $\mathcal{N}(P)$, $\mathcal{S}(P)$, $\mathcal{E}(P)$, and $\mathcal{W}(P)$ is $O(n)$. Thus, we apply the same techniques as in Subsect. 3.3 with the difference of the complexities of the envelopes and get the result.

5 Concluding Remarks

We showed how to compute and maintain $\mathcal{RCH}_\theta(P)$ as θ changes from 0 to 2π, for the cases where P is a collection of line segments, polygons, or polygonal lines in the plane. In all cases, our algorithms are worst-case optimal in both time and space.

From our algorithms we actually obtain a general approach to compute the rectilinear convex hull with arbitrary orientation of a collection of geometric objects. The efficiency bottleneck is the complexity of the lower envelope of the collection. We are currently working on a couple of cases in which we are about to successfully adapt our approach, to obtain algorithms that are worst-case optimal as well. The first case is a collection of circles. The lower envelope of such a collection has $O(n)$ complexity [11], and can be computed in $O(n \log n)$ time and $O(n)$ space [5]. The second case is a collection of (non-necessarily equally oriented) ellipses. From the results of [21], we know that the lower envelope of a collection of n ellipses has $O(n\alpha(n))$ complexity. On the other hand, since a line segment can be thought of as an almost-flat ellipse, the lower bound construction from [23] for line segments seems to imply that the complexity bound is tight.

Acknowledgements. Justin Dallant is supported by the French Community of Belgium via the funding of a FRIA grant. Pablo Pérez-Lantero was partially supported by project DICYT 042332PL Vicerrectoría de Investigación, Desarrollo e Innovación USACH (Chile). Carlos Seara is supported by Project PID2019-104129GB-I00/AEI/10.13039/501100011033 of the Spanish Ministry of Science and Innovation.

References

1. Alegría, C., Orden, D., Seara, C., Urrutia, J.: Separating bichromatic point sets in the plane by restricted orientation convex hulls maintenance of maxima of 2D point sets. J. Global Optim. **85**, 1–34 (2022). https://doi.org/10.1007/s10898-022-01238-9

2. Biedl, T., Genç, B.: Reconstructing orthogonal Polyhedra from putative vertex sets. Comput. Geomet. Theor. Appl. **44**(8), 409–417 (2011). https://doi.org/10.1016/j.comgeo.2011.04.002

3. Chazelle, B.: An optimal convex hull algorithm in any fixed dimension. Discr. Comput. Geom. **10**(4), 377–409 (1993). https://doi.org/10.1007/bf02573985

4. Davari, M.J., Edalat, A., Lieutier, A.: The convex hull of finitely generable subsets and its predicate transformer. In: 2019 34th Annual ACM/IEEE Symposium on Logic in Computer Science (LICS). IEEE (2019). https://doi.org/10.1109/lics.2019.8785680

5. Devillers, O., Golin, M.J.: Incremental algorithms for finding the convex hulls of circles and the lower envelopes of parabolas. Inf. Process. Lett. **56**(3), 157–164 (1995). https://doi.org/10.1016/0020-0190(95)00132-V

6. Díaz-Bañez, J.M., López, M.A., Mora, M., Seara, C., Ventura, I.: Fitting a two-joint orthogonal chain to a point set. Comput. Geom. **44**(3), 135–147 (2011). https://doi.org/10.1016/j.comgeo.2010.07.005

7. Fink, E., Wood, D.: Restricted-orientation Convexity. Monographs in Theoretical Computer Science (An EATCS Series). 1st Edn. Springer, Heidelberg (2004). https://doi.org/10.1007/978-3-642-18849-7

8. Franěk, V.: On algorithmic characterization of functional D-convex hulls, Ph. D. thesis, Faculty of Mathematics and Physics, Charles University in Prague (2008)

9. Franěk, V., Matoušek, J.: Computing D-convex hulls in the plane. Comput. Geomet. Theor. Appl. **42**(1), 81–89 (2009). https://doi.org/10.1016/j.comgeo.2008.03.003

10. Hershberger, J.: Finding the upper envelope of n line segments in $O(n \log n)$ time. Inf. Process. Lett. **33**, 169–174 (1989). https://doi.org/10.1016/0020-0190(89)90136-1

11. Kedem, K., Livne, R., Pach, J., Sharir, M.: On the union of Jordan regions and collision-free translational motion amidst polygonal obstacles. Discr. Comput. Geomet. **1**(1), 59–71 (1986). https://doi.org/10.1007/BF02187683

12. Melkman, A.A.: On-line construction of the convex hull of a simple polyline. Inf. Process. Lett. **25**(1), 11–12 (1987). https://doi.org/10.1016/0020-0190(87)90086-X

13. Metelskii, N.N., Martynchik, V.N.: Partial convexity. Math. Notes **60**(3), 300–305 (1996). https://doi.org/10.1007/BF02320367

14. Nicholl, T.M., Lee, D.T., Liao, Y.Z., Wong, C.K.: On the X-Y convex hull of a set of X-Y polygons. BIT Numer. Math. **23**(4), 456–471 (1983). https://doi.org/10.1007/BF01933620

15. Ottmann, T., Soisalon-Soininen, E., Wood, D.: On the definition and computation of rectilinear convex hulls. Inf. Sci. **33**(3), 157–171 (1984). https://doi.org/10.1016/0020-0255(84)90025-2

16. Pérez-Lantero, P., Seara, C., Urrutia, J.: A fitting problem in three dimension. In: Book of Abstracts of the XX Spanish Meeting on Computational Geometry, pp. 21–24. EGC 2023 (2023)

17. Preparata, F.P., Shamos, M.I.: Computational geometry: an introduction. Text and Monographs in Computer Science. 1st Edn. Springer, NY (1985). https://doi.org/10.1007/978-1-4612-1098-6

18. Rawlins, G.J.E.: Explorations in restricted-orientation geometry, Ph. D. thesis, School of Computer Science, University of Waterloo (1987)

19. Rawlins, G.J., Wood, D.: Ortho-convexity and its generalizations. In: Toussaint, G.T. (ed.) Computational Morphology, Machine Intelligence and Pattern Recognition, vol. 6, pp. 137–152. North-Holland (1988). https://doi.org/10.1016/B978-0-444-70467-2.50015-1

20. Schuierer, S., Wood, D.: Restricted-orientation visibility. Tech. Rep. 40, Institut für Informatik, Universität Freiburg (1991)

21. Sharir, M., Agarwal, P.K.: Davenport-Schinzel Sequences and Their Geometric Applications. Cambridge University Press (1995)

22. Wang, C., Zhou, R.G.: A quantum search algorithm of two-dimensional convex hull. Commun. Theoret. Phys. **73**(11), 115102 (2021). https://doi.org/10.1088/1572-9494/ac1da0

23. Wiernik, A., Sharir, M.: Planar realizations of nonlinear Davenport-Schinzel sequences by segments. Discr. Comput. Geomet. **3**(1), 15–47 (1988). https://doi.org/10.1007/BF02187894

Domino Snake Problems on Groups

Nathalie Aubrun[ID] and Nicolas Bitar[(⊠)][ID]

Université Paris-Saclay, CNRS, LISN, 91190 Gif-sur-Yvette, France
{nathalie.aubrun,nicolas.bitar}@lisn.fr

Abstract. In this article we study domino snake problems on finitely generated groups. We provide general properties of these problems and introduce new tools for their study. The first is the use of symbolic dynamics to understand the set of all possible snakes. Using this approach we solve many variations of the infinite snake problem including the geodesic snake problem for certain classes of groups. Next, we introduce a notion of embedding that allows us to reduce the decidability of snake problems from one group to another. This notion enable us to establish the undecidability of the infinite snake and ouroboros problems on nilpotent groups for any generating set, given that we add a well-chosen element. Finally, we make use of monadic second order logic to prove that domino snake problems are decidable on virtually free groups for all generating sets.

Keywords: Domino Snake Problems · Computability Theory · Symbolic Dynamics · Combinatorial Group Theory · MSO logic

1 Introduction

Since their introduction more than 60 years ago [27], domino problems have had a long history of providing complexity lower bounds and undecidability of numerous decision problems [6,11,15,16]. The input to these problems is a set of *Wang tiles*: unit square tiles with colored edges and fixed orientation. The decision problems follow the same global structure; given a finite set of Wang tiles, is there an algorithm to determine if they tile a particular shape or subset of the infinite grid such that adjacent tiles share the same color along their common border? An interesting variant of this general formula are domino snake problems. First introduced by Myers in 1979 [24], snake problems ask for threads –or snakes– of tiles that satisfy certain constraints. In particular, three of them stand-out. The infinite snake problem asks is there exists a tiling of a self-avoiding bi-infinite path on the grid, the ouroboros problem asks if there exists a non-trivial cycle on the grid, and the snake reachability problem asks if there exists a tiling of a self-avoiding path between two prescribed points. Adjacency rules are only required to be respected along the path. These problems have had their computability completely classified [1,9,10,13,14,19] (see Theorem 1). In this article, we expand the scope of domino snake problems to finitely generated

H. Fernau and K. Jansen (Eds.): FCT 2023, LNCS 14292, pp. 46–59, 2023.
https://doi.org/10.1007/978-3-031-43587-4_4

groups, as has been done for other domino problems [3], to understand how the underlying structure affects computability.

We present three novel ways in which to approach these problems. The first is the use of symbolic dynamics to understand the set of all possible snakes. Theorem 3 states that when this set is defined through a regular language of forbidden patterns, the infinite snake problem becomes decidable. Using this approach we solve many variations of the infinite snake problem including the geodesic snake problem for some classes of groups. Next, we introduce a notion of embedding that allows us to reduce the decidability of snake problems from one group to another. This notion enable us to establish the undecidability of the infinite snake and ouroboros problems on a large class of groups –that most notably include nilpotent groups– for any generating set, provided that we add a central torsion-free element. Finally, to tackle virtually free groups, we express the three snake problems in the language of Monadic Second Order logic. Because for this class of groups this fraction of logic is decidable, we show that our three decision problems are decidable independently of the generating set.

2 Preliminaries

Given a finite alphabet A, we denote by A^n the set of words on A of length n, and A^* the set of all finite length words including the empty word ϵ. Furthermore, we denote by $A^+ = A^* \setminus \{\epsilon\}$ the set of non-empty finite words over A. A factor v of a word w is a contiguous subword; we denote this by $v \sqsubseteq w$. We denote discrete intervals by $[\![n, m]\!] = \{n, n + 1, ..., m - 1, m\}$. We also denote the free group defined from the free generating set S by $\mathbb{F}_S$. The proofs missing in the text are present in [4].

2.1 Symbolic Dynamics

Given a finite alphabet A, we define the *full-shift* over A as the set of configurations $A^{\mathbb{Z}} = \{x : \mathbb{Z} \to A\}$. There is a natural $\mathbb{Z}$-action on the full-shift called the *shift*, $\sigma : A^{\mathbb{Z}} \to A^{\mathbb{Z}}$, given by $\sigma(x)_i = x_{i+1}$. The full-shift is also endowed with the prodiscrete topology, making it a compact space.

Let $F \subseteq \mathbb{Z}$ be a finite subset. A *pattern* of support F is an element $p \in A^F$. We say a pattern $p \in A^F$ appears in a configuration $x \in A^{\mathbb{Z}}$, denoted $p \sqsubseteq x$, if there exists $k \in \mathbb{Z}$ such that $x_{k+i} = p_i$ for all $i \in F$. Given a set of patterns $\mathcal{F}$, we can define the set of configurations where no pattern from $\mathcal{F}$ appears,

$$X_{\mathcal{F}} := \{x \in A^{\mathbb{Z}} \mid \forall p \in \mathcal{F}, \ p \text{ does not appear in } x\}.$$

A *subshift* is a subset of the full-shift $X \subseteq A^{\mathbb{Z}}$ such that there exists a set of patterns $\mathcal{F}$ that verifies $X = X_{\mathcal{F}}$. A classic result shows subshifts can be equivalently defined as closed σ-invariant subsets of the full-shift. We say a subshift $X_{\mathcal{F}}$ is

– a *subshift of finite type* (SFT) if $\mathcal{F}$ is finite,

- *sofic* if $\mathcal{F}$ is a regular language,
- *effective* if $\mathcal{F}$ is a decidable language.

Each class is strictly contained within the next. Sofic subshifts can be equivalently defined as the set of bi-infinite walks on a labeled finite graph. It is therefore decidable, given the automaton or graph that defines the subshift, to determine if the subshift is empty. Similarly, given a Turing machine for the forbidden language of an effective subshift, we have a semi-algorithm to determine if the subshift is empty.

Lastly, we say a configuration $x \in X$ is *periodic* if there exists $k \in \mathbb{Z}^*$ such that $\sigma^k(x) = x$. We say the subshift X is *aperiodic* if it contains no periodic configurations. For a comprehensive introduction to one-dimensional symbolic dynamics we refer the reader to [21].

2.2   Combinatorial Group Theory

Let G be a finitely generated (f.g.) group and S a finite generating set. Elements in the group are represented as words on the alphabet $S \cup S^{-1}$ through the evaluation function $w \mapsto \overline{w}$. Two words w and v represent the same element in G when $\overline{w} = \overline{v}$, and we denote it by $w =_G v$. We say a word is *reduced* if it contains no factor of the form ss^{-1} or $s^{-1}s$ with $s \in S$.

Definition 1. *Let G be a group, S a subset of G and R a language on $S \cup S^{-1}$. We say (S, R) is a presentation of G, denoted $G = \langle S \mid R \rangle$, if the group is isomorphic to $\langle S \mid R \rangle = \mathbb{F}_S / \langle\langle R \rangle\rangle$, where $\langle\langle R \rangle\rangle$ is the normal closure of R, i.e. the smallest normal subgroup containing R. We say G is recursively presented if there exists a presentation (S, R) such that S is finite and R is recursively enumerable.*

For a group G and a generating set S, we define:

$$\text{WP}(G, S) := \{ w \in (S \cup S^{-1})^* \mid \overline{w} = 1_G \}.$$

Definition 2. *The word problem (WP) of a group G with respect to a set of generators S asks to determine, given a word $w \in (S \cup S^{-1})^*$, if $w \in \text{WP}(G, S)$.*

We say a word $w \in (S \cup S^{-1})^+$ is *G-reduced* if w contains no factor in $\text{WP}(G, S)$. We say a word $w \in (S \cup S^{-1})^*$ is a *geodesic* if for all words $v \in (S \cup S^{-1})^*$ such that $\overline{w} = \overline{v}$ we have $|w| \leq |v|$. For a given group G and generating set S, we denote its language of geodesics by $\text{Geo}(G, S)$.

We say an element $g \in G$ has *torsion* if there exists $n \geq 1$ such that $g^n = 1_G$. If there is no such n, we say g is *torsion-free*. Analogously, we say G is a torsion group if all of its elements have torsion. Otherwise if the only element of finite order is 1_G we say the group is torsion-free.

Finally, let $\mathcal{P}$ be a class of groups (for example abelian groups, free groups, etc.). We say a group G is *virtually* $\mathcal{P}$, if there exists a finite index subgroup $H \leq G$ that is in $\mathcal{P}$.

3   Snake Behaviour

Although the original snake problems were posed for the Wang tile model, we make use of a different, yet equivalent, formalism (see [4] for a proof of the equivalence of the two models).

Definition 3. *A tileset graph (or tileset) for a f.g. group (G, S) is a finite multi-graph $\Gamma = (A, B)$ such that each edge is uniquely determined by (a, a', s) where $s \in S \cup S^{-1}$ and $a, a' \in A$ are its initial and final vertices respectively, and if $(a, a', s) \in B$ then $(a', a, s^{-1}) \in B$.*

In what follows, I denotes $\mathbb{Z}$, $\mathbb{N}$, or a discrete interval $[\![n, m]\!]$, depending on the context.

Definition 4. *Let (G, S) be a f.g. group, and $\Gamma = (A, B)$ a tileset for the pair. A snake or Γ-snake is a pair of functions (ω, ζ), where $\omega : I \to G$ is an injective function, referred to as the snake's skeleton and $\zeta : I \to A$ the snake's scales. These pairs must satisfy that $d\omega_i := \omega(i)^{-1}\omega(i+1) \in S \cup S^{-1}$ and $(\zeta(i), \zeta(i+1))$ must be an edge in Γ labeled by $d\omega_i$ (Fig. 1).*

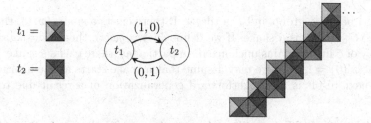

Fig. 1. Two Wang tiles that do not tile $\mathbb{Z}^2$, the corresponding graph Γ for the generating set $\{(1,0), (0,1)\}$ (edges labeled with generator inverses are omitted for more readability) and a Γ-snake with $I = \mathbb{N}$. In the Wang tile model, two tiles can be placed next to each other if they have the same color on the shared side.

We say a snake (ω, ζ) connects the points $p, q \in G$ if there exists a $n \in \mathbb{N}$ such that (ω, ζ) is defined over $[\![0, n]\!]$, $\omega(0) = p$ and $\omega(n) = q$. We say a snake is bi-infinite if its domain is $\mathbb{Z}$. A Γ-ouroboros is a Γ-snake defined over $[\![0, n]\!]$, with $n \geq 2$, that is injective except for $\omega(0) = \omega(n)$. In other words, a Γ-ouroboros is a well-tiled simple non-trivial cycle. We study the following three decision problems:

Definition 5. *Let (G, S) be a f.g. group. Given a tileset Γ for (G, S) and two points $p, q \in G$,*

- *the* infinite snake problem *asks if there exists a bi-infinite Γ-snake,*
- *the* ouroboros problem *asks if there exists a Γ-ouroboros,*

– *the* snake reachability problem *asks if there exists a Γ-snake connecting p and q.*

We can also talk about the *seeded* variants of these three problems. In these versions, we add a selected tile $a_0 \in A$ to our input and ask for the corresponding snake/ouroboros to satisfy $\zeta(0) = a_0$.

All of these problems have been studied and classified for $\mathbb{Z}^2$ with its standard generating set $\{(1,0),(0,1)\}$.

Theorem 1. *Let S be the standard generating set for $\mathbb{Z}^2$. Then,*

1. *The snake reachability problem for $(\mathbb{Z}^2, S)$ is PSPACE-complete [14],*
2. *The infinite snake problem for $(\mathbb{Z}^2, S)$ is Π_1^0-complete [1],*
3. *The ouroboros problem for $(\mathbb{Z}^2, S)$ is Σ_1^0-complete [9,19].*

In addition, the seeded variants of these problems are undecidable [9].

Our aim is to extend these results to larger classes of groups and different generating sets.

3.1   General Properties

Let (G, S) be a f.g. group and Γ a tileset. If there exists a snake (ω, ζ), then for every $g \in G$, $(g\omega, \zeta)$ is a snake. If we define $\tilde{\omega}(i) = g\omega(i)$, then $d\tilde{\omega} = d\omega$, as the adjacency of ζ in Γ remains unchanged. In particular, there exists a snake (ω', ζ) such that $\omega'(0) = 1_G$, i.e. we may assume that a snake starts at the identity 1_G.

The next result is a straightforward generalization of a result due to Kari [19] for $\mathbb{Z}^2$.

Proposition 1. *Let Γ be a tileset for a f.g. group (G, S). Then, the following are equivalent:*

1. *Γ admits a bi-infinite snake,*
2. *Γ admits a one-way infinite snake,*
3. *Γ admits a snake of every length.*

This result implies that a tileset that admits no snakes will fail to tile any snake bigger than a certain length. Therefore, if we have a procedure to test snakes of increasing length, we have a semi-algorithm to test if a tileset does not admit an infinite snake.

Corollary 1. *If G has decidable WP, the infinite snake problem is in Π_1^0.*

A similar process can be done for the ouroboros problem.

Proposition 2. *If G has decidable WP, the ouroboros problem is in Σ_1^0.*

Proof. Let Γ be a tileset graph for (G, S). For each $n \geq 1$, we test each word of length n to see if it defines a simple loop and if it admits a valid tiling. More precisely, for $w \in (S \cup S^{-1})^n$, we use the word problem algorithm to check if w is G-reduced and evaluates to 1_G. If it is reduced, we test all possible tilings by Γ of the path defined by following the generators in w. If we find a valid tiling, we accept. If not, we keep iterating with the next word length n and eventually with words of length $n + 1$.

If there is a Γ-ouroboros, this process with halt and accept. Similarly, if the process halts we have found a Γ-ouroboros. Finally, if there is no Γ-ouroboros the process continues indefinitely.

We also state reductions when working with subgroups or adding generators.

Lemma 1. *Let* (G, S) *be a f.g. group,* (H, T) *a f.g. subgroup of* G *and* $w \in (S \cup S^{-1})^+$. *Then,*

- *The infinite snake, ouroboros and reachability problems in* (H, T) *many one-reduce to their respective analogues in* $(G, S \cup T)$.
- *The infinite snake, ouroboros and reachability problems in* (G, S) *many one-reduce to their respective analogues in* $(G, S \cup \{w\})$.

Proof. Any tileset graph for (H, T) is a tileset graph for $(G, S \cup T)$, and any tileset graph for (G, S) is a tileset graph for $(G, S \cup \{w\})$.

4   Ossuary

Much of the complexity of snakes comes from the paths they define on the underlying group. It stands to reason that understanding the structure of all possible injective bi-infinite paths on the group can shed light on the computability of the infinite snake problem. Let G be a f.g. group with S a set of generators. The *skeleton* subshift of the pair (G, S) is defined as

$$X_{G,S} := \{x \in (S \cup S^{-1})^{\mathbb{Z}} \mid \forall w \sqsubseteq x, \ w \notin \mathrm{WP}(G, S)\}.$$

This subshift is the set of all possible skeletons: recall from Definition 4 that for any skeleton ω, we can define $d\omega : \mathbb{Z} \to S \cup S^{-1}$ as $d\omega_i = \omega(i)^{-1}\omega(i+1)$. Thus, for any infinite snake (ω, ζ): $d\omega \in X_{G,S}$.

This formalism allows us to introduce variations of the infinite snake problem where we ask for additional properties on the skeleton. We say a subset $Y \subseteq X_{G,S}$ is *skeletal* if it is shift-invariant. In particular, all subshifts of $X_{G,S}$ are skeletal.

Definition 6. *Let* Y *be a skeletal subset. The* Y-*snake problem asks, given a tileset* Γ, *does there exist a bi-infinite* Γ-*snake* (ω, ζ) *such that* $d\omega \in Y$?

4.1   Skeletons and Decidability

A snake (ω, ζ) can be seen as two walks that follow the same labels: ω is a self-avoiding walk on the Cayley graph of the group, and ζ a walk on the tileset graph. The next result, which is a direct consequence of the definitions, uses this fact to construct a single object that captures both walks.

For (G, S) a f.g. group, Γ a tileset graph, denote $X_\Gamma \subseteq (S \cup S^{-1})^{\mathbb{Z}}$ the subshift whose configurations are the labels of bi-infinite paths over Γ. This implies X_Γ is a sofic subshift.

Proposition 3. *Let (G, S) be a f.g. group, Γ a tileset graph and Y a non-empty skeletal subset. Then $X = Y \cap X_\Gamma$ is non-empty if and only if there is a bi-infinite Γ-snake (ω, ζ) with $d\omega \in Y$. In addition, if Y is an effective/sofic subshift, then X is an effective/sofic subshift.*

The previous result reduces the problem of finding an infinite Y-snake, to the problem of emptiness of the intersection of two one-dimensional subshifts. As previously stated determining if a subshift is empty is co-recursively enumerable for effective subshifts, and decidable for sofics. Therefore, we can provide a semi-algorithm when the skeleton is effective. This is true for the class of recursively presented groups. Because these groups have recursively enumerable word problem, $\mathrm{WP}(G, S)$ is recursively enumerable for all finite generating sets. This enumeration gives us an enumeration of the forbidden patterns of our subshift.

Proposition 4. *Let G be a recursively presented group. Then $X_{G,S}$ is effective for every finite generating set S.*

This allows us to state the following proposition.

Proposition 5. *Let Y be a skeletal subshift. Then, if Y is sofic (resp. effective) the Y-snake problem is decidable (resp. in Π_1^0). In particular, if G is recursively presented, the infinite snake problem for any generating set is in Π_1^0.*

We now identify where the undecidability of the infinite snake problem comes from. If we restrict the directions in which the snake moves to 2 or 3, the problem becomes decidable. This means that the ability to make "space-filling"-like curves in $\mathbb{Z}^2$ is, heuristically speaking, required for the proof of the undecidability of the infinite snake problem.

Theorem 2. *The infinite snake problem in $\mathbb{Z}^2$ restricted to 2 or 3 directions among $(1, 0), (0, 1), (-1, 0), (0, -1)$ is decidable.*

Proof. The set of skeletons of snakes restricted to 3 directions, for instance left, right and up (denoted by a^{-1}, a and b respectively), is the subshift $Y_3 \subseteq \{a, a^{-1}, b\}^{\mathbb{Z}}$ where the only forbidden words are aa^{-1} and $a^{-1}a$. As Y_3 is a skeletal SFT of $X_{\mathbb{Z}^2, \{a,b\}}$, by Proposition 5, the Y_3-snake problem is decidable. The case of two directions is analogous as Y_2 is the full shift on the two generators a and b.

A natural variation of the infinite snake problem, from the point of view of group theory, is asking if there is an infinite snake whose skeleton defines a geodesic. These skeletons are captured by the geodesic skeleton subshift; a subshift of $X_{G,S}$ comprised exclusively of bi-infinite geodesic rays. Formally,

$$X_{G,S}^g = \{x \in X_{G,S} \mid \forall w \sqsubseteq x, w' =_G w : |w| \leq |w'|\}.$$

This subshift can be equivalently defined through the set of forbidden patterns given by $\mathrm{Geo}(G, S)$. Then, by the definition of a sofic shift, we can state the following proposition.

Proposition 6. *Let (G, S) be a f.g. group. If $\mathrm{Geo}(G, S)$ is regular, then $X_{G,S}^g$ is sofic.*

$\mathrm{Geo}(G, S)$ is known to be regular for all generating sets in abelian groups [25] and hyperbolic groups [12], and for some generating sets in many other classes of groups [2,7,17,18,25]. Theorem 3 implies that the geodesic infinite snake problem is decidable for all such (G, S); most notably for $\mathbb{Z}^2$ with its standard generating set.

Theorem 3. *The geodesic infinite snake problem is decidable for any f.g. group (G, S) such that $\mathrm{Geo}(G, S)$ is regular. In particular, it is decidable for abelian and hyperbolic groups for all generating sets.*

What happens with skeletal subsets that are not closed and/or not effective? In these cases, Ebbinghaus showed that the problem can be undecidable outside of the arithmetical hierarchy [10]. If we define Y to be the skeletal subset of $(\mathbb{Z}^2, \{a, b\})$ of skeletons that are not eventually a straight line, then, Y is not closed and deciding if there exists a Y-skeleton snake is Σ_1^1-complete. Similarly, if we take the Y to be the set of non-computable skeletons of $\mathbb{Z}^2$, the Y-skeleton problem is also Σ_1^1-complete.

5   Snake Embeddings

Let us introduce a suitable notion of embedding, that guarantees a reduction of snake problems. To do this, we will make use of a specific class of finite-state transducer called *invertible-reversible transducer*, that will translate generators of one group to another in an automatic manner.

Definition 7. *An invertible-reversible transducer $\mathcal{M}$ is a tuple $(Q, S, T, q_0, \delta, \eta)$ where,*

- *Q is a finite set of states,*
- *S, T are finite alphabets,*
- *$q_0 \in Q$ is an initial state,*
- *$\delta : Q \times S \to Q$ is a transition function,*
- *$\eta : Q \times S \to T$ is such that $\eta(q, \cdot)$ is an injective function for all $q \in Q$,*

such that for all $q \in Q$ and $s \in S$ there exists a unique q' such that $\delta(q', s) = q$.

We extend both η and δ to manage inverses of S by setting $\eta(q, s^{-1}) = \eta(q', s)^{-1}$ and $\delta(q, s^{-1}) = q'$, where q' is the unique state satisfying $\delta(q', s) = q$. Furthermore, we denote by q_w the state of $\mathcal{M}$ reached after reading the word $w \in (S \cup S^{-1})^*$ starting from q_0. We introduce the function $f_{\mathcal{M}} : (S \cup S^{-1})^* \to (T \cup T^{-1})^*$ recursively defined as $f_{\mathcal{M}}(\epsilon) = \epsilon$ and $f_{\mathcal{M}}(ws^{\pm 1}) = f_{\mathcal{M}}(w)\eta(q_w, s^{\pm 1})$.

Definition 8. *Let (G, S) and (H, T) be two f.g. groups. A map $\phi : G \to H$ is called a* snake embedding *if there exists a transducer $\mathcal{M}$ such that $\phi(g) = f_{\mathcal{M}}(w)$ for all $w \in (S \cup S^{-1})^*$ such that $\bar{w} = g$, and $f_{\mathcal{M}}(w) =_H f_{\mathcal{M}}(w')$ if and only if $w =_G w'$.*

Remark 1. Snake-embeddings are a strictly stronger form of a translation-like action. Such an action is a right group action $* : G \to H$ that is free, i.e. $h * g = h$ implies $g = 1_G$, and $\{d(h, h * g) \mid h \in H\}$ is bounded for all $g \in G$. A straightforward argument shows that if $\phi : (G, S) \to (H, T)$ is a snake-embedding, then $h * g = h\phi(g)$ is a translation-like action. The converse is not true: there are translation-like actions that are not defined by snake-embeddings. For instance, from Definition 8 we see that there is a snake embedding from $\mathbb{Z}$ to a group G if and only if $\mathbb{Z}$ is a subgroup of G. Nevertheless, infinite torsion groups admit translation-like actions from $\mathbb{Z}$, as shown by Seward in [26], but do not contain $\mathbb{Z}$ as a subgroup.

Proposition 7. *Let (G, S) and (H, T) be two f.g. groups such that there exists a snake-embedding $\phi : G \to H$. Then, the infinite snake (resp. ouroboros) problem on (G, S) many-one reduces to the infinite snake (resp. ouroboros) problem on (H, T).*

The reduction consists in taking a tileset for (G, S) and using the transducer to create a tileset for (H, T) that is consistent with the structure of G. Because the transducer is invertible-reversible, we have a computable way to transform a bi-infinite snake from one group to the other.

Using snake-embeddings we can prove that non-$\mathbb{Z}$ f.g. free abelian groups have undecidable snake problems.

Proposition 8. *The infinite snake and ouroboros problems on $\mathbb{Z}^d$ with $d \geq 2$ are undecidable for all generating sets.*

Proof. Let $S = \{v_1, ..., v_n\}$ be a generating set for $\mathbb{Z}^d$. As S generates the group, there are two generators v_{i_1} and v_{i_2}, such that $v_{i_1}\mathbb{Z} \cap v_{i_2}\mathbb{Z} = \{1_{\mathbb{Z}^d}\}$. Then, $H = \langle v_{i_1}, v_{i_2} \rangle \simeq \mathbb{Z}^2$ and there is clearly a snake-embedding from $\mathbb{Z}^2$ to H. Finally, by Lemma 1, the infinite snake and ouroboros problems are undecidable for $(\mathbb{Z}^d, S)$.

6   Virtually Nilpotent Groups

Through the use of snake-embeddings and skeleton subshifts, we extend undecidability results from abelian groups to the strictly larger class of virtually nilpotent groups. For any group G we define $Z_0(G) = \{1_G\}$ and

$$Z_{i+1}(G) = \{g \in G \mid ghg^{-1}h^{-1} \in Z_i(G), \forall h \in G\}.$$

The set $Z_1(G)$ is called the *center* of G, and by definition is the set of elements that commute with every element in G. We say a group is *nilpotent* if there exists $i \geq 0$ such that $Z_i(G) = G$.

The next Lemma is stated for a larger class of groups that contain nilpotent groups, that will allow us to prove the undecidability results on the latter class.

Lemma 2. *Let (G, S) be a f.g. group that contains an infinite order element g in its center, such that $G/\langle g \rangle$ is not a torsion group. Then, there is a snake embedding from $(\mathbb{Z}^2, \{a, b\})$ into $(G, S \cup \{g\})$, where $\{a, b\}$ is the standard generating set for $\mathbb{Z}^2$.*

The proof consists in finding a distorted copy of $\mathbb{Z}^2$ within $(G, S \cup \{g\})$. One of the copies of $\mathbb{Z}$ is given by $\langle g \rangle \simeq \mathbb{Z}$. The other is obtained through the following result.

Proposition 9. *Let (G, S) be a f.g. group. Then, G is a torsion group if and only if $X_{G,S}$ is aperiodic.*

Using g and a periodic point from this proposition we construct the snake-embedding.

Proposition 10. *Let (G, S) be a f.g. group that contains a infinite order element g in its center and $G/\langle g \rangle$ is not a torsion group. Then, $(G, S \cup \{g\})$ has undecidable infinite snake and ouroboros problems.*

Proof. By Lemma 2, there is a snake-embedding from $\mathbb{Z}^2$ to $(G, S \cup \{g\})$. Combining Proposition 7 and Theorem 1, we conclude that both problems are undecidable on $(G, S \cup \{g\})$.

Theorem 4. *Let (G, S) be a f.g. infinite, non-virtually $\mathbb{Z}$, nilpotent group. Then there exists g such that $(G, S \cup \{g\})$ has undecidable infinite snake and ouroboros problems.*

Proof. Let G be a f.g. infinite nilpotent group that is not virtually cyclic. Because G is nilpotent, there exists a torsion-free element $g \in Z_1(G)$. Furthermore no quotient of G is an infinite torsion group [8]. In addition, as G is not virtually cyclic $G/\langle g \rangle$ is not finite. Therefore, by Proposition 10, both problems are undecidable on $(G, S \cup \{g\})$.

Through Lemma 1 we obtain undecidability for virtually nilpotent groups.

Corollary 2. *Let G be a f.g. infinite, non virtually $\mathbb{Z}$, virtually nilpotent group. Then there exists a finite generating set S such that (G, S) has undecidable infinite snake and ouroboros problems.*

7   Snakes and Logic

We want to express snake problems as a formula that can be shown to be satisfied for a large class of Cayley graphs. To do this we use Monadic Second-Order (MSO) logic, as has been previously been done for the domino problem. Our formalism is inspired by [5]. Let $\Lambda = (V, E)$ be an S-labeled graph with root v_0. This fraction of logic consists of variables $P, Q, R, \dots$ that represent subsets of vertices of Λ, along with the constant set $\{v_0\}$; as well as an operation for each $s \in S$, $P \cdot s$, representing all vertices reached when traversing an edge labeled by s from a vertex in P. In addition, we can use the relation $\subseteq$, Boolean operators $\wedge, \vee, \neg$ and quantifiers $\forall, \exists$. For instance, we can express set equality by the formula $(P = Q) \equiv (P \subseteq Q \wedge Q \subseteq P)$ and emptiness by $(P = \varnothing) \equiv \forall Q(P \subseteq Q)$. We can also manipulate individual vertices, as being a singleton is expressed by $(|P| = 1) \equiv P \neq \varnothing \wedge \forall Q \subseteq P(Q = \varnothing \vee P = Q)$. For example, $\forall v \in P$ is shorthand notation for the expression $\forall Q(Q \subseteq P \wedge |Q| = 1)$. Notably, we can express non-connectivity of a subset $P \subseteq V$ by the formula $\mathrm{NC}(P)$ defined as

$$\exists Q \subseteq P, \exists v, v' \in P(v \in Q \wedge v' \notin Q \wedge \forall u, w \in P(u \in Q \wedge \mathrm{edge}(u, w) \Rightarrow w \in Q)),$$

where $\mathrm{edge}(u, w) \equiv \bigvee_{s \in S} u \cdot s = w$. The set of formulas without free variables obtained with these operations is denoted by $\mathrm{MSO}(\Lambda)$. We say Λ has *decidable* MSO logic, if the problem of determining if given a formula in $\mathrm{MSO}(\Lambda)$ is satisfied is decidable.

The particular instance we are interested in is when Λ is the Cayley graph of a f.g. group G labeled by S a symmetric finite set of generators, that is, $S = S^{-1}$ (we take such a set to avoid cumbersome notation in this section). In this case, the root of our graph is the identity $v_0 = 1_G$. A landmark result in the connection between MSO logic and tiling problems comes from Muller and Schupp [22,23], as well as Kuskey and Lohrey [20], who showed that virtually free groups have decidable MSO logic. Because the Domino Problem can be expressed in MSO, it is decidable on virtually free groups. Our goal is to obtain an analogous result for domino snake problems. To express infinite paths and loops, given a tileset graph $\Gamma = (A, B)$, we will partition a subset $P \subseteq V$ into subsets indexed by S and A, such that $P_{s,a}$ will contain all vertices with the tile a that point through s to the continuation of the snake. We denote the disjoint union as $P = \coprod_{s \in S, a \in A} P_{s,a}$. First, we express the property of always having a successor within P as

$$N(P, \{P_{s,a}\}) \equiv \bigwedge_{s \in S, a \in A} (P_{s,a} \cdot s \subseteq P).$$

We also want for this path to not contain any loops, by asking for a unique predecessor for each vertex:

$$\mathrm{up}(v) \equiv \exists! s \in S, a \in A : v \in P_{s,a} \cdot s,$$

$$\equiv \left(\bigvee_{\substack{s \in S \\ a \in A}} v \in P_{s,a} \cdot s \right) \wedge \left(\bigwedge_{\substack{a, a' \in A \\ a \neq a'}} \bigwedge_{\substack{s, t \in S \\ s \neq t}} \neg((v \in P_{s,a} \cdot s) \wedge (v \in P_{t,a'} \cdot t)) \right).$$

Then, for a one-way infinite path

$$UP(P, \{P_{s,a}\}) \equiv \forall v \in P \left((v = v_0 \wedge \bigwedge_{s \in S, a \in A} v \notin P_{s,a} \cdot s) \vee (v \neq v_0 \wedge up(v)) \right),$$

Thus, we state the property of having an infinite path as follows:

$$\infty RAY(P, \{P_{s,a}\}) \equiv \left(v_0 \in P \wedge P = \coprod_{\substack{s \in S \\ a \in A}} P_{s,a} \wedge N(P, \{P_{s,a}\}) \wedge UP(P, \{P_{s,a}\}) \right).$$

In fact, we can do a similar procedure to express the property of having a simple loop within P by slightly changing the previous expressions. The only caveat comes when working with S a symmetric finite generating set of some group, as we must avoid trivial loops such as ss^{-1}.

$$\ell(P, \{P_{s,a}\}) \equiv \forall v \in P, up(v) \wedge \bigwedge_{s \in S, a, a' \in A} \left(P_{s,a} \cdot s \cap P_{s^{-1}, a'} = \varnothing \right).$$

This way, admitting a simple loop is expressed as

$$LOOP(P, \{P_{s,a}\}) \equiv \left(v_0 \in P \wedge P = \coprod_{s \in S, a \in A} P_{s,a} \wedge N(P, \{P_{s,a}\}) \wedge \ell(P, \{P_{s,a}\}) \right)$$

$$\wedge \forall Q \subseteq P, \forall \{Q_{s,a}\} \left(\neg \infty RAY(Q, \{Q_{s,a}\}) \right).$$

Lemma 3. *Let $P \subseteq V$. Then,*

1. *If there exists a partition $\{P_{s,a}\}_{s \in S, a \in A}$ such that $\infty RAY(P, \{P_{s,a}\})$ is satisfied, P contains an infinite injective path. Conversely, if P is the support of an injective infinite path rooted at v_0, there exists a partition $\{P_{s,a}\}_{s \in S, a \in A}$ such that $\infty RAY(P, \{P_{s,a}\})$ is satisfied.*
2. *If there exists a partition $\{P_{s,a}\}_{s \in S, a \in A}$ such that $LOOP(P, \{P_{s,a}\})$ is satisfied, P contains a simple loop. Conversely, if P is the support of a simple loop based at v_0, there exists a partition $\{P_{s,a}\}_{s \in S, a \in A}$ such that $LOOP(P, \{P_{s,a}\})$ is satisfied.*

With these two structure-detecting formulas, we can simply add the additional constraint that P partitions in a way compatible with the input tileset graph of the problem, in the direction of the snake. This is captured by the formula

$$D_\Gamma(\{P_{s,a}\}) \equiv \bigwedge_{(a,a',s) \notin B} \bigwedge_{s' \in S} P_{a,s} \cdot s \cap P_{a',s'} = \varnothing.$$

Theorem 5. *Let Λ be a Cayley graph of generating set S. The infinite snake problem, the reachability problem and the ouroboros problem can be expressed in $MSO(\Lambda)$.*

Proof. Let $\Gamma = (A, B)$ be a tileset graph for Λ. By Lemma 3, it is clear that

$$\infty\text{-SNAKE}(\Gamma) \equiv \exists P \exists \{P_{s,a}\} \left(\infty\text{RAY}(P, \{P_{s,a}\}) \wedge D_\Gamma(\{P_{s,a}\})\right),$$

$$\text{OUROBOROS}(\Gamma) \equiv \exists P \exists \{P_{s,a}\} \left(\text{LOOP}(P, \{P_{s,a}\}) \wedge D_\Gamma(\{P_{s,a}\})\right),$$

exactly capture the properties of admitting a one-way infinite Γ-snake and Γ-ouroboros respectively. Remember that Proposition 1 tells us that admitting a one-way infinite snake is equivalent to admitting a bi-infinite snake. We finish by noting that for reachability in a Cayley graph, we can take $p = v_0$. Then, verifying the formula $\text{REACH}(\Gamma, q)$ defined as

$$\exists P \exists \{P_{s,a}\} \left(q \in P \wedge \neg\text{NC}(P) \wedge P = \coprod_{\substack{s \in S \\ a \in A}} P_{s,a} \wedge UP(P, \{P_{s,a}\}) \wedge D_\Gamma(\{P_{s,a}\}) \right),$$

is equivalent to P containing the support of a Γ-snake that connects p to q.

As previously mentioned, virtually free groups have decidable MSO logic for all generating sets. Thus, we can state the following corollary.

Corollary 3. *Both the normal and seeded versions of the infinite snake, reachability and ouroboros problems are decidable on virtually free groups, independently of the generating set.*

Proof. Let $\Gamma = (A, B)$ be a tileset graph with $a_0 \in A$ the targeted tile. Then adding the clause $\bigvee_{s \in S} v_0 \in P_{s,a_0}$ to the formulas of any of the problems in question, we obtain a formula that expresses its corresponding seeded version.

Acknowledgments. We would like to thank Pierre Guillon and Guillaume Theyssier for helping with Lemma 3. We would also like to thank David Harel and Yael Etzion for providing a copy of [13]. We are grateful to the anonymous referees for their useful remarks.

References

1. Adleman, L., Kari, J., Kari, L., Reishus, D.: On the decidability of self-assembly of infinite ribbons. In: Proceedings of the 43rd Annual IEEE Symposium on Foundations of Computer Science 2002, pp. 530–537. IEEE (2002)
2. Antolin, Y., Ciobanu, L.: Finite generating sets of relatively hyperbolic groups and applications to geodesic languages. Trans. Am. Math. Soc. **368**(11), 7965–8010 (2016)
3. Aubrun, N., Barbieri, S., Jeandel, E.: About the domino problem for subshifts on groups. In: Berthé, V., Rigo, M. (eds.) Sequences, Groups, and Number Theory. TM, pp. 331–389. Springer, Cham (2018). https://doi.org/10.1007/978-3-319-69152-7_9
4. Aubrun, N., Bitar, N.: Domino snake problems on groups. arXiv preprint arXiv:2307.12655 (2023)

5. Bartholdi, L.: Monadic second-order logic and the domino problem on self-similar graphs. Groups Geom. Dyn. **16**, 1423–1459 (2022)
6. Berger, R.: The Undecidability of the Domino Problem. No. 66. American Mathematical Soc. (1966)
7. Charney, R., Meier, J.: The language of geodesics for Garside groups. Math. Zeitschrift **248**(3), 495–509 (2004)
8. Clement, A.E., Majewicz, S., Zyman, M.: Introduction to nilpotent groups. In: The Theory of Nilpotent Groups, pp. 23–73. Springer, Cham (2017). https://doi.org/ 10.1007/978-3-319-66213-8_2
9. Ebbinghaus, H.D.: Undecidability of some domino connectability problems. Math. Logic Q. **28**(22–24), 331–336 (1982)
10. Ebbinghaus, H.-D.: Domino threads and complexity. In: Börger, E. (ed.) Computation Theory and Logic. LNCS, vol. 270, pp. 131–142. Springer, Heidelberg (1987). https://doi.org/10.1007/3-540-18170-9_161
11. van Emde Boas, P.: The convenience of tilings. In: Complexity, Logic, and Recursion Theory, pp. 331–363. CRC Press (2019)
12. Epstein, D.B.: Word Processing in Groups. A K Peters/CRC Press, Massachusetts (1992). https://doi.org/10.1201/9781439865699
13. Etzion, Y.: On the Solvability of Domino Snake Problems. Master's thesis, Dept. of Applied Math. and Computer Science, Wiezmann Institute of Science, Rehovot, Israel (1991)
14. Etzion-Petruschka, Y., Harel, D., Myers, D.: On the solvability of domino snake problems. Theor. Comput. Sci. **131**(2), 243–269 (1994)
15. Grädel, E.: Domino games and complexity. SIAM J. Comput. **19**(5), 787–804 (1990)
16. Harel, D.: Recurring dominoes: making the highly undecidable highly understandable. In: North-Holland Mathematics Studies, vol. 102, pp. 51–71. Elsevier (1985)
17. Holt, D.F., Rees, S.: Artin groups of large type are shortlex automatic with regular geodesics. Proc. London Math. Soc. **104**(3), 486–512 (2012)
18. Howlett, R.B.: Miscellaneous facts about Coxeter groups. University of Sydney, School of Mathematics and Statistics (1993)
19. Kari, J.: Infinite snake tiling problems. In: Ito, M., Toyama, M. (eds.) DLT 2002. LNCS, vol. 2450, pp. 67–77. Springer, Heidelberg (2003). https://doi.org/10.1007/ 3-540-45005-X_6
20. Kuske, D., Lohrey, M.: Logical aspects of Cayley-graphs: the group case. Ann. Pure Appl. Logic **131**(1–3), 263–286 (2005)
21. Lind, D., Marcus, B.: An Introduction to Symbolic Dynamics and Coding. Cambridge University Press, Cambridge (2021)
22. Muller, D.E., Schupp, P.E.: Groups, the theory of ends, and context-free languages. J. Comput. Syst. Sci. **26**(3), 295–310 (1983)
23. Muller, D.E., Schupp, P.E.: The theory of ends, pushdown automata, and second-order logic. Theor. Comput. Sci. **37**, 51–75 (1985)
24. Myers, D.: Decidability of the tiling connectivity problem. Abstract 79T–E42. Notices Am. Math. Soc. **195**(26), 177–209 (1979)
25. Neumann, W.D., Shapiro, M.: Automatic structures, rational growth, and geometrically finite hyperbolic groups. Inventiones Math. **120**(1), 259–287 (1995)
26. Seward, B.: Burnside's problem, spanning trees and tilings. Geom. Topology **18**(1), 179–210 (2014). https://doi.org/10.2140/gt.2014.18.179
27. Wang, H.: Proving theorems by pattern recognition-II. Bell Syst. Tech. J. **40**(1), 1–41 (1961)

Parsing Unranked Tree Languages, Folded Once

Martin Berglund$^{(\boxtimes)}$ [iD], Henrik Björklund[iD], and Johanna Björklund[iD]

Umeå University, Umeå 90836, Sweden
{mbe,henrikb,johanna}@cs.umu.se

Abstract. A regular unranked tree folding consists of a regular unranked tree language and a folding operation that merges, i.e., *folds*, selected nodes of a tree to form a graph; the combination is a formal device for representing graph languages. If, in the process of folding, the order among edges is discarded so that the result is an unordered graph, then two applications of a fold operation is enough to make the associated parsing problem NP-complete. However, if the order is kept, then the problem is solvable in non-uniform polynomial time. In this paper we address the remaining case where only one fold operation is applied, but the order among edges is discarded. We show that under these conditions, the problem is solvable in non-uniform polynomial time.

1 Introduction

Graphs are one of the most commonly used data structures in computer science. Whether we are conducting social network analysis [12], defining the semantics of programming languages [10], or devising a better method for training deep neural networks [13], we are likely to operate on some form of graph representation. Practical applications of formal graph languages typically require that the parsing problem is efficiently solvable. This means that given a graph g, we can decide whether the graph adheres to the formalism in polynomial time, and also produce a certificate attesting the veracity of our decision. In the case of so-called order-preserving graph grammars (ODPGs) for example, we can decide in linear time if a given graph g is well-formed with respect to a particular grammar G, and provide a unique derivation tree for g in G as proof [3,4].

Significant effort has been devoted to finding graph formalisms that combine expressiveness with parsing efficiency, see e.g. [9,11]. Most of these are restrictions of hyperedge replacement grammars (HRGs) [8], a natural generalisation of context-free grammars, in which nonterminals are replaced by labelled hyperedges that provide restricted access to the intermediate graphs. The previously mentioned ODPGs is one of the most easily parsed restrictions of HRGs. They are designed to be just strong enough to represent so-called abstract meaning

J. Björklund—Supported by the Swedish Research Council under Grant Number 2020-03852, and by the Wallenberg AI, Autonomous Systems and Software Program through the NEST project *STING*.

H. Fernau and K. Jansen (Eds.): FCT 2023, LNCS 14292, pp. 60–73, 2023.
https://doi.org/10.1007/978-3-031-43587-4_5

representations (AMRs), i.e., a semantic representation based on a limited type of directed acyclic graphs. At the more powerful end of the spectrum we have s-grammars [9]. In this formalism, terms over a small set of operators and a finite set of elementary graphs are evaluated in the domain of node-labelled graphs. The membership problem for HRGs and s-grammars require exponential time in general, and HRGs can generate languages for which the associated parsing problem is NP-complete [1].

In [5], the authors introduced *regular tree foldings*, a generative device consisting of a (finite representation of a) regular tree language and a new type of folding operation. The tree language is non-standard in that there is an auxiliary set of symbols Δ, which can be used to mark already labelled nodes, or to label nodes under which sits a single subtree. The folding operation then translates each tree t in the language into a graph by processing t bottom up: every time evaluation reaches a node labelled $\alpha \in \Delta$, it merges all nodes in the subtree sitting below v that carries the mark α into a single node, and clears it from the mark α (see Fig. 1). Similar to ODPGs, regular tree-foldings are suitable to model AMRs, but can also accommodate cyclic graphs and more varied types of node-sharing. On the down-side, the parsing problem is more difficult: If the relative order among the edges attaching to a node is preserved by the folding, then parsing can be done in polynomial time in the size of the input graph [2]. If however this order is relaxed, so that the output graph is considered to be unordered, then the parsing problem is NP-complete [2]. An analysis of the proof in [2] yields that the problem is NP-hard already when only two applications of the folding operation are allowed. The case when there is a single application of a folding operation is not addressable by either of the proof techniques in [2], and was therefore not solved.

In this paper, we show that the parsing problem for unranked regular tree languages folded once under an order-cancelling fold semantics (see [2] for the contrasting order-preserving case) is solvable in non-uniform polynomial time, i.e. in polynomial time when considering the automaton constant.

2   Preliminaries

This section recalls relevant formal language theory and fixes notation [2,5].

The set of non-negative integers is denoted by $\mathbb{N}$. For $i \in \mathbb{N}$, $[i] = \{1, \ldots, i\}$ and $[0] = \emptyset$. A *multiset* S' is a set in which elements can have multiple instances.

Graphs. An *alphabet* is a finite nonempty set of symbols. Let Σ be an alphabet. A (directed and rooted) *graph* over Σ is a tuple $g = (V, E, src, tar, lab, r)$ consisting of: Finite sets V and E of *nodes* and *edges*, respectively; *source* and *target mappings* $src \colon E \to V$ and $tar \colon E \to V$ assigning to each edge e its source $src(e)$ and target $tar(e)$; a *labelling* $lab \colon V \to \Sigma$; and a *root* node $r \in V$. The set of *incoming edges* of $u \in V$ is $in(u)$, and the set of *outgoing edges* is $out(u)$. A node with out-degree 0 is a *leaf*. The *size* of the graph g is $|g| = |V| + |E|$. The set of all graphs with node labels in Σ is denoted by $\mathbb{G}_\Sigma$. We write $root(g)$ for the root node r, and $nodes(g)_\sigma$ for the set of nodes in g that are labelled $\sigma \in \Sigma$.

An *edge-ordered graph* over Σ is a pair $(g, <)$ where $g = (V, E, src, tar, lab, r)$ in $\mathbb{G}_\Sigma$ and $<$ is a binary relation on E that becomes a total order when restricted to any set $in(v)$ or $out(v)$, for $v \in V$. From here on, we leave the second component implicit and refer to it as $<_g$ when needed. The set of all edge-ordered graphs over Σ is denoted by $\mathbb{G}_\Sigma^<$.

A *path* in the graph $g = (V, E, src, tar, lab, r)$ is a finite and possibly empty sequence $p = e_0 e_1 \cdots e_k$ of edges such that for each $i \in [k]$, the target of e_{i-1} is the source of e_i. Here, we say that p is a path from $src(e_0)$ to $tar(e_k)$. The path p is a *cycle* if $src(e_0) = tar(e_k)$.

Trees. Let Σ be an alphabet. An *ordered unranked tree over* Σ is a tuple $t = ((V, E, src, tar, lab, r), <) \in \mathbb{G}_\Sigma^<$ such that (i) t is connected, (ii) r has no incoming edges, and (iii) every node except for r has exactly one incoming edge. We denote the set of all ordered unranked trees over Σ by T_Σ.

Let $\sigma \in \Sigma$, $k \in \mathbb{N}$, and $t_1, \ldots, t_k$ be trees over Σ with $t_i = (g_i, <_i)$ and $g_i = (V_i, E_i, src_i, tar_i, lab_i, v_i)$, such that the node sets V_i, $i \in [k]$, are mutually disjoint, and similarly for the edge sets E_i, $i \in [k]$. The *top-concatenation* of $t_1, \ldots, t_k$ with σ, denoted by $\sigma[t_1, \ldots, t_k]$ is the tree obtained by attaching the trees $t_1, \ldots, t_k$ as children underneath a new root node with label σ.

Top-concatenation is analogously defined for single-rooted graphs, so we may write $\sigma[g_1, \ldots, g_k]$ without risk of confusion. In the case that the graphs are ordered, so are the edges $e_1, \ldots, e_k$ that attach the subgraphs $g_1, \ldots, g_k$; otherwise they are unordered.

Let X be a set of variables such that $X \cap \Sigma = \emptyset$. A *context* is a tree $c \in \mathrm{T}_{\{x\} \cup \Sigma}$, for $x \in X$, such that c contains exactly one occurrence of x, and this occurrence is a leaf. Given such a context and a tree t, we let $c[\![t]\!]$ denote the tree obtained from c by replacing the node labelled x with t. Formally, $c[\![t]\!] = t$ if $c = x$, and otherwise $c[\![t]\!] = \sigma[s_1, \ldots, s_{i-1}, s_i[\![t]\!], s_{i+1}, \ldots, s_n]$, where $c = \sigma[s_1, \ldots, s_n]$ and $s_i \in \mathrm{T}_{\{x\} \cup S}$ is the unique context among $s_1, \ldots, s_n$.

Automata for Unranked Trees. A *Z-automaton* [6] is a tuple $A = (\Sigma, Q, R, F)$ consisting of a finite *input alphabet* Σ; a finite set Q of *states* which is disjoint from Σ; a finite set R of *transition rules*, each of the form $s \to q$ consisting of a left-hand side $s \in \mathrm{T}_{\Sigma \cup Q}$ and a right-hand side $q \in Q$; and a finite set $F \subseteq Q$ of accepting states. Henceforth, we write m for $|Q|$.

Let $t \in \mathrm{T}_{\Sigma \cup Q}$. A transition rule r of the form $s \to q$ is *applicable* to t, if t can be written as $t = c[\![\sigma[t_1, \ldots, t_n]]\!]$, such that $s = \sigma[t_1, \ldots, t_k]$ for some $k \leq n$. If so, then there is a *computation step* $t \to_A c[\![q[t_{k+1}, \ldots, t_n]]\!]$. A *computation* of A on a tree $t \in \Sigma$ is a sequence of computation steps $t \to_A^* q$, for some $q \in Q$. The computation is *accepting* if $q \in F$. A tree $t \in \mathrm{T}_\Sigma$ is *accepted*, or *recognised*, by A if there is an accepting computation of A on t. The *language* accepted by A, denoted by $\mathcal{L}(A)$, is the set of all trees in T_Σ that A accepts.

As shown in [6], Z-automata recognise the same class of languages as unranked tree automata [7]. We use a normal form, in which all transition rules are of the form $\sigma \to r$, $\sigma[q] \to r$, or $p[q] \to r$, for $\sigma \in \Sigma$ and $p, q, r \in Q$.

(a)   (b)   (c)   (d)

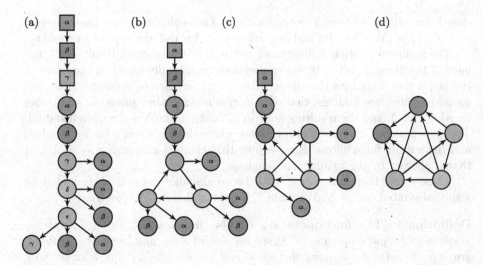

Fig. 1. In the tree in Subfigure (a), round nodes with and without annotation denote a label in $\Sigma \times \Delta$ and Σ, respectively, and square nodes denote a label in Δ. Arrows indicate edges, pointing from the source node of each edge to its target. When the tree is evaluated bottom-up, nodes with labels in $\Sigma \cup (\Sigma \times \Delta)$ are copied to the output graph until the transformation reaches a node with a label $\alpha \in \Delta$. Here, all nodes below with a label in $\Sigma \times \{\alpha\}$ are merged into a single node, and the square node labelled α is removed. The result is the graph in Subfigure (b). The process continues upwards until all Δ-labelled nodes have been cleared, yielding the graph in Subfigure (d).

A run of a Z-automaton $A = (\Sigma, Q, R, F)$ in normal form on a tree t can be seen as assigning states to *nodes(t)*. For each state (q, σ) in $Q \times \Sigma$, there is a regular language $r_{q,\sigma}$ such that in a run, the sequence w of states assigned to the children of a node that has label σ and is assigned q must be a string in $r_{q,\sigma}$ [6]. We denote by *parikh*(q, σ) the Parikh image of $r_{q,\sigma}$.

3   Regular Tree Foldings

The purpose of the folding operation is to turn an unranked tree t over an alphabet Σ into a graph g by merging nodes. The folding is done by marking nodes with symbols from an auxiliary alphabet Δ, meaning that some nodes will have labels in $\Sigma \times \Delta$, and then, for each $\alpha \in \Delta$ merging all nodes marked with symbols in $\Sigma \times \{\alpha\}$ into a single node. By itself, this formalism can only produce output graphs where at most $|\Delta|$ nodes have more than one incoming edge. For this reason, the merging is divided into *scopes* by allowing nodes in t that have labels from Δ: A node u in t labelled $\alpha \in \Delta$ is an instruction that all nodes with labels in $\Sigma \times \{\alpha\}$ *below it* are to be merged, whereupon the node u itself is to be deleted. The tree is then evaluated bottom-up, so that Δ-labeled nodes lower in the tree have their corresponding operations performed earlier. To keep the result well-defined, the tree language must force nodes with labels in Δ to have exactly one direct subtree. The combination of a regular unranked tree language over $\Sigma \cup (\Sigma \times \Delta) \cup \Delta$ and a fold operation over Δ is called a *regular tree folding*.

The folding operation is illustrated in Fig. 1. In the original definition [5], the label of the merged node v is chosen non-deterministically based on the labels of the nodes that went into the merger. However, there is a (less compact) normal form of regular tree foldings, that only merges nodes if they share the same label $(\sigma, \alpha) \in \Sigma \times \Delta$, and the resulting node in the output graph is then labelled σ [2]. This paper is only concerned with the case where there is a single folding symbol, and the normal form allows us to assume that there is a unique $\langle \sigma, \alpha \rangle \in \Sigma \cup \Delta$ that is allowed by the regular tree language.

Thus, throughout this paper, let Σ be an alphabet, let $\sigma \in \Sigma$, and let α be a special symbol not in Σ. We write Γ for the alphabet $\Sigma \cup \{\langle \sigma, \alpha \rangle, \alpha\}$.

Definition 1 (The fold operation F). *The function $[\![\alpha]\!] : \mathbb{G}_\Gamma \to \mathbb{G}_\Gamma$ takes a single-rooted input graph $g = (V, E, src, tar, lab, r) \in \mathbb{G}_\Gamma$ and computes an output graph $h = [\![\alpha]\!](g)$ by merging the set of nodes $nodes(g)_{\langle \sigma, \alpha \rangle}$ into a single node u, and assigning u the label σ. If $root(g) \in nodes(g)_{\langle \sigma, \alpha \rangle}$, then $root(h) = u$; otherwise $root(h) = root(g)$. The fold operation $F : \mathrm{T}_\Gamma \to \mathbb{G}_\Gamma$ is defined for every tree $t = \gamma[t_1, \ldots, t_k] \in \mathrm{T}_\Gamma$ by $F(t) = [\![\gamma]\!](F(t_1))$ if $\gamma = \alpha$ (when k is always 1), and $\gamma[F(t_1), \ldots, F(t_k)]$ otherwise. It is extended to sets of trees in the expected way: For $L \subseteq \mathrm{T}_\Gamma$, $F(L) = \bigcup_{t \in L} \{F(t)\}$.*

Definition 2 (Regular Tree Folding (with one folding symbol)). *A regular tree folding (RTF) over Γ is defined through a Z-automaton A over the same alphabet, such that for every $t \in \mathcal{L}(A)$ and every node v in t, it holds that if the label of v is α, then v has exactly one direct subtree. The folded graph language with respect to A is $\mathcal{L}_F(A) = F(\mathcal{L}(A))$.*

Example 1. Figure 2 shows two trees annotated with folding symbols, along with the corresponding graphs they fold into. In the first tree (located on the left in the illustration), the two nodes labeled $\alpha \in \Delta$ (represented as blue squares) appear

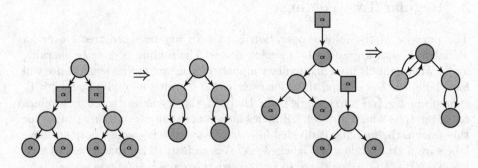

Fig. 2. In the above figure, we see two examples of trees decorated with folding symbols and the graphs they fold into. As in Fig. 1, round nodes with and without annotation denote a label in $\Sigma \times \Delta$ and Σ, respectively, and square nodes denote a label in Δ.

side by side. There is no interaction between their scopes: Removing either of these nodes would not affect the outcome of the application of the other. Moving to the second tree (two steps to the right in the same illustration), the lower square node labelled α shadows the scope of the upper one. Had it not been present, then all blue round nodes would have been merged into a single node.

The remainder of this paper is devoted to the membership problem for a fixed RTF represented by a Z-automaton A over Γ. It asks: *Given a graph g, is g in $\mathcal{L}_F(A)$?* In the special case where folding is only applied once, the problem can be restated as one of combining tree fragments into a single tree in a target language. From here on, $x \notin \Gamma$ is a fixed but arbitrary variable symbol.

A *tree fragment* is an unordered, unranked tree t with the following properties: (i) The root has exactly one child. (ii) Some leafs may have label x, while all other nodes have labels from Γ. We call the unique child of the root the *prior* of t, denoted by $prior(t)$. A *tree fragment* is an unordered, unranked tree t with the following properties: (i) The root has exactly one child. (ii) Some leaves may have label x, while all other nodes have labels from Γ. If t has more than one node, we call the unique child of the root the *prior* of t, denoted by $prior(t)$.

A *substitution* is an operation that takes a tree or a tree fragment t and $k = |nodes(t)_x|$ sets $T_1, \ldots, T_k$ of tree fragments. It assigns to each node $v_i \in nodes(t)_x$ a unique set T_i. For each i, the roots of the fragments in T_i are then identified with v_i. Finally, v_i is labelled $\langle \sigma, \alpha \rangle$.

We can thus view any tree as composed from a tree and a number of tree fragments through substitution. Taking this idea further, we note that a single application of a tree folding has the effect of turning the input tree into a set of tree fragments, all but at most one rooted at the merged node, and some of which have leaves attaching to the merged node (see Fig. 3). The merged nodes hide how these tree fragments originally fit together, and solving the membership problem is tantamount to recapturing this information. In the following, we denote by $order(t)$ the set of ordered trees that can be obtained from an unordered tree t by attaching an order to its edges.

Remark 1. Constructing the tree fragments is trivial in the interesting cases, i.e., the graph will have a single node which is obviously the merged node as it has more than one incoming edge. The tree fragments are then obtained by giving each edge incident with the merged node its own copy of that node, as is shown in Fig. 3. The other cases, where zero or one node is "merged" can easily be avoided by rewriting the automaton. That is: use the states to track and verify a nondeterministic guess whether zero, one, or more than one, nodes will get merged by a folding operator. For zero, skip generating the folding operator node (that would do nothing), for one node instead generate it with its resulting label *and* inhibit generating the folding operator node. For two or more, simply operate the same as the original automaton (but check the guess). Refer to Lemma 4.1 in [2] for a detailed construction easily modified for this case. We begin by dealing with the case of only a single fold, but in Corollary 1 we sketch the straightforward steps needed to reintroduce multiple folds.

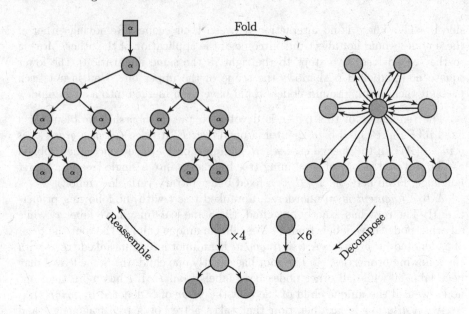

Fig. 3. To parse a folded graph (top right), we first decompose it into a number of tree fragments attaching to the merged node (bottom row), and then search for a way of reassembling the fragments into a tree in the folded tree language (top left).

Definition 3 (Membership problem for one folding). *Let A be a fixed but arbitrary Z-automaton over Γ. Given a multiset of unordered tree fragments $\{T_1, \ldots, T_n\}$ is there a sequence of substitutions that uses each tree exactly once, and produces a tree t such that $\alpha[t'] \in \mathcal{L}(A)$ for some $t' \in order(t)$?*

4   Unfolding Folded Trees

Since the input trees we are working with are unordered (see Definition 3), we extend A's behaviour to the unordered case.

Definition 4 (Unordered runs). *A run of a Z-automaton $A = (\Sigma, Q, R, F)$ on an unranked, unordered tree $t = (V, E, src, tar, lab, r)$ is a mapping $\rho : nodes(t) \to Q$ such that for each node v, the states assigned to the children of the node v, when viewed as a multiset, belongs to $parikh(\rho(v), lab(v))$. For a tree fragment s, a partial run is a run, except that the condition on the children does not apply to the nodes in $nodes(s)_x$.*

We can now define the *signature* of a tree fragment in terms of which partial runs it can realize. The intuition is that for each x-labeled node and possible assignment of a state to it, we have to find a set of trees to attach that can evaluate to states in the appropriate Parikh image.

Definition 5 (Signature). *Let $A = (\Sigma, Q, R, F)$ be a Z-automaton with $m = |Q|$, and let t be a tree fragment with $k = |nodes(t)_x|$. The signature of t with* *respect to A, denoted by $sig(t)$, is a set of tuples of the form (q, S), where $q \in Q$ and S is a multiset of elements from Q, defined as follows. Let $v_1, \ldots, v_k$ be the nodes in $nodes(t)_x$. Then (q, S) belongs to the signature iff there is a partial run ρ on t and a partitioning of S into $S_1, \ldots, S_k$ such that $\rho(prior(t)) = q$ and, for each i, $S_i \in parikh(\rho(v_i), \langle \sigma, \alpha \rangle)$.*

The intuition of the above definition is that if (q, S) is in the signature of t if t can "accept" $|S|$ tree fragments whose priors have been assigned the states in S and then "deliver" a state q at its prior.

Given the input set $\{T_1, \ldots, T_n\}$, we only need to consider a polynomial number of signatures. Since there are n input trees, we only consider signatures where $|S| \leq n$. The number of such signatures is bounded from above by mn^m. In other words, the number of possible signatures for all input trees is polynomial. The signature for each input fragment can be computed in polynomial time using a CYK-like dynamic programming algorithm.

Given a set of tree fragments we compute their signatures, and then reassemble them as in Fig. 3, leading to the final theorem.

Theorem 1. *The non-uniform membership problem for tree languages folded once under an order-cancelling semantics is decidable in polynomial time.*

To prove Theorem 1, we rely on the signatures to tell us what multisets of states each tree fragment can "consume", and what state it then "produces". Finding a way of puzzling the fragments together consistently is a combinatorial problem which we will solve by reducing it to reachability in a restricted form of vector addition system. We next present these systems and prove the relevant complexity bound. We then explain the reduction in the proof of Theorem 1.

For vectors u and v let $(u; v)$ denote their concatenation. Let $\mathbb{0}$ denote the set of all vectors of zeros, and for all $k \geq 1$ let $\mathbb{1}_{[k]} = \{(z_1; 1; z_2) \mid z_1, z_2 \in \mathbb{0}, |z_1| = k - 1\}$, i.e., the unit vectors with a 1 in position k. Let $\mathbb{1} = \cup_{k \geq 1} \mathbb{1}_{[k]}$, i.e., all unit vectors. We may treat $\mathbb{0}$ and $\mathbb{1}$ as vectors when the length is implied by context.

Definition 6. *A* Vector Addition System *of dimension $k \in \mathbb{N}$ (a k-VAS) is a finite set $V = \{(p, p') \mid p \in \{-v \mid v \in \mathbb{N}^k\}, p' \in \mathbb{N}^k\}$. We call these the operations. An operation sequence in V is denoted $s_0 \to_{(p_1, p_1')} s_1 \to_{(p_2, p_2')} \cdots \to_{(p_n, p_n')} s_n$ for $s_0, \ldots, s_n \in \mathbb{N}^k$, and for each $1 \leq i \leq n$*

$$1.\ (p_i, p_i') \in V, \qquad 2.\ s_{i-1} + p_i \geq \mathbb{0}, \qquad and, \qquad 3.\ s_i = s_{i-1} + p_i + p_i' .$$

A vector s_n is reachable *from s_0 if and only if such an operation sequence exists.*

For any $k, l \in \mathbb{N}$ we call a $(k+l)$-VAS V a (k, l)-VAS to differentiate the first k elements of the vectors from the rest, writing $((u; v), (u'; v')) \in V$ to signify that $|u| = |u'| = k$ and $|v| = |v'| = l$.

While we define reachability in terms of going from a vector s to a vector t, we are primarily interested in the special case of $\mathbb{0}$ being reachable, since the numbers will represent tree fragments, all of which must be used.

Definition 7. *A (k, l)-VAS V is metered if all $((s; b), (s'; b')) \in V$ have $s, b \in -\mathbb{1}$ and $b' \in \mathbb{0}$.*

That is, in a metered (k, l)-VAS every operation takes precisely one unit from each of the two parts of the vector, and never adds anything to the second part. We will use the first part, the s vector, to represent a *multiset of states*, while the second, the b vector, represents a *budget*.

Definition 8. *For a (k, l)-VAS V an operation sequence $(s_0; b_0) \rightarrow_{p_1} \cdots \rightarrow_{p_n} (s_n; b_n)$ visits a position $i \in [k]$ if at least one vector $s \in \{s_0, \ldots, s_n\}$ is nonzero in position i. The set $I \subseteq [k]$ is visited if all elements are visited.*

Visits only concern the first (state) part of the vector, which makes sense for a metered VAS, as a position in the second part is visited iff it is visited in b_0. We will use *visits* to correspond to uses of states in a run of an automaton.

We next show that if there exists an operation sequence which visits a set I starting from a vector v_0 then there is a *short subsequence* that does the same.

Lemma 1. *For any metered (k, l)-VAS V and operation sequence $v_0 \rightarrow_{p_1} \cdots \rightarrow_{p_n} v_n$ which visits I there exists an operation sequence $v_0' \rightarrow_{p_1'} \cdots \rightarrow_{p_m'} v_m'$ such that: (i) this sequence also visits I; (ii) $v_0' = v_0$; (iii) the sequence $p_1' \cdots p_m'$ forms a subsequence of $p_1 \cdots p_n$; (iv) and $m \leq k|I|$.*

Proof. Let $(s_0; b_0) \rightarrow_{p_1} \cdots \rightarrow_{p_n} (s_n; b_n)$ be an operation sequence visiting I, for an arbitrary $n \in \mathbb{N}$. Let us abbreviate it as $p_1 \cdots p_n$, leaving the vectors implicit. First consider the singleton case where $I = \{i\}$, when the following procedure constructs the indicated subsequence. Define $P(p_1 \cdots p_n, i)$ recursively as:

1. Let j be the smallest index such that with $(s_0; b_0) \rightarrow_{p_1} \cdots \rightarrow_{p_j} (s_j; b_j)$ the vector s_j is nonzero at position i.
2. If $j = 0$ return the empty sequence.
3. Letting $p_j = ((v_1; v_2), (v_1'; v_2'))$, take i' to be the unique (as V is metered) position in v_1 which equals -1.
4. Return $P(p_1 \cdots p_{j-1}, i') \cdot p_j$. That is, the sequence constructed by finding a short visit of i' in steps 1 through $j - 1$ (one must exist as step p_j subtracted 1 from that position) followed by the operation p_j (which visits i).

Now $P(p_1 \cdots p_n, i)$, starting from $(s_0; b_0)$, forms an operation sequence of length at most k which visits i. The visit to i is straight-forward (the final "p_j"), it is of length at most k as each level of the recursion has a distinct i (as the current one is eliminated from the sequence used in the recursive call) and adds one operation. Finally, it is an operation sequence as each p_j appended in step 4 has its required visit provided by the step immediately prior. The budget part of the vector is unproblematic, as it can only be increased by shortening the sequence.

Generalizing this to an arbitrary set $I \subseteq [k]$ amounts to an iteration of the above argument. Here, elements of I may compete for the same operations, but this can be handled by having Step 1 in the above procedure pick nondeterministically among the k first such indices j, and then letting the computation fail if two elements of i reuse an operation. In the worst case, each $i \in I$ introduces k new operations to the subsequence, for a total of $k|I|$ operations. $\square$

Next we define a relaxed operation sequence, which contains less information, but the existence of which implies the existence of an operation sequence.

Definition 9. *For a metered* (k, l)*-VAS* V*, a vector* $v \in \mathbb{N}^{k+l}$*, and any* $I \subseteq [k]$*, a destructured sequence is a tuple* (v, I, P, S) *where* $P = p_1 \cdots p_{k|I|} \in V^*$*, where*

1. $v \to_{p_1} \cdots \to_{p_{k|I|}} v'$ *is an operation sequence visiting* I*,*
2. S *is a multiset over* V *such that for all* $((s; b), (s'; b')) \in S$ *the vectors* s *and* s' *are zero in all positions* $z \notin I$*, and,*
3. $v + \left(\sum_{(s,s') \in S} s + s' \right) + \left(\sum_{(p,p') \in P} p + p' \right) = \mathbb{0}.$

The key then becomes that we only need to consider destructured sequences to demonstrate reachability, as they turn out equivalent to full operation sequences in this setting. That is, past a certain length and ensuring a certain subsequence exists, we can disregard the order of operations.

Lemma 2. *For a metered* (k, l)*-VAS* V *and vector* $v \in \mathbb{N}^{k+l}$*, there exists an operation sequence* $v \to \cdots \to \mathbb{0}$ *iff there exists a destructured sequence* (v, I, P, S)*.*

Proof. The "only-if" direction: Let $v \to_{p_1} \cdots \to_{p_n} \mathbb{0}$ be an operation sequence, and take $I \subseteq [k]$ be the largest set visited by this sequence. Then apply Lemma 1 to construct P, and let $S = \{p_1, \ldots, p_n\} \setminus P$. Then (v, I, P, S) is a destructured sequence, fulfilling the requirements of Definition 9: Cond. 1 by Lemma 1, Cond. 2 by construction, and, Cond. 3 as the original sequence reaches $\mathbb{0}$, so the sum across all its operators, plus the initial vector v, has to be zero.

The "if" direction: given a destructured sequence (v, I, P, S) we can construct an operation sequence which reaches $\mathbb{0}$ from v. The only thing keeping any particular sequencing of the operations P and S from being an operation sequence from v to $\mathbb{0}$ is Condition 2 in Definition 6. That is, some position in the first (state) part of the vector may turn negative during the application of an operation. Only the first (state) part going negative can cause problems, as the second (budget) part is used up correctly by all orders, and all orders do reach $\mathbb{0}$ by the definition of a destructured sequence. We now demonstrate how to intersperse the operations in S in P to create a valid operation sequence reaching $\mathbb{0}$ from v.

Phase 1: Place Cycles at First Visit. An ordered submultiset $\{p_1, \ldots, p_n\} \subseteq S$ is a cycle on i if there are vectors $s, t \in \mathbb{N}^k$, $s', t' \in \mathbb{N}^l$ such that $(s; s') \to_{p_1} \cdots \to_{p_n} (t; t')$ with $s \in \mathbb{1}_{[i]}$, $t \geq s$. If such a cycle exists, split $P = P_1 P_2$ such that P_1 ends with the first visit to i (must exist by the definition of a destructured sequence), and construct a new operation sequence $P_1 \cdot p_1 \cdots p_n \cdot P_2$, and a new set of operations $S \setminus \{p_1, \ldots, p_n\}$. Take these to be P and S and iterate this procedure until no cycles remain in S. This retains all the properties of the destructured sequence (v, I, P, S), fulfilling all the conditions except the length of P. Specifically, the spliced sequence is an operation sequence because the visit to i allows it to be applied (the budget as usual irrelevant to the reordering), and as $t \geq s$ it cannot cause any later operation to fail.

Phase 2: Order Remainder Topologically. If S is cycle-free then there is some i such that every $((s; b), (s'; b')) \in S$ has s' zero in position i. Otherwise, a visit to i can be made by some operation that requires a visit to i', but i' can be visited by some operation that requires a visit to i'', etc. But no position may repeat in this chain, as that would be a cycle, and there are only finitely many positions, which causes a contradiction. Pick such an i, let $\{p_1, \ldots, p_n\} \subseteq S$ be all such operations p which have $p = ((-\mathbb{1}_{[i]}; b), (s'; b'))$ (for any b, s' and b') and construct a new sequence $P \cdot p_1 \cdots p_n$ and a new multiset $S \setminus \{p_1, \ldots, p_n\}$. Take these to be P and S and iterate this procedure until S is empty. This produces an operation sequence, since we maintain the destructured sequence invariant that summing P, S and v produces $\mathbb{0}$, and as the i picked in each step is not generated by any operation in S, all needed visits must already be in P. After these steps, S is empty and P is an operation sequence reaching $\mathbb{0}$ from v.   □

Theorem 2. *For a metered (k, l)-VAS V and vectors $s_0 \in \mathbb{N}^k$ and $b_0 \in \mathbb{N}^l$ it is decidable in time $\mathcal{O}(|V|^{k^2+1}(\sum b_0))$ whether $\mathbb{0}$ is reachable from $(s_0; b_0)$.*

Proof. All relevant tuples (v, I, P, S) can be evaluated to see if they are a destructured sequence as in Definition 9, if there is one, then by Lemma 2 there exists an operation sequence reaching $\mathbb{0}$ from v, which is the definition of reachability.

First, for a vector v define the *next smaller* vector v' as the one formed by decrementing the *first* nonzero element of v (for example, for $(0, 1, 2)$ the next vector is $(0, 0, 2)$, and the next smaller of that is $(0, 0, 1)$).

To see that the bound holds, regard what tuples we need to test. There are $|V|^{k^2}$ ways of picking P by the bound on its length, and I is entirely determined by P. Try the following for all such P.

Given v, I and P compute the v' reached from v using the operations in P. Then construct S by searching for a path from v' to $\mathbb{0}$ in the following graph.

1. The vector v' is a node in the graph.
2. If the vector $(s; b)$ is a node in the graph and there exists an operation $p = ((u; v), (u'; v')) \in V$ such that
    - $s' = s + u + u'$, $b' = b + v + v'$ (vector elements can be negative),
    - the next smaller vector than b is b', and,
    - u and v are zero in all positions not in I,
   then there is a node $(s'; b')$ and an edge from $(s; b)$ to $(s'; b')$ labeled p.

Then let S be a multiset of operations used (i.e. edges traversed) finding a path from v' to $\mathbb{0}$ in this graph. This procedure is sound and complete.

- An S found this way does make (v, I, P, S) a valid destructured sequence, as by construction $v + \sum_{(s,s') \in S} s + s' + \sum_{(p,p') \in P} p + p' = \mathbb{0}$ and S only visits I. All other needed properties derive from an exhaustive enumeration of possible operation sequences P.
- If a destructured sequence does exist this procedure will find it. Note that the only real pruning happening is requiring the budget to decrease according to the *next smaller* order. This is necessary to limit the effect l has on the size of the graph, but as S is itself unordered requiring the budget to be used in a certain order is not a real restriction.

Finally, all paths in this graph is of length at most $\sum b_0$ and each node has at most $|V|$ outgoing edges. Combining trying all P with exhaustive search on the graph gives a bound of $\mathcal{O}(|V|^{k^2})\mathcal{O}(|V|(\sum b_0)) = \mathcal{O}(|V|^{k^2+1}(\sum b_0))$. $\square$

Proof (of Theorem 1). By Definition 3, we have a fixed Z-automaton A and are given as input a graph, which we can decompose into a set of tree fragments $T_1, \ldots, T_n$. Assume that the root node was folded, i.e. the graph has no node with zero incoming edges. This causes no loss of generality: if the graph has a distinguished root node r, give it a parent marked by a new symbol "*dummy*", and give that node an incoming edge from the folded node. Then modify A with the necessary additional transitions (such that where it would have previously accepted $\alpha[t]$ it now accepts $\alpha[\langle \sigma, \alpha \rangle [dummy[t]]]$.) Let $Q = \{q_1, \ldots, q_m\}$ be the states of A, and assume, without loss of generality, that q_m is the only accepting state, and that it occurs on no left-hand-side of a rule in A.

Construct the signatures $sig(T_1), \ldots, sig(T_n)$ and from these construct a metered $(m, n+1)$-VAS V by giving it precisely the following operations: (i) for each $i \in [n]$ and $(q_j, S) \in sig(T_i)$, V has the operation $((-1_{[j]}; -1_{[i]}), (\bar{S}; 0))$, where $\bar{S}$ is S turned into a vector of length m, letting position k be the number of occurrences of q_k in S; and; (ii) for each S such that $S \in parikh(q_m, \langle \sigma, \alpha \rangle)$ and $|S| \leq n$, V has the operation $((-1_{[m]}; -1_{[n+1]}), (\bar{S}; 0))$. Finally, the initial vector is $v = (s; b)$, where $s \in 1_{[m]}$ (with $|s| = m$) and $b = 1 \cdots 1$. Intuitively, V simulates reassembling the tree from the fragments. In each step there is a current vector $(s'; b')$, where s' describes the multiset of states which still need to be replaced by a tree fragment, and b describes which tree fragments have already been used. Operations of type (i) attach a tree fragment using one of the present states, where type (ii) initializes the multiset of states to one from which A can accept by going to q_m.

Then 0 is reachable from $(s; b)$ if and only if the tree fragments can be reassembled into a tree accepted by A. By induction on the length of a VAS operation sequence $(s; b) \rightarrow \cdots \rightarrow 0$, relating each step to a part of some final tree t such that $\alpha[t'] \in \mathcal{L}(A)$ for some $t' \in order(t)$. That is, the first step (the only operation of type (ii) by construction) establishes the root and a multiset of states the children must produce. The second operation attaches some tree fragment T_i as one of those children by: picking some $(q, S) \in sig(T_i)$, removing one q from the state multiset represented, removing the tree fragment itself from the budget, and providing a new set of unaccounted-for children with state multiset S. This maintains the invariant that the part of the tree already constructed can be accepted by A given that the multiset of states currently tracked are provided by the remainder of the procedure. Since 0 is reached no further states are needed, and all tree fragments have been placed.

Finally, we can check whether 0 is reachable from $(s; b)$ by applying Theorem 2, observing that this $(m, n+1)$-VAS has m constant (as A is assumed fixed) and both n and $|V|$ polynomial. Substituting these into the bound of Theorem 2 yields a polynomial bound. Observe the role Lemma 2 and Theorem 2 play here; in effect the destructured sequence corresponds to constructing a small tree t which visits all necessary states without exhibiting any loop. Once this is in place, the remaining tree fragments can be added without keeping record of precisely where they are placed, producing a proper tree.   □

We have thus shown that for a fixed regular tree language $\mathcal{L}$, the question of whether a graph g could have been produced by a single application of the order-cancelling fold operation on a tree in $\mathcal{L}$ is solvable in polynomial time. As it turns out, we can extend this to any number of folding operation applications.

Corollary 1. *The non-uniform membership problem for tree languages folded using only a single folding symbol under an order-cancelling semantics is decidable in polynomial time.*

Proof. This follows from a helpful separability of graphs using a single folding symbol: If the graph contains no edge which would bisect the graph if removed, then the graph contains at most one folded node. This must be the case as the two groups of nodes merged must have been in scope of different instances of the folding operator (or they would have *all* been merged). If the folding operators were on the same path, removing the edge placed where the lower folding operator once was would then bisect the graph. If instead they were on different paths, removing either again bisects the graph.

For this reason, the general case can be checked through the following steps.

1. If the graph contains a single merged node (by Remark 1 we can assume this without loss of generality), apply Theorem 1 and halt with that result.
2. If there is no edge e which can bisect the graph in a way that separates two merged nodes, reject it. As argued above it cannot be in the language.
3. Pick an edge e which bisects the graph into subgraphs g_1 and g_2 which both contain merged nodes, with g_1 having e outgoing and g_2 having e incoming. Additionally, pick e such that the node e is outgoing from has no incoming edge which would bisect the graph in this way. Observe that such a choice always exists if any bisecting edge does, as we can otherwise pick that incoming edge, and doing so cannot lead us into a cycle (as removing an edge from a cycle would not bisect the graph). This condition ensures that the folding operator creating the merged node in g_2 is also in g_2.
4. Try each rule in the tree automaton to find one generating e:
   (a) Recursively apply this procedure to check if g_1 is a member of the language when e points to the state generated by the rule.
   (b) Recursively apply this procedure to check if g_2 is a member of the language which results from making the state, consumed by the rule, the only accepting state.
   (c) If 4a and 4b accept g_1 and g_2, respectively, accept the graph g.
5. If all rule options have been tried without accepting, reject.

This procedure is correct, as every graph in the language has to either contain only a single application of the folding operator (checked in Step 1) or it can be bisected guessing the rule applied (all enumerated by Step 4).

Moreover, the procedure runs in polynomial time, since it applies the polynomial time algorithm of Theorem 1 a polynomial number of times, as the recursion eliminates one edge each step.   □

Acknowledgements. We would like to express our sincere gratitude to the anonymous reviewers for their constructive feedback, which greatly improved the quality of this manuscript.

References

1. Arnborg, S., Corneil, D., Proskurowski, A.: Complexity of finding embeddings in a k-tree. SIAM J. Algebraic Discrete Methods **8**(2), 277–284 (1987)
2. Berglund, M., Björklund, H., Björklund, J., Boiret, A.: Transduction from trees to graphs through folding. Available at SSRN 4291269 (2022)
3. Björklund, H., Björklund, J., Ericson, P.: On the regularity and learnability of ordered DAG languages. In: Carayol, A., Nicaud, C. (eds.) CIAA 2017. LNCS, vol. 10329, pp. 27–39. Springer, Cham (2017). https://doi.org/10.1007/978-3-319-60134-2_3
4. Björklund, H., Drewes, F., Ericson, P.: Parsing weighted order-preserving hyperedge replacement grammars. In: 16th Meeting on the Mathematics of Language, MOL 2019, Toronto, Canada, pp. 1–11. ACL (2019)
5. Björklund, J.: Tree-to-graph transductions with scope. In: Hoshi, M., Seki, S. (eds.) DLT 2018. LNCS, vol. 11088, pp. 133–144. Springer, Cham (2018). https://doi.org/10.1007/978-3-319-98654-8_11
6. Björklund, J., Drewes, F., Satta, G.: Z-Automata for compact and direct representation of unranked tree languages. In: Hospodár, M., Jirásková, G. (eds.) CIAA 2019. LNCS, vol. 11601, pp. 83–94. Springer, Cham (2019). https://doi.org/10.1007/978-3-030-23679-3_7
7. Brüggemann-Klein, A., Murata, M., Wood, D.: Regular tree and regular hedge languages over unranked alphabets: Version 1. Technical report HKUST-TCSC-2001-0, The Hongkong University of Science and Technology (2001)
8. Drewes, F., Kreowski, H.J., Habel, A.: Hyperedge replacement graph grammars. In: Rozenberg, G. (ed.) Handbook of Graph Grammars and Computing by Graph Transformation, pp. 95–162. World Scientific, River Edge, NJ, USA (1997)
9. Koller, A.: Semantic construction with graph grammars. In: Proceedings of the 14th International Conference on Computational Semantics, IWCS. London (2015)
10. Plump, D.: The graph programming language GP. In: Bozapalidis, S., Rahonis, G. (eds.) CAI 2009. LNCS, vol. 5725, pp. 99–122. Springer, Heidelberg (2009). https://doi.org/10.1007/978-3-642-03564-7_6
11. Quernheim, D., Knight, K.: DAGGER: a toolkit for automata on directed acyclic graphs. In: Proceedings of the 10th International Workshop Finite-State Methods and Natural Language Processing, FSMNLP 2012, pp. 40–44. ACL (2012)
12. Tang, L., Liu, H.: Graph mining applications to social network analysis. In: Aggarwal, C., Wang, H. (eds.) Managing and Mining Graph Data. Advances in Database Systems, vol. 40, pp. 487–513. Springer, Boston (2010). https://doi.org/10.1007/978-1-4419-6045-0_16
13. You, J., Leskovec, J., He, K., Xie, S.: Graph structure of neural networks. In: International Conference on Machine Learning, pp. 10881–10891. PMLR (2020)

The Impact of State Merging
on Predictive Accuracy in Probabilistic
Tree Automata: Dietze's Conjecture
Revisited

Johanna Björklund[✉][iD]

Umeå University, Umeå, Sweden
johanna@cs.umu.se

Abstract. Dietze's conjecture concerns the problem of equipping a tree automaton M with weights to make it probabilistic, in such a way that the resulting automaton N predicts a given corpus C as accurately as possible. The conjecture states that the accuracy cannot increase if the states in M are merged with respect to an equivalence relation $\sim$ so that the result is a smaller automaton $M^\sim$. Put differently, merging states can never improve predictions. This is under the assumption that both M and $M^\sim$ are bottom-up deterministic and accept every tree in C. We prove that the conjecture holds, using a construction that turns any probabilistic version $N^\sim$ of $M^\sim$ into a probabilistic version N of M, such that N assigns at least as great a weight to each tree in C as $N^\sim$ does.

Keywords: Tree automata · Statistical ML · Probability distributions

1   Introduction

Supervised learning is concerned with recovering or approximating a target function, based on a limited sample of input-output pairs called a *corpus*. The difficulty of the problem depends on the class of functions admitted and on various characteristics of the corpus. In this work, we focus on the supervised learning of *probabilistic tree languages* [4], in particular, on probability distributions computable by *weighted tree automata (wta)* over the probability semiring [7]. Some of the earliest results are due to Klein and Manning, who extracted a simple grammar from a corpus and then manually refined it based on linguistic patterns in the data [9]. Petrov et al. later translated their approach into an automated method that alternately splits and merges nonterminals to maximize the likelihood of the target corpus [10].

Björklund, J—Supported by the Swedish Research Council under Grant Number 2020-03852, and by the Wallenberg AI, Autonomous Systems and Software Program through the NEST project *STING — Synthesis and analysis with Transducers and Invertible Neural Generators.*

H. Fernau and K. Jansen (Eds.): FCT 2023, LNCS 14292, pp. 74–87, 2023.
https://doi.org/10.1007/978-3-031-43587-4_6

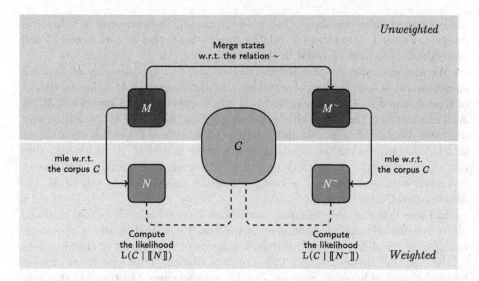

Fig. 1. By merging the states of the unweighted tree automaton M with respect to the equivalence relation $\sim$ on the states of M, we obtain the smaller automaton $M^\sim$. The probabilistic automata N and $N^\sim$ are derived from M and $M^\sim$, respectively, by computing the maximum likelihood estimation with respect to the corpus C. The conjecture states that if all automata are deterministic and contain C in their support, then $L(C \mid [\![N^\sim]\!])$ cannot exceed $L(C \mid [\![N]\!])$.

Dietze [5] formalised the ideas of Petrov et al. [10] into an algorithmic framework for supervised learning of weighted tree automata: First, the *read-off automaton* is extracted from a given corpus C. This is a wta over the Boolean semiring whose states and transitions are in one-to-one correspondence with the subtrees of C. The states of this automaton are then iteratively merged or split, and after each pass, the resulting wta M is equipped with weights to optimise prediction accuracy with respect to the weight of the trees in C. The resulting probabilistic automaton N is a *maximum-likelihood estimation* (mle) of M with respect to C. Intuitively, this means that among all probabilistic automata that can be derived from M, the automaton N is that which makes the weights of the trees in C as likely as possible. The learning process terminates when the predictions produced by N are sufficiently well-aligned with C, or when the number of merge-split cycles has reached some maximal limit.

A question that arises from this approach is how the split and merge operations affect the automaton's ability to predict the corpus. Merging states has the advantage that it allows the automaton to accept more trees, thereby generalising what has been learnt from data. However, Dietze conjectures that if we have a determinsitic wta (dta) M that already accepts every tree in the corpus and merge the states of M with respect to some equivalence relation $\sim$ to form a dta $M^\sim$, then the mle $N^\sim$ of $M^\sim$ cannot yield a better prediction of C than the mle N of M [5, Conjecture 6.2.2]. This means that further progress cannot be made by merging states alone (but possibly through a succession of splits and merges). Figure 1 illustrates the relationship between the various weighted and unweighted automata involved in the conjecture.

We are now able to give a constructive proof of the veracity of Dietze's conjecture. It is based on the observation that the *support* of $M^\sim$, that is, the set of trees mapped to a nonzero value by $M^\sim$, is a superset of the support of M [5]. As a consequence, the probability mass assigned by the mle of $M^\sim$ is diluted over a greater number of trees. We find that it is possible to reshape this distribution into a distribution over the smaller support of the mle of M, in such a way that the weight of every tree in the support either increases or remains unchanged, but never decreases. Since C is contained in the support of the mle of M, the likelihood of C also increases.

In these times of deep learning, statistical approaches such as that outlined above remain relevant for a number of reasons. First, their training process is transparent, enabling a straightforward analysis of the impact of training data on the resultant model. Second, they are data efficient, rendering them suitable for deployment in low-data régimes. Finally, with an energy footprint that is generally smaller than corresponding neural approaches, they are a feasible choice for embedded systems and internet-of-things applications.

Finally, suppose that we are interested in supervised learning of structures that can be uniquely represented as terms in some algebra. We can then parse each element in the corpus into a term, i.e., a tree, and then apply the above learning framework. Examples of such structures are, e.g., strings, hedges [3], and certain types of directed acyclic graphs [1,2]. This means that learning results for tree languages have immediate consequences for countless other domains. For further reading on tree-based generation and processing, see e.g. [6,8].

Outline. This paper is organized as follows. Section 2 provides automata-theoretic definitions and fixes notation. Section 3 recalls Dietze's conjecture and proves its correctness. Section 4 gives pointers for future work.

2   Preliminaries

Sets, Numbers and Strings. The set of *natural numbers* (excluding zero) is denoted by $\mathbb{N}$, and the set of *non-negative reals* is denoted by $\mathbb{R}_{\geq 0}$.

The *power set* of a set S is written 2^S, the *size* of S is denoted by $|S|$, and the *empty set* is written $\emptyset$. A *partition* of a set S is a set of pairwise disjoint, sets $\{S_1, \ldots, S_k\} \subseteq 2^S$ whose union is S. A *probability distribution* over S is a function $p: S \to \mathbb{R}_{\geq 0}$ such that $\sum_{s \in S} p(s) = 1$.

A binary relation on S is an equivalence relation if it is reflexive, symmetric and transitive. Let $\sim$ be such a relation. The *equivalence class* of an element s in S with respect to s is the set $[s]_\sim = \{s' \in S \mid s \sim s'\}$. Since $[s]_\sim$ and $[s']_\sim$ are equal if $s \sim s'$ and disjoint otherwise, the relation $\sim$ induces a partition $(S/\sim) = \{[s]_\sim \mid s \in S\}$ of S.

An *alphabet* Σ is a finite nonempty set. A *string* is an ordered sequence of zero or more symbols from Σ. The set of all strings over Σ is denoted by Σ^* and the *empty string* is denoted by ϵ.

Semirings. A *monoid* is a structure $(A, \oplus, \mathbb{0})$ where A is a set, $\oplus$ a binary operation on A, and $\mathbb{0}$ is an identity element with respect to $\oplus$ (so $a \oplus \mathbb{0} = \mathbb{0} \oplus a = a$ for every $a \in A$). If, for every $a, b \in A$, we have that $a \oplus b = b \oplus a$ then the monoid is *commutative*.

A *(commutative) semiring* is a structure $\mathcal{A} = (A, \oplus, \odot, \mathbb{0}, \mathbb{1})$ such that $(A, \oplus, \mathbb{0})$ and $(A, \odot, \mathbb{1})$ are commutative monoids, $\odot$ distributes over $\oplus$, and $\mathbb{0}$ is an absorbing element with respect to $\odot$ (that is, $a \odot \mathbb{0} = \mathbb{0} \odot a = \mathbb{0}$ for every $a \in A$). The *Boolean semiring* is the semiring $\mathbb{B} = (\{0, 1\}, \vee, \wedge, 0, 1)$ and the *probability semiring* is the semiring $\mathbb{P} = (\mathbb{R}_{\geq 0} \cup \{\infty\}, +, \cdot, 0, 1)$. Let S be a set, $\mathcal{A} = (A, \oplus, \odot, \mathbb{0}, \mathbb{1})$ a semiring, and $f : S \to A$ a function. The *support* of f is the set $sup(f) = \{a \in A \mid f(a) \neq \mathbb{0}\}$.

Trees. An *unranked tree* over the alphabet Σ is a function $t : D \to \Sigma$ where $D \subseteq \mathbb{N}^*$ is such that $\epsilon \in D$ and, for every $v \in D$, there exists a $k \in \mathbb{N}$ with $\{i \in \mathbb{N} \mid vi \in D\} = [k]$. We call D the *domain* of t and denote it by $dom(t)$. An element v of $dom(t)$ is called a *node of t*, and k is the *rank of v*. The *subtree* of $t \in T_\Sigma$ rooted at v is the tree t/v defined by $dom(t/v) = \{u \in \mathbb{N}^* \mid vu \in dom(t)\}$ and $t/v(u) = t(vu)$ for every $u \in D$. If $t(\epsilon) = f$ and $t/i = t_i$ for all $i \in [k]$, where k is the rank of ϵ in t, then we denote t by $f[t_1, \ldots, t_k]$. If $k = 0$, then $f[]$ is usually abbreviated as f. The *height* of t is $height(t) = \max\{|v| \mid v \in dom(t)\}$.

A *ranked alphabet* is an alphabet $\Sigma = \bigcup_{k \in \mathbb{N}} \Sigma_k$ which is partitioned into subsets Σ_k. For every $k \in \mathbb{N}$ and $f \in \Sigma_k$, the *rank* of f is $rank(f) = k$. A *tree over Σ* is an unranked tree t over Σ such that the rank of every node $v \in dom(t)$ coincides with the rank of $t(v)$. The set of all (ranked) trees over Σ is denoted by T_Σ. A *tree language over Σ* is a subset of T_Σ. A *weighted tree language* over Σ and a semiring $\mathcal{A}$ is a mapping $T_\Sigma \to \mathcal{A}$.

Weighted Automata. A *weighted tree automaton* (wta) over the semiring $\mathcal{A}$ is a tuple $M = (Q, \Sigma, \mathcal{A}, F, \delta)$ where

- Q is an alphabet of *states*,
- Σ is a ranked alphabet of *input symbols*,
- $\mathcal{A}$ is a communtative semiring,
- $F : Q \to \mathcal{A}$ is a *final weight mapping*,
- $\delta = (\delta_k)_{k \in \mathbb{N}}$ is a family of functions such that $\delta_k : Q^k \times \Sigma_k \times Q \to \mathcal{A}$, for every $k \in \mathbb{N}$.

The wta is *(bottom-up) deterministic* if for every choice of $q_1 \cdots q_k \in Q^k$ and $\sigma \in \Sigma$, there is at most one $q \in Q$ such that $\delta_k(q_1 \cdots q_k, \sigma, q) \neq \mathbb{0}$.

A *run* of M on $t \in T_\Sigma$ is a function $dom(t) \to Q$. The set of all runs of M on t is denoted by $runs_M(t)$. The weight function $w_M : runs_M(t) \to \mathcal{A}$ is defined for every $r \in runs_M(t)$ by

$$w_M(r) = \bigodot_{v \in dom(t)} \delta_k(r(v1) \cdots r(vk), t(v), r(v)) \ ,$$

where $k = rank(v)$. If M is deterministic, then there is at most one run $r \in runs_M(t)$ such that $w_M(t) \neq 0$. If it exists, we denote by q_t the state $r(\epsilon)$.

The $\mathcal{A}$-*weighted tree language* computed by M is the function $[\![M]\!] : T_\Sigma \to \mathcal{A}$ that is defined, for every $t \in T_\Sigma$, by

$$[\![M]\!](t) = \bigoplus_{r \in runs_M(t)} w_M(r) \odot F(r(\epsilon)) \ .$$

For every $q \in Q$, we denote by M_q the wta $(Q, \Sigma, \mathcal{A}, F', \delta)$, where $F'(q) = \mathbb{1}$ and $F'(p) = \mathbb{0}$ for every $p \in Q \setminus \{q\}$.

Definition 1 (Probabilistic automata, cf. Sect. 4.3 of [5]). *Let $M = (Q, \Sigma, \mathbb{P}, F, \delta)$ be a weighted tree automaton over the probability semiring.. The automaton M is:*

– out-probabilistic *if for every $q \in Q$,*

$$\sum_{k \in \mathbb{N}, \sigma \in \Sigma_k, q_1 \cdots q_k \in Q^k} \delta_k(q_1 \cdots q_k, \sigma, q) = 1 \ ,$$

– end-probabilistic *if*

$$\sum_{q \in Q} F(q) = 1 \ , \ and$$

– consistent *if*

$$\sum_{t \in T_\Sigma} [\![M]\!](t) = 1 \ .$$

The wta M is probabilistic *if it is out- and end-probabilistic, and consistent.*

Corpora. In [5], a *corpus* is a mapping $C : T_\Sigma \to \mathbb{R}_{\geq 0}$ such that $sup(C)$ is finite and nonempty. The size of C is $|C| = \sum_{t \in T_\Sigma} C(t)$. Let p be a probability distribution over T_Σ. The *likelihood* of C under p is:

$$L(C \mid p) = \prod_{t \in T_\Sigma} p(t)^{C(t)} \ .$$

Let $M = (Q, \Sigma, \mathbb{B}, F, \delta)$ be a wta and let C be a corpus. We denote by $prob(M)$ the set of probabilistic wta $(Q, \Sigma, \mathbb{P}, F', \delta')$ such that $sup(F') \subseteq sup(F)$ and $sup(\delta') \subseteq sup(\delta)$. Finally,

$$mle_C(M) = \operatorname*{argmax}_{N \in prob(M)} L(C \mid [\![N]\!]) \ .$$

3   Dietze's Conjecture

Dietze's conjecture concerns the impact of state merging on the inference of probabilistic tree languages from corpora. In essence, the conjecture says that when attempting to transform a dta M into a probabilistic wta N to maximise predictive accuracy with respect to a given corpus C, where M already accepts every tree in C, no predictive power can be gained by merging states of M to obtain a smaller dta $M^\sim$ and subsequently endowing $M^\sim$ with weights. To formalise this assertion, we recall the concept of merging states in a weighted tree automaton based on an equivalence relation.

Definition 2. *Let $M = (Q, \Sigma, \mathcal{A}, F, \delta)$ be a wta and let $\sim$ be an equivalence relation on Q. We denote by $M^\sim$ the wta $(Q^\sim, \Sigma, \mathcal{A}, F^\sim, \delta^\sim)$ with the components*

- *$Q^\sim = (Q/\sim)$,*
- *$F^\sim \colon Q^\sim \to \mathcal{A}$, where for every $p = [q]_\sim \in Q^\sim$,*

$$F^\sim(p) = \bigoplus_{q' \in [q]_\sim} F(q') \text{ , and}$$

- *$\delta^\sim = (\delta_k^\sim)_{k \in \mathbb{N}}$ where $\delta_k^\sim \colon (Q/\sim)^k \times \Sigma \times (Q/\sim) \to \mathcal{A}$ and for every $p = [q]_\sim, p_1 = [q_1]_\sim, \ldots, p_k = [q_k]_\sim \in Q^\sim$ and $\sigma \in \Sigma$, we have*

$$\delta_k^\sim(p_1 \cdots p_k, \sigma, p) = \bigoplus_{\substack{q' \in [q]_\sim, \\ q_i' \in [q_i]_\sim, i \in [k]}} \delta_k(q_1' \cdots q_k', \sigma, q') \text{ .}$$

Recall from [5, Thm 5.1.1.] that if M is a wta over the Boolean semiring, then $sup(\llbracket M \rrbracket) \subseteq sup(\llbracket M^\sim \rrbracket)$. Dietze's conjecture can now be stated as follows.

Theorem 1 (cf. Conjecture 6.2.2 of [5]). *Let $M = (Q, \Sigma, \mathbb{B}, F, \delta)$ be a wta over the Boolean semiring, $\sim$ an equivalence relation on Q, and C a corpus such that $sup(C) \subseteq sup(\llbracket M \rrbracket)$. If M and $M^\sim$ are bottom-up deterministic, then*

$$L(C \mid \llbracket mle_C(M) \rrbracket) \geq L(C \mid \llbracket mle_C(M^\sim) \rrbracket) \text{ .}$$

We note that the condition that M accepts every tree in C is necessary for the statement to hold true. If it was not, state-merging could be used to expand the support of M so that it encompasses a larger part of the corpus C: Consequently, the likelihood of C with respect to probabilistic wta derived from the smaller automaton may increase [5].

To prove Theorem 1, we show that for every $N^\sim = ((Q/\sim), \Sigma, \mathbb{P}, F^\sim \delta^\sim)$ in $prob(M^\sim)$, we can construct a wta $N = (Q, \Sigma, \mathbb{P}, F', \delta')$ in $prob(M)$ that has the following property:

For every $t \in sup(\llbracket M \rrbracket) \cap sup(\llbracket N^\sim \rrbracket)$, it holds that $\llbracket N \rrbracket(t) \geq \llbracket N^\sim \rrbracket(t)$.   (P1)

Since $C \subseteq sup(\llbracket M \rrbracket)$, this is enough to validate the conjecture.

The construction of N is based on the following observation. Recall that every state p in $N^\sim$ is a set of states $\{q_1, \ldots, q_k\}$ in M, and let T_p be the set of trees that are mapped by $N^\sim$ to p with nonzero weight. Since $N^\sim$ is probabilistic, by a straightforward induction the weights assigned to the trees in T_p by $N^\sim$ sum to 1. When p is split into its constituent states $q_1, \ldots, q_k$, the set T_p is partitioned over the states $q_1, \ldots, q_k$ into sets $T_{q_1}, \ldots, T_{q_k}$, such that the state q_i is reached by M by precisely the trees in T_{q_i}. When we add weights to M to form N, the weights assigned by N to the trees in T_{q_i} must also sum to 1 for every $i \in [k]$. We can therefore compute a value $\theta(q_i)$ for each q_i that corresponds to the probability mass given by $N^\sim$ to the trees in T_{q_i}, relative to the probability mass given by $N^\sim$ to all of T_p. Note that in Definition 3 below, we do not explicitly divide by the sum of the latter because it is always 1. We define a mapping θ on Q such that $\theta(q_i)$ allows us to compute the weights of the transitions entering q_i, by scaling the weights of the transitions entering p in $N^\sim$ by a factor of $\theta(q_i)^{-1}$. If we also scale the accepting weights of $N^\sim$, then we obtain a wta N with the property (P1).

Definition 3 (The mapping θ). *Let $\theta\colon Q \to [0,1]$ be a mapping, defined for every $q \in Q$ by*

$$\theta(q) = \sum_{t \in sup([M_q])} [\![N^\sim_{[q]_\sim}]\!](t) \ .$$

In the upcoming proofs we need an expanded form of θ. To this end, we introduce a sequence of functions $\theta_0, \theta_1, \theta_2, \ldots$ which converge to θ. In the following recursive definition, the value of θ_0 is zero for all states, while the value of θ_1 is nonzero only for states that can be reached on symbols of rank 0.

Definition 4. *For every $q \in Q$, $\theta_0(q) = 0$ and for every $i \geq 1$,*

$$\theta_i(q) = \sum_{(q_1 \cdots q_k, \sigma, q) \in sup(\delta)} \delta^\sim([q_1]_\sim \cdots [q_k]_\sim, \sigma, [q]_\sim) \cdot \prod_{j \in [k]} \theta_{i-1}(q_j) \ . \tag{1}$$

Lemma 1. *For every $q \in Q$ and $i = 0, 1, 2, \ldots$*

$$\theta_i(q) = \sum_{\substack{t \in sup([M_q]) \\ and\ height(t) < i}} [\![N^\sim_{[q]_\sim}]\!](t) \ .$$

Proof. The proof is by induction on i. The base case, where $i = 0$, is trivially true. Assume that $i \geq 1$ and that Lemma 1 holds for $i-1$. We compute as follows:

$\theta_i(q)$

= by Definition 4 of θ_i

$$\sum_{(q_1\cdots q_k,\sigma,q)\in sup(\delta)} \delta_k^\sim([q_1]_\sim\cdots[q_k]_\sim,\sigma,[q]_\sim) \cdot \prod_{j\in[k]}\theta_{i-1}(q_j)$$

= by the induction hypothesis

$$\sum_{(q_1\cdots q_k,\sigma,q)\in sup(\delta)} \delta_k^\sim([q_1]_\sim\cdots[q_k]_\sim,\sigma,[q]_\sim) \cdot \prod_{j\in[k]}\sum_{\substack{t_j\,\in\,sup(M_{q_j}),\\ height(t_j)\,<\,i-1}} [\![N^\sim_{[q_j]_\sim}]\!](t_j)$$

= because multiplication distributes over addition

$$\sum_{(q_1\cdots q_k,\sigma,q)\in sup(\delta)} \delta_k^\sim([q_1]_\sim\cdots[q_k]_\sim,\sigma,[q]_\sim) \cdot$$
$$\sum_{\substack{t_1\cdots t_k\,\in\,sup(M_{q_1})\times\cdots\times sup([M_{q_k}])\\ \text{and } height(t_j)\,<\,i-1,\text{ for } j\,\in\,[k]}} [\![N^\sim_{[q_1]_\sim}]\!](t_1)\cdot\cdots\cdot[\![N^\sim_{[q_k]_\sim}]\!](t_k)$$

= because multiplication distributes over addition

$$\delta_k^\sim([q_1]_\sim\cdots[q_k]_\sim,\sigma,[q]_\sim)\cdot$$
$$\sum_{\substack{(q_1\cdots q_k,\sigma,q)\,\in\,sup(\delta),\\ t_1\cdots t_k\,\in\,sup(M_{q_1})\times\cdots\times sup([M_{q_k}])\\ \text{and } height(t_j)\,<\,i-1,\text{ for } j\,\in\,[k]}} [\![N^\sim_{[q_1]_\sim}]\!](t_1)\cdot\cdots\cdot[\![N^\sim_{[q_k]_\sim}]\!](t_k)$$

= by the definition of wta semantics in Sect. 2

$$\sum_{\substack{t\,\in\,sup(M_q)\\ \text{and } height(t)\,<\,i}} [\![N^\sim_{[q]_\sim}]\!](t)$$

□

Lemma 2. *For every* $q \in Q$, $\theta(q) = \lim_{i\to\infty}\theta_i(q)$.

Proof. To show that the value of $\theta_i(q)$ converges, we argue that there is a constant c such that for all $i > c$, there is a $j < i$ such that

$$\max\{[\![N^\sim]\!](t) \mid height(t) = i\} < \max\{[\![N^\sim]\!](t) \mid height(t) = j\} \ .$$

The claim holds because the products of the weight used in each loop in $N^\sim$ must be strictly less than 1. If this was not the case, then $N^\sim$ would not be probabilistic, because at the state marking the beginning and end of the loop, there must be one transition leading into the loop and one transition exiting the loop, both of which must have non-zero weight. Once a tree has been constructed by pumping the loop several times, there is a smaller tree with strictly greater weight that was built by pumping the loop fewer times. In combination with Lemma 1, this means that the sequence $\theta_i(q)$ converges to $\theta(q)$ for all $q \in Q$. □

We can now use the function θ to rescale the weight mappings used in $N^\sim$ to compute suitable weights for M, such that the resulting probabilistic tree automaton N assigns at least as great a weight to every tree in C as $N^\sim$ does.

Definition 5 (The automaton N). *Let $M = (Q, \Sigma, \mathbb{B}, F, \delta)$ be a wta over the Boolean semiring, $\sim$ an equivalence relation on Q, and $N^\sim = ((Q/\sim), \Sigma, \mathbb{P}, F^\sim \delta^\sim) \in prob(M^\sim)$. We define N as the wta $(Q, \Sigma, \mathbb{P}, F', \delta')$, where for every $q \in Q$*

$$F'(q) = F^\sim([q]_\sim) \cdot \frac{\theta(q)}{\sum_{p \in [q]_\sim} \theta(p)} \,, \tag{2}$$

and for every $(q_1 \cdots q_k, \sigma, q) \in sup(\delta)$,

$$\delta'_k(q_1 \cdots q_k, \sigma, q) = \frac{\delta^\sim_k([q_1]_\sim \cdots [q_k]_\sim, \sigma, [q]_\sim) \cdot \prod_{i \in [k]} \theta(q_i)}{\theta(q)} \,, \tag{3}$$

and 0 otherwise.

Example 1. Consider the automata in Fig. 2. If we momentarily ignore the weights, we can take these to be the wta M and $M^\sim$ over the Boolean semiring. We see that $M^\sim$ can be obtained from M by merging the states with respect to the equivalence relation $\sim$, which is such that $(Q/\sim) = \{p_1, p_2\}$, $p_1 = \{q_1, q_2\}$, and $p_2 = \{q_3, q_4, q_5, q_6\}$. Assume then that $N^\sim$ is the wta in $prob(M^\sim)$ that results from equipping $M^\sim$ with weights as shown on the right in the figure. The table in the lower right of the figure lists the values of θ for the states in M, and from these the weights that turn M into the wta N are computed according to Definition 5. As we can see,

$$sup(\llbracket M \rrbracket) = \{f[h[a], h[b]], \; f[h[b], h[a]], \; g[h[a], h[b]], \; g[h[b], h[a]]\} \,.$$

It is easy to verify that for every tree t in this set $N(t) \geq N^\sim(t)$. For example, $N(f[h[a], h[b]]) = \frac{1}{2 \cdot 3}$ and $N^\sim(f[h[a], h[b]]) = \frac{1}{3^5}$. ◇

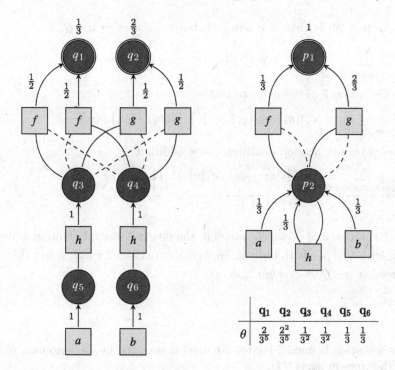

Fig. 2. The automata discussed in Example 1. The circles and boxes denote states and transitions, respectively. In a transition of the form $(q_3 q_4, f, q_1)$, the first and second argument (q_3 and q_4) are indicated with a solid and dashed line, respectively, while the target state q_1 is indicated with an arrow. Accepting states are drawn with double strokes, and final weights are written above them.

We note that if a tree in $sup(\llbracket M \rrbracket)$ is assigned a nonzero weight by $N^\sim$, then it is also assigned a nonzero weight by N.

Lemma 3. *If* $t \in sup(\llbracket M \rrbracket) \cap sup(\llbracket N^\sim \rrbracket)$, *then* $t \in sup(\llbracket N \rrbracket)$.

Proof. Let $t \in sup(\llbracket M \rrbracket) \cap sup(\llbracket N^\sim \rrbracket)$. There is then a unique run $r \in runs_M(t)$. Intuitively, $\theta(q)$ records the quotient of the weight assigned by $N^\sim$ to the trees that reach q in M, divided by the weight assigned by $N^\sim$ to the trees that reach $[q]$. It follows that all nominators and denominators in the Eqs. 2 and 3 of Definition 5 are nonzero when the weights are computed for every transition $r(v)$, $v \in dom(t)$, and the final weight $F'(r(\epsilon))$. Hence, $t \in sup(\llbracket N \rrbracket)$. □

For N be useful in the main proof, we need to verify that it is a probabilistic wta as per Defintion 1.

Lemma 4. *The automaton N is an end- and out-probabilistic wta.*

Proof. The definition of the final weight function F' contains a normalisation factor in the denominator which ensures that the weights $F'(q), q \in Q$, sum up to 1. This means that N is end-probabilistic.

Let us now verify that N is out-probabilistic. For every $q \in Q$,

$$\sum_{k \in \mathbb{N}, \sigma \in \Sigma_k, q_1 \cdots q_k \in Q^k} \delta'_k(q_1 \cdots q_k, \sigma, q)$$

= by Definition 5 of the automaton N

$$\sum_{(q_1 \cdots q_k, \sigma, q) \in sup(\delta)} \delta^{\sim}_k([q_1]_{\sim} \cdots [q_k]_{\sim}, \sigma, [q]_{\sim}) \cdot \theta(q)^{-1} \cdot \prod_{i \in [k]} \theta(q_i)$$

= because multiplication distributes over addition

$$\frac{\sum_{(q_1 \cdots q_k, \sigma, q) \in sup(\delta)} \delta^{\sim}_k([q_1]_{\sim} \cdots [q_k]_{\sim}, \sigma, [q]_{\sim}) \cdot \prod_{i \in [k]} \theta(q_i)}{\theta(q)}$$

= by Definition 3 of θ (which occurs in the denominator), Definition 4 of θ_i (which corresponds to the nominator), and Lemma 2 which states that for every $q \in Q$, $\theta(q) = \lim_{i \to \infty} \theta_i(q)$

1 .

$\square$

The following Lemma 5 relates the weights assigned by the automata N and $N^{\sim}$ to the trees in $sup(\llbracket M \rrbracket)$.

Lemma 5. *For every* $t \in sup(\llbracket M \rrbracket) \cap sup(\llbracket N^{\sim} \rrbracket)$,

$$\llbracket N \rrbracket(t) = \frac{1}{\sum_{p \in [q]_{\sim}} \theta(p)} \cdot \llbracket N^{\sim} \rrbracket(t) \ .$$

Proof. Let $t \in sup(\llbracket M \rrbracket) \cap sup(\llbracket N^{\sim} \rrbracket)$. By Lemma 3, there is exactly one run $r \in runs_N(t)$ and one run $r' \in runs_{N^{\sim}}(t)$ with non-zero weights. Let $q = r(\epsilon)$. We compute the weight of r through the following equations (We refer to this computation as the series of equations *):

$w_N(r)$

= by the definition of w_N in Sect. 2

$$\prod_{v \in dom(t)} \delta'_k(r(v1) \cdots r(vk), t(v), r(v))$$

= by Definition 5 of δ'

$$\prod_{v \in dom(t)} \frac{\delta^{\sim}_k([r(v1)]_{\sim} \cdots [r(vk)]_{\sim}, t(v), [r(v)]_{\sim}) \cdot \prod_{i \in [k]} \theta(r(vi))}{\theta(r(v))}$$

= by simplification through cancellation at each v, where the weights divided away at the nodes vi, $i \in [k]$, are multiplied back in at the node v

$$\frac{\prod_{v \in dom(t)} \delta^{\sim}_k([r(v1)]_{\sim} \cdots [r(vk)]_{\sim}, t(v), [r(v)]_{\sim})}{\theta(q)}$$

= by Definition 2 of merged wta, and because $N^\sim$ is deterministic

$$\frac{w_{N^\sim}(r')}{\theta(q)}$$

Using this relation between the weight functions, we find that:

$$[\![N]\!](t)$$

= by the definition of wta semantics in Sect. 2

$$F'(q) \cdot w_N(r)$$

= by Definition 5 of F'

$$F^\sim([q]_\sim) \cdot \frac{\theta(q)}{\sum_{p\in[q]_\sim} \theta(p)} \cdot w_N(r)$$

= by the sequence of equations (*) above

$$F^\sim([q]_\sim) \cdot \frac{\theta(q)}{\sum_{p\in[q]_\sim} \theta(p)} \cdot \frac{w_{N^\sim}(r')}{\theta(q)}$$

= by cancelling and reordering terms

$$\frac{1}{\sum_{p\in[q]_\sim} \theta(p)} \cdot F^\sim([q]_\sim) \cdot w_{N^\sim}(r')$$

= by the definition of wta semantics in Sect. 2

$$\frac{1}{\sum_{p\in[q]_\sim} \theta(p)} \cdot [\![N^\sim]\!](t) .$$

□

To ensure that N is a probablistic wta, it remains to verify consistency.

Lemma 6. *The automaton N is consistent.*

Proof.

$$\sum_{t\in T_\Sigma} [\![N]\!](t)$$

= by associativity and commutativity

$$\sum_{P\in Q^\sim} \sum_{p\in P} \sum_{t\in sup([N_p])} [\![N]\!](t)$$

= by Lemma 5 which relates $[\![N]\!](t)$ to $[\![N^\sim]\!](t)$

$$\sum_{P\in Q^\sim} \sum_{p\in P} \sum_{t\in sup([N_p])} \frac{1}{\sum_{p\in P} \theta(p)} \cdot [\![N^\sim]\!](t)$$

= by distributivity

$$\sum_{P\in Q^\sim} \frac{1}{\sum_{p\in P} \theta(p)} \sum_{p\in P} \sum_{t\in sup([N_p])} [\![N^\sim]\!](t)$$

= by construction of θ and by changing the range of the final sum

$$\sum_{P \in Q^\sim} \frac{1}{\sum_{p \in P} \theta(p)} \sum_{p \in P} \theta(p) \sum_{t \in sup([N_P^\sim])} [\![N^\sim]\!](t)$$

= by distributivity

$$\sum_{P \in Q^\sim} \frac{1}{\sum_{p \in P} \theta(p)} \left(\sum_{p \in P} \theta(p) \right) \sum_{t \in sup([N_P^\sim])} [\![N^\sim]\!](t)$$

= by cancellation

$$\sum_{P \in Q^\sim} \sum_{t \in sup([N_P^\sim])} [\![N^\sim]\!](t)$$

= since $N^\sim$ is consistent

$$1 \ . \hspace{4cm} \square$$

$$\square$$

We now connect the intermediate results into a proof of Dietze's conjecture.

Proof (of Theorem 1). By Lemma 4 and Lemma 6, the wta N is a probabilistic wta and by construction, it is in $prob(M)$. For every $t \in sup([\![M]\!]) \cap sup([\![N^\sim]\!])$,

$$[\![N]\!](t) = \frac{1}{\sum_{p \in [q_t]_\sim} \theta(q_t)} \cdot [\![N^\sim]\!](t) \ .$$

By Def. 3 of θ, the sum $\sum_{p \in [q_t]_\sim} \theta(p)$ cannot exceed 1, and it is assumed that $C \subseteq sup([\![M]\!])$, so for every $t \in C$,

$$[\![N]\!](t) \geq [\![N^\sim]\!](t) \ .$$

By definition of the maximum likelihood estimation,

$$L(C \mid [\![mle_C(M)]\!]) \geq L(C \mid [\![N]\!])$$

from which it follows that

$$L(C \mid [\![mle_C(M)]\!]) \geq L(C \mid [\![N^\sim]\!]) \ ,$$

including when $[\![N^\sim]\!] = mle_C(M^\sim)$. $\hspace{3cm} \square$

4 Conclusion

We have seen that given an automaton M and an equivalence relation $\sim$ on the states of M, we can derive from every wta $N^\sim \in prob(M^\sim)$ a wta $N \in prob(M)$ such that N assigns a weight to every tree in $sup([\![M]\!]) \cap sup([\![N^\sim]\!])$ that is greater or equal to that assigned by $N^\sim$. From this it follows that if $C \subseteq sup([\![M]\!])$, then

$$L(C \mid [\![mle_C(M)]\!]) \geq L(C \mid [\![mle_C(M^\sim)]\!]) \ .$$

Now that we know that Dietze's conjecture holds, we can try to relax the assumptions made in the statement. As mentioned earlier, the conjecture cannot be true if $C \nsubseteq sup(\llbracket M \rrbracket)$. However, it remains uncertain whether the restriction to deterministic automata is essential. The proof presented in this paper incorporates the determinism of the automata while defining the mapping θ. Nonetheless, a similar argument could potentially be applicable if we shift our focus to the weight of runs rather than the weight of trees. This is due to the monotonic nature of addition, implying that increasing the likelihood of every run of a tree would result in an increased likelihood of the tree itself.

Acknowledgements. I would like to thank the reviewers for their constructive criticism that helped improved this article. I am also grateful to Frank Drewes for his willingness to act as a sounding board throughout the work.

References

1. Björklund, H., Björklund, J., Ericson, P.: On the regularity and learnability of ordered DAG languages. In: Carayol, A., Nicaud, C. (eds.) CIAA 2017. LNCS, vol. 10329, pp. 27–39. Springer, Cham (2017). https://doi.org/10.1007/978-3-319-60134-2_3
2. Björklund, H., Drewes, F., Ericson, P.: Between a rock and a hard place - uniform parsing for hyperedge replacement DAG grammars. In: 10th International Conference on Language and Automata Theory and Applications (LATA 2016), Prague, Czech Republic, 2016, pp. 521–532 (2016)
3. Bruggemann-Klein, A., Murata, M., Wood, D.: Regular tree and regular hedge languages over unranked alphabets. Technical report (2001)
4. Collins, M.: Head-driven statistical models for natural language parsing. Comput. Linguist. **29**(4), 589–637 (2003)
5. Dietze, T.: A Formal View on Training of Weighted Tree Automata by Likelihood-driven State Splitting and Merging. Technische Universität Dresden (2004)
6. Drewes, F.: Grammatical Picture Generation. TTCSAES, Springer, Heidelberg (2006). https://doi.org/10.1007/3-540-32507-7
7. Fülöp, Z., Vogler, H.: Weighted Tree Automata and Tree Transducers. In: Droste, M., Kuich, W., Vogler, H. (eds.) Handbook of Weighted Automata. Monographs in Theoretical Computer Science. An EATCS Series. Springer, Berlin, Heidelberg (2009). https://doi.org/10.1007/978-3-642-01492-5_9
8. Fülöp, Z., Vogler, H.: Syntax-directed semantics: Formal models based on tree transducers. Springer Science & Business Media (2012). https://doi.org/10.1007/978-3-642-72248-6
9. Klein, D., Manning, C.D.: Accurate unlexicalized parsing. In: Proceedings of the 41st Annual Meeting of the Association for Computational Linguistics, pp. 423–430, Sapporo, Japan (2003). Association for Computational Linguistics
10. Petrov, S., Barrett, L., Thibaux, R., Klein, D.: Learning accurate, compact, and interpretable tree annotation. In: Proceedings of the 21st International Conference on Computational Linguistics and 44th Annual Meeting of the Association for Computational Linguistics, pp. 433–440 (2006)

Computing Subset Vertex Covers in H-Free Graphs

Nick Brettell[1] , Jelle J. Oostveen[2]([⊠]) , Sukanya Pandey[2],
Daniël Paulusma[3] , and Erik Jan van Leeuwen[2]

[1] Victoria University of Wellington, Wellington, New Zealand
nick.brettell@vuw.ac.nz
[2] Utrecht University, Utrecht, The Netherlands
{j.j.oostveen,s.pandey1,e.j.vanleeuwen}@uu.nl
[3] Durham University, Durham, UK
daniel.paulusma@durham.ac.uk

Abstract. We consider a natural generalization of VERTEX COVER:
the SUBSET VERTEX COVER problem, which is to decide for a graph
$G = (V, E)$, a subset $T \subseteq V$ and integer k, if V has a subset S of size
at most k, such that S contains at least one end-vertex of every edge
incident to a vertex of T. A graph is H-free if it does not contain H as
an induced subgraph. We solve two open problems from the literature
by proving that SUBSET VERTEX COVER is NP-complete on subcubic
(claw,diamond)-free planar graphs and on 2-unipolar graphs, a subclass
of $2P_3$-free weakly chordal graphs. Our results show for the first time
that SUBSET VERTEX COVER is computationally harder than VERTEX
COVER (under $P \neq NP$). We also prove new polynomial time results. We
first give a dichotomy on graphs where $G[T]$ is H-free. Namely, we show
that SUBSET VERTEX COVER is polynomial-time solvable on graphs G,
for which $G[T]$ is H-free, if $H = sP_1 + tP_2$ and NP-complete otherwise.
Moreover, we prove that SUBSET VERTEX COVER is polynomial-time
solvable for $(sP_1 + P_2 + P_3)$-free graphs and bounded mim-width graphs.
By combining our new results with known results we obtain a partial
complexity classification for SUBSET VERTEX COVER on H-free graphs.

1 Introduction

We consider a natural generalization of the classical VERTEX COVER problem:
the SUBSET VERTEX COVER problem, introduced in [5]. Let $G = (V, E)$ be a
graph and T be a subset of V. A set $S \subseteq V$ is a T-vertex cover of G if S contains
at least one end-vertex of every edge incident to a vertex of T. We note that T
itself is a T-vertex cover. However, a graph may have much smaller T-vertex
covers. For example, if G is a star whose leaves form T, then the center of G
forms a T-vertex cover. We can now define the problem; see also Fig. 1.

Oostveen, J—was supported by the NWO grant OCENW.KLEIN.114 (PACAN).

H. Fernau and K. Jansen (Eds.): FCT 2023, LNCS 14292, pp. 88–102, 2023.
https://doi.org/10.1007/978-3-031-43587-4_7

SUBSET VERTEX COVER
> *Instance:* A graph $G = (V, E)$, a subset $T \subseteq V$, and a positive integer k.
> *Question:* Does G have a T-vertex cover S_T with $|S_T| \leq k$?

If we set $T = V$, then we obtain the VERTEX COVER problem. Hence, as VERTEX COVER is NP-complete, so is SUBSET VERTEX COVER.

To obtain a better understanding of the complexity of an NP-complete graph problem, we may restrict the input to some special graph class. In particular, *hereditary* graph classes, which are the classes closed under vertex deletion, have been studied intensively for this purpose. It is readily seen that a graph class $\mathcal{G}$ is hereditary if and only if $\mathcal{G}$ is characterized by a unique minimal set of forbidden induced subgraphs $\mathcal{F}_G$. Hence, for a systematic study, it is common to first consider the case where $\mathcal{F}_G$ has size 1. This is also the approach we follow in this paper. So, for a graph H, we set $\mathcal{F}_G = \{H\}$ for some graph H and consider the class of H-free graphs (graphs that do not contain H as an induced subgraph). We now consider the following research question:

For which graphs H is SUBSET VERTEX COVER, *restricted to H-free graphs, still* NP-*complete and for which graphs H does it become polynomial-time solvable?*

We will also address two open problems posed in [5] (see Sect. 2 for any undefined terminology):

Q1. What is the complexity of SUBSET VERTEX COVER for claw-free graphs?
Q2. Is SUBSET VERTEX COVER is NP-complete for P_t-free graphs for some t?

The first question is of interest, as VERTEX COVER is polynomial-time solvable even on $rK_{1,3}$-free graphs for every $r \geq 1$ [4], where $rK_{1,3}$ is the disjoint union of r claws (previously this was known for rP_3-free graphs [13] and $2P_3$-free graphs [14]). The second question is of interest due to some recent quasi-polynomial-time results. Namely, Gartland and Lokshtanov [9] proved that for every integer t, VERTEX COVER can be solved in $n^{O(\log^3 n)}$-time for P_t-free graphs. Afterwards, Pilipczuk, Pilipczuk and Rzążewski [18] improved the running time to $n^{O(\log^2 n)}$ time. Even more recently, Gartland et al. [10] extended the results of [9,18] from P_t-free graphs to H-free graphs where every connected component of H is a path or a subdivided claw.

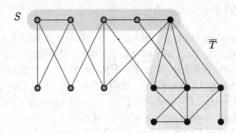

Fig. 1. An instance (G, T, k) of SUBSET VERTEX COVER, where T consists of the orange vertices, together with a solution S (a T-vertex cover of size 5). Note that S consists of four vertices of T and one vertex of $\overline{T} = V \setminus T$.

Grötschel, Lovász, and Schrijver [11] proved that VERTEX COVER can be solved in polynomial time for the class of perfect graphs, which includes well-known graph classes, such as bipartite graphs and (weakly) chordal graphs. Before we present our results, we first briefly discuss the relevant literature.

Existing Results and Related Work

Whenever VERTEX COVER is NP-complete for some graph class $\mathcal{G}$, then so is the more general problem SUBSET VERTEX COVER. Moreover, SUBSET VERTEX COVER can be polynomially reduced to VERTEX COVER: given an instance (G, T, k) of the former problem, remove all edges not incident to a vertex of T to obtain an instance (G', k) of the latter problem. Hence, we obtain:

Proposition 1. *The problems* VERTEX COVER *and* SUBSET VERTEX COVER *are polynomially equivalent for every graph class closed under edge deletion.*

For example, the class of bipartite graphs is closed under edge deletion and VERTEX COVER is polynomial-time solvable on bipartite graphs. Hence, by Proposition 1, SUBSET VERTEX COVER is polynomial-time solvable on bipartite graphs. However, a class of H-free graphs is only closed under edge deletion if H is a complete graph, and VERTEX COVER is NP-complete even for triangle-free graphs [19]. This means that there could still exist graphs H such that VERTEX COVER and SUBSET VERTEX COVER behave differently if the former problem is (quasi)polynomial-time solvable on H-free graphs. The following well-known result of Alekseev [1] restricts the structure of such graphs H.

Theorem 1 ([1]). *For every graph H that contains a cycle or a connected component with two vertices of degree at least* 3, VERTEX COVER, *and thus* SUBSET VERTEX COVER, *is* NP-*complete for H-free graphs.*

Due to Theorem 1 and the aforementioned result of Gartland et al. [10], every graph H is now either classified as a quasi-polynomial case or NP-hard case for VERTEX COVER. For SUBSET VERTEX COVER the situation is much less clear. So far, only one positive result is known, which is due to Brettell et al. [5].

Theorem 2 ([5]). *For every $s \geq 0$,* SUBSET VERTEX COVER *is polynomial-time solvable on $(sP_1 + P_4)$-free graphs.*

Subset variants of classic graph problems are widely studied, also in the context of H-free graphs. Indeed, Brettell et al. [5] needed Theorem 2 as an auxiliary result in complexity studies for SUBSET FEEDBACK VERTEX SET and SUBSET ODD CYCLE TRANSVERSAL restricted to H-free graphs. The first problem is to decide for a graph $G = (V, E)$, subset $T \subseteq V$ and integer k, if G has a set S of size at most k such that S contains a vertex of every cycle that intersects T. The second problem is similar but replaces "cycle" by "cycle of odd length". Brettell et al. [5] proved that both these subset transversal problems are polynomial-time solvable on $(sP_1 + P_3)$-free graphs for every $s \geq 0$. They also showed that ODD CYCLE TRANSVERSAL is polynomial-time solvable for P_4-free graphs and NP-complete for split graphs, which form a subclass of $2P_2$-free graphs, whereas NP-completeness for SUBSET FEEDBACK VERTEX SET on split graphs was shown by Fomin et al. [8]. Recently, Paesani et al. [17] extended the result of [5] for SUBSET FEEDBACK VERTEX SET from $(sP_1 + P_3)$-free graphs to $(sP_1 + P_4)$-free graphs for every integer $s \geq 0$. If H contains a cycle or claw, NP-completeness for both subset transversal problems follows from corresponding results for FEEDBACK VERTEX SET [16,19] and ODD CYCLE TRANSVERSAL [6].

Combining all the above results leads to the following theorems (see also [5, 17]). Here, we write $F \subseteq_i G$ if F is an induced subgraph of G.

Theorem 3. *For a graph H, SUBSET FEEDBACK VERTEX SET on H-free graphs is polynomial-time solvable if $H \subseteq_i sP_1 + P_4$ for some $s \geq 0$, and NP-complete otherwise.*

Theorem 4. *For a graph $H \neq sP_1 + P_4$ for some $s \geq 1$, SUBSET ODD CYCLE TRANSVERSAL on H-free graphs is polynomial-time solvable if $H = P_4$ or $H \subseteq_i sP_1 + P_3$ for some $s \geq 0$, and NP-complete otherwise.*

Our Results

In Sect. 3 we prove two new hardness results, using the same basis reduction, which may have a wider applicability. We first answer Q1 by proving that SUBSET VERTEX COVER is NP-complete even for subcubic planar line graphs of triangle-free graphs, or equivalently, subcubic planar (claw, diamond)-free graphs.

We then answer Q2 by proving that SUBSET VERTEX COVER is NP-complete even for a 2-unipolar graphs, which are $2P_3$-free (and thus P_7-free).

Our hardness results show a sharp contrast with VERTEX COVER, which can be solved in polynomial time for both weakly chordal graphs [11] and $rK_{1,3}$-free graphs for every $r \geq 1$ [4]. Hence, SUBSET VERTEX COVER may be harder than VERTEX COVER for a graph class closed under vertex deletion (if P $\neq$ NP). This is in contrast to graph classes closed under edge deletion (see Proposition 1).

In Sect. 3 we also prove that SUBSET VERTEX COVER is NP-complete for inputs (G, T, k) if the subgraph $G[T]$ of G induced by T is P_3-free. On the other hand, our first positive result, shown in Sect. 4, shows that the problem is polynomial-time solvable if $G[T]$ is sP_2-free for any $s \geq 2$. In Sect. 4 we also prove that SUBSET VERTEX COVER can be solved in polynomial time for $(sP_1 + P_2 + P_3)$-free graphs for every $s \geq 1$. Our positive results generalize known results for VERTEX COVER. The first result also implies that SUBSET VERTEX COVER is polynomial-time solvable for split graphs, contrasting our NP-completeness result for 2-unipolar graphs, which are generalized split, $2P_3$-free, and weakly chordal. Combining our new results with Theorem 2 gives us a partial classification and a dichotomy, both of which are proven in Sect. 5.

Theorem 5. *For a graph $H \neq rP_1+sP_2+P_3$ for any $r \geq 0$, $s \geq 2$; $rP_1+sP_2+P_4$ for any $r \geq 0$, $s \geq 1$; or $rP_1 + sP_2 + P_t$ for any $r \geq 0$, $s \geq 0$, $t \in \{5,6\}$, SUBSET VERTEX COVER on H-free graphs is polynomial-time solvable if $H \subseteq_i sP_1 + P_2 + P_3$, sP_2, or $sP_1 + P_4$ for some $s \geq 1$, and NP-complete otherwise.*

Theorem 6. *For a graph H, SUBSET VERTEX COVER on instances (G, T, k), where $G[T]$ is H-free, is polynomial-time solvable if $H \subseteq_i sP_2$ for some $s \geq 1$, and NP-complete otherwise.*

Theorems 3–6 show that SUBSET VERTEX COVER on H-free graphs can be solved in polynomial time for infinitely more graphs H than SUBSET FEEDBACK VERTEX SET and SUBSET ODD CYCLE TRANSVERSAL. This is in line with the behaviour of the corresponding original (non-subset) problems.

In Sect. 6 we discuss some directions for future work, which naturally originate from the above results and our final new result, which is proven in the full version of our paper[1], and which states that SUBSET VERTEX COVER is polynomial-time solvable on every graph class of bounded mim-width, such as the class of circular-arc graphs.

2   Preliminaries

Let $G = (V, E)$ be a graph. The *degree* of a vertex $u \in V$ is the size of its *neighbourhood* $N(u) = \{v \mid uv \in E\}$. We say that G is *subcubic* if every vertex of G has degree at most 3. An independent set I in G is *maximal* if there exists no independent set I' in G with $I \subsetneq I'$. Similarly, a vertex cover S of G is *minimal* if there no vertex cover S' in G with $S' \subsetneq S$. For a graph H we write $H \subseteq_i G$ if H is an *induced* subgraph of G, that is, G can be modified into H by a sequence of vertex deletions. If G does not contain H as an induced subgraph, G is *H-free*. For a set of graphs $\mathcal{H}$, G is *$\mathcal{H}$-free* if G is H-free for every $H \in \mathcal{H}$. If $\mathcal{H} = \{H_1, \ldots, H_p\}$ for some $p \geq 1$, we also write that G is *$(H_1, \ldots, H_p)$-free*.

The *line graph* of a graph $G = (V, E)$ is the graph $L(G)$ that has vertex set E and an edge between two vertices e and f if and only if e and f share a common end-vertex in G. The *complement* $\overline{G}$ of a graph $G = (V, E)$ has vertex set V and an edge between two vertices u and v if and only if $uv \notin E$.

For two vertex-disjoint graphs F and G, the *disjoint union* $F+G$ is the graph $(V(F) \cup V(G), E(F) \cup E(G))$. We denote the disjoint union of s copies of the same graph G by sG. A *linear forest* is a disjoint union of one or more paths.

Let C_s be the cycle on s vertices; P_t the path on t vertices; K_r the complete graph on r vertices; and $K_{1,r}$ the star on $(r+1)$ vertices. The graph $C_3 = K_3$ is the *triangle*; the graph $K_{1,3}$ the *claw*, and the graph $\overline{2P_1 + P_2}$ is the *diamond* (so the diamond is obtained from the K_4 after deleting one edge). The *subdivision* of an edge uv replaces uv with a new vertex w and edges uw, wv. A *subdivided claw* is obtained from the claw by subdividing each of its edges zero or more times.

---

[1] The full version is available on arXiv, see https://arxiv.org/abs/2307.05701.

A graph is *chordal* if it has no induced C_s for any $s \geq 4$. A graph is *weakly chordal* if it has no induced C_s and no induced $\overline{C_s}$ for any $s \geq 5$. A cycle C_s or an anti-cycle $\overline{C_s}$ is *odd* if it has an odd number of vertices. By the Strong Perfect Graph Theorem [7], a graph is *perfect* if it has no odd induced C_s and no odd induced $\overline{C_s}$ for any $s \geq 5$. Every chordal graph is weakly chordal, and every weakly chordal graph is perfect. A graph $G = (V, E)$ is *unipolar* if V can be partitioned into two sets V_1 and V_2, where $G[V_1]$ is a complete graph and $G[V_2]$ is a disjoint union of complete graphs. If every connected component of $G[V_2]$ has size at most 2, then G is *2-unipolar*. Unipolar graphs form a subclass of *generalized split graphs*, which are the graphs that are unipolar or their complement is unipolar. It can also be readily checked that every 2-unipolar graph is weakly chordal (but not necessarily chordal, as evidenced by $G = C_4$).

For an integer r, a graph G' is an *r-subdivision* of a graph G if G' can be obtained from G by subdividing every edge of G r times, that is, by replacing each edge $uv \in E(G)$ with a path from u to v of length $r + 1$.

3   NP-Hardness Results

In this section we prove our hardness results for SUBSET VERTEX COVER, using the following notation. Let G be a graph with an independent set I. We say that we *augment* G by adding a (possibly empty) set F of edges between some pairs of vertices of I. We call the resulting graph an *I-augmentation* of G.

The following lemma forms the basis for our hardness gadgets.

Lemma 1. *Every vertex cover of a graph $G = (V, E)$ with an independent set I is a $(V \setminus I)$-vertex cover of every I-augmentation of G, and vice versa.*

Proof. Let G' be an I-augmentation of G. Consider a vertex cover S of G. For a contradiction, assume that S is not a $(V \setminus I)$-vertex cover of G'. Then $G' - S$ must contain an edge uv with at least one of u, v belonging to $V \setminus I$. As $G - S$ is an independent set, uv belongs to $E(G') \setminus E(G)$ implying that both u and v belong to I, a contradiction.

Now consider a $(V \setminus I)$-vertex cover S' of G'. For a contradiction, assume that S' is not a vertex cover of G. Then $G - S'$ must contain an edge uv (so $uv \in E$). As G' is a supergraph of G, we find that $G' - S'$ also contains the edge uv. As S' is a $(V \setminus I)$-vertex cover of G', both u and v must belong to I. As $uv \in E$, this contradicts the fact that I is an independent set. $\square$

To use Lemma 1 we need one other lemma, which follows directly from an observation due to Poljak [19].

Lemma 2 ([19]). *For an integer r, a graph G with m edges has an independent set of size k if and only if the $2r$-subdivision of G has an independent set of size $k + rm$.*

We are now ready to prove our first two hardness results. Recall that a graph is (claw, diamond)-free if and only if it is a line graph of a triangle-free graph. Hence, the result in particular implies NP-hardness of SUBSET VERTEX COVER for line graphs. Recall also that we denote the claw and diamond by $K_{1,3}$ and $\overline{2P_1 + P_2}$, respectively.

Theorem 7. SUBSET VERTEX COVER *is* NP-*complete for* $(K_{1,3}, \overline{2P_1 + P_2})$-*free subcubic planar graphs.*

Proof. We reduce from VERTEX COVER, which is NP-complete even for cubic planar graphs [15]. As an n-vertex graph has a vertex cover of size at most k if and only if it has an independent set of size at least $n - k$, we find that VERTEX COVER is NP-complete even for subcubic planar graphs that are 4-subdivisions due to an application of Lemma 2 with $r = 2$ (note that subdividing an edge preserves both maximum degree and planarity). So, let (G, k) be an instance of VERTEX COVER, where $G = (V, E)$ is a subcubic planar graph that is a 4-subdivision of some cubic planar graph G^*, and k is an integer.

In G, we let $U = V(G^*)$ and W be the subset of $V(G) \setminus U$ that consists of all neighbours of vertices of U. Note that W is an independent set in G. We construct a W-augmentation G' as follows.

For every vertex $u \in U$ of degree 3 in G, we pick two arbitrary neighbours of u (which both belong to W) and add an edge between them. It is readily seen that G' is $(K_{1,3}, \overline{2P_1 + P_2})$-free, planar and subcubic. By Lemma 1, it holds that G has a vertex cover of size at most k if and only if G' has a $(V \setminus W)$-vertex cover of size at most k.   □

See the full version of our paper for the proof of our second hardness result. It can be readily checked that 2-unipolar graphs are $(2C_3, C_5, C_6, C_3 + P_3, 2P_3, \overline{P_6}, \overline{C_6})$-free graphs, and thus are $2P_3$-free weakly chordal.

Theorem 8. SUBSET VERTEX COVER *is* NP-*complete for instances* (G, T, k), *for which* G *is* 2-*unipolar and* $G[T]$ *is a disjoint union of edges.*

4   Polynomial-Time Results

In this section, we prove our polynomial-time results. We start with the case where $H = sP_2$ for some $s \geq 1$. For this case we need the following two well-known results. The *delay* of an enumeration algorithm is the maximum of the time taken before the first output and that between any pair of consecutive outputs.

Theorem 9 ([2]). *For every constant* $s \geq 1$, *the number of maximal independent sets of an* sP_2-*free graph on* n *vertices is at most* $n^{2s} + 1$.

Theorem 10 ([20]). *For every constant* $s \geq 1$, *it is possible to enumerate all maximal independent sets of an* sP_2-*free graph* G *on* n *vertices and* m *edges with a delay of* $O(nm)$.

We show a slightly stronger result than proving that SUBSET VERTEX COVER is polynomial-time solvable for sP_2-free graphs. The idea behind the algorithm is to remove any edges between vertices in $V \setminus T$, as these edges are irrelevant. As a consequence, we may leave the graph class, but this is not necessarily an obstacle. For example, if $G[T]$ is a complete graph, or T is an independent set, we can easily solve the problem. Both cases are generalized by the result below.

Theorem 11. *For every $s \geq 1$, SUBSET VERTEX COVER can be solved in polynomial time for instances (G, T, k) for which $G[T]$ is sP_2-free.*

Proof. Let $s \geq 1$, and let (G, T, k) be an instance of SUBSET VERTEX COVER where $G = (V, E)$ is a graph such that $G[T]$ is sP_2-free. Let $G' = (V, E')$ be the graph obtained from G after removing every edge between two vertices of $V \setminus T$, so $G'[V \setminus T]$ is edgeless. We observe that G has a T-vertex cover of size at most k if and only if G' has a T-vertex cover of size at most k. Moreover, $G'[T]$ is sP_2-free, and we can obtain G' in $O(|E(G)|)$ time. Hence, from now on, we consider the instance (G', T, k).

We first prove the following two claims, see Fig. 2 for an illustration.

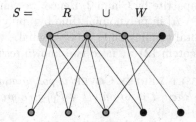

$$S = \quad R \quad \cup \quad W$$

Fig. 2. An example of the $2P_2$-free graph G' of the proof of Theorem 11. Here, T consists of the orange vertices. A solution S can be split up into a minimal vertex cover R of $G'[T]$ and a vertex cover W of $G[V \setminus R]$.

Claim 1. A subset $S \subseteq V(G')$ is a T-vertex cover of G' if and only if $S = R \cup W$ for a minimal vertex cover R of $G'[T]$ and a vertex cover W of $G'[V \setminus R]$.

We prove Claim 1 as follows. Let $S \subseteq V(G')$. First assume that S is a T-vertex cover of G'. Let $I = V \setminus S$. As S is a T-vertex cover, $T \cap I$ is an independent set. Hence, S contains a minimal vertex cover R of $G'[T]$. As $G'[V \setminus T]$ is edgeless, S is a vertex cover of G, or in other words, I is an independent set. In particular, this means that $W \setminus R$ is a vertex cover of $G'[V \setminus R]$.

Now assume that $S = R \cup W$ for a minimal vertex cover R of $G'[T]$ and a vertex cover W of $G'[V \setminus R]$. For a contradiction, suppose that S is not a T-vertex cover of G'. Then $G' - S$ contains an edge $uv \in E'$, where at least one of u, v belongs to T. First suppose that both u and v belong to T. As R is a vertex cover of $G'[T]$, at least one of u, v belongs to $R \subseteq S$, a contradiction. Hence, exactly one of u, v belongs to T, say $u \in T$ and $v \in V \setminus T$, so in particular, $v \notin R$. As $R \subseteq S$, we find that $u \notin R$. Hence, both u and v belong to $V \setminus R$. As W is a vertex cover of $V \setminus R$, this means that at least one of u, v belongs to $W \subseteq S$, a contradiction. This proves the claim. ◇

Claim 2. For every minimal vertex cover R of $G'[T]$, the graph $G'[V \setminus R]$ is bipartite.

We prove Claim 2 as follows. As R is a vertex cover of $G'[T]$, we find that $T \setminus R$ is an independent set. As $G'[V \setminus T]$ is edgeless by construction of G', this means that $G'[V \setminus R]$ is bipartite with partition classes $T \setminus R$ and $V \setminus T$. ◇

We are now ready to give our algorithm. We enumerate the minimal vertex covers of $G'[T]$. For every minimal vertex cover R, we compute a minimum vertex cover W of $G'[V \setminus R]$. In the end, we return the smallest $S = R \cup W$ that we found.

The correctness of our algorithm follows from Claim 1. It remains to analyse the running time. As $G'[T]$ is sP_2-free, we can enumerate all maximal independent sets I of $G'[T]$ and thus all minimal vertex covers $R = T \setminus I$ of $G'[T]$ in $(n^{2s} + 1) \cdot O(nm)$ time due to Theorems 9 and 10. For a minimal vertex cover R, the graph $G'[V \setminus R]$ is bipartite by Claim 2. Hence, we can compute a minimum vertex cover W of $G'[V \setminus R]$ in polynomial time by applying König's Theorem. We conclude that the total running time is polynomial. □

For our next result (Theorem 12) we need two known results as lemmas.

Lemma 3 ([5]). *If* SUBSET VERTEX COVER *is polynomial-time solvable on H-free graphs for some H, then it is so on $(H + P_1)$-free graphs.*

Lemma 4 ([4]). *For every $r \geq 1$,* VERTEX COVER *is polynomial-time solvable on $rK_{1,3}$-free graphs.*

We are now ready to prove our second polynomial-time result.

Theorem 12. *For every integer s,* SUBSET VERTEX COVER *is polynomial-time solvable on $(sP_1 + P_2 + P_3)$-free graphs.*

Proof. Due to Lemma 3, we can take $s = 0$, so we only need to give a polynomial-time algorithm for $(P_2 + P_3)$-free graphs. Hence, let (G, T, k) be an instance of SUBSET VERTEX COVER, where $G = (V, E)$ is a $(P_2 + P_3)$-free graph.

First compute a minimum vertex cover of G. As G is $(P_2 + P_3)$-free, and thus $2K_{1,3}$-free, this takes polynomial time by Lemma 4. Remember the solution S_{vc}.

We now compute a minimum T-vertex cover S of G that is not a vertex cover of G. Then $G - S$ must contain an edge between two vertices in $G - T$. We branch by considering all $O(n^2)$ options of choosing this edge. For each chosen edge uv we do as follows. As both u and v will belong to $G - S$ for the T-vertex cover S of G that we are trying to construct, we first add every neighbour of u or v that belongs to T to S.

Let T' consist of all vertices of T that are neither adjacent to u nor to v. As G is $(P_2 + P_3)$-free and $uv \in E$, we find that $G[T']$ is P_3-free and thus a disjoint union of complete graphs. We call a connected component of $G[T']$ *large* if it has at least two vertices; else we call it *small* (so every small component of $G[T']$ is an isolated vertex). See also Fig. 3 for an illustration.

Case 1. The graph $G[T']$ has at most two large connected components.
Let D_1 and D_2 be the large connected components of $G[T']$ (if they exist). As $V(D_1)$ and $V(D_2)$ are cliques in $G[T]$, at most one vertex of D_1 and at most one vertex of D_2 can belong to $G - S$. We branch by considering all $O(n^2)$ options of choosing at most one vertex of D_1 and at most one vertex of D_2 to be these vertices. For each choice of vertices we do as follows. We add all other vertices of D_1 and D_2 to S. Let T^* be the set of vertices of T that we have not added to S. Then T^* is an independent set.

We delete every edge between any two vertices in $G - T$. Now the graph G^* induced by the vertices of $T^* \cup (V \setminus T)$ is bipartite (with partition classes T^* and $V \setminus T$). It remains to compute a minimum vertex cover S^* of G^*. This can be done in polynomial time by applying König's Theorem. We let S consist of S^* together with all vertices of T that we had added in S already.

For each branch, we remember the output, and in the end we take a smallest set S found and compare its size with the size of S_{vc}, again taking a smallest set as the final solution.

Case 2. The graph $G[T']$ has at least three large connected components.
Let $D_1, \ldots, D_p$, for some $p \geq 3$, be the large connected components of $G[T']$. Let A consists of all the vertices of the small connected components of $G[T']$.

We first consider the case where $G - S$ will contain a vertex $w \in V \setminus T$ with one of the following properties:

1. for some i, w has a neighbour and a non-neighbour in D_i; or
2. for some i, j with $i \neq j$, w has a neighbour in D_i and a neighbour in D_j; or
3. for some i, w has a neighbour in D_i and a neighbour in A.

We say that a vertex w in $G - S$ is *semi-complete* to some D_i if w is adjacent to all vertices of D_i except at most one. We show the following claim that holds if the solution S that we are trying to construct contains a vertex $w \in V \setminus (S \cup T)$ that satisfies one of the three properties above. See Fig. 3 for an illustration.

Claim. Every vertex $w \in V \setminus (S \cup T)$ that satisfies one of the properties 1–3 is semi-complete to every $V(D_j)$.

We prove the Claim as follows. Let $w \in V \setminus (S \cup T)$. First assume w satisfies Property 1. Let x and y be vertices of some D_i, say D_1, such that $wx \in E$ and $wy \notin E$. For a contradiction, assume w is not semi-complete to some D_j. Hence, D_j contains vertices y' and y'', such that $wy' \notin E$ and $wy'' \notin E$. If $j \geq 2$, then $\{y', y'', w, x, y\}$ induces a $P_2 + P_3$ (as D_1 and D_j are complete graphs). This contradicts that G is $(P_2 + P_3)$-free. Hence, w is semi-complete to every $V(D_j)$ with $j \geq 2$. Now suppose $j = 1$. As $p \geq 3$, the graphs D_2 and D_3 exist. As w

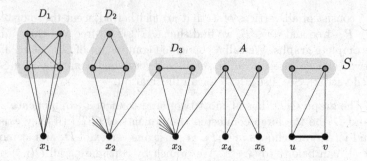

Fig. 3. An illustration of the graph G in the proof of Theorem 12, where T consists of the orange vertices, and $p = 3$. Edges in $G[V \setminus T]$ are not drawn, and for x_2 and x_3 some edges are partially drawn. None of x_1, x_4, x_5 satisfy a property; x_2 satisfies Property 1 for D_2 and Property 2 for D_2 and D_3; and x_3 satisfies Property 3 for D_3.

is semi-complete to every $V(D_j)$ for $j \geq 2$ and every D_j is large, there exist vertices $x' \in V(D_2)$ and $x'' \in V(D_3)$ such that $wx' \in E$ and $wx'' \in E$. However, now $\{y', y'', x', w, x''\}$ induces a $P_2 + P_3$, a contradiction.

Now assume w satisfies Property 2, say w is adjacent to $x_1 \in V(D_1)$ and to $x_2 \in V(D_2)$. Suppose w is not semi-complete to some $V(D_j)$. If $j \geq 3$, then the two non-neighbours of w in D_j, together with x_1, w, x_2, form an induced $P_2 + P_3$, a contradiction. Hence, w is semi-complete to every $V(D_j)$ for $j \geq 3$. If $j \in \{1, 2\}$, say $j = 1$, then let y, y' be two non-neighbours of w in D_1 and let x_3 be a neighbour of w in D_3. Now, $\{y, y', x_2, w, x_3\}$ induces a $P_2 + P_3$, a contradiction. Hence, w is semi-complete to $V(D_1)$ and $V(D_2)$ as well.

Finally, assume w satisfies Property 3, say w is adjacent to $z \in A$ and $x_1 \in V(D_1)$. If w not semi-complete to $V(D_j)$ for some $j \geq 2$, then two non-neighbours of w in D_j, with z, w, x_1, form an induced $P_2 + P_3$, a contradiction. Hence, w is semi-complete to every $V(D_j)$ with $j \geq 2$. As before, by using a neighbour of w in D_2 and one in D_3, we find that w is also semi-complete to $V(D_1)$.   ◇

We now branch by considering all $O(n)$ options for choosing a vertex $w \in V \setminus (S \cup T)$ that satisfies one of the properties 1–3. For each chosen vertex w, we do as follows. We remove all its neighbours in T, and add them to S. By the above Claim, the remaining vertices in T form an independent set. We delete any edge between two vertices from $V \setminus T$, so $V \setminus T$ becomes an independent set as well. It remains to compute, in polynomial time by König's Theorem, a minimum vertex cover in the resulting bipartite graph and add this vertex cover to S. For each branch, we store S. After processing all of the $O(n)$ branches, we keep a smallest S, which we denote by S^*.

We are left to compute a smallest T-vertex cover S of G over all T-vertex covers that contain every vertex from $V \setminus T$ that satisfy one of the properties 1–3. We do this as follows. First, we put all vertices from $V \setminus T$ that satisfy one of the three properties 1–3 to the solution S that we are trying to construct. Let G^* be the remaining graph. We do not need to put any vertex from any connected component of G^* that contains no vertex from T in S.

Now consider the connected component D_1' of G^* that contains the vertices from D_1. As D_1' contains no vertices from $V \setminus T$ satisfying properties 2 or 3, we find that D_1' contains no vertices from A or from any D_j with $j \geq 2$, so $V(D_1') \cap T = V(D_1)$. Suppose there exists a vertex v in $V(D_1') \setminus V(D_1)$, which we may assume has a neighbour in D_1 (as D_1' is connected). Then, v is complete to D_1 as it does not satisfy Property 1. Then, we must put at least $|V(D_1)|$ vertices from D_1' in S, so we might just as well put every vertex of D_1 in S. As $V(D_1') \cap T = V(D_1)$, this suffices. If $D_1' = D_1$, then we put all vertices of D_1 except for one arbitrary vertex of D_1 in S.

We do the same as we did for D_1 for the connected components $D_2', \ldots, D_p'$ of G^* that contain $V(D_2), \ldots V(D_p)$, respectively.

Now, it remains to consider the induced subgraph F of G^* that consists of connected components containing the vertices of A. Recall that A is an independent set. We delete every edge between two vertices in $V \setminus T$, resulting in another independent set. This changes F into a bipartite graph and we can compute a minimum vertex cover S_F of F in polynomial time due to König's Theorem. We put S_F to S and compare the size of S with the size of S^* and S_{vc}, and pick the one with smallest size as our solution.

The correctness of our algorithm follows from the above description. The number of branches is $O(n^4)$ in Case 1 and $O(n^3)$ in Case 2. As each branch takes polynomial time to process, this means that the total running time of our algorithm is polynomial. This completes our proof. $\square$

5   The Proof of Theorems 5 and 6

We first prove Theorem 5, which we restate below.

Theorem 5 (restated). *For a graph $H \neq rP_1 + sP_2 + P_3$ for any $r \geq 0$, $s \geq 2$; $rP_1 + sP_2 + P_4$ for any $r \geq 0$, $s \geq 1$; or $rP_1 + sP_2 + P_t$ for any $r \geq 0$, $s \geq 0$, $t \in \{5, 6\}$, SUBSET VERTEX COVER on H-free graphs is polynomial-time solvable if $H \subseteq_i sP_1 + P_2 + P_3$, sP_2, or $sP_1 + P_4$ for some $s \geq 1$, and NP-complete otherwise.*

Proof. Let H be a graph not equal to $rP_1 + sP_2 + P_3$ for any $r \geq 0$, $s \geq 2$; $rP_1 + sP_2 + P_4$ for any $r \geq 0$, $s \geq 1$; or $rP_1 + sP_2 + P_t$ for any $r \geq 0$, $s \geq 0$, $t \in \{5, 6\}$. If H has a cycle, then we apply Theorem 1. Else, H is a forest. If H has a vertex of degree at least 3, then the class of H-free graphs contains all $K_{1,3}$-free graphs, and we apply Theorem 7. Else, H is a linear forest. If H contains an induced $2P_3$, then we apply Theorem 8. If not, then $H \subseteq_i sP_1 + P_2 + P_3$, sP_2, or $sP_1 + P_4$ for some $s \geq 1$. In the first case, apply Theorem 12; in the second case Theorem 11; and in the third case Theorem 2. $\square$

We now prove Theorem 6, which we restate below.

Theorem 6 (restated). *For a graph H, SUBSET VERTEX COVER on instances (G, T, k), where $G[T]$ is H-free, is polynomial-time solvable if $H \subseteq_i sP_2$ for some $s \geq 1$, and NP-complete otherwise.*

Proof. First suppose $P_3 \subseteq_i H$. As a graph that is a disjoint union of edges is P_3-free, we can apply Theorem 8. Now suppose H is P_3-free. Then $H \subseteq_i sP_2$ for some $s \geq 1$, and we apply Theorem 11. □

6   Conclusions

Apart from giving a dichotomy for SUBSET VERTEX COVER restricted to instances (G, T, k) where $G[T]$ is H-free (Theorem 6), we gave a partial classification of SUBSET VERTEX COVER for H-free graphs (Theorem 5). Our partial classification resolved two open problems from the literature and showed that for some hereditary graph classes, SUBSET VERTEX COVER is computationally harder than VERTEX COVER (if P ≠ NP). This is in contrast to the situation for graph classes closed under edge deletion. Hence, SUBSET VERTEX COVER is worth studying on its own, instead of only as an auxiliary problem (as in [5]).

Our results raise the question whether there exist other hereditary graph classes on which SUBSET VERTEX COVER is computationally harder than VERTEX COVER. Recall that VERTEX COVER is polynomial-time solvable for perfect graphs [11], and thus for weakly chordal graphs and chordal graphs. On the other hand, we showed that SUBSET VERTEX COVER is NP-complete for 2-unipolar graphs, a subclass of $2P_3$-free weakly chordal graphs. Hence, as the first candidate graph class to answer this question, we propose the class of chordal graphs. A standard approach for VERTEX COVER on chordal graphs is dynamic programming over the clique tree of a chordal graph. However, a naive dynamic programming algorithm over the clique tree does not work for SUBSET VERTEX COVER, as we may need to remember an exponential number of subsets of a bag (clique) and the bags can have arbitrarily large size. In the full version of our paper, we show that SUBSET VERTEX COVER can be solved in polynomial time on graphs of bounded mim-width. Using known results, this immediately implies the following:

Corollary 1. SUBSET VERTEX COVER *can be solved in polynomial time on interval and circular-arc graphs.*

Corollary 1 makes the open question of the complexity of SUBSET VERTEX COVER on chordal graphs, a superclass of the class of interval graphs, even more pressing. Recall that SUBSET FEEDBACK VERTEX SET, which is also solvable in polynomial time for graphs of bounded mim-width [3], is NP-complete for split graphs and thus for chordal graphs [8].

We note that our polynomial algorithms for SUBSET VERTEX COVER for sP_2-free graphs and $(P_2 + P_3)$-free graphs can easily be adapted for WEIGHTED SUBSET VERTEX COVER for sP_2-free graphs and $(P_2 + P_3)$-free graphs. every $s \geq 1$ [5] (see also Theorem 4).

Finally, to complete the classification of SUBSET VERTEX COVER for H-free graphs we need to solve the open cases where $H = sP_2 + P_3$ for $s \geq 2$; or $H = sP_2 + P_4$ for $s \geq 1$; or $H = sP_2 + P_t$ for $s \geq 0$ and $t \in \{5, 6\}$. Brettell et al. [5] asked what the complexity of SUBSET VERTEX COVER is for P_5-free graphs. In contrast, VERTEX COVER is polynomial-time solvable even for P_6-free graphs [12]. However, the open cases where $H = sP_2+P_t$ ($s \geq 1$ and $t \in \{4,5,6\}$) are even open for VERTEX COVER on H-free graphs (though a quasi-polynomial time algorithm is known [9,18]). So for those cases we may want to first restrict ourselves to VERTEX COVER instead of SUBSET VERTEX COVER.

References

1. Alekseev, V.E.: The effect of local constraints on the complexity of determination of the graph independence number. In: Combinatorial-Algebraic Methods in Applied Mathematics, pp. 3–13 (1982). (in Russian)
2. Balas, E., Yu, C.S.: On graphs with polynomially solvable maximum-weight clique problem. Networks **19**(2), 247–253 (1989)
3. Bergougnoux, B., Papadopoulos, C., Telle, J.A.: Node Multiway Cut and Subset Feedback Vertex Set on graphs of bounded mim-width. Algorithmica **84**(5), 1385–1417 (2022)
4. Brandstädt, A., Mosca, R.: Maximum Weight Independent Set for ℓclaw-free graphs in polynomial time. Discrete Appl. Math. **237**, 57–64 (2018)
5. Brettell, N., Johnson, M., Paesani, G., Paulusma, D.: Computing subset transversals in H-free graphs. Theor. Comput. Sci. **902**, 76–92 (2022)
6. Chiarelli, N., Hartinger, T.R., Johnson, M., Milanič, M., Paulusma, D.: Minimum connected transversals in graphs: new hardness results and tractable cases using the price of connectivity. Theor. Comput. Sci. **705**, 75–83 (2018)
7. Chudnovsky, M., Robertson, N., Seymour, P., Thomas, R.: The strong perfect graph theorem. Ann. Math. **164**, 51–229 (2006)
8. Fomin, F.V., Heggernes, P., Kratsch, D., Papadopoulos, C., Villanger, Y.: Enumerating minimal subset feedback vertex sets. Algorithmica **69**, 216–231 (2014)
9. Gartland, P., Lokshtanov, D.: Independent Set on P_k-free graphs in quasi-polynomial time. Proc. FOCS **2020**, 613–624 (2020)
10. Gartland, P., Lokshtanov, D., Masařík, T., Pilipczuk, M., Pilipczuk, M., Rzążewski, P.: Maximum Weight Independent set in graphs with no long claws in quasi-polynomial time. CoRR arXiv:2305.15738 (2023)
11. Grötschel, M., Lovász, L., Schrijver, A.: Polynomial algorithms for perfect graphs. Ann. Discrete Math. **21**, 325–356 (1984)
12. Grzesik, A., Klimošová, T., Pilipczuk, M., Pilipczuk, M.: Polynomial-time algorithm for Maximum Weight Independent Set on P_6-free graphs. ACM Trans. Algorithms **18**, 4:1–4:57 (2022)
13. Lozin, V.V.: From matchings to independent sets. Discrete Appl. Math. **231**, 4–14 (2017)
14. Lozin, V.V., Mosca, R.: Maximum regular induced subgraphs in $2P_3$-free graphs. Theor. Comput. Sci. **460**, 26–33 (2012)
15. Mohar, B.: Face covers and the genus problem for apex graphs. J. Comb. Theor. Ser. B **82**(1), 102–117 (2001)
16. Munaro, A.: On line graphs of subcubic triangle-free graphs. Discrete Math. **340**, 1210–1226 (2017)
17. Paesani, G., Paulusma, D., Rzążewski, P.: Classifying Subset Feedback Vertex Set for H-Free Graphs. In: Bekos, M.A., Kaufmann, M. (eds.) Graph-Theoretic Concepts in Computer Science. WG 2022. Lecture Notes in Computer Science. vol. 13453. Springer, Cham (2022). https://doi.org/10.1007/978-3-031-15914-5_30

18. Pilipczuk, M., Pilipczuk, M., Rzążewski, P.: Quasi-polynomial-time algorithm for Independent Set in P_t-free graphs via shrinking the space of induced paths. Proc. SOSA **2021**, 204–209 (2021)
19. Poljak, S.: A note on stable sets and colorings of graphs. Commentationes Mathematicae Universitatis Carolinae **15**, 307–309 (1974)
20. Tsukiyama, S., Ide, M., Ariyoshi, H., Shirakawa, I.: A new algorithm for generating all the maximal independent sets. SIAM J. Comput. **6**, 505–517 (1977)

On Computing Optimal Temporal Branchings

Daniela Bubboloni[1], Costanza Catalano[1(✉)], Andrea Marino[2], and Ana Silva[3]

[1] Dipartimento di Matematica e Informatica, Università degli Studi di Firenze, Firenze, Italy
{daniela.bubboloni,costanza.catalano}@unifi.it
[2] Dipartimento di Statistica, Informatica, Applicazioni, Università degli Studi di Firenze, Firenze, Italy
andrea.marino@unifi.it
[3] Departamento de Matematica, Universidade Federal do Ceará, Fortaleza, Brazil
anasilva@mat.ufc.br

Abstract. The computation of out/in-branchings spanning the vertices of a digraph (also called directed spanning trees) is a central problem in theoretical computer science due to its application in reliable network design. This concept can be extended to temporal graphs, which are graphs where arcs are available only at prescribed times and paths make sense only if the availability of the arcs they traverse is non-decreasing. In this context, the paths of the out-branching from the root to the spanned vertices must be valid temporal paths. While the literature has focused only on minimum weight temporal out-branchings or the ones realizing the earliest arrival times to the vertices, the problem is still open for other optimization criteria. In this work we define four different types of optimal temporal out-branchings (TOB) based on the optimization of the travelling time (ST-TOB), of the travel duration (FT-TOB), of the number of transfers (MT-TOB) or of the departure time (LD-TOB). For D ∈ {ST, MT, LD}, we provide necessary and sufficient conditions for the existence of spanning D-TOBs; when those do not exist, we characterize the maximum vertex set that a D-TOB can span. Moreover, we provide a log linear algorithm for computing such D-TOBs. Oppositely, we show that deciding the existence of an FT-TOB spanning all the vertices is NP-complete. This is quite surprising, as all the above distances, including FT, can be computed in polynomial time, meaning that computing temporal distances is inherently different from computing D-TOBs. Finally, we show that the same results hold for optimal temporal in-branchings.

Keywords: Temporal graph · temporal network · link stream · optimal branching · optimal temporal walk

Daniela Bubboloni is partially supported by GNSAGA of INdAM (Italy). Daniela Bubboloni, Costanza Catalano and Andrea Marino are partially supported by Italian PNRR CN4 Centro Nazionale per la Mobilità Sostenibile, NextGeneration EU - CUP, B13C22001000001. Ana Silva is partially supported by: FUNCAP MLC-0191-00056.01.00/22 and PNE-0112-00061.01.00/16, CNPq 303803/2020-7 (Brazil).

H. Fernau and K. Jansen (Eds.): FCT 2023, LNCS 14292, pp. 103–117, 2023.
https://doi.org/10.1007/978-3-031-43587-4_8

1   Introduction

A temporal graph is a graph where arcs are active only at certain time instants, with a possible *delay* or *travelling time* indicating the time it takes to traverse an arc. There is not a unified terminology in the literature to call these objects, as they are also known as stream graphs [14], dynamic networks [18], and time-varying graphs [13] to name a few. Important categories of temporal graphs are those of transport networks, where arcs are labeled by the times of bus/train/flight departures and arrivals [9], and communication networks as phone calls and emails networks, where each arc represents the interaction between two parties [19]. Temporal graphs find application in a vast number of fields such as neural, ecological and social networks, distributed computing, epidemiology etc.; we refer the reader to [10] for a survey on temporal graphs. Fundamental properties of static graphs, such as the fact that concatenation of walks is a walk, do not necessarily hold in temporal graphs. For instance, a public transports route can happen only at increasing time instants, since a person cannot catch a bus that already left. This often makes temporal graphs much harder to handle: e.g. computing strongly connected components takes linear time in a static graph, but is an NP-complete problem in a temporal graph [17], and the same happens to Eulerian walks [16], and many other problems. We will see in the next section that this is also the case for temporal branchings.

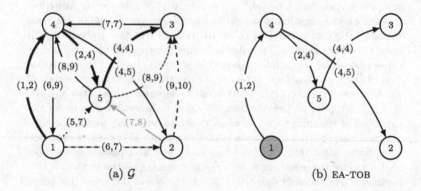

(a) $\mathcal{G}$ (b) EA-TOB

Fig. 1. (a) Temporal graph with different walks from vertex 1 to 3, each one represented by a style (two-style arcs belong to two walks). **Bold**: walk realizing EA(1, 3). Dotted: walk realizing MT(1, 3). Dashed: walk realizing both ST(1, 3) and LD(1, 3). Grey: walk realizing FT(1, 3). (b) An EA-TOB of $\mathcal{G}$ with root 1.

Background on Temporal Graphs. We denote by $\mathbb{N}$ the set of positive integers. Given $n \in \mathbb{N}$, we set $[n] := \{x \in \mathbb{N} : x \le n\}$. A temporal graph $\mathcal{G}$ is a triple (V, A, τ), where V is the set of vertices, $\tau \in \mathbb{N}$ is the *lifetime*, and $A \subseteq \{(u, v, s, t) : u, v \in V, u \ne v \text{ and } s, t \in [\tau], s \le t\}$ is the set of *temporal arcs*. We set $|A| := m$ and $|V| := n$. Given $a \in A$, we write $a = (\mathrm{t}(a), \mathrm{h}(a), t_s(a), t_a(a))$,

Table 1. Computational time of single source shortest paths in a temporal graph.

EA	MT	ST	LD	FT
$O(m)$ [11,20]	$O(m \log n)$ [2]	$O(m \log m)$ [1,20,21]	$O(m \log m)$ [1]	$O(m \log n)$ [2]

where $t(a)$ and $h(a)$ are, respectively, the *tail* and *head* vertices of the temporal arc a, and $t_s(a)$ and $t_a(a)$ are, respectively, the *starting time* and the *arrival time* of a. These functions are easily interpreted: $t_s(a)$ is the time at which it is possible to begin a trip along a from vertex $t(a)$ to vertex $h(a)$, and $t_a(a)$ is the arrival time of that trip. The temporal graph $\mathcal{G}$ has the multidigraph $\mathcal{D}_g = (V, A, t, h)$ as underlying structure. Figure (1a) presents an example of temporal graph, where every arc a is labeled by the ordered pair $(t_s(a), t_a(a))$. Each arc has an *elapsed time* $el(a) := t_a(a) - t_s(a)$. In temporal graphs, walks make sense only if they are time-consistent. More precisely, a *temporal* (u, v)-*walk of length* $k \in \mathbb{N}$ in $\mathcal{G}$ is a (u, v)-walk $W = (u, a_1, v_1, \ldots, v_{k-1}, a_k, v)$ in the underlying multidigraph such that $t_a(a_i) \leq t_s(a_{i+1})$ for all $i \in [k-1]$; in this case we also say that v is *temporally reachable* from u. For the walk W, we consider the *starting time* $t_s(W) := t_s(a_1)$ and the *arrival time* $t_a(W) := t_a(a_k)$. The *travelling time* of W is $tt(W) := \sum_{i=1}^{k} el(a_i)$ and the *duration* of W is $dur(W) := t_a(W) - t_s(W)$. The *length* of W is denoted by $\ell(W)$. Given $u, v \in V$, $\mathcal{W}_g(u, v)$ is the set of temporal walks from u to v in $\mathcal{G}$. We consider the following optimization criteria.

Earliest Arrival time: $\text{EA}_g(u, v) := \min\{t_a(W) : W \in \mathcal{W}_g(u, v)\}$;

Latest Departure time: $\text{LD}_g(u, v) := \max\{t_s(W) : W \in \mathcal{W}_g(u, v)\}$;

Minimum Transfers: $\text{MT}_g(u, v) := \min\{\ell(W) : W \in \mathcal{W}_g(u, v)\}$;

Fastest Time: $\text{FT}_g(u, v) := \min\{dur(W) : W \in \mathcal{W}_g(u, v)\}$;

Shortest Travelling time: $\text{ST}_g(u, v) := \min\{tt(W) : W \in \mathcal{W}_g(u, v)\}$.

Consistently with the literature [3], we refer to the above definitions as *distances*.[1] All these concepts are widely used (see [1,2,9,11,20,21]), although sometimes they appear with different names. For any $\text{D} \in \{\text{EA}, \text{LD}, \text{MT}, \text{FT}, \text{ST}\}$, we say that a temporal (u, v)-walk *realizes* $\text{D}_g(u, v)$ if it attains the minimum (or maximum if $\text{D}=\text{LD}$) of the functions in the corresponding definition of $\text{D}_g(u, v)$. Figure (1a) shows, for each D, a temporal walk from vertex 1 to 3 realizing $\text{D}_g(u, v)$. Each distance is computable in polynomial-time: Table 1 reports the time to compute $\text{D}_g(r, v)$ from a given vertex r to all the other vertices v.[2]

Optimal temporal branchings. In static directed graphs, spanning branchings are well-studied objects; they represent a minimal set of arcs that connect a special vertex called the root to any other vertex (out-branching), or any vertex to the root (in-branching). They are also called arborescences or spanning directed trees, since their underlying structure is a tree. Spanning branchings

---

[1] They do not necessarily satisfy the triangle inequality.

[2] Notice that [2] deals with waiting-time constrains. Nonetheless, to the best of our knowledge, their algorithms provide the best running time for distances such as MT and FT also when there are no time-constrains or restrictions on the elapsed times.

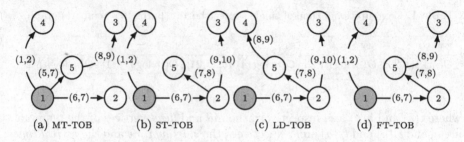

Fig. 2. Example of D-TOBs of the temporal graph $\mathcal{G}$ in Figure (1a) for different distances. The grey vertex is the root of the TOB.

representing shortest distances are also well-studied. Their existence is guaranteed simply by the reachability of any vertex from/to the root and they can be computed in $O(m \log m)$ time by Dijkstra's algorithm [7]. Branchings are, to cite a few, important for engineering applications and in social networks in relation to information dissemination and spreading. We can similarly define spanning branchings in temporal graphs, here called spanning TOBs (Temporal Out-Branchings) and TIBs (Temporal In-Branchings), representing the minimal set of temporal arcs that temporally connect any vertex from/to the root. This definition of TOB has already appeared in the literature [11, 12].[3] In the context of urban mobility, suppose that a concert is just finished in a remote location X, and you want to guarantee that every person can go back home via public transports, while optimizing the number of bus/train rides. This problem can be solved by a spanning TOB with root X. We also may ask this TOB to arrive the earliest possible in every point of interest of the city, or to use the least number of transfers, or optimize any of the distances that we have introduced before. It is then natural to extend the notion of shortest distance branchings to the temporal framework. For each D $\in$ {EA, LD, MT, FT, ST}, we call spanning D-TOB a spanning TOB representing the distance D, i.e. for every vertex v, the unique (r, v)-walk within the branching realizes $\mathrm{D}(r, v)$. We define similarly spanning D-TIBs. Figure (1b) and Fig. 2 show, for each distance, a spanning D-TOB with root 1 of the temporal graph in Figure (1a). Notice that the MT-TOB can be modified by adding the arc $(2, 3, 9, 10)$ and by deleting the arc $(5, 3, 8, 9)$ while still obtaining a spanning MT-TOB. Thus, in general, D-TOBs are not unique.

In [11] the authors prove that a spanning TOB as well as a EA-TOB exists iff every vertex is temporally reachable from the root and provide an algorithm to compute them in $O(m)$ time. Nonetheless, for all the other distances but EA, the problem of computing optimal branching is still open and seems to be a more difficult task. We start observing that for D $\neq$ EA, the temporal reachability from the root to any vertex is no longer sufficient for the existence of a spanning D-TOB; this is showed in Fig. 3 where for each D $\in$ {LD, MT, ST, FT} we present a temporal graph that does not admit a spanning D-TOB even if every vertex is

---

[3] Notice that [12] proposes it in a simplified context, while the conditions listed in the definition of [11] are not all necessary to describe the concept.

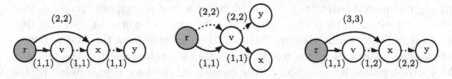

(a) No spanning MT,LD-TOB   (b) No spanning FT-TOB   (c) No spanning ST-TOB

Fig. 3. Examples of temporal graphs that do not admit a spanning D-TOB with root r. Solid arcs represent a maximum D-TOB.

temporally reachable from r. In Figures (3a) and (3c), observe that there is a unique temporal path from r to y; call it P. This is clearly the only spanning TOB of the temporal graphs under consideration. However, P does not realize $D(r, x)$, which is realized by the temporal arc from r to x. Therefore, P is not a D-TOB. We emphasize that adding the arc from r to x to P would no longer form a TOB (the underlying graph would not be a branching). As for Figure (3b), notice that the temporal path $(r, (r, v, 1, 1), v, (v, x, 1, 1), x)$ is the only temporal path realizing $FT(r, x)$. Similarly, the temporal path $(r, (r, v, 2, 2), v, (v, y, 2, 2), y)$ is the only one realizing $FT(r, y)$. This implies that a possible spanning D-TOB must be equal to the graph itself, which clearly is not a branching. Notice that in the examples $\tau = 2$ for D $\in \{MT, LD\}$, which is the smallest value possible, as when $\tau = 1$ the temporal graph reduces to a static graph. When D = ST, we have that $\tau = 3$: it can be proven that this is again the smallest value possible. Notice also that in all the examples, we can always find a D-TOB on the vertex set $\{r, v, x\}$, with D chosen accordingly; this TOB is highlighted by solid arcs in the figures. In Figure (3b), also the dotted arcs form an FT-TOB on the vertex set $\{r, v, y\}$. The following questions naturally arise:

1. When does a spanning D-TOB exist?
2. If it does not exist, can we identify the maximum set of vertices that can be spanned by a D-TOB (*maximum* D-TOB)?
3. Can we compute a maximum D-TOB in polynomial time?
4. Can we answer to all the above questions for D-TIBs?

Our contribution. In this paper we solve all the above problems. For each D $\in \{ST, MT, LD\}$, we provide a necessary and sufficient condition for the existence of a spanning D-TOB in a temporal graph; this property is based on the concept of optimal substructure. Moreover, we characterize the vertex set of a maximum D-TOB, which turns out to be uniquely identified; this property is crucial to find efficient polynomial-time algorithms for computing a maximum D-TOB (Sect. 4). In particular, our algorithms compute a D-TOB whose path from the root arrives the earliest possible in every vertex. The characterization does not hold for D = FT, and in fact we show that computing an FT-TOB is an NP-complete problem (Sect. 5). Finally, we show that the same results hold for optimal temporal in-branchings (Sect. 3). A summary of our results and of the computational time of our algorithms can be found in Table 2. We underline that any algorithm computing $D(r, v)$ for all vertices v of a temporal graph cannot suffice by itself to find a D-TOB. Indeed we have seen in Figure (3a) and (3c) that $D(r, y)$ is well-defined because y is temporally reachable from the root r, but no D-TOBs can span y. In other words, there are no guarantees that the union of the shortest paths, with respect to the considered distance D, computed by the aforementioned algorithms would form a TOB. In addition, for D $=$FT we have the extreme case where computing $FT(r, v)$ is polynomial-time, but finding an FT-TOB is NP-complete. Also, applying Dijkstra's algorithm on the static expansion of a temporal graph returns a branching on the static expansion, but does not guarantee to obtain a TOB in the original temporal graph.

Table 2. Our contribution: summary results. Second column refers to the time to compute any TOB/TIB, the others refer to the time to compute any D-TOB/D-TIB.

	any	EA-	MT-	ST-	LD-	FT-
TOB	$O(m)$ [11]	$O(m)$ [11]	$O(m \log n)$	$O(m \log m)$	$O(m \log m)$	NP-c
TIB	$O(m)$	$O(m \log m)$	$O(m \log n)$	$O(m \log m)$	$O(m)$	NP-c

Further Related Results. We already mentioned the results of [11], where they also show that finding minimum weight spanning TOBs is NP-hard. Kuwata et. al. [13] are interested in the temporal reachability from the root that realizes EA, and they obtain it by making use of Dijsktra's algorithm on the static expansion of the temporal graph, which we already observed does not translate into a TOB in the original temporal graph. Different versions of the problem of finding arc-disjoint TOBs in temporal graphs are investigated in [4,12].

2   Preliminaries

We set $\mathbb{N}_0 = \mathbb{N} \cup \{0\}$, $[n] := \{x \in \mathbb{N} : x \le n\}$ and $[n]_0 := \{x \in \mathbb{N}_0 : x \le n\}$, for $n \in \mathbb{N}_0$. Given a set $\mathcal{X}$ and a property $\mathcal{P}$, we say that $\mathcal{X}$ is *minimal* for property $\mathcal{P}$ if $\mathcal{X}$ has property $\mathcal{P}$, and for all $\mathcal{Y} \subsetneq \mathcal{X}$, $\mathcal{Y}$ does not have property $\mathcal{P}$. A digraph is a pair $D = (V, A)$ where V is the nonempty, finite set of vertices, and $A \subseteq V \times V$ is the set of arcs. Informally, a multidigraph is a digraph where multiple arcs are allowed; it is formalized by a quadruple $\mathcal{D} = (V, A, t, h)$, where V is the set of vertices, A the set of arcs and $t, h : A \to V$ are respectively the *head* and the *tail* function where $\forall a \in A$, $t(a) \ne h(a)$, i.e. no selfloops are allowed. The in-degree and out-degree of v are defined respectively as $d_{\mathcal{D}}^-(v) := |\{a \in A : h(a) = v\}|$, $d_{\mathcal{D}}^+(v) := |\{a \in A : t(a) = v\}|$. A (u, v)-*walk* of length $k \in \mathbb{N}_0$ in $\mathcal{D}$ is an alternating ordered sequence $W = (u, a_1, v_1, \ldots, v_{k-1}, a_k, v_k = v)$ of vertices $u, v_1, \ldots, v_k \in V$ and arcs $a_1, \ldots, a_k \in A$ such that $t(a_1) = u$, $h(a_k) = v$ and $h(a_i) = v_i = t(a_{i+1})$ for all $i \in [k-1]$. The set of vertices of W is denoted by $V(W)$ and the set of arcs of W by $A(W)$. A *path* is a walk where the vertices are all distinct. The concatenation of two walks W_1 and W_2 is denoted by $W_1 + W_2$.

For $h \in [k]$ the v_h-prefix of W is the subwalk of W given by $(u, a_1, v_1, \ldots, v_h)$; the v_h-suffix of W is the subwalk of W given by $(v_h, a_{h+1}, \ldots, a_k, v_k)$. Note that, for a fixed $z \in V(W)$, there are, in general, many z-prefixes and many z-suffixes of W; they are unique if W is a path. A digraph $\mathcal{D} = (V, A)$ is called an *out-branching* (resp. *in-branching*) with root $r \in V$ if for every $v \in V$ there exists a unique (r, v)-*walk* (resp. (v, r)-*walk*) in $\mathcal{D}$. In a branching, every walk is a path. A temporal graph $\mathcal{G}' = (V', A', \tau')$ is a *temporal subgraph* of $\mathcal{G} = (V, A, \tau)$ if $V' \subseteq V$, $A' \subseteq A$ and $\tau' \leq \tau$. For $V' \subseteq V$, we denote by $\mathcal{G}[V']$ the temporal subgraph of $\mathcal{G}$ on vertex set V'. When the temporal graph is clear from the context, we usually omit the subscripts.

3 Temporal Branching and Preliminary Results

3.1 Temporal Out-Branching

In this section, we present the formal notion of temporal out-branching and define related optimization problems.

Definition 1. *A temporal graph* $\mathcal{B} = (V, A, \tau)$ *is called a* temporal out-branching (TOB) *with root* $r \in V$ *if* A *is a minimal set of temporal arcs such that for all* $v \in V$, *there exists a temporal* (r, v)-*walk in* $\mathcal{B}$.

The following lemma provides characterizations of a TOB, which are crucial for the proofs of the results of Sect. 4.

Lemma 1. *Let* $\mathcal{B} = (V, A, \tau)$ *be a temporal graph. The following facts are equivalent:*

1. $\mathcal{B}$ *is a TOB with root* r;
2. *For all* $v \in V$ *there is a temporal* (r, v)-*walk in* $\mathcal{B}$. *Additionally,* $d_{\mathcal{B}}^-(r) = 0$ *and, for all* $v \in V \setminus \{r\}$, $d_{\mathcal{B}}^-(v) = 1$;
3. *For all* $v \in V$ *there is a temporal* (r, v)-*walk in* $\mathcal{B}$, *and* $|A| = |V| - 1$;
4. *The underlying digraph* $\mathcal{D}_{\mathcal{B}}$ *of* $\mathcal{B}$ *is an out-branching with root* r *and for all* $v \in V$, *the unique* (r, v)-*walk in* $\mathcal{B}$ *is temporal.*

In a TOB with root r, the unique temporal walk from r to v is a temporal path.

Definition 2. *Let* $\mathcal{G} = (V, A, \tau)$ *be a temporal graph and* $\mathcal{B} = (V_{\mathcal{B}}, A_{\mathcal{B}}, \tau_{\mathcal{B}})$ *a* TOB *with root* r. *We say that* $\mathcal{G}$ *admits the* TOB $\mathcal{B}$ *if* $\mathcal{B}$ *is a temporal subgraph of* $\mathcal{G}$. $\mathcal{B}$ *is also said to be a* TOB *of* $\mathcal{G}$. $\mathcal{B}$ *is called a* spanning TOB *of* $\mathcal{G}$ *if* $V_{\mathcal{B}} = V$; *a* maximum TOB *of* $\mathcal{G}$ *if* $|V_{\mathcal{B}}|$ *is the largest possible.*

We now expand the concept of TOB to the various distances considered in the introduction. The idea is that we are not only interested in temporally reaching the maximum number of vertices from the root, but we want also to *minimize* their distance from the root, which can translate into arriving the earliest possible, the fastest possible, by starting the journey the latest possible, by travelling the shortest time possible or by making the least number of transfers possible, depending on the preferences and needs.

Definition 3. *Let* $D \in \{EA, LD, MT, FT, ST\}$ *and let* $\mathcal{B} = (V_B, A_B, \tau_B)$ *be a* TOB *with root* r *of a temporal graph* $\mathcal{G} = (V, A, \tau)$. *We say that* $\mathcal{B}$ *is a* D-TOB *of* $\mathcal{G}$ *if* $D_B(r, v) = D_G(r, v)$ *for every* $v \in V_B$. $\mathcal{B}$ *is a* spanning D-TOB *of* $\mathcal{G}$ *if* $V_B = V$; *is a* maximum D-TOB *of* $\mathcal{G}$ *if* $|V_B|$ *is the largest possible.*

Problem 1 (Maximum D-TOB). Let $D \in \{EA, LD, MT, ST, FT\}$ and $\mathcal{G}$ be a temporal graph. Find a maximum D-TOB of $\mathcal{G}$.

Problem 1 has already been solved for $D = EA$ in [11]. Their result also implies that a maximum EA-TOB spans all the vertices that are temporally reachable from the root. We will see that also for every $D \in \{LD, MT, ST\}$, the vertex set of a maximum D-TOB of a temporal graph is uniquely determined, which is key for the polynomiality of the related problems. However, the property of being temporally reachable from the root is not sufficient anymore, as showed in Fig. 3. Instead for FT, we show that the related problem is NP-complete. As we will see, in the polynomial cases we can constrain ourselves to the earliest arrival paths that realize the distances.

Definition 4. *Given a temporal graph* $\mathcal{G} = (V, A, \tau)$, *for every* $u, v \in V$ *and* $D \in \{MT, ST, LD, FT\}$, *we define* $EAD_G(u, v) := \min\{t_a(W) : W \text{ realizes } D_G(u, v)\}$. *A* TOB $\mathcal{B} = (V_B, A_B, \tau_B)$ *of* $\mathcal{G}$ *with root* r *is called an* EAD-TOB *if* $\mathcal{B}$ *is a* D-TOB *and, for every* $v \in V_B$, $EAD_B(r, v) = EAD_G(r, v)$. $\mathcal{B}$ *is called* spanning *if* $V_B = V$; *maximum if* $|V_B|$ *is the largest possible.*

3.2 Temporal In-Branching

In this section, we present definitions of temporal in-branchings and prove that the related problems are computationally equivalent to TOBs.

Definition 5. *A temporal graph* $\mathcal{B} = (V_B, A_B, \tau_B)$ *is said a* temporal in-branching (TIB) *with root* r *if* A_B *is a minimal set of temporal arcs such that for all* $v \in V$, *there exists a temporal* (v, r)-*walk in* $\mathcal{B}$. *A temporal graph* $\mathcal{G} = (V, A, \tau)$ *admits the* TIB $\mathcal{B}$ *with root* r *if* $\mathcal{B}$ *is a temporal subgraph of* $\mathcal{G}$; *we also say that* $\mathcal{B}$ *is a* TIB *of* $\mathcal{G}$. $\mathcal{B}$ *is* spanning *if* $V_B = V$; $\mathcal{B}$ *is* maximum *if* $|V_B|$ *is the largest possible. Given* $D \in \{EA, LD, MT, FT, ST\}$ *and* $\mathcal{B}$ *a* TIB *of* $\mathcal{G}$ *with root* r, *we say that* $\mathcal{B}$ *is a* D-TIB *of* $\mathcal{G}$ *if* $\forall v \in V_B$, $D_B(v, r) = D_G(v, r)$. $\mathcal{B}$ *is called a* spanning D-TIB *if* $V_B = V$; *a* maximum D-TIB *if* $|V_B|$ *is the largest possible.*

Problem 2 (Maximum D-TIB). Let $D \in \{EA, LD, MT, ST, FT\}$ and $\mathcal{G}$ be a temporal graph. Find a maximum D-TIB of $\mathcal{G}$.

The next proposition shows that finding maximum TIBs can be reduced to finding maximum TOBs in an auxiliary temporal graph. We define the *reversal* of a temporal graph $\mathcal{G} = (V, A, \tau)$ as the temporal graph $\mathcal{G}^\circlearrowleft = (V, A^\circlearrowleft, \tau)$ where the order of the timesteps is reversed as well as the direction of the arcs. Formally, $A^\circlearrowleft = \{(h(a), t(a), \tau - t_a(a) + 1, \tau - t_s(a) + 1) : a \in A\} := \{a^\circlearrowleft : a \in A\}$. A similar transformation has been used in [3].

Proposition 1. *Given a temporal graph $\mathcal{G}$, it holds that:*

- $\mathcal{B}$ *is a maximum* EA-TIB *of* $\mathcal{G}$ *iff* $\mathcal{B}^{\circlearrowleft}$ *is a maximum* LD-TOB *of* $\mathcal{G}^{\circlearrowleft}$;
- $\mathcal{B}$ *is a maximum* LD-TIB *of* $\mathcal{G}$ *iff* $\mathcal{B}^{\circlearrowleft}$ *is a maximum* EA-TOB *of* $\mathcal{G}^{\circlearrowleft}$;
- *For each* D $\in \{$MT, ST, FT$\}$, $\mathcal{B}$ *is a maximum* D-TIB *of* $\mathcal{G}$ *iff* $\mathcal{B}^{\circlearrowleft}$ *is a maximum* D-TOB *of* $\mathcal{G}^{\circlearrowleft}$.

4   Computing Maximum D-TOBs for D $\in \{$MT, ST, LD$\}$

The following concept allows us to establish a necessary and sufficient condition for the existence of a spanning D-TOB with root r in a temporal graph.

Definition 6. *Let $\mathcal{G}$ be a temporal graph and W be a temporal (u, v)-walk in $\mathcal{G}$. For every* D $\in \{$LD, MT, FT, ST$\}$ *we say that:*

- W *is* D*-prefix-optimal if* $\forall x \in V(W)$, *any x-prefix of W realizes* $\mathrm{D}_{\mathcal{G}}(u, x)$;
- W *is* EAD*-prefix-optimal if it is* D*-prefix-optimal and* $\forall x \in V(W)$, *any x-prefix of W realizes* $\mathrm{EAD}_{\mathcal{G}}(u, x)$.

Theorem 1. *Let $\mathcal{G} = (V, A, \tau)$ be a temporal graph, $r \in V$ and* D $\in \{$LD, MT, ST$\}$. *Then $\mathcal{G}$ admits a spanning* D-TOB *with root r if and only if for all $v \in V$, there exists a* D*-prefix-optimal temporal (r, v)-path in $\mathcal{G}$.*

Notice that Theorem 1 does not hold for D = FT. Indeed the temporal graph in Figure (3b) has an FT-prefix-optimal path from r to any other vertex, but does not admit a spanning FT-TOB as previously observed. We are now ready to characterize the vertex set of a maximum D-TOB.

Corollary 1. *Let $\mathcal{G} = (V, A, \tau)$ be a temporal graph, $r \in V$, and* D $\in \{$LD, MT, ST$\}$. *Then a maximum* D-TOB *with root r of $\mathcal{G}$ has vertex set:*

$$V_{\mathcal{B}} = \{v \in V : \text{there exists a D-prefix-optimal } (r, v)\text{-path in } \mathcal{G}\}. \qquad (1)$$

Proof. Consider $\mathcal{G}[V_{\mathcal{B}}]$. Let $v \in V_{\mathcal{B}}$ and W a D-prefix-optimal (r, v)-temporal walk in $\mathcal{G}$. By definition of D-prefix-optimal walk, $\forall u \in V(W)$, $u \in V_{\mathcal{B}}$, so W is also a D-prefix-optimal walk in $\mathcal{G}[V_{\mathcal{B}}]$. Hence by Theorem 1, $\mathcal{G}[V_{\mathcal{B}}]$ admits a spanning D-TOB $\mathcal{B}$, which is a D-TOB of $\mathcal{G}$. We now show that $\mathcal{B}$ is maximum. It suffices to prove that if $V' \subseteq V$ is such that $V' \setminus V_{\mathcal{B}} \neq \emptyset$, then $\mathcal{G}[V']$ does not admit a spanning D-TOB with root r. Let $u \in V' \setminus V_{\mathcal{B}}$. By hypothesis there does not exist a D-prefix-optimal temporal (r, u)-walk in $\mathcal{G}$, hence there does not exist one in $\mathcal{G}[V']$. Thus, by Theorem 1, $\mathcal{G}[V']$ does not admit a spanning D-TOB. $\square$

Corollary 1 shows that, even if v is temporally reachable from r, if none of the walks that realize $\mathrm{D}(r, v)$ is D-prefix-optimal, then no D-TOB can span v. The next sections present algorithms for finding D-TOBs of a given temporal graph in polynomial time when D $\in \{$MT, ST, LD$\}$. In particular, we show that these algorithms always return an EAD-TOB. This implies that for

$D \in \{MT, ST, LD\}$, the existence of a D-prefix-optimal (r,v)-path in $\mathcal{G}$ is equivalent to the existence of an EAD-prefix-optimal (r,v)-path in $\mathcal{G}$. For $D = FT$ this is no longer true: indeed consider Figure (3b). The only FT-prefix-optimal (r,y)-path is $W = (r,(r,v,2),v,(v,y,2),y)$, but it is not EAFT-prefix-optimal: in fact, $EAFT(r,v) = 1$ since the path $(r,(r,v,1),v)$ realizes $FT(r,v)$ and arrives in v at time 1, while W arrives in v at time 2. This difference will be crucial for showing that computing an FT-TOB is an NP-complete problem (Sect. 5).

4.1   Algorithm for MT

A maximum MT-TOB of a temporal graph can be computed in polynomial time. Due to space constraints, we do not report the whole algorithm here, while providing an informal description of it instead.

Theorem 2. *A maximum EAMT-TOB of a temporal graph, for a chosen root, can be computed in $O(m \log n)$ time. In particular, a maximum MT-TOB can be computed in $O(m \log n)$ time.*

First observe that, given an MT-prefix-optimal temporal (r,v)-walk $W = (r = v_0, a_1, v_1, \ldots, a_k, v_k = v)$, we have that $MT(r,v_i) = MT(r,v_{i+1}) - 1 < MT(r,v_{i+1})$ for all $i \in [k-1]$, i.e. the sequence of distances in any MT-prefix-optimal walk is strictly increasing. The main idea of the algorithm is to compute a priori the MT-distances of all vertices from the root, and then build the MT-TOB guided by these computed distances, using their strict monotonicity property. More specifically, given $h = \max\{MT(r,v) : v \in V\}$, the algorithm grows an MT-TOB starting from the root and adding, at step $i \in [h]$, all the vertices at distance i. During this process, when adding some vertex v, we choose, among its neighbors at distance $i - 1$, which one can be the parent of v. To choose the right parent, we look at the incoming temporal arcs having tail in vertices at distance $i - 1$ and we consider only the arcs $a' = (u', v, s', t')$ such that, if $W_{u'}$ is the unique temporal (r, u')-path in the MT-TOB built so far, then $s' \geq t_a(W_{u'})$, i.e. the new arc can be concatenated with $W_{u'}$ to obtain a temporal (r,v)-path. Among the arcs fulfilling these constraints, we choose a' minimizing t', the arrival time in v; such arc a' exists if and only if there exists an MT-prefix-optimal (r,v)-path in $\mathcal{G}$. We prove that such choice of a' ensures that we are actually representing in the TOB a temporal (r,v)-path that realizes the distance $MT(r,v)$ and has the earliest arrival time among the paths realizing such distance, i.e. we are computing an EAMT-TOB. The algorithm takes $O(m \log n)$ time to compute all the initial MT distances (see Table 1), while the remaining part of the algorithm takes $O(m)$ time as it requires only one scan of each temporal arc.

4.2   Algorithm for LD and ST

Algorithm 1 computes a maximum D-TOB $\mathcal{B}$ with root r for a given temporal graph when $D \in \{LD, ST\}$.   The issue is that if $W = (r = v_0, a_1, \ldots, a_k, v_k = v)$ is a D-prefix-optimal walk, then it is possible to have $D(r, v_{i-1}) = D(r, v_i)$ for some

---

Algorithm 1: Computing a maximum D-TOB, with D $\in \{$LD, ST$\}$.

Input: A temporal graph $\mathcal{G} = (V, A, \tau)$, a vertex $r \in V$, D $\in \{$LD, ST$\}$.
Output: A maximum D-TOB $\mathcal{B} = (V_{\mathcal{B}}, A_{\mathcal{B}}, \tau_{\mathcal{B}})$ of $\mathcal{G}$ with root r.

1 $\mathcal{EA}(r) \leftarrow 0; \forall v \in V \setminus \{r\}, \mathcal{EA}(v) \leftarrow +\infty$;
2 $d(r) \leftarrow 0; \forall v \in V \setminus \{r\}, d(v) \leftarrow$ D$_{\mathcal{G}}(r, v)$;
3 $\langle d_1, \dots, d_h \rangle \leftarrow$ list of the elements of the set $\{d(v) : v \in V, d(v) < +\infty\}$ in increasing order;
4 $V_{\mathcal{B}} \leftarrow \{r\}; A_{\mathcal{B}} \leftarrow \emptyset; \tau_{\mathcal{B}} \leftarrow 0, D_0 \leftarrow \{r\}$;
5 **for** $i = 1, \dots, h$ **do**
6      $D_i \leftarrow \{v \in V \setminus \{r\} : d(v) = d_i\}$;
7   **if** D = LD **then** enqueue all $(r, v, s, t) \in A$ such that $s = d_i$ in a min priority queue Q with weight t;
8   **if** D = ST **then** enqueue all $(u, v, s, t) \in A$ such that $u \in D_0 \cup \dots \cup D_{i-1}$ and $v \in D_i$ in a min priority queue Q with weight t;
9   **while** $Q \neq \emptyset$ **do**
10   dequeue $a \leftarrow (u, v, s, t)$ from Q;
11   **while** $s < \mathcal{EA}(u)$ *or* $t \geq \mathcal{EA}(v)$ *or* (D = ST *and* $t - s \neq d_i - d(u)$) **do**
12   **if** $Q = \emptyset$ **then** go to Line 5 with next value of i;
13   dequeue $a \leftarrow (u, v, s, t)$ from Q;
14   **end**
        /* $a = (u, v, s, t)$ is s.t. $a \in \arg\min_{(u', v', s', t') \in Q} t'$, $s \geq \mathcal{EA}(u)$, $t < \mathcal{EA}(v) = +\infty$, and if D = ST, $t - s = d_i - d(u)$.   */
15          $\mathcal{EA}(v) \leftarrow t, V_{\mathcal{B}} \leftarrow V_{\mathcal{B}} \cup \{v\}, A_{\mathcal{B}} \leftarrow A_{\mathcal{B}} \cup \{a\}, \tau_{\mathcal{B}} \leftarrow \max\{\tau_{\mathcal{B}}, t\}$;
16   enqueue all $(v, v', s', t') \in A$ such that $v' \in D_i$ in Q with weight t';
17   **end**
18 **end**

$i \in [k]$. Indeed, if D = LD, then all the vertices in the walk share the same latest departure time, i.e. $t_s(W) = $ LD(r, v_i) for all $i \in [k]$. If D = ST and $el(a_i) = 0$ for some $i \in [k]$, then ST$(r, v_{i-1}) = $ ST(r, v_i). However, in any case we have that D$(r, v_{i-1}) \leq$ D(r, v_i) for all $i \in [k]$. This implies that, by letting D_i denote the set of vertices at distance d_i from r with the distances d_i being in increasing order, to choose the parent of each vertex of D_i in $\mathcal{B}$, we cannot look only at vertices in $D_0 \cup \dots \cup D_{i-1}$, but also at the ones in D_i itself (in particular, only at the ones in D_i when D = LD). Note that this gives us an additional difficulty as we cannot simply choose an arbitrary vertex $v \in D_i$ to be the next one to be added to $\mathcal{B}$, as it might happen that the good parent of v (i.e. the in-neighbor of v within an EAD-prefix-optimal (r, v)-walk) has not been added to $\mathcal{B}$ yet. To overcome this, we add vertices in D_i to $\mathcal{B}$ in increasing order of the value of EAD(r, v). Observe however that EAD(r, v) is not known a priori, so to do that we use a queue that keeps the outgoing temporal arcs from vertices in $\mathcal{B}$ in increasing order of their arrival time. These ideas are formalized below. At step i of the **for** loop at lines 5–18, Algorithm 1 adds to $\mathcal{B}$ the vertices of D_i that are reachable by a D-prefix-optimal walk. To this aim, it uses a min priority queue Q for temporal arcs a with head vertices in D_i with weight $t_a(a)$. For D = LD, Q is initialized with all the outgoing temporal arcs from r with starting time d_i, as they are the only arcs that can realize a latest departure time equal to d_i. For $D = \text{ST}$, Q is initialized with all the temporal arcs with tail in $D_0 \cup \ldots \cup D_{i-1}$ and head in D_i. The vector $\mathcal{E}\mathcal{A}$ in the algorithm, initialized at $+\infty$ for all the vertices but the root, keeps track of the arrival time in the vertices every time they are added to the TOB. In the **while** loop at lines 9–17, we dequeue temporal arcs from Q that cannot possibly be within an EAD-prefix-optimal walk. Formally, if such loop is not broken in line 12, then at the end we are left with an arc $a = (u, v, s, t) \in \arg\min_{(u',v',s',t')\in Q} t'$, i.e. an arc that minimizes the arrival time in the queue, satisfying:

- $s \geq \mathcal{E}\mathcal{A}(u)$, so that a is temporally compatible with the temporal (r, u)-walk W_u that is already present in the TOB, i.e. $W_u + (u, a, v)$ is a temporal walk;
- $t < \mathcal{E}\mathcal{A}(v)$, which ensures that we add to the TOB a new vertex each time;
- $t - s = d_i - d(u)$ if $D = \text{ST}$, ensuring that $W_u + (u, a, v)$ realizes $\text{ST}(r, v)$.

We then add v and the temporal arc a to the TOB and we update the arrival time in v to $\mathcal{E}\mathcal{A}(v) = t$, which is equal to $\text{EAD}(r, v)$ and will be no longer updated until the end of the algorithm. Finally, we add to Q all the outgoing arcs from v with head vertices in D_i. When at distance d_i there are no arcs satisfying these constraints, i.e. the queue Q at line 12 is empty, we go to the next distance d_{i+1}, as it means that we have already spanned all the possible vertices in D_i. The initial computation of all $D(r, v)$ requires $O(m \log m)$ by Table 1. Concerning the remaining part of the algorithm, the i-th iteration of the **for** loop considers only arcs whose head is in D_i, hence each arc is considered only in one of the iterations of the **for** loop. Moreover, each arc is dequeued from Q at most once. As the dequeue from Q costs $O(\log m)$ we obtain a running time of $O(m \log m)$.

Theorem 3. *For any* $D \in \{\text{LD}, \text{ST}\}$, *Algorithm 1 returns a maximum* D-TOB *of a temporal graph, for a chosen root, in* $O(m \log m)$ *time. Besides, the output is an* EAD-TOB.

5   Computing Maximum FT-TOBs

As previously observed, Theorem 1 does not hold for $D = \text{FT}$. Indeed for FT the problem becomes NP-complete even in the following very constrained situations: when $el(a) = 0$ for all $a \in A$, also called *nonstrict* temporal graphs, and when $el(a) = 1$ for all $a \in A$, also called *strict* temporal graphs (see e.g. [5,8]). The nonstrict model is used when the time-scale of the measured phenomenon is relatively big: this is the case in a disease-spreading scenario [22] where the spreading speed might be unclear, or in time-varying graphs [17], where a single snapshot corresponds e.g. to all the streets available within a day.[4]

---

[4] The literature often focused on nonstrict/strict variations to provide stronger negative results. In this paper, we have used the more general model to provide stronger positive results, while using the nonstrict/strict when providing negative ones.

Theorem 4. *Let $\mathcal{G} = (V, A, \tau)$ be a temporal graph and $r \in V$. Deciding whether $\mathcal{G}$ admits a spanning FT-TOB with root r is NP-complete, even if $\tau = 2$ and $el(a) = 0$ for every $a \in A$, or if $\tau = 3$ and $el(a) = 1$ for every $a \in A$.*

Sketch of the proof. The problem is in NP, since computing $\mathrm{FT}_\mathcal{G}(r, v)$ for every vertex v can be done in polynomial time (Table 1), and because testing whether a given temporal subgraph $\mathcal{B}$ is a TOB can be done in polynomial time. To prove hardness, we make a reduction from 3-SAT, largely known to be NP-complete [6,15]. For this, consider a formula ϕ in CNF form on variables $X = \{x_1, \ldots, x_n\}$ and on clauses $C = \{c_1, \ldots, c_m\}$. We first construct $\mathcal{G} = (V, A, \tau)$ for the case where every arc has elapsed time 0 (observe Figure (4a) to follow the construction). First, let $V = X \cup C \cup \{r\}$. For each variable x_i, add to A the temporal arcs $(r, x_i, 1, 1)$ and $(r, x_i, 2, 2)$. Then, for each clause c_j and each variable x_i appearing in c_j, add temporal arc $(x_i, c_j, 1, 1)$ if x_i appears in c_j positively, while add the temporal arc $(x_i, c_j, 2, 2)$ if x_i appears in c_j negatively. It is possible to prove that ϕ is satisfiable if and only if there exists a spanning FT-TOB rooted in r. In the case where $el(a) = 1$ for every arc a, the reduction is similar to the previous one. Specifically, for each $x_i \in X$, we add arcs $(r, x_i, 1, 2)$ and $(r, x_i, 2, 3)$. For each clause c_j, if x_i appears positively in c_j we add the temporal arc $(x_i, c_j, 2, 3)$, while if x_i appears negatively in c_j we add the temporal arc $(x_i, c_j, 3, 4)$, see also Figure (4b). Similar correctness proof applies. $\square$

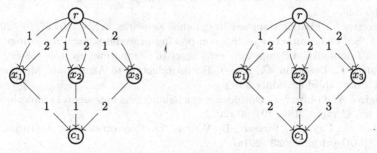

(a) All temporal arcs have elapsed time 0. (b) All temporal arcs have elapsed time 1.

Fig. 4. Example of the construction in the proof of Theorem 4. Clause c_1 is equal to $(x_1 \vee x_2 \vee \neg x_3)$. The value on top of each arc represents the starting time.

The gaps left by the above theorem are when $\tau = 1$ or when $\tau = 2$ and $el(a) \geq 1$ for all $a \in A$. In the former case, the temporal graph reduces to a static graph, so the problem is solvable in polynomial time by Dijkstra's algorithm. In the latter case, one can see that the maximum FT-TOB rooted in r contains exactly r and every $u \in V$ such that $(r, u, 1, 2)$ is an arc in $\mathcal{G}$.

6   Conclusions and Future Work

We have showed that for $\mathrm{D} \in \{\mathrm{MT}, \mathrm{ST}, \mathrm{LD}\}$, a spanning D-TOB does not always exist, but computing a D-TOB that spans the maximal number of vertices is polynomial-time. When D=FT, also finding a maximum FT-TOB becomes NP-complete. The fact that not all the vertices can be spanned by a maximum D-TOB could be an issue, for example, in a public transports setting, where we still want to reach all possible places. A natural follow-up of our work would be to relax the definition of spanning D-TOB, by asking to find a subgraph that reaches all the vertices from the root with a path realizing the distance, while having the least amount of arcs possible. Preliminary results suggest that this might become a much harder problem.

References

1. Himmel, A.-S., Bentert, M., Nichterlein, A., Niedermeier, R.: Efficient computation of optimal temporal walks under waiting-time constraints. In: Cherifi, H., Gaito, S., Mendes, J.F., Moro, E., Rocha, L.M. (eds.) COMPLEX NETWORKS 2019. SCI, vol. 882, pp. 494–506. Springer, Cham (2020). https://doi.org/10.1007/978-3-030-36683-4_40

2. Brunelli, F., Viennot, L.: Minimum-cost temporal walks under waiting-time constraints in linear time. arXiv:2211.12136 (2023)

3. Calamai, M., Crescenzi, P., Marino, A.: On computing the diameter of (weighted) link streams. ACM J. Exp. Algorithmics 27, 4.3:1–4.3:28 (2022)

4. Campos, V., Lopes, R., Marino, A., Silva, A.: Edge-disjoint branchings in temporal graphs. Electronic J. Combinatorics 28 (2020). https://doi.org/10.1007/978-3-030-48966-3_9

5. Casteigts, A.: Finding structure in dynamic networks. arXiv:1807.07801 (2018)

6. Cook, S.: The complexity of theorem-proving procedures. In: Proceedings of the Third Annual ACM Symposium on Theory of Computing, pp. 151–158 (1971)

7. Cormen, T., Leiserson, C., Rivest, R.: Introduction to Algorithms. McGraw-Hill, MIT Press, third ed. edn. (2001)

8. Deligkas, A., Potapov, I.: Optimizing reachability sets in temporal graphs by delaying. Inf. Comput. 285, 104890 (2022)

9. Dibbelt, J., Pajor, T., Strasser, B., Wagner, D.: Connection scan algorithm. ACM J. Exp. Algorithmics 23 (2018)

10. Holme, P., Saramäki, J.: Temporal networks. Phys. Rep. 519(3), 97–125 (2012)

11. Huang, S., Fu, A.W.C., Liu, R.: Minimum spanning trees in temporal graphs. In: ACM SIGMOD International Conference on Management of Data, pp. 419–430 (2015)

12. Kamiyama, N., Kawase, Y.: On packing arborescences in temporal networks. Inf. Process. Lett. 115(2), 321–325 (2015)

13. Kuwata, Y., Blackmore, L., Wolf, M., Fathpour, N., Newman, C., Elfes, A.: Decomposition algorithm for global reachability analysis on a time-varying graph with an application to planetary exploration. In: Intelligent Robot and System, pp. 3955–3960 (2009)

14. Latapy, M., Viard, T., Magnien, C.: Stream graphs and link streams for the modeling of interactions over time. Soc. Netw. Anal. 8(1), 611–6129 (2018)

15. Levin, L.: Universal sequential search problems. Problemy peredachi informatsii 9(3), 115–116 (1973)

16. Marino, A., Silva, A.: Eulerian walks in temporal graphs. Algorithmica 85, 805–830 (2023)

17. Nicosia, V., Tang, J., Musolesi, M., Russo, G., Mascolo, C., Latora, V.: Components in time-varying graphs. Chaos: Interdisc. J. Nonlinear Sci. **22**(2) (2012)
18. Ranshous, S., Shen, S., Koutra, D., Harenberg, S., Faloutsos, C., Samatova, N.: Anomaly detection in dynamic networks: a survey. WIREs Comput. Stat. **7**(3), 223–247 (2015)
19. Tang, J.K., Mascolo, C., Musolesi, M., Latora, V.: Exploiting temporal complex network metrics in mobile malware containment. In: 2011 IEEE International Symposium on a World of Wireless, Mobile and Multimedia Networks, pp. 1–9 (2010)
20. Wu, H., Cheng, J., Huang, S., Ke, Y., Lu, Y., Xu, Y.: Path problems in temporal graphs. Proc. VLDB Endow. **7**(9), 721–732 (2014)
21. Wu, H., Cheng, J., Ke, Y., Huang, S., Huang, Y., Wu, H.: Efficient algorithms for temporal path computation. Knowl. Data Eng. **28**(11), 2927–2942 (2016)
22. Zschoche, P., Fluschnik, T., Molter, H., Niedermeier, R.: The complexity of finding small separators in temporal graphs. J. Comp. Syst. Sci. **107**, 72–92 (2020)

Contracting Edges to Destroy a Pattern: A Complexity Study

Dipayan Chakraborty[1,2] and R. B. Sandeep[3(✉)]

[1] LIMOS, Université Clermont Auvergne, Aubière, France
dipayan.chakraborty@uca.fr
[2] Department of Mathematics and Applied Mathematics,
University of Johannesburg, Johannesburg, South Africa
[3] Department of Computer Science and Engineering, Indian Institute of Technology
Dharwad, Dharwad, India
sandeeprb@iitdh.ac.in

Abstract. Given a graph G and an integer k, the objective of the Π-CONTRACTION problem is to check whether there exists at most k edges in G such that contracting them in G results in a graph satisfying the property Π. We investigate the problem where Π is 'H-free' (without any induced copies of H). It is trivial that H-FREE CONTRACTION is polynomial-time solvable if H is a complete graph of at most two vertices. We prove that, in all other cases, the problem is NP-complete. We then investigate the fixed-parameter tractability of these problems. We prove that whenever H is a tree, except for seven trees, H-FREE CONTRACTION is W[2]-hard. This result along with the known results leaves behind only three unknown cases among trees.

Keywords: Edge contraction problem · H-free · NP-completeness · W[2]-hardness · Trees

1 Introduction

Let Π be any graph property. Given a graph G and an integer k, the objective of the Π-CONTRACTION problem is to check whether G contains at most k edges so that contracting them results in a graph with property Π. This is a vertex partitioning problem in disguise: Find whether there is a partition $\mathcal{P}$ of the vertices of G such that each set in $\mathcal{P}$ induces a connected subgraph of G, $G/\mathcal{P}$ (the graph obtained by contracting each set in $\mathcal{P}$ into a vertex) has property

This work is partly sponsored by SERB (India) grants "Complexity dichotomies for graph modification problems" (SRG/2019/002276), and "Algorithmic study on hereditary graph properties" (MTR/2022/000692), and a public grant overseen by the French National Research Agency as part of the "Investissements d'Avenir" through the IMobS3 Laboratory of Excellence (ANR-10-LABX-0016), and the IDEX-ISITE initiative CAP 20-25 (ANR-16-IDEX-0001). We also acknowledge support of the ANR project GRALMECO (ANR-21-CE48-0004).

H. Fernau and K. Jansen (Eds.): FCT 2023, LNCS 14292, pp. 118–131, 2023.
https://doi.org/10.1007/978-3-031-43587-4_9

Π, and $n - |\mathcal{P}| \leq k$. These problems, for various graph properties Π, have been studied for the last four decades. Asano and Hirata [2] proved that the problem is NP-complete if Π is any of the following classes - planar, series-parallel, outerplanar, chordal. When Π is a singleton set $\{H\}$, then the problem is known as H-CONTRACTION. Brouwer and Veldman [5] proved that H-CONTRACTION is polynomial-time solvable if H is a star, and NP-complete if H is a connected triangle-free graph other than a star graph. Belmonte, Heggernes, and van 't Hof [3] proved that it is polynomial-time solvable when H is a split graph. Golovach, Kaminski, Paulusma, and Thilikos [13] studied the problem when Π is 'minimum degree at least d' and proved that the problem is NP-complete even for $d = 14$ and W[1]-hard when parameterized by k. Heggernes, van 't Hof, Lokshtanov, and Paul [16] proved that the problem is fixed parameter tractable when Π is the class of bipartite graphs. Guillemot and Marx [14] obtained a faster FPT algorithm for the problem. Cai, Guo [8], and Lokshtanov, Misra, and Saurabh [20] proved that the problem is W[2]-hard when Π is the class of chordal graphs. Garey and Johnson [11] mentioned that, given two graphs G and H, the problem of checking whether H can be obtained from G by edge contractions is NP-complete. Edge contraction has applications in Graph minor theory (see [21]), Hamiltonian graph theory [17], and geometric model simplification [12].

We consider the H-FREE CONTRACTION problem: Given a graph G and an integer k, find whether G can be transformed, by at most k edge contractions, into a graph without any induced copies of H. The parameter we consider is k. Unlike graph contraction problems, other major graph modification problems are well-understood for these target graph classes. In particular, P versus NP-complete dichotomies are known for H-FREE EDGE EDITING, H-FREE EDGE DELETION, H-FREE EDGE COMPLETION [1], and H-FREE VERTEX DELETION [19] (here, the allowed operations are edge editing, edge deletion, edge completion, and vertex deletion respectively). It is also known that all these problems are in FPT for every graph H [6]. The picture is far from complete for H-FREE CONTRACTION. See Table 1. It is trivial to note that H-FREE CONTRACTION is polynomial-time solvable if H is a complete graph of at most 2 vertices. Cai, Guo [8,15], and Lokshtanov, Misra, and Saurabh [20] proved the following results for H-FREE CONTRACTION.

- FPT when H is a complete graph
- If H is a path or a cycle, then the problem is FPT when H has at most 3 edges, and W[2]-hard otherwise.
- W[2]-hard when H is 3-connected but not complete, or a star graph on at least 5 vertices, or a diamond.

The W[2]-hardness results mentioned above also imply NP-completeness of the problems. Guo [15] proved that the problem is NP-complete when H is a complete graph on t vertices, for every $t \geq 3$. Eppstein [10] proved that the Hadwiger number problem (find whether the size of a largest clique minor of a graph is at least k) is NP-complete. This problem is essentially $2K_1$-FREE CONTRACTION, if we ignore the parameter. This result implies that the problem is NP-complete when H is a P_3. We build on these results and prove the following.

Table 1. Complexities of various graph modification problems where the target property is H-free. The number of vertices, the number of edges, and the number of nonedges in H are denoted by n, m, m' respectively.

Problem	P	NPC	FPT	W-hard
Edge Editing	$n \leq 2$ [trivial]	otherwise [1]	For all H [6]	
Edge Deletion	$m \leq 1$ [trivial]	otherwise [1]	For all H [6]	
Edge Completion	$m' \leq 1$ [trivial]	otherwise [1]	For all H [6]	
Vertex Deletion	$n \leq 1$ [trivial]	otherwise [19]	For all H [6]	
Edge Contraction	K_1, K_2 [trivial]	otherwise [**Theorem 1**]	K_t $(t \geq 3)$ [15,20], $P_3, P_4, K_2 + K_1$ ([8,20], MSO_1 expressibility)	W[2]-hard for 3-connected non-complete graphs, diamond [8,15], C_t $(t \geq 4)$ [8,20], all trees except 7 trees [**Theorem 2**]

- H-FREE CONTRACTION is NP-complete if H is not a complete graph on at most 2 vertices.
- H-FREE CONTRACTION is W[2]-hard if H is a tree which is neither a star on at most 4 vertices (o ,⊙ ,⊙⊙ ,⊙⊙⊙) nor a bistar in {⊙⊙ ,⊙⊙⊙ ,⊙⊙⊙⊙ }.

Our W[2]-hardness results, along with known positive results, leaves behind only three open cases among trees - ⊙⊙⊙ , ⊙⊙⊙ , ⊙⊙⊙⊙ .

Due to space constraints, many proofs are moved to a full version of the paper.

2 Preliminaries

Graphs. All graphs considered in this paper are simple and undirected. A complete graph and a path on t vertices are denoted by K_t and P_t respectively. A *universal vertex* of a graph is a vertex adjacent to every other vertex of the graph. An *isolated* vertex is a vertex with degree 0. For an integer $t \geq 0$, a *star* on $t + 1$ vertices, denoted by $K_{1,t}$, is a tree with a single universal vertex and t degree-1 vertices. The universal vertex in $K_{1,t}$ is also called the *center* of the star. For integers t, t' such that $t \geq t' \geq 0$, a *bistar* on $t + t' + 2$ vertices, denoted by $T_{t,t'}$, is a tree with two adjacent vertices v and v', where t degree-1 vertices are attached to v and t' degree-1 vertices are attached to v'. The bistar $T_{t,0}$ is the star $K_{1,t+1}$ and the bistar $T_{1,1}$ is P_4. We say that vv' is the *central edge* of the bistar. By $G_1 + G_2$ we denote the disjoint union of the graphs G_1 and G_2. A graph is H-free if it does not contain any induced copies of H. In a graph G, *replacing a vertex* v with a graph H is the graph obtained from G by removing v, introducing a copy of H, and adding edges between every vertex of the H and every neighbor of v in G. A *separator* S of a connected graph G is a subset of its vertices such that $G - S$ (the graph obtained from G by removing the vertices in S) is disconnected. A separator is *universal* if every vertex of the separator is adjacent to every vertex outside the separator. We will be using the term 'universal K_1 (resp. K_2) separator' to denote a universal separator which induces a K_1 (resp. K_2). Let V' be a subset of vertices of a graph H. By $H[V']$ we denote the graph induced by V' in H. For a graph G and two subsets A and B of vertices of G, by $E[A, B]$ we denote the set of edges in G, where each edge in the set is having one end point in A and the other end point in B.

Contraction. *Contracting* an edge uv in a graph G is the operation in which the vertices u and v are identified to be a new vertex w such that w is adjacent to every vertex adjacent to either u or v. Given a graph G and a subset F of edges of G, the graph G/F obtained by contracting the edges in F does not depend on the order in which the edges are contracted. Every vertex w in G/F represents a subset W of vertices (which are contracted to w) of G such that W induces a connected graph in G. Let G_F be the subgraph of G containing all vertices of G and the edges in F. There is a partition $\mathcal{P}$ of vertices of G implied by F: Every set in $\mathcal{P}$ corresponds to the vertices of a connected component in G_F. We note that many subsets of edges may imply the same partition - it does not matter which all edges of a connected subgraph are contracted to get a single vertex. The graph G/F is nothing but the graph in which there is a vertex corresponding to every set in $\mathcal{P}$ and two vertices in G/F are adjacent if and only if there is at least one edge in G between the corresponding sets in $\mathcal{P}$. The graph G/F is equivalently denoted by $G/\mathcal{P}$. Assume that $\mathcal{P}'$ is a partition of a subset of vertices of G. Then by $G/\mathcal{P}'$ we denote the graph obtained from G by contracting each set in $\mathcal{P}'$ into a single vertex. The *cost* of a set P in $\mathcal{P}$ is the number $|P| - 1$, which is equal to the minimum number of edges required to *form* the set P. The cost of $\mathcal{P}$, denoted by $\text{cost}(\mathcal{P})$, is the sum of costs of the sets in $\mathcal{P}$. We observe that $|\mathcal{P}| + \text{cost}(\mathcal{P}) = n$, where n is the number of vertices of G. We say that F *touches* a subset W of vertices of G, if there is at least one edge uv in F such that either u or v is in W. Let u, v be two non-adjacent vertices of G. *Identifying* u and v in G is the operation of removing u and v, adding a new vertex w, and making w adjacent to every vertex adjacent to either u or v.

Fixed Parameter (in)tractability. A parameterized problem is fixed-parameter tractable (FPT) if it can be solved in time $f(k)|I|^{O(1)}$-time, where f is a computable function and (I, k) is the input. Parameterized problems fall into different levels of complexities which are captured by the W-hierarchy. A parameterized reduction from a parameterized problem Q' to a parameterized problem Q is an algorithm which takes as input an instance (I', k') of Q' and outputs an instance (I, k) of Q such that the algorithm runs in time $f(k')|I'|^{O(1)}$ (where f is a computable function), and (I', k') is a yes-instance of Q' if and only if (I, k) is a yes-instance of Q, and $k \leq g(k')$ (for a computable function g). We use parameterized reductions to transfer fixed-parameter intractability. For more details on these topics, we refer to the textbook [9]. The problem that we deal with in this paper is defined as follows.

H-FREE CONTRACTION: Given a graph G and an integer k, can G be modified into an H-free graph by at most k edge contractions?

3   NP-Completeness

In this section we prove that H-FREE CONTRACTION is NP-complete whenever H is not a complete graph of at most two vertices. First we obtain reductions for the cases when H is connected but does not have any universal K_1 separator and universal K_2 separator. Then we deal with non-star graphs with universal K_1 separator or universal K_2 separator. Next we resolve the case of stars. Then the cases when H is a $2K_2$ or a $K_2 + K_1$ are handled. These come as base cases in the inductive proof of the main result of the section. We crucially use the following results.

Proposition 1 ([10,15]). *H-FREE CONTRACTION is NP-complete when H is a $2K_1$, or a P_3, or a K_t, for any $t \geq 3$.*

3.1   A General Reduction

Inspired by a reduction by Asano and Hirata [2], we introduce the following reduction from VERTEX COVER which handles connected graphs H without any universal K_1 separator and universal K_2 separator.

Construction 1. *Let G' be a graph without any isolated vertices, and H be a connected non-complete graph. We obtain a graph G from G' and H as follows.*

- *Subdivide each edge of G' once, i.e., for every edge uv, introduce a new vertex and make it adjacent to both u and v, and delete the edge uv.*
- *Replace each new vertex by a copy H.*
- *Let w be a non-universal vertex of H (the existence of w is guaranteed as H is not a complete graph). Identify w of every copy of H (introduced in the previous step) to be a single vertex named w.*

Let the resultant graph be G. The vertices in G copied from G' form the set V', which forms an independent set in G. For each edge uv in G', the vertices, except w, of the copy of H is denoted by W_{uv}. By W we denote any such set. We note that w is adjacent to every vertex in V', as G' does not have any isolated vertices. This completes the construction. An example is shown in Fig. 1.

Let H be a connected non-complete graph with h vertices and without any universal K_1 or K_2 separator. Let G' be a graph without any isolated vertices. Let G be obtained from (G', H) by Construction 1. Assume that (G', k) is a yes-instance of VERTEX COVER and let T be any vertex cover of size at most k of G'. Let $F = \{wu : u \in T\}$. Let w itself denote the vertex obtained by contracting the edges in F. It can be seen that w is a universal vertex in G/F. If we remove w from G/F, then the resultant graph is a disjoint union of W's

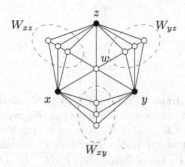

Fig. 1. Construction of G from (G', H) by Construction 1, where G' is a triangle and H is a P_4. The vertices of G' in G are darkened.

and graphs obtained by adding universal vertices to W. This helps us to prove that G/F is H-free. Let F be a solution of (G, k). Then we can create a vertex cover T of G' as follows: If F touches u for some vertex $u \in V'$, then add u to T. If F touches an edge in the graph induced by $W_{uv} \cup \{w\}$, then arbitrarily add either u or v in T. Since H is induced by $W_{uv} \cup \{w\}$ for every uv in G', the graph induced by $W_{uv} \cup \{u, v, w\}$ is touched by F. Therefore, T contains either u or v. Thus we obtain Lemma 1.

Lemma 1. *Let H be a connected non-complete graph with neither a universal K_1 separator nor a universal K_2 separator. Then H-FREE CONTRACTION is NP-complete.*

3.2 Graphs with Universal Clique Separators

Now we handle the graphs H with either a universal K_1 separator (except stars) or with a universal K_2 separator. We note that H cannot have both a universal K_1 separator and a universal K_2 separator. Further H cannot have more than one such separator.

Construction 2. *Let G', H be any graphs and let V' be any subset of vertices of H. Let b, c, k be positive integers. We obtain a graph G from (G', H, V', b, c, k) as follows. For every set S of vertices of G', where S induces a clique on b vertices in G', do the following: Introduce $k + c$ copies of $H[V']$ and make every vertex of the copies adjacent to every vertex of S. Let W_S denote the set of new vertices introduced for S, and let W be the set of all new vertices.*

Let H be a graph with a universal K_1 separator or a universal K_2 separator. Let the set of vertices of the separator be denoted by K. Assume that there exists at least two non-isomorphic components in $H - K$ (therefore, H cannot be a star). Let J be a component in $H - K$ with minimum number of vertices. Let c be the number of times J appears (as a component) in $H - K$. Let V' be the set of vertices of a copy of J. Let H' be the graph obtained from H by removing the vertices of every component isomorphic to J in $H - K$. Let (G', k) be an instance

Fig. 2. Construction of G from (G', H, V', b, c, k) by Construction 2, where G' is a triangle and H is a paw. Since paw has a universal K_1 separator (denote it by K), $b = |K| = 1$. Since we get a K_1 and a K_2 after removing the universal K_1 separator from paw, the smallest component is K_1. Therefore V' contains a single vertex. Since there are only one copy of K_1 (as a component) in $H - K$, $c = 1$. Assume that $k = 1$. The vertices of G' in G are darkened.

of H'-FREE CONTRACTION. Let G be obtained from $(G', H, V', b = |K|, c, k)$ by Construction 2. An example of the construction is shown in Fig. 2.

Let F' be a solution of (G', k). If there is an induced H in G/F' then every vertex in W can only act as a vertex in J in the induced H. Therefore, there will be an induced H' in G'/F', which is a contradiction. Unmanageable number of copies of J attached to the cliques of size b in G' ensures that one has to kill all H's in G' to kill all H in G. This gives us Lemma 2.

Lemma 2. *Let H be a graph with a universal K_1 separator or a universal K_2 separator, denoted by K. Assume that $H - K$ has at least two components which are not isomorphic. Let J be a component in $H - K$ with minimum number of vertices. Let H' be obtained from H by removing all components of $H - K$ isomorphic to J. Then there is a polynomial-time reduction from H'-FREE CONTRACTION to H-FREE CONTRACTION.*

What remains to handle is the case when H has a universal K_1 separator or a universal K_2 separator K such that $H - K$ is a disjoint union of a graph J. The diamond graph is an example. For this we need the concept of an enforcer - a structure to forbid contraction of certain edges. Enforcers are used widely in connection with proving hardness results for edge modification problems (see [7,15,22]). We use enforces to come up with a reduction which gives us Lemma 3.

Lemma 3. *Let H be a graph with a universal K_1 separator or a universal K_2 separator, denoted by K. Assume that $H - K$ is a disjoint union of t copies of a graph J, for some $t \geq 2$. Let H be not a star and let H' be tJ. Then there is a polynomial-time reduction from H'-FREE CONTRACTION to H-FREE CONTRACTION.*

3.3 Stars and Small Graphs

Here, we resolve the cases of $2K_2$, $K_2 + K_1$, and star graphs. We start with a reduction for $2K_2$.

Construction 3. *Let G' be a graph and k be an integer. We obtain a graph G by attaching $k + 1$ pendant vertices, denoted by a set Z_u, to every vertex u in G'. Let Z denote the set of all newly added vertices.*

Let (G', k) be an instance of $2K_1$-FREE CONTRACTION. We obtain an instance (G, k) of $2K_2$-FREE CONTRACTION by applying Construction 3 on (G', k).

Lemma 4. *Let* (G', k) *be a yes-instance of* $2K_1$-FREE CONTRACTION. *Then* (G, k) *is a yes-instance of* $2K_2$-FREE CONTRACTION.

Proof. Let $\mathcal{P}'$ be a partition of vertices of G' such that $\text{cost}(\mathcal{P}') \leq k$ and $G'/\mathcal{P}'$ is H-free. We obtain a partition $\mathcal{P}$ of vertices of G from $\mathcal{P}'$ by introducing singleton sets corresponding to the vertices in Z. Clearly, $\text{cost}(\mathcal{P}) = \text{cost}(\mathcal{P}') \leq k$. Since there is no edge induced by the sets corresponds to vertices in Z, we obtain that if $G/\mathcal{P}$ is not $2K_2$-free, then there is an induced $2K_1$ in $G'/\mathcal{P}'$, which is a contradiction.

Lemma 5. *Let* (G, k) *be a yes-instance of* $2K_2$-FREE CONTRACTION. *Then* (G', k) *is a yes-instance of* $2K_1$-FREE CONTRACTION.

Proof. Let $\mathcal{P}$ be a partition of vertices of G such that $G/\mathcal{P}$ is $2K_2$-free and $\text{cost}(\mathcal{P}) \leq k$. We obtain a partition $\mathcal{P}'$ of vertices of G' as follows: For every set $P \in \mathcal{P}$, include $P \setminus Z$ in $\mathcal{P}'$. Since P induces a connected graph in G, $P \setminus Z$ induces a connected graph in G'. Assume that there is a $2K_1$ induced by $P'_u, P'_v \in \mathcal{P}'$. Let $u \in P'_u$ and $v \in P'_v$. Since there is a set Z_u of $k+1$ pendant vertices attached to u and a set Z_v of $k+1$ pendant vertices attached to v, at least one vertex from Z_u and at least one vertex from Z_v form singleton sets in $\mathcal{P}$. Then, those two sets along with the sets containing P'_u and the set containing P'_v in $\mathcal{P}$ induces a $2K_2$ in $G/\mathcal{P}$, which is a contradiction.

Now, the NP-completeness of $2K_2$-FREE CONTRACTION follows from that of $2K_1$-FREE CONTRACTION (Proposition 1) and Lemmas 4 and 5.

Lemma 6. $2K_2$-FREE CONTRACTION *is NP-complete.*

The hardness of $K_2 + K_1$-FREE CONTRACTION can be proved by a reduction from DOMATIC NUMBER. Domatic number of a graph is the size of a largest set of disjoint dominating sets of the graph, which partitions the vertices of the graph. For example, the domatic number of a complete graph is the number of vertices of it, and that of a star graph is 2. The DOMATIC NUMBER problem is to find whether the domatic number of the input graph is at least k or not. It is known that DOMATIC NUMBER is NP-complete [11] even for various classes of graphs [4, 18]. Recall that $K_2 + K_1$-free graphs are exactly the class of complete multipartite graphs. The reduction that we use is exactly the same as the reduction for the NP-completeness of Hadwiger number problem (which is equivalent to $2K_1$-FREE CONTRACTION) described by Eppstein [10]. The proof requires some adaptation.

Lemma 7. $K_2 + K_1$-FREE CONTRACTION *is NP-complete.*

A reduction from $2K_1$-FREE CONTRACTION resolves the case of star graph of at least 4 vertices.

Lemma 8. *For* $t \geq 3$, $K_{1,t}$-FREE CONTRACTION *is NP-complete.*

3.4 Putting Them Together

Recall that the reduction from VERTEX COVER does not handle disconnected graph. This is the main ingredient that remains to be added to obtain the main result of the section. This turns out to be easy. Guo [15] has a reduction for transferring the hardness of H'-FREE CONTRACTION to H-FREE CONTRACTION, where H' is any component of H.

Proposition 2 ([15]). *Let H be a disconnected graph. Let H' be any component of it. Then there is a polynomial-time reduction from H'-FREE CONTRACTION to H-FREE CONTRACTION.*

Proposition 2 does not help us to prove the hardness when every component of H is either a K_1 or a K_2. But there are simple reductions to handle them.

Lemma 9. *Let H be a disconnected graph with an isolated vertex v. There is a polynomial-time reduction from $(H - v)$-FREE CONTRACTION to H-FREE CONTRACTION.*

Repeated application of Lemma 9 implies that there is a polynomial-time reduction from $2K_1$-FREE CONTRACTION to tK_1-FREE CONTRACTION, for every $t \geq 3$. Then the NP-Completeness of $2K_1$-FREE CONTRACTION (Proposition 1) implies Lemma 10.

Lemma 10. *For every $t \geq 3$, tK_1-FREE CONTRACTION is NP-complete.*

Now, we handle the case when H is a disjoint union of t copies of K_2.

Lemma 11. *Let $H = tK_2$, for any integer $t \geq 3$ and let H' be $(t - 1)K_2$. There is a polynomial-time reduction from H'-FREE CONTRACTION to H-FREE CONTRACTION.*

Repeated application of Lemma 11 and the NP-completeness of $2K_2$-FREE CONTRACTION (Lemma 6) give us the following Lemma.

Lemma 12. *For every $t \geq 3$, tK_2-FREE CONTRACTION is NP-complete.*

Now we are ready to prove the main result of this section.

Theorem 1. *Let H be any graph other than K_1 and K_2. Then H-FREE CONTRACTION is NP-complete.*

Proof. We prove this by induction on n, the number of vertices of H. The base cases are when $n = 2$ and $n = 3$, i.e., when H is $2K_1$ or P_3 or triangle (Proposition 1), or $3K_1$ (Lemma 10), or $K_2 + K_1$ (Lemma 7). Assume that $n \geq 4$.

Let H be a disconnected graph. Assume that H has a component H' with at least three vertices. By Proposition 2, there is a polynomial-time reduction from H'-FREE CONTRACTION to H-FREE CONTRACTION. Then we are done by induction hypothesis. Assume that every component of H is either a K_2 or a K_1. If H has an isolated vertex, then we are done by Lemma 9. If there are no isolated vertex in H, then H is isomorphic to tK_2, for $t \geq 2$. Then we are done by Lemma 12.

Let H be a connected graph. If H is complete, then Proposition 1 is sufficient. Assume that H is non-complete. Therefore, there is a non-universal vertex in H. Assume that H has neither a universal K_1 separator nor a universal K_2 separator. Then we are done by Lemma 1. Assume that H has either a universal K_1 separator or a universal K_2 separator, denoted by K. Further assume that H is not a star. Let $H - K$ has at least two non-isomorphic components. Let J be any component in $H - K$ with least number of vertices. Let H' be obtained from H by removing all copies of J in $H - K$. Then by Lemma 2, there is a polynomial-time reduction from H'-FREE CONTRACTION to H-FREE CONTRACTION, and we are done. Assume that $H - K$ is disjoint union of t copies of a graph J, for some $t \geq 2$. Then Lemma 3 gives us a polynomial-time reduction from tJ-FREE CONTRACTION to H-FREE CONTRACTION. For the last case, assume that H is a star graph. Then we are done by Lemma 8.

4   W[2]-Hardness

In this section, we prove that H-FREE CONTRACTION is W[2]-hard whenever H is a tree, except for 7 trees. All our reductions are from DOMINATING SET, which is well-known to be a W[2]-hard problem. First we obtain a reduction for all trees which are neither stars nor bistars. Then we come up with a reduction for a subset of bistars, a corner case of the same proves the case of stars. Then we come up with a reduction which handles the remaining bistars.

Recall that a dominating set D of a graph G is a subset of vertices of G such that every vertex of G is either in D or adjacent to a vertex in D. The objective of the DOMINATING SET problem is to check whether a graph has a dominating set of size at most k or not.

4.1   A General Reduction for Trees

Here we handle all trees which are neither stars nor bistars. The reduction that we use is an adapted version of a reduction used in [8] (to handle 3-connected graphs) and a reduction used in [20] (to handle cycles).

Construction 4. *Let G', H be graphs and k be an integer. Let $\{v_1, v_2, \ldots, v_n\}$ be the set of vertices of G'. We construct a graph G from (G', H, k) as follows.*

- *Introduce a clique $X = \{x_1, x_2, \ldots, x_n\}$.*
- *Introduce n copies of H denoted by $H_1, H_2, \ldots, H_n$. Let V_i denote the set of vertices of H_i, for $1 \leq i \leq n$.*
- *Let w be any vertex in H. Identify w's of all copies of H. Let the vertex obtained so be denoted by w. Let the remaining vertices in each copy H_i be denoted by W_i, i.e., $W_i \cup \{w\}$ induces H, for $1 \leq i \leq n$.*
- *Make w adjacent to every vertex in X.*

– *Make x_i adjacent to every vertex in W_j if and only if $i = j$ or v_i is adjacent to v_j in G'.*

This completes the construction. An example is shown in Fig. 3.

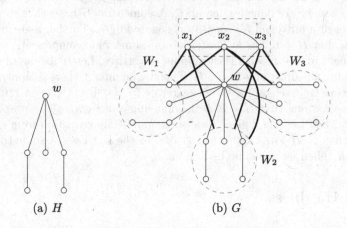

(a) H (b) G

Fig. 3. Construction of G from $(G' = P_3, H)$ by Construction 4

Let (G', k) be an instance of DOMINATING SET. Let H be a tree which is neither a star nor a bistar. We obtain a graph G from (G', H, k) by applying Construction 4.

Lemma 13. *Let (G', k) be a yes-instance of* DOMINATING SET. *Then (G, k) is a yes-instance of H-FREE CONTRACTION.*

Proof. Let D be a dominating set of size at most k of G'. Let $F = \{wx_i | v_i \in D\}$. Clearly, $|F| = |D| \leq k$. We claim that G/F is H-free. Let w itself denote the vertex obtained by contracting the edges in F. To get a contradiction, assume that there is an H induced by a set U of vertices of G/F. We observe that w is a universal vertex in G/F, due to the fact that D is a dominating set of G'. Since H does not contain a universal vertex (a tree has a universal vertex if and only if it is a star), w cannot be in U. Since H is triangle-free, U can have at most two vertices from X. Assume that U has no vertex in X. Then U must be a subset of W_i, which is a contradiction (observe that W_i and W_j are nonadjacent in G/F for $i \neq j$). Assume that U has exactly one vertex, say x_i, from X. Then the rest of the vertices are from W_js adjacent to x_i. Recall that x_i is adjacent to either all or none of vertices in W_j. Since H is triangle-free, the vertices in U from W_js adjacent to x_i form an independent set. Then H is a star, which is a contradiction. If U has exactly two vertices from X, then with similar arguments, we obtain that H is a bistar, which is a contradiction.

Lemma 14. *Let (G, k) be a yes-instance of H-FREE CONTRACTION. Then (G', k) is a yes-instance of* DOMINATING SET.

Proof. Let F' be a subset of edges of G such that G/F' is H-free and $|F'| \leq k$. We construct a new solution F from F' as follows. If F' contains an edge touching W_j, then we replace that edge with the edge wx_j in F. Clearly $|F| \leq |F'| \leq k$. Since an edge touching W_j kills only the H induced by $W_j \cup \{w\}$, which is killed by wx_j, we obtain that G/F is H-free. Let $\mathcal{P}$ be the partition of vertices of G corresponds to F. Let P_w be the set in $\mathcal{P}$ containing w. Let $D = P_w \cap X$. Clearly, $|D| \leq |\text{cost}(\mathcal{P})| \leq k$. We claim that $D' = \{v_i | x_i \in D\}$ is a dominating set of G'. Assume that there is a vertex v_j in G' not dominated by D'. Then $W_j \cup \{w\}$ induces an H in G/F, which is a contradiction.

Lemmas 14 and 13 imply that there is a parameterized reduction from DOM-INATING SET to H-FREE CONTRACTION.

Lemma 15. *Let H be a tree which is neither a star nor a bistar. Then H-FREE CONTRACTION is W[2]-hard.*

4.2 Stars and Bistars

First we generalize a reduction given in [15] for $K_{1,4}$-FREE CONTRACTION. This generalized reduction covers all bistars $T_{t,t'}$ such that $t \geq 3$ and $t > t' \geq 0$. As a boundary case (when $t' = 0$) we obtain hardness result for all stars of at least 5 vertices.

Construction 5. *Let G' be a graph with a vertex set $V' = \{v_1, v_2, \ldots, v_n\}$ and let k, t, t' be integers. We construct a graph G from (G', k, t, t') as follows.*

- *Create two cliques $X = \{x_1, x_2, \ldots, x_n\}, Y = \{y_1, y_2, \ldots, y_n\}$.*
- *Make x_i adjacent to y_j if and only if $i = j$ or $v_i v_j$ is an edge in G'.*
- *Create a vertex w and two cliques A and B of $k + 1$ vertices each.*
- *Make w adjacent to all vertices of $X \cup A \cup B$. Make $A \cup B \cup Y$ a clique.*
- *Introduce $t - 1$ cliques with $k + 1$ vertices each and make them adjacent to A. Let A' denote the set of these vertices.*
- *Introduce $t - 1$ cliques with $k + 1$ vertices each and make them adjacent to B. Let B' denote the set of these vertices.*
- *Introduce $t - 1$ cliques with $k + 1$ vertices each and make them adjacent to X. Let X' denote the set of these vertices.*
- *Introduce a clique of $k + 1$ vertices and make it adjacent to Y. Let Y' denote the set of these vertices.*
- *For every vertex in $A' \cup B' \cup X' \cup Y'$, attach t' degree-1 vertices.*

Let H be $T_{t,t'}$ for $t > t' \geq 0$ and $t \geq 3$. Let (G', k) be an instance of DOMINATING SET. We obtain G from (G', t, t', k) by Construction 5. An example is shown in Fig. 4. Let D be a dominating set of size at most k of G'. Let $F = \{wx_i | v_i \in D\}$. We can prove that G/F is H-free. For the other direction, let F be a minimal subset of edges of G such that G/F is H-free and $|F| \leq k$. We can prove that F does not contain any edges of G other than those from $E[w, X] \cup E[X, X] \cup E[Y, Y]$. Then we can come up with a dominating set of size at most k for G'.

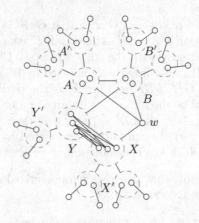

Fig. 4. Construction of G from $(G' = P_3, k = 1, t = 3, t' = 1)$ by Construction 5. Dashed circles denote cliques. This corresponds to the reduction for $T_{3,1}$-FREE CONTRACTION.

Lemma 16. *Let $t > t' \geq 0$ and $t \geq 3$. Then $T_{t,t'}$-FREE CONTRACTION is W[2]-hard.*

Now, we are left with the bistars $T_{t,t'}$ where $t = t'$. For this, we come up with a reduction that handles more than this case and obtain the following Lemma.

Lemma 17. *Let $t \geq t' \geq 3$. Then $T_{t,t'}$-FREE CONTRACTION is W[2]-hard.*

Now, Lemmas 15, 16, and 17 imply the main result of this section.

Theorem 2. *Let T be a tree which is neither a star of at most 4 vertices ($\{K_1, K_2, P_3, K_{1,3}\}$) nor a bistar in $\{T_{1,1}, T_{2,1}, T_{2,2}\}$. Then $T_{t,t'}$-FREE CONTRACTION is W[2]-hard.*

We believe that our W[2]-hardness result on trees will be a stepping stone for an eventual parameterized complexity classification of H-FREE CONTRACTION. The most challenging hurdle for such a complete classification can be the graphs H where each component is of at most 2 vertices, and the case of claw, the usual trouble-maker for other graph modification problems to H-free graphs.

We conclude with some folklore observations. As noted in a version of [20], the property that "there exists at most k edges contracting which results in an H-free graph" can be expressed in MSO_1. The length of the corresponding MSO_1 formula will be a function of k. Then, there exists FPT algorithms for H-FREE CONTRACTION, whenever H-free graphs have bounded rankwidth (See Chap. 7 of the textbook [9]). This, in particular, implies that $K_2 + K_1$-FREE CONTRACTION can be solved in FPT time. It is known that every component of a paw-free graph is either triangle-free or complete multipartite [23], where where paw is the graph . Then the existance of FPT algorithms for K_3-FREE CONTRACTION and $K_2 + K_1$-FREE CONTRACTION imply that there exists an FPT algorithm for *paw*-FREE CONTRACTION.

References

1. Aravind, N.R., Sandeep, R.B., Sivadasan, N.: Dichotomy results on the hardness of H-free edge modification problems. SIAM J. Disc. Math. **31**(1), 542–561 (2017)
2. Asano, T., Hirata, T.: Edge-contraction problems. J. Comput. Syst. Sci. **26**(2), 197–208 (1983)
3. Belmonte, R., Heggernes, P., van't Hof, P.: Edge contractions in subclasses of chordal graphs. Disc. Appl. Math. **160**(7–8), 999–1010 (2012)
4. Bonuccelli, M.A.: Dominating sets and domatic number of circular arc graphs. Disc. Appl. Math. **12**(3), 203–213 (1985)
5. Brouwer, A.E., Veldman, H.J.: Contractibility and np-completeness. J. Graph Theory **11**(1), 71–79 (1987)
6. Cai, L.: Fixed-parameter tractability of graph modification problems for hereditary properties. Inf. Process. Lett. **58**(4), 171–176 (1996)
7. Cai, L., Cai, Y.: Incompressibility of H-free edge modification problems. Algorithmica **71**(3), 731–757 (2015). https://doi.org/10.1007/s00453-014-9937-x
8. Cai, L., Guo, C.: Contracting few edges to remove forbidden induced subgraphs. In: Gutin, G., Szeider, S. (eds.) IPEC 2013. LNCS, vol. 8246, pp. 97–109. Springer, Cham (2013). https://doi.org/10.1007/978-3-319-03898-8_10
9. Cygan, M., et al.: Parameterized Algorithms. Springer, Heidelberg (2015). https://doi.org/10.1007/978-3-319-21275-3
10. Eppstein, D.: Finding large clique minors is hard. J. Graph Algor. Appl. **13**(2), 197–204 (2009)
11. Garey, M.R., Johnson, D.S.: Computers and Intractability: A Guide to the Theory of NP-Completeness. W. H Freeman, New York City (1979)
12. Garland, M., Heckbert, P.S.: Surface simplification using quadric error metrics. In: Proceedings of the 24th Annual Conference on Computer Graphics and Interactive Techniques, pp. 209–216 (1997)
13. Golovach, P.A., Kaminski, M., Paulusma, D., Thilikos, D.M.: Increasing the minimum degree of a graph by contractions. Theor. Comput. Sci. **481**, 74–84 (2013)
14. Guillemot, S., Marx, D.: A faster FPT algorithm for bipartite contraction. Inf. Process. Lett. **113**(22–24), 906–912 (2013)
15. Guo, C.: Parameterized complexity of graph contraction problems. Ph.D. thesis, Chinese University of Hong Kong, Hong Kong (2013)
16. Heggernes, P., van 't Hof, P., Lokshtanov, D., Paul, C.: Obtaining a bipartite graph by contracting few edges. SIAM J. Disc. Math. **27**(4), 2143–2156 (2013)
17. Hoede, C., Veldman, H.J.: Contraction theorems in hamiltonian graph theory. Disc. Math. **34**(1), 61–67 (1981)
18. Kaplan, H., Shamir, R.: The domatic number problem on some perfect graph families. Inf. Process. Lett. **49**(1), 51–56 (1994)
19. Lewis, J.M., Yannakakis, M.: The node-deletion problem for hereditary properties is NP-complete. J. Comput. Syst. Sci. **20**(2), 219–230 (1980)
20. Lokshtanov, D., Misra, N., Saurabh, S.: On the hardness of eliminating small induced subgraphs by contracting edges. In: Gutin, G., Szeider, S. (eds.) IPEC 2013. LNCS, vol. 8246, pp. 243–254. Springer, Cham (2013). https://doi.org/10.1007/978-3-319-03898-8_21
21. Lovász, L.: Graph minor theory. Bull. Am. Math. Soc. **43**(1), 75–86 (2006)
22. Marx, D., Sandeep, R.B.: Incompressibility of H-free edge modification problems: towards a dichotomy. J. Comput. Syst. Sci. **125**, 25–58 (2022)
23. Olariu, S.: Paw-fee graphs. Inf. Process. Lett. **28**(1), 53–54 (1988)

Distance-Based Covering Problems
for Graphs of Given Cyclomatic Number

Dibyayan Chakraborty[1] , Florent Foucaud[2] , and Anni Hakanen[2,3]([✉])

[1] Univ Lyon, CNRS, ENS de Lyon, Université Claude Bernard Lyon 1,
LIP UMR5668, Lyon, France
[2] Université Clermont Auvergne, CNRS, Clermont Auvergne INP,
Mines Saint-Étienne, LIMOS, 63000 Clermont-Ferrand, France
[3] Department of Mathematics and Statistics, University of Turku,
20014 Turku, Finland
anehak@utu.fi

Abstract. We study a large family of graph covering problems, whose
definitions rely on distances, for graphs of bounded cyclomatic number
(that is, the minimum number of edges that need to be removed from
the graph to destroy all cycles). These problems include (but are not
restricted to) three families of problems: (i) variants of metric dimension,
where one wants to choose a small set S of vertices of the graph such that
every vertex is uniquely determined by its ordered vector of distances to
the vertices of S; (ii) variants of geodetic sets, where one wants to select
a small set S of vertices such that any vertex lies on some shortest path
between two vertices of S; (iii) variants of path covers, where one wants
to select a small set of paths such that every vertex or edge belongs to
one of the paths. We generalize and/or improve previous results in the
area which show that the optimal values for these problems can be upper-
bounded by a linear function of the cyclomatic number and the degree 1-
vertices of the graph. To this end, we develop and enhance a technique
recently introduced in [C. Lu, Q. Ye, C. Zhu. Algorithmic aspect on the
minimum (weighted) doubly resolving set problem of graphs, *Journal of
Combinatorial Optimization* 44:2029–2039, 2022] and give near-optimal
bounds in several cases. This solves (in some cases fully, in some cases
partially) some conjectures and open questions from the literature. The
method, based on breadth-first search, is of algorithmic nature and thus,
all the constructions can be computed in linear time. Our results also
imply an algorithmic consequence for the computation of the *optimal*
solutions: they can all be computed in polynomial time for graphs of
bounded cyclomatic number.

Research funded by the French government IDEX-ISITE initiative 16-IDEX-0001 (CAP
20–25) and by the ANR project GRALMECO (ANR-21-CE48-0004).
A. Hakanen—Research supported by the Jenny and Antti Wihuri Foundation and
partially by Academy of Finland grant number 338797.

H. Fernau and K. Jansen (Eds.): FCT 2023, LNCS 14292, pp. 132–146, 2023.
https://doi.org/10.1007/978-3-031-43587-4_10

1   Introduction

Distance-based covering problems in graphs are a central class of problems in graphs, both from a structural and from an algorithmic point of view, with numerous applications. Our aim is to study such problems for graphs of bounded cyclomatic number. The latter is a measure of sparsity of the graph that is popular in both structural and algorithmic graph theory. We will mainly focus on three types of such problems, as follows.

Metric Dimension and Its Variants. In these concepts, introduced in the 1970s [16,33], the aim is to distinguish elements in a graph by using distances. A set $S \subseteq V(G)$ is a *resolving set* of G if for all distinct vertices $x, y \in V(G)$ there exists $s \in S$ such that $d(s, x) \neq d(s, y)$. The smallest possible size of a resolving set of G is called the *metric dimension* of G (denoted by $\dim(G)$). During the last two decades, many variants of resolving sets and metric dimension have been introduced. In addition to the original metric dimension, we consider the edge and mixed metric dimensions of graphs. A set $S \subseteq V(G)$ is an *edge resolving set* of G if for all distinct edges $x, y \in E(G)$ there exists $s \in S$ such that $d(s, x) \neq d(s, y)$, where the distance from a vertex v to an edge $e = e_1 e_2$ is defined as $\min\{d(v, e_1), d(v, e_2)\}$ [18]. A mixed resolving set is both a resolving set and an edge resolving set, but it must also distinguish vertices from edges and vice versa; a set $S \subseteq V(G)$ is a *mixed resolving set* of G if for all distinct $x, y \in V(G) \cup E(G)$ there exists $s \in S$ such that $d(s, x) \neq d(s, y)$ [17]. The *edge metric dimension* $\edim(G)$ (resp. *mixed metric dimension* $\mdim(G)$) is the smallest size of an edge resolving set (resp. mixed resolving set) of G. More on the different variants of metric dimension and their applications (such as detection problems in networks, graph isomorphism, coin-weighing problems or machine learning) can be found in the recent surveys [22,34].

Geodetic Numbers. A *geodetic set* of a graph G is a set S of vertices such that any vertex of G lies on some shortest path between two vertices of S [15]. The *geodetic number* of G is the smallest possible size of a geodetic set of G. The version where the edges must be covered is called an *edge-geodetic set* [3]. "Strong" versions of these notions have been studied. A *strong (edge-) geodetic set* of graph G is a set S of vertices of G such that we can assign for any pair x, y of vertices of S a shortest xy-path such that each vertex (edge) of G lies on one of the chosen paths [2,24]. Recently, the concept of *monitoring edge-geodetic set* was introduced in [14] as a strengthening of a strong edge-geodetic set: here, for every edge e, there must exist two solution vertices x, y such that e lies on *all* shortest paths between x and y. These concepts have numerous applications related to the field of convexity in graphs, see the book [27].

We also consider the concept of *distance-edge-monitoring-sets* introduced in [12,13], which can be seen as a relaxation of monitoring edge-geodetic sets. A set S is a distance-edge-monitoring-set if, for every edge e of G, there is a vertex x of S and a vertex y of G such that e lies on all shortest paths between x and y. The minimum size of such a set is denoted $\dem(G)$.

Path Covering Problems. In this type of problem, one wishes to cover the vertices (or edges) of a graph using a small number of paths. A *path cover* is a set of paths of a graph G such that every vertex of G belongs to one of the paths. If one path suffices, the graph is said to be Hamiltonian, and deciding this property is one of the most fundamental algorithmic complexity problems. The paths may be required to be shortest paths, in which case we have the notion of an *isometric path cover* [5,11]; if they are required to be chordless, we have an *induced path cover* [25]. The edge-covering versions have also been studied [1]. This type of problems has numerous applications, such as program and circuit testing [1,26], or bioinformatics [4].

Our Goal. Our objective is to study the three above classes of problems, on graphs of bounded cyclomatic number. (See Fig. 1 for a diagram showing the relationships between the optimal solution sizes of the studied problems.) A *feedback edge set* of a graph G is a set of edges whose removal turns G into a forest. The smallest size of such a set, denoted by $c(G)$, is the *cyclomatic number* of G. It is sometimes called the *feedback edge (set) number* or the *cycle rank* of G. For a graph G on n vertices, m edges and k connected components, it is not difficult to see that we have $c(G) = m - n + k$, since a forest on n vertices with k components has $n - k$ edges. In this paper, we assume all our graphs to be connected. To find an optimal feedback edge set of a connected graph, it suffices to consider a spanning tree; the edges not belonging to the spanning tree form a minimum-size feedback edge set.

Graphs whose cyclomatic number is constant have a relatively simple structure. They are sparse (in the sense that they have a linear number of edges). They also have bounded treewidth (indeed the treewidth is at most the cyclomatic number), a parameter that plays a central role in the area of graph algorithms, see for example Courcelle's celebrated theorem [8]. Thus, they are studied extensively from the perspective of algorithms (for example for the metric dimension [10], the geodetic number [19] or other graph problems [7,35]). They are also studied from a more structural angle [30–32].

Conjectures Addressed in this Paper. In order to formally present the conjectures, we need to introduce some structural concepts and notations. A *leaf* of a graph G is a vertex of degree 1, and the number of leaves of G is denoted by $\ell(G)$. Consider a vertex $v \in V(G)$ of degree at least 3. A *leg* attached to the vertex v is a path $p_1 \ldots p_k$ such that p_1 is adjacent to v, $\deg_G(p_k) = 1$ and $\deg_G(p_i) = 2$ for all $i \neq k$. The number of legs attached to the vertex v is denoted by $l(v)$.

A set $R \subseteq V(G)$ is a *branch-resolving set* of G, if for every vertex $v \in V(G)$ of degree at least 3 the set R contains at least one element from at least $l(v) - 1$ legs attached to v. The minimum cardinality of a branch-resolving set of G is denoted by $L(G)$, and we have

$$L(G) = \sum_{v \in V(G),\, \deg(v) \geq 3,\, l(v) > 1} (l(v) - 1).$$

It is well-known that for any tree T with at least one vertex of degree 3, we have $\dim(T) = L(T)$ (and if T is a path, then $\dim(T) = 1$) [6,16,20,33]. This has motivated the following conjecture.

Conjecture 1 ([32]). Let G be a connected graph with $c(G) \geq 2$. Then $\dim(G) \leq L(G) + 2c(G)$ and $\text{edim}(G) \leq L(G) + 2c(G)$.

The restriction $c(G) \geq 2$ is missing from the original formulation of Conjecture 1 in [32]. However, Sedlar and Škrekovski have communicated to us that this restriction should be included in the conjecture. Conjecture 1 holds for cacti with $c(G) \geq 2$ [32]. The bound $\dim(G) \leq L(G) + 18c(G) - 18$ was shown in [10] (for $c(G) \geq 2$), and is the first bound established for the metric dimension in terms of $L(G)$ and $c(G)$.

Conjecture 2 ([31]). If $\delta(G) \geq 2$ and $G \neq C_n$, then $\dim(G) \leq 2c(G) - 1$ and $\text{edim}(G) \leq 2c(G) - 1$.

In [31], Sedlar and Škrekovski showed that Conjecture 2 holds for graphs with minimum degree at least 3. They also showed that if Conjecture 2 holds for all 2-connected graphs, then it holds for all graphs G with $\delta(G) \geq 2$. Recently, Lu at al. [23] addressed Conjecture 2 and showed that $\dim(G) \leq 2c(G) + 1$ when G has minimum degree at least 2.

Conjecture 3 ([30]). Let G be a connected graph. If $G \neq C_n$, then $\text{mdim}(G) \leq \ell(G) + 2c(G)$.

Conjecture 3 is known to hold for trees [17], cacti and 3-connected graphs [30], and balanced theta graphs [29].

The following conjecture was also posed recently.

Conjecture 4 ([12,13]). For any graph G, $\text{dem}(G) \leq c(G) + 1$.

The original authors of the conjecture proved the bound when $c(G) \leq 2$, and proved that the bound $\text{dem}(G) \leq 2c(G) - 2$ holds when $c(G) \geq 3$ [12]. The conjectured bound would be tight [12,13].

Our Contributions. In this paper, we are motivated by Conjectures 1–4, which we address. We will show that both $\dim(G)$ and $\text{edim}(G)$ are bounded from above by $L(G) + 2c(G) + 1$ for all connected graphs G. Moreover, we show that if $L(G) \neq 0$, then the bounds of Conjecture 1 hold.

We show that Conjecture 3 is true when $\delta(G) = 1$, and when $\delta(G) \geq 2$ and G contains a cut-vertex. We also show that $\text{mdim}(G) \leq 2c(G) + 1$ in all other cases. We also consider the first part of Conjecture 1, that $\dim(G) \leq L(G) + 2c(G)$ from [32], in the case where $\delta(G) = 1$, and we show that it is true when $L(G) \geq 1$ and otherwise we have $\dim(G) \leq 2c(G) + 1$. We also consider the conjecture that $\text{edim}(G) \leq L(G) + 2c(G)$ from [32], and we show that it is true when $\delta(G) = 1$ and $L(G) \geq 1$, and when $\delta(G) \geq 2$ and G contains a cut-vertex. We also show that $\text{edim}(G) \leq 2c(G) + 1$ in all other cases.

Thus, our results yield significant improvements towards the Conjectures 1–3, since they are shown to be true in most cases, and are approximated by an additive term of 1 for all graphs. Moreover, we also resolve in the affirmative Conjecture 4.

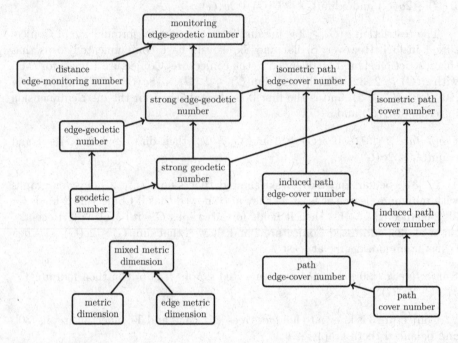

Fig. 1. Relations between the parameters discussed in the paper. If a parameter A has a directed path to parameter B, then for any graph, the value of A is upper-bounded by a linear function of the value of B.

To obtain the above results, we develop a technique from [23], who introduced it in order to study a strengthening of metric dimension called *doubly resolving sets* in the context of graphs of minimum degree 2. We notice that the technique can be adapted to work for all graphs and in fact it applies to many types of problems: (variants of) metric dimension, (variants of) geodetic sets, and path-covering problems. For all these problems, the technique yields upper bounds of the form $a \cdot c(G) + f(\ell(G))$, where $\ell(G)$ is the number of leaves of G, f is a linear function that depends on the respective problem, and a is a small constant.

The technique is based on a breadth-first-search rooted at a specific vertex, that enables to compute an optimal feedback edge set F by considering the edges of the graph that are not part of the breadth-first-search spanning tree. We then select vertices of the edges of F (or neighbouring vertices); the way to select these vertices depends on the problem. For the metric dimension and path-covering problems, a pre-processing is done to handle the leaves of the graph (for the geodetic set variants, all leaves must be part of the solution). Our results demonstrate that the techniques used by most previous works to handle graphs of bounded cyclomatic number were not precise enough, and the simple technique we employ is much more effective. We believe that it can be used with sucess in similar contexts in the future.

Algorithmic Applications. For all the considered problems, our method in fact implies that the optimal solutions can be computed in polynomial time for graphs with bounded cyclomatic number. In other words, we obtain XP algorithms with respect to the cyclomatic number. This was already observed in [10] for the metric dimension (thanks to our improved bounds, we now obtain a better running time, however it should be noted that in [10] the more general weighted version of the problem was considered).

Organisation. We first describe the general method to compute the special feedback edge set in Sect. 2. We then use it in Sect. 3 for the metric dimension and its variants. We then turn to geodetic sets and its variants in Sect. 4, and to path-covering problems in Sect. 5. We describe the algorithmic consequence in Sect. 6, and conclude in Sect. 7.

2 The General Method

The *length* of a path P, denoted by $|P|$, is the number of its vertices minus one. A path is *induced* if there are no graph edges joining non-consecutive vertices. A path is *isometric* if it is a shortest path between its endpoints. For two vertices u, v of a graph G, $d(u, v)$ denotes the length of an isometric path between u and v. Let r be a vertex of G. An edge $e = uv \in E(G)$ is a *horizontal* edge *with respect to* r if $d(u, r) = d(v, r)$ (otherwise, it is a *vertical* edge). For a vertex u of G, let $B_r(u)$ denote the set of edges $uv \in E(G)$ such that $d(u, r) = d(v, r) + 1$. A set F of edges of G is *good with respect to* r if F contains all horizontal edges with respect to r and for each $u \neq r$, $|B_r(u) \cap F| = |B_r(u)| - 1$. A set F of edges is simply *good* if F is good with respect to some vertex $r \in V(G)$. For a set F of good edges of a graph G, let T_F denote the subgraph of G obtained by removing the edges of F from G.

Lemma 5. *For any connected graph G with n vertices and m edges and a vertex $r \in V(G)$, a good edge set with respect to r can be computed in $O(n + m)$ time.*

Proof. By doing a Breadth First Search on G from r, distances of r from u for all $u \in V(G)$ can be computed in $O(n + m)$ time. Then the horizontal and vertical edges can be computed in $O(m)$ time. Then the sets $B_r(u)$ for all $u \in V(G)$ can be computed in $O(n + m)$ time. Hence the set of good edges with respect to r can be computed in $O(n + m)$ time. □

Lemma 6. *For a set F of good edges with respect to a vertex r of a connected graph G, the subgraph T_F is a tree rooted at r. Moreover, every path from r to a leaf of T_F is an isometric path in G.*

Proof. First observe that T_F is connected, as each vertex u has exactly one edge $uv \in E(T_F)$ with $d(u, r) = d(v, r) + 1$. Now assume for contradiction that T_F has a cycle C. Let $v \in V(C)$ be a vertex that is furthest from r among all vertices of C. Formally, v is a vertex such that $d(r, v) = \max\{d(r, w) : w \in V(C)\}$. Let E' denote the set of edges in T_F incident to v. Observe that $|E'|$ is at least two. Hence either E' contains an horizontal edge, or $E' \cap B_r(v)$ contains at least two edges. Either case contradicts that F is a good edge set with respect to r. This proves the first part of the observation.

Now consider a path P from r to a leaf v of T_F and write it as $u_1 u_2 \ldots u_k$ where $u_1 = r$ and $u_k = v$. By definition, we have $d(r, u_i) = d(r, u_{i-1}) + 1$ for each $i \in [2, k]$. Hence, P is an isometric path in G. □

Observation 7. *Any set F of good edges of a connected graph G is a feedback edge set of G with minimum cardinality.*

Proof. Due to Lemma 6 we have that T_F is a tree and therefore $|F| = m - n + 1$ which is same as the cardinality of a feedback edge set of G with minimum cardinality. □

The *base graph* [10] G_b of a graph G is the graph obtained from G by iteratively removing vertices of degree 1 until there remain no such vertices. We use the base graph in some cases where preprocessing the leaves and other tree-like structures is needed.

3   Metric Dimension and Variants

In this section, we consider three metric dimension variants and conjectures regarding them and the cyclomatic number. We shall use the following result.

Distinct vertices x, y are *doubly resolved* by $v, u \in V(G)$ if $d(v, x) - d(v, y) \neq d(u, x) - d(u, y)$. A set $S \subseteq V(G)$ is a *doubly resolving set* of G if every pair of distinct vertices of G are doubly resolved by a pair of vertices in S. Lu et al. [23] constructed a doubly resolving set of G with $\delta(G) \geq 2$ by finding a good edge set with respect to a root $r \in V(G)$ using breadth-first search. We state a result obtained by Lu et al. [23] using the terminologies of this paper.

Theorem 8 ([23]). *Let G be a connected graph such that $\delta(G) \geq 2$ and $r \in V(G)$. Let $S \subseteq V(G)$ consist of r and the endpoints of the edges of a good edge set with respect to r.*

(i) The set S is a doubly resolving set of G.
(ii) If r is a cut-vertex, then the set $S \setminus \{r\}$ is a doubly resolving set of G.
(iii) We have $|S| \leq 2c(G) + 1$.

A doubly resolving set of G is also a resolving set of G, and thus $\dim(G) \leq 2c(G) + 1$ when $\delta(G) \geq 2$ due to Theorem 8. Moreover, if G contains a cut-vertex and $\delta(G) \geq 2$, we have $\dim(G) \leq 2c(G)$. Therefore, Conjecture 1 holds for the metric dimension of a graph with $\delta(G) \geq 2$ and at least one cut-vertex.

A doubly resolving set is not necessarily an edge resolving set or a mixed resolving set. Thus, more work is required to show that edge and mixed resolving sets can be constructed with good edge sets. A *layer* of G is a set $L_d = \{v \in V(G) \mid d(r, v) = d\}$ where r is the chosen root and d is a fixed distance.

Proposition 9. *Let G be a graph with $\delta(G) \geq 2$, and let $r \in V(G)$. If the set S contains r and the endpoints of a good edge set F with respect to r, then S is an edge resolving set.*

Proof. Suppose to the contrary that there exist distinct edges $e = e_1 e_2$ and $f = f_1 f_2$ that are not resolved by S. In particular, we have $d(r, e) = d(r, f)$. Due to this, say, e_1 and f_1 are in the same layer L_d, and e_2 and f_2 are in $L_d \cup L_{d+1}$. If e is a horizontal edge with respect to r, then $e_1, e_2 \in S$ and e and f are resolved. Thus, neither e nor f is a horizontal edge with respect to r and we have $e_2, f_2 \in L_{d+1}$.

If $e_2 = f_2$, then $e, f \in B_r(e_2)$. Thus, we have $e_2 \in S$ and at least one of e_1 and f_1 is also in S. Now e and f are resolved by e_1 or f_1. Therefore, we have $e_2 \neq f_2$ and $e_2, f_2 \notin S$.

Let $w \in V(G)$ be a leaf in T_F such that e_2 lies on a path between w and r in T_F. Since $\delta(G) \geq 2$, the vertex w is an endpoint of some edge in F, and thus $w \in S$. Since e and f are not resolved by S, we have $d(w, f_2) = d(w, e_2) = d' - d - 1$, where $w \in L_{d'}$, due to the path between w and r being isometric (Lemma 6). Let P_f be a shortest path $w - f_2$ in G, and assume that P_f is such that it contains an element of S as close to f_2 as possible. Denote this element of S by s. We have $s \in L_i$ for some $d + 1 < i \leq d'$ (notice that we may have $s = w$). As the edges e and f are not resolved by S, we have $d(s, e) = d(s, f)$, which implies that $d(s, e_2) = d(s, f_2) = i - d - 1$. Let P_e' and P_f' be shortest paths $s - e_2$ and $s - f_2$, respectively. The paths P_e' and P_f' are internally vertex disjoint, since otherwise the vertex after which the paths diverge is an element of S which contradicts the choice of P_f and s. Let v_e and v_f be the vertices adjacent to s in P_e' and P_f', respectively. Now, we have $sv_e, sv_f \in B_r(s)$, and thus $v_e \in S$ (otherwise, $v_f \in S$, which contradicts the choice of P_f and s). If $d(v_e, e_2) < d(v_e, f_2)$, then v_e resolves e and f, a contradiction. Thus, we have $d(v_e, e_2) \geq d(v_e, f_2)$, but now there exists a shortest path $w - f_2$ that contains v_e, which is closer to f_2 than s is, a contradiction. $\square$

Proposition 10. *Let G be a graph with $\delta(G) \geq 2$, and let $r \in V(G)$. If the set S contains r and the endpoints of a good edge set F with respect to r, then S is a mixed resolving set.*

Proof. The set S resolves all pairs of distinct vertices by Theorem 8 and all pairs of distinct edges by Proposition 9. Therefore we only need to show that all pairs consisting of a vertex and an edge are resolved.

Suppose to the contrary that $v \in V(G)$ and $e = e_1 e_2 \in E(G)$ are not resolved by S. In particular, the root r does not resolve v and e, and thus $v, e_1 \in L_d$ for some $d \geq 1$. If e is a horizontal edge, then $e_1, e_2 \in S$ and e and v are resolved. Thus, assume that $e_2 \in L_{d+1}$. Let $w \in V(G)$ be a leaf in T_F such that e_2 lies on a path between w and r in T_F. Since $\delta(G) \geq 2$, the vertex w is an endpoint of some edge in F, and thus $w \in S$. We have $d(w, e_2) = d' - d - 1$, where $w \in L_{d'}$. However, now $d(w, v) \geq d' - d > d(w, e_2)$, and w resolves v and e, a contradiction. $\square$

As pointed out in [23], if R is a doubly resolving set that contains a cut-vertex v, then the set $R \setminus \{v\}$ is also a doubly resolving set. The following observation states that the same result holds for mixed resolving sets, and with certain constraints to (edge) resolving sets.

Observation 11. *Let G be a connected graph with a cut-vertex v.*

(i) *Let $R \subseteq V(G)$ be such that there are at least two connected components in $G - v$ containing elements of R. If $d(v, x) \neq d(v, y)$ for some $x, y \in V(G) \cup E(G)$, then there exists an element $s \in R$, $s \neq v$, such that $d(s, x) \neq d(s, y)$.*

(ii) *If $R \subseteq V(G)$ is a mixed resolving set of G, then every connected component of $G - v$ contains at least one element of R.*

(iii) *If $R \subseteq V(G)$ is a resolving set or edge resolving set of G, then at most one connected component of $G - v$ does not contain any elements of R, and that component is isomorphic to P_n for some $n \geq 1$.*

The following corollary follows from Propositions 9 and 10, and Observation 11.

Corollary 12. *Let G be a graph with $\delta(G) \geq 2$.*

(i) *If G contains a cut-vertex, then $\mathrm{edim}(G) \leq 2c(G)$ and $\mathrm{mdim}(G) \leq 2c(G)$.*

(ii) *If G does not contain a cut-vertex, then $\mathrm{edim}(G) \leq 2c(G) + 1$ and $\mathrm{mdim}(G) \leq 2c(G) + 1$.*

We then turn our attention to graphs with $\delta(G) = 1$. We will show that a good edge set can be used to construct a (edge, mixed) resolving set also in this case. Moreover, we show that Conjecture 3 holds, and Conjecture 1 holds when $L(G) \geq 1$. We also show that $\dim(G)$ and $\mathrm{edim}(G)$ are at most $2c(G) + 1$ when $L(G) = 0$. We use the following results on trees in our proof.

Proposition 13 ([17]). *Let T be a tree, and let $R \subseteq V(T)$ be the set of leaves of T. The set R is a mixed metric basis of T.*

Proposition 14 ([18,20]). *Let T be a tree that is not a path. If $R \subseteq V(T)$ is a branch-resolving set of T, then it is a resolving set and an edge resolving set.*

Theorem 15. *Let G be a connected graph that is not a tree such that $\delta(G) = 1$. Let $r \in V(G_b)$, and let $S \subseteq V(G_b)$ contain r and the endpoints of a good edge set $F \subseteq E(G_b)$ with respect to r. If R is a branch resolving set of G, then the set $R \cup S$ is a resolving set and an edge resolving set of G. If R is the set of leaves of G, then the set $R \cup S$ is a mixed resolving set of G.*

Proof. Let R be either a branch-resolving set of G (for the regular and edge resolving sets) or the set of leaves of G (for mixed metric dimension). We will show that the set $R \cup S$ is a (edge, mixed) resolving set of G.

The graph $G - E(G_b)$ is a forest (note that some of the trees might be isolated vertices) where each tree contains a unique vertex of G_b. Let us denote these trees by T_v, where $v \in V(G_b)$.

Consider distinct $x, y \in V(G) \cup E(G)$. We will show that x and y are resolved by $R \cup S$.

- Assume that $x, y \in V(T_v) \cup E(T_v)$ for some $v \in V(G_b)$. Denote $R_v = (V(T_v) \cap R) \cup \{v\}$. The set R_v is a (edge, mixed) resolving set of T_v by Propositions 14 and 13. If x and y are resolved by some element in R_v that is not v, then we are done. If x and y are resolved by v, then they are resolved by any element in $S \setminus \{v\}$. Since G is not a tree, the set $S \setminus \{v\}$ is clearly nonempty, and x and y are resolved in G.
- Assume that $x, y \in V(G_b) \cup E(G_b)$. Now x and y are resolved by S due to Theorem 8, Proposition 9 or Proposition 10.
- Assume that $x \in V(T_v) \cup E(T_v)$ and $y \in V(T_w) \cup E(T_w)$ where $v, w \in V(G_b)$, $v \neq w$. The set S is a doubly resolving set of G_b according to Theorem 8. Thus, there exist distinct $s, t \in S$ such that $d(s, v) - d(s, w) \neq d(t, v) - d(t, w)$. Suppose to the contrary that $d(s, x) = d(s, y)$ and $d(t, x) = d(t, y)$. Now we have

$$d(w, y) - d(v, x) = d(s, v) - d(s, w) \neq d(t, v) - d(t, w) = d(w, y) - d(v, x),$$

  a contradiction. Thus, s or t resolves x and y.
- Assume that $x \in V(T_v) \cup E(T_v)$ for some $v \in V(G_b)$, $v \neq x$, and $y = y_1 y_2 \in E(G_b)$. Suppose that $d(r, x) = d(r, y)$. Without loss of generality, we may assume that $d(r, y) = d(r, y_1) = d$. Now $y_1 \in L_d$ and $v \in L_{d - d_x}$, where $d_x = d(v, x) \in \{0, \ldots, d\}$. If $y_2 \in L_d$, then y is a horizontal edge and $y_1, y_2 \in S$. Now x and y are resolved by y_1 or y_2. So assume that $y_2 \in L_{d+1}$. Let $z \in V(G_b)$ be a leaf in T_F such that y_2 lies on a path from r to z in T_F. Since $\delta(G_b) \geq 2$, the vertex z is an endpoint of some edge in F, and thus $z \in S$. Now $z \in L_{d'}$ for some $d' > d+1$ and $d(z, y_2) = d' - d - 1$ by Lemma 6. Consequently,

$$d(z, x) = d(z, v) + d_x \geq d' - (d - d_x) + d_x = 2d_x + 1 + d(z, y_2) > d(z, y).$$

$\square$

Since the root r can be chosen freely, we can choose the root to be a cut-vertex in G whenever G contains cut-vertices. The bounds in the next corollary then follow from Observations 7 and 11, and Theorem 15.

Corollary 16. *Let G be a connected graph that is not a tree such that $\delta(G) = 1$. We have $\dim(G) \leq \lambda(G) + 2c(G)$, $\mathrm{edim}(G) \leq \lambda(G) + 2c(G)$, and $\mathrm{mdim}(G) \leq \ell(G) + 2c(G)$, where $\lambda(G) = \max\{L(G), 1\}$.*

The relationship of metric dimension and edge metric dimension has garnered a lot of attention since the edge metric dimension was introduced. Zubrilina [36] showed that the ratio $\frac{\text{edim}(G)}{\dim(G)}$ cannot be bounded from above by a constant, and Knor et al. [21] showed the same for the ratio $\frac{\dim(G)}{\text{edim}(G)}$. Inspired by this, Sedlar and Škrekovski [28] conjectured that for a graph $G \neq K_2$, we have $|\dim(G) - \text{edim}(G)| \leq c(G)$. This bound, if true, is tight due to the construction presented in [21]. It is easy to see that $\dim(G) \geq \lambda(G)$ and $\text{edim}(G) \geq \lambda(G)$ (the fact that $\dim(G) \geq L(G)$ is shown explicitly in [6], for example). Thus, we now obtain the bound $|\dim(G) - \text{edim}(G)| \leq 2c(G)$ due to the bounds established in Corollaries 12 and 16.

4   Geodetic Sets and Variants

We now address the problems related to geodetic sets, and show that the same method can be applied in this context as well. Note that all leaves of a graph belong to any of its geodetic sets. Due to lack of space, we only present the constructions of the solution sets.

4.1   Geodetic Sets

Theorem 17 (*). *Let G be a connected graph. If G has a cut-vertex then $g(G) \leq 2c(G) + \ell(G)$. Otherwise, $g(G) \leq 2c(G) + 1$.*

Proof (Sketch). We construct a good set F of edges of G by Lemma 5 (if G has a cut-vertex then the root r shall be a cut-vertex). We select as solution vertices, all leaves of G, all endpoints of edges of F, and r (only if G is biconnected).   □

The upper bound of Theorem 17 is tight when there is a cut-vertex, indeed, consider the graph formed by a disjoint union of k odd cycles and l paths, all identified via a single vertex. This graph has cyclomatic number k, l leaves, and geodetic number $2k + l$. When there is no cut-vertex, any odd cycle has geodetic number 3 and cyclomatic number 1, so the bound is tight in this case too.

4.2   Monitoring Edge-Geodetic Sets

It was proved in [14] that $\text{meg}(G) \leq 9c(G) + \ell(G) - 8$ for every graph G, and some graphs were constructed for which $\text{meg}(G) = 3c(G) + \ell(G)$. We next improve the former upper bound, therefore showing that the latter construction is essentially best possible.

Theorem 18 (*). *For any graph G, we have $\text{meg}(G) \leq 3c(G) + \ell(G) + 1$. If G contains a cut-vertex, then $\text{meg}(G) \leq 3c(G) + \ell(G)$.*

Proof (Sketch). We construct a good set F of edges of G by Lemma 5, by choosing r as a vertex belonging to a cycle, if possible. The solution set contains r (if G is biconnected), all leaves of G, and for each edge of F, both its endpoints. Moreover, for each vertex u of G with $|B_r(u)| \geq 2$, we add all endpoints of the edges of $B_r(u)$.   □

4.3   Distance-edge-monitoring-sets

We now prove Conjecture 4.

Theorem 19 (*). *For any connected graph G,* $\text{dem}(G) \leq c(G) + 1$.

Proof (Sketch). We construct a good set F of edges of the base graph G_b of G by Lemma 5, and select as solution vertices the root r, one arbitrary endpoint of each horizontal edges of F, as well as each vertex v with $|B_r(v)| \geq 2$.   □

5   Path Covers and Variants

In this section, we consider the path covering problems. We focus on isometric path edge-covers (sets of isometric paths that cover all edges of the graph), indeed those have the most restrictive definition and the bound thus holds for all other path covering problems from Fig. 1.

Theorem 20 (*). *For any graph G,* $\text{ipec}(G) \leq 3c(G) + \lceil (\ell(G) + 1)/2 \rceil$.

Proof (Sketch). We construct a good set F of edges of the base graph G_b of G by Lemma 5, and select as solution paths the horizontal edges of F; for each vertex v with $|B_r(v)| \geq 2$, we add to S, $|B_r(v)|$ shortest paths from v to r, each starting with a different edge from $B_r(v)$. This covers the edges of G_b. To cover the edges of $G - E(G_b)$, we carefully construct (using an iterative procedure) a pairing of the leaves of G and connect each paired pair by a shortest path.   □

The upper bound of Theorem 20 is nearly tight, indeed, consider (again) the graph formed by a disjoint union of k odd cycles and l paths, all identified via a single vertex. The obtained graph has cyclomatic number k, l leaves, and isometric path edge-cover number $3k + \lceil l/2 \rceil$.

6   Algorithmic Consequences

Theorem 21 (*). *For all the problems considered here, if we have an upper bound on the solution size of $a \cdot c(G) + f(\ell(G))$ for some $a \in \mathbb{N}$, we obtain an algorithm with running time $O(n^{a \cdot c(G)})$ on graphs G of order n.*

Proof (Sketch). One needs to be able to compute the optimal number of leaves required in a solution, using the methods described in the proofs of the theorems. Then, a simple brute-force algorithm trying all subsets of size $a \cdot c(G)$ completes the algorithm.   □

7   Conclusion

We have demonstrated that a simple technique based on breadth-first-search is very efficient to obtain bounds for various distance-based covering problems, when the cyclomatic number and the number of leaves are considered. This resolves or advances several open problems and conjectures from the literature on this type of problems. There remain some gaps between the obtained bounds and the conjectures or known constructions, that still need to be closed.

A refinement of the cyclomatic number of a (connected) graph G is called its *max leaf number*, which is the maximum number of leaves in a spanning tree of G. It is known that the cyclomatic number is always upper-bounded by the max leaf number plus the number of leaves [9], so, all our bounds also imply bounds using the max leaf number only.

Regarding the algorithmic applications, we note that the XP algorithms described in Theorem 21 can sometimes be improved to obtain an FPT algorithm. This is the case for geodetic sets [19], but whether this is possible for the metric dimension remains a major open problem [9,19] (this is however shown to be possible for the larger parameter "max leaf number" [9]).

References

1. Andreatta, G., Mason, F.: Path covering problems and testing of printed circuits. Disc. Appl. Math. **62**(1), 5–13 (1995)
2. Arokiaraj, A., Klavžar, S., Manuel, P.D., Thomas, E., Xavier, A.: Strong geodetic problems in networks. Discussiones Mathematicae Graph Theory **40**(1), 307–321 (2020)
3. Atici, M.: On the edge geodetic number of a graph. Int. J. Comput. Math. **80**(7), 853–861 (2003)
4. Cáceres, M., Cairo, M., Mumey, B., Rizzi, R., Tomescu, A.I.: Sparsifying, shrinking and splicing for minimum path cover in parameterized linear time. In: Proceedings of SODA 2022, pp. 359–376. SIAM (2022)
5. Chakraborty, D., Dailly, A., Das, S., Foucaud, F., Gahlawat, H., Ghosh, S.K.: Complexity and algorithms for isometric path cover on chordal graphs and beyond. In: Proceedings of ISAAC 2022, LIPIcs, vol. 248, pp. 12:1–12:17 (2022)
6. Chartrand, G., Eroh, L., Johnson, M.A., Oellermann, O.: Resolvability in graphs and the metric dimension of a graph. Disc. Appl. Math. **105**(1–3), 99–113 (2000)
7. Coppersmith, D., Vishkin, U.: Solving NP-hard problems in 'almost trees': vertex cover. Disc. Appl. Math. **10**(1), 27–45 (1985)
8. Courcelle, B.: The monadic second-order logic of graphs. i. recognizable sets of finite graphs. Inf. Comput. **85**(1), 12–75 (1990)
9. Eppstein, D.: Metric dimension parameterized by max leaf number. J. Graph Algor. Appl. **19**(1), 313–323 (2015)
10. Epstein, L., Levin, A., Woeginger, G.J.: The (weighted) metric dimension of graphs: Hard and easy cases. Algorithmica **72**(4), 1130–1171 (2015)
11. Fisher, D.C., Fitzpatrick, S.L.: The isometric number of a graph. J. Comb. Math. Comb. Comput. **38**(1), 97–110 (2001)
12. Foucaud, F., Kao, S., Klasing, R., Miller, M., Ryan, J.: Monitoring the edges of a graph using distances. Disc. Appl. Math. **319**, 424–438 (2022)

13. Foucaud, F., Klasing, R., Miller, M., Ryan, J.: Monitoring the edges of a graph using distances. In: Changat, M., Das, S. (eds.) CALDAM 2020. LNCS, vol. 12016, pp. 28–40. Springer, Cham (2020). https://doi.org/10.1007/978-3-030-39219-2_3

14. Foucaud, F., Narayanan, K., Sulochana, L.R.: Monitoring edge-geodetic sets in graphs. In: Bagchi, A., Muthu, R. (eds.) Proceedings of CALDAM 2023. Lecture Notes in Computer Science, vol. 13947, pp. 245–256. Springer, Heidelberg (2023). https://doi.org/10.1007/978-3-031-25211-2_19

15. Harary, F., Loukakis, E., Tsouros, C.: The geodetic number of a graph. Math. Comput. Model. **17**(11), 89–95 (1993)

16. Harary, F., Melter, R.: On the metric dimension of a graph. Ars Combinatoria **2**, 191–195 (1976)

17. Kelenc, A., Kuziak, D., Taranenko, A., Yero, I.G.: Mixed metric dimension of graphs. Appl. Math. Comput. **314**, 429–438 (2017)

18. Kelenc, A., Tratnik, N., Yero, I.G.: Uniquely identifying the edges of a graph: the edge metric dimension. Disc. Appl. Math. **251**, 204–220 (2018)

19. Kellerhals, L., Koana, T.: Parameterized complexity of geodetic set. J. Graph Algor. Appl. **26**(4), 401–419 (2022)

20. Khuller, S., Raghavachari, B., Rosenfeld, A.: Landmarks in graphs. Disc. Appl. Math. **70**(3), 217–229 (1996)

21. Knor, M., Majstorović, S., Masa Toshi, A.T., Škrekovski, R., Yero, I.G.: Graphs with the edge metric dimension smaller than the metric dimension. Appl. Math. Comput. **401**, 126076 (2021)

22. Kuziak, D., Yero, I.G.: Metric dimension related parameters in graphs: a survey on combinatorial, computational and applied results. arXiv preprint arXiv:2107.04877 (2021)

23. Lu, C., Ye, Q., Zhu, C.: Algorithmic aspect on the minimum (weighted) doubly resolving set problem of graphs. J. Comb. Optim. **44**, 2029–2039 (2022)

24. Manuel, P., Klavžar, S., Xavier, A., Arokiaraj, A., Thomas, E.: Strong edge geodetic problem in networks. Open Math. **15**(1), 1225–1235 (2017)

25. Manuel, P.D.: Revisiting path-type covering and partitioning problems. arXiv preprint arXiv:1807.10613 (2018)

26. Ntafos, S., Hakimi, S.: On path cover problems in digraphs and applications to program testing. IEEE Trans. Softw. Eng. SE **5**(5), 520–529 (1979)

27. Pelayo, I.M.: Geodesic Convexity in Graphs. Springer, Heidelberg (2013). https://doi.org/10.1007/978-1-4614-8699-2

28. Sedlar, J., Škrekovski, R.: Bounds on metric dimensions of graphs with edge disjoint cycles. Appl. Math. Comput. **396**, 125908 (2021)

29. Sedlar, J., Škrekovski, R.: Extremal mixed metric dimension with respect to the cyclomatic number. Appl. Math. Comput. **404**, 126238 (2021)

30. Sedlar, J., Škrekovski, R.: Mixed metric dimension of graphs with edge disjoint cycles. Disc. Appl. Math. **300**, 1–8 (2021)

31. Sedlar, J., Škrekovski, R.: Metric dimensions vs. cyclomatic number of graphs with minimum degree at least two. Appl. Math. Comput. **427**, 127147 (2022)

32. Sedlar, J., Škrekovski, R.: Vertex and edge metric dimensions of cacti. Disc. Appl. Math. **320**, 126–139 (2022)

33. Slater, P.J.: Leaves of trees. In: Proceedings of the Sixth Southeastern Conference on Combinatorics, Graph Theory, and Computing (Florida Atlantic Univ., Boca Raton, Fla., 1975), pp. 549–559. Congressus Numerantium, No. XIV. Utilitas Math., Winnipeg, Man. (1975)

34. Tillquist, R.C., Frongillo, R.M., Lladser, M.E.: Getting the lay of the land in discrete space: a survey of metric dimension and its applications. arXiv preprint arXiv:2104.07201 (2021)
35. Uhlmann, J., Weller, M.: Two-layer planarization parameterized by feedback edge set. Theor. Comput. Sci. **494**, 99–111 (2013)
36. Zubrilina, N.: On the edge dimension of a graph. Disc. Math. **341**(7), 2083–2088 (2018)

An Efficient Computation of the Rank Function of a Positroid

Lamar Chidiac[1]([✉])[iD], Santiago Guzmán-Pro[2][iD], Winfried Hochstättler[1][iD],
and Anthony Youssef[3]

[1] Fakultät für Mathematik und Informatik, FernUniversität in Hagen,
Hagen, Germany
{lamar.chidiac,winfried.hochstaettler}@fernuni-hagen.de
[2] Institut für Algebra, TU Dresden, Dresden, Germany
sanguzpro@ciencias.unam.mx
[3] Reply S.p.A, London, UK

Abstract. Positroids are a class of matroids in bijection with several combinatorial objects. In particular, every positroid can be constructed from a decorated permutation or from a Le-graph.

In this paper, we present two algorithms, one that computes the rank of a subset of a positroid using its representation as a Le-graph and the other takes as input a decorated permutation σ and outputs the Le-graph that represent the same positroid as σ. These two algorithms combined form an improvement to Mcalmon and Oh's result on the computation of the rank function of a positroid from the decorated permutation.

Keywords: Decorated permutation · Le diagrams · Positroid · Algorithm

1 Introduction

A matroid is a combinatorial object that unifies several notions of independence. It can be defined in many different but equivalent way. In this work we use the definition of matroids in terms of bases as it is more convenient for studying positroids. A matroid M is a pair $M = (E, \mathcal{B})$ where E is the set of elements of M and is finite, and $\mathcal{B}$ is its collection of bases which satisfies the following two conditions: $\mathcal{B}$ is not empty and if B_1 and B_2 are members of $\mathcal{B}$ and $x \in B_1 - B_2$, then there is an element y of $B_2 - B_1$ such that $(B_1 - x) \cup y \in \mathcal{B}$.

Positroids are a class of matroids, introduced by Postnikov in [7] as the column sets of a matrix's nonzero maximal minors, with all maximal minors ($d \times d$ submatrices) being non-negative. In other terms, a positroid is a representable matroid $M(A)$ associated with a $(d \times n)$-matrix A of rank d with all real entries such that all its maximal minors are non-negative. Positroids seemed to have several interesting combinatorial characteristics. In particular, Postnikov [7] demonstrated that positroids are in bijection with a number of interesting combinatorial objects, including Grassmann necklaces, decorated permutations,

This work was carried out during a visit of Guzmán-Pro at FernUniversität in Hagen, supported by DAAD grant 57552339.

Le-graphs, and plabic graphs. Mcalmon and Oh [5] showed how to compute the rank of an arbitrary subset of a positroid from the associated decorated permutation using non-crossing partitions. It is not clear whether this algorithm runs in polynomial time. This fact, together with the high technicality involved to prove its correctness, encouraged us to search for a simpler approach. For that purpose, we first notice that the rank of a set in a positroid P can be easily computed using a Right-first-search variant of the Depth-first-search on a Le-graph representing P. Secondly, we propose an efficient construction of the Le-graph representing the same positroid from the given decorated permutation σ using what we call the Le-graph-construction algorithm. Thus, by composing both algorithms we obtain a polynomial-time algorithm that computes the rank of a given set of a positroid represented by a decorated permutation.

The rest of this work is structured as follows. In Sect. 2, we define positroids using Le-graphs and present all the background needed. In Sect. 3, we present the Rank-Function algorithm along with an example and prove its correctness. In Sect. 4, we recall the algorithm in [3] that computes the decorated permutation from a Le-diagram and in Sect. 5, we present the algorithm that computes the Le-diagram of a decorated permutation for the reverse direction. We conclude with a small discussion in the last section.

2   Preliminaries

We assume familiarity with matroid theory, a reference is [6]. We start by defining the Young diagram which is needed for the definition of Le-diagrams.

Definition 1 (Young diagram). *A Young diagram Y_λ (also called a Ferrers diagram) is a finite collection of boxes arranged in left-justified rows, with the row lengths (number of boxes in a row) in non-increasing order, where λ is a partition of the total number of boxes of the diagram and the sets of λ are the lengths of the rows. The Young diagram is said to be of shape λ, and it carries the same information as that partition.*

Definition 2 (Le-diagram). *Fix d and n. A Le-diagram D of shape λ and type (d, n) is a Young diagram Y_λ contained in a $d \times (n - d)$ rectangle, where some of its boxes are filled with dots in such a way that the λ-property is satisfied: there is no empty box which has a dot above it in the same column and a dot to its left in the same row. Figure 1 is an example of a Le-diagram.*

Given a Le-diagram we can always construct a Le-graph as follows.

Definition 3 (Le-graph). *A Le-graph is constructed from a Le-diagram as follows.*

- *Every dot of the Le-diagram is a vertex of the corresponding Le-graph.*
- *Add a vertex in the middle of each step of the east-south boundary and label them from 1 to n starting at the north-east corner of the path and ending at the south-west corner. We refer to these vertices as* vertical *and* horizontal *vertices, depending on whether they lie on a vertical or horizontal step.*

– *Add edges directed to the right between any two vertices that lie in the same row and which have no other vertices between them in the same row, and edges directed upwards between any two nodes that lie in the same column and which have no other vertices between them in the same column.*

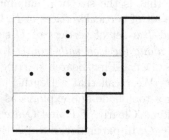

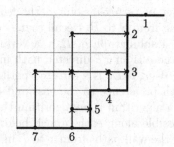

Fig. 1. A Le-diagram with $\lambda = (3,3,2)$, $d = 3$ and $n = 7$

Fig. 2. The corresponding Le-graph obtained from Fig. 1

We always consider a Le-graph Γ to be embedded in the Le-diagram from which Γ is constructed. This embedding is a planar embedding.

We define now a positroid using Le-graphs.

Definition 4 (Positroid). *Let Γ be a Le-graph and B be the set of nodes labelling vertical vertices of the boundary path of Γ and $r = |B|$. We define the collection $\mathcal{I}$ of independent sets of the positroid $P(\Gamma)$ to be all sets $I \in \binom{[n]}{k}$ where $k \leq r$, such that there exist k pairwise vertex-disjoint paths from I to B in Γ. A matroid $M = ([n], \mathcal{I})$ is a positroid if there is a Le-graph Γ such that $M \cong P(\Gamma)$. Note that if $x \in I \cap B$ then x is the trivial path from I to B and therefore every element of B is an independent set on its own.*

As it is commonly done, we identify a matroid with its collection of basis. These are the independent sets of size r. For example, the positroid built from the Le-graph in Fig. 2 is

$$\mathcal{P} = \{235, 236, 245, 246, 256, 257, 267, 356, 357, 367, 456, 457, 467\}.$$

Notice that given a Le-graph Γ, the loops in $P(\Gamma)$ correspond to isolated horizontal vertices, as 1 in Fig. 2; while coloops correspond to isolated vertical vertices.

We state without a proof the following proposition.

Proposition 1 ([3]). *Positroids are closed under taking duals and minors.*

3   The Rank-Function Algorithm

Our algorithm takes a subset $S \subseteq [n]$ of a positroid $\mathcal{P}$ of n elements and return its rank. A *routing* from S to B in Γ is a set of vertex-disjoint paths with initial vertices in S and end vertices in B. The rank of a subset S is the size of the biggest independent set that it contains. Thus, our goal is to find the maximal number of vertex-disjoint paths from S to B, that is the size of a maximum routing from S to B. Recall that B denotes the set of vertical vertices of Γ. Let D be a subdigraph of Γ, x a vertex of D, and T a set of vertices of D such that there is an xT-directed path in D. An *all along the wall path* from x to T in D, $W(x, T, D)$ is a directed path obtained by Depth-First-Search starting in x, with preference to the east and ending in T. We recall that a Depth-First-Search algorithm is an algorithm that starts at a root node and explores as far as possible along each branch before backtracking. Clearly, if P and Q are all along the wall paths from x to T in D, then $P = Q$. In particular, if $x \in T$ then $W(x, T, D)$ is the trivial path x.

Let $\mathcal{P}$ be a positroid on n elements, represented by a Le-graph D, and S a subset of $\mathcal{P}$ whose rank we want to compute. The following is a pseudocode of the Rank-function algorithm.

```
Arrange the set S in increasing order
Create B, the set of all vertical vertices of D
r = |B| (r is the rank of the positroid)
rk(S) = 0.

For i in S
If rk(S) = r
Break
If W(i,B,D) exists
rk(S) = rk(S) + 1
D = D - W(i,B,D)   (Delete all vertices of the path W(i,B,D))
Else
U = All visited vertices during the search for W(i,B,D)
D = D - U   (Delete all visited vertices during the search for W(i,B,D))
print rk(S)
```

Notice that since we are looking for vertex-disjoint paths, we can delete all vertices of $W(i, B, D)$ once it is found. Similarly, if $W(i, B, D)$ does not exist, that is if no path is found from i to B in D, we can delete all vertices traversed during the search; that is to avoid visiting them again knowing they do not lead to successful paths. It is important to mention that $W(i, B, D)$ can be constructed in linear time with respect to the size of the edge set of D, which is $\mathcal{O}(n^2)$. In the following example we use a marker to emphasis the steps done by the algorithm. We also color all visited vertices in red. Note that all edges are directed upwards and from left to right, but for the simplicity of graphs we draw simple edges instead.

Remark 1. Positroids form a subclass of the gammoids (see e.g. [2]). While rank computation even in represented gammoids can be hard as it involves the computation of a maximal set of vertex disjoint paths, the simple structure of the underlying digraph makes the computation a lot easier in the case of Le-graphs.

Example 1. Let us find the rank of $S = \{3, 4, 5, 8, 9\}$, a subset of the positroid $\mathcal{P}(D)$ using its Le-diagram D shown below. We first create the set of verticals $B = \{2, 3, 6, 8\}$. Let $rk(S) = 0$ and $r = |B| = 4$. We start by placing the marker on 3. Since 3 is in B, then $W(3, B, D)$ is the trivial path 3. We delete the vertex 3, increase the rank of S by one and move on to the second element in S, and so on. After finding the trivial path 8 we move the marker to the last element 9 and continue our search but one can notice that $W(9, B, D - U)$ does not exist, where $D - U$ is the diagram shown in the last row of this example. After completing the Rank-function algorithm we get that the rank of the subset S is $rk(S) = 3$.

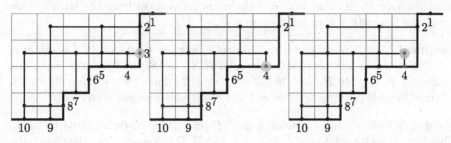

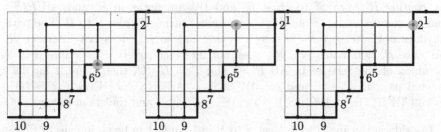

We next prove that the paths $W(i, B, D)$ found in the Rank-Function algorithm form a maximum routing from S to B, which we call the *all along the wall routing*. Given a set $S \subseteq [n]$, we recursively define the all along the wall routing of S, $AWR(S)$ as follows. We define the base case $AWR(\emptyset) = \emptyset$. When $S \neq \emptyset$, let $s = \max(S)$, U be the union of vertices of the paths in $AWR(S - s)$ and $T = B \setminus U$.

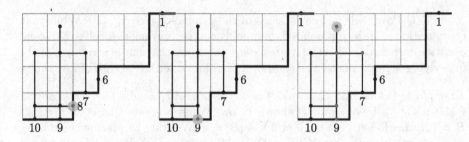

$$AWR(S) = \begin{cases} AWR(S-s) & \text{if there is no } sT\text{-directed path in } D-U, \\ AWR(S-s) \cup \{W(s,T,D-U)\} & \text{otherwise.} \end{cases}$$

For the proof, the following lemma is needed. Let $i \in [n]$ and consider a pair P and Q of iB-directed paths in D. We say that P *dominates* Q if Q is never strictly above P, in other words, if there are no vertices from the path Q that are above the path P and not touching P.

Lemma 1. *Let Γ be a Le-graph, B its set of vertical vertices and S a subset of $[n]$. If S is an independent set, then $|AWR(S)| = |S|$, and for any maximum routing R from S to B, we always have that R_i dominates W_i for all $i \in S$, where R_i and W_i are paths of R and $AWR(S)$ whose initial vertex is i.*

Proof. Let $i \in [n]$ and consider a pair P and Q of iB-directed paths in Γ. Consider a positive integer j, $i < j \leq n$. If P dominates Q, then any jB-directed path that intersects Q also intersects P. Now we prove the lemma by induction over $|S|$ where the base case trivially holds. Let $m = \max(S)$ and $S' = S - m$. By induction we have that $|S'| = |AWR(S')|$ and that for any routing R' from S' to B, if R'_i and W'_i are paths in R' and $AWR(S')$ then R'_i dominates W'_i. Consider a maximum routing R from S to B. The path R_m does not intersect R_i for any $i < m$, and so R_m does not intersect R'_i and since R'_i dominates W'_i, R_m does not intersect any W'_i. Therefore if U is the union of the vertices in $AWR(S')$ and $T = B \setminus U$, then there is an mT-directed path in $\Gamma - U$, and so, $AWR(S) = AWR(S') \cup \{W(m, T, \Gamma - U)\}$. Hence $|AWR(S)| = |AWR(S')| + 1 = |S'| + 1 = |S|$. and the claim follows. ☐

Consider a routing R from a set S to B in Γ and let m be an integer in S. We denote by $R_{|m}$ the subset of R that consists of those paths whose initial vertex belongs to $S \cap [m]$. In other words, paths whose initial vertex are $i \in S$ and $i \leq m$. By the order in which we process the elements of S when we construct $AWR(S)$, the equality $AWR(S \cap [m]) = AWR(S)_{|m}$ holds. This simple observation will be used in the proof of the following proposition.

Proposition 2. *Let Γ be a Le-graph and S a subset of $[n]$. The all along the wall routing of S, $AWR(S)$, is a maximum routing from S to B in Γ, where B is the set of all vertical vertices of Γ.*

Proof. We proceed by induction over $|S|$ where we consider the base case to be when S is an independent set, which holds by Lemma 1. Suppose that S is a dependent set and let I be the set of initial vertices of the paths in $AWR(S)$. Let $m = \min(S \setminus I)$ and $I' = I \cap [m]$. We first prove that $I' \cup \{m\}$ is a dependent set. Proceeding by contradiction, suppose that $I' \cup \{m\}$ is an independent set. By the choice of m, we know that $I' \cup \{m\} = S \cap [m]$. By the observation above this paragraph and by the fact that $m \notin I$, the following equation holds; $|AWR(S \cap [m])| = |AWR(S)_{|m}| = |I'| = |I' \cup \{m\}| - 1$. On the other hand, since $I' \cup \{m\}$ is an independent set, by Lemma 1, we have that $|AWR(I' \cup \{m\})| = |I' \cup \{m\}|$. This contradiction implies that $I' \cup \{m\}$ must be a dependent set. So $rk(S - m) = rk(S)$, and thus any maximum routing of $S - m$ is a maximum routing of S. Since $m \in S \setminus I$, by the recursive rule for constructing all along the wall routings, we know that $AWR(S) = AWR(S - m)$. Therefore, by our induction hypothesis we conclude that $AWR(S)$ is a maximum routing of $S - m$, and thus it is a maximum routing of S.   □

We highlight that with an adequate implementation, the all along the wall routing of any set $S \subseteq [n]$ can be constructed in linear time with respect to the edge set of Γ which is $\mathcal{O}(n^2)$. Moreover, since the rank of a set $S \subseteq [n]$ in a positroid $P(\Gamma)$, is the size of a maximum routing from S to B in Γ, by Proposition 2 the following statement holds.

Corollary 1. *Let n be a positive integer and Γ a Le-graph with boundary vertices $[n]$. The rank of a given set of a positroid $P(\Gamma)$ can be computed in $\mathcal{O}(n^2)$-time.*

4   From Le-Diagram to Decorated Permutation

As previously mentioned, positroids are also in bijection with decorated permutations. Before defining them, let us first recall that the *i-order* $<_i$ on the set $[n]$ is the total order

$$i <_i i + 1 <_i \cdots <_i n <_i 1 <_i \cdots <_i i - 2 <_i i - 1.$$

Definition 5 (Decorated Permutation). *A decorated permutation of the set $[n]$ is a bijection $\sigma : [n] \to [n]$ whose fixed points are colored either "clockwise" or "counterclockwise." We denote a clockwise fixed point by $\sigma(j) = j$ and a counterclockwise fixed point by $\sigma(j) = \bar{j}$. A weak i-excedance of the decorated permutation σ is an element $j \in [n]$ such that either $j <_i \sigma(j)$ or $\sigma(j) = \bar{j}$. The number of weak i-excedances of σ is the same for any $i \in [n]$; we will simply call it the number of weak excedances of σ. A simple way to represent a permutation σ is by the vector $(\sigma(1), \sigma(2), \ldots, \sigma(n))$.*

Postnikov proved in [7] that there is a bijection β between decorated permutation and Le-diagrams such that for every decorated permutation σ, the positroid represented by σ, $P(\sigma)$ is the same as $P(\beta(\sigma))$. Later on, in 2013, Ardila et al. [3] presented a simple algorithm that, given a Le-graph Γ, efficiently computes

$\beta^{-1}(\Gamma)$ (Lemma 2). However, we are unaware of a published algorithm that takes as an input a decorated permutation σ and outputs $\beta(\sigma)$. Although the existence of such a bijection β was proven by Postnikov, he did not present an explicit algorithm that computes it. In Sect. 5 we propose an algorithm that computes β in polynomial time. For a clear understanding of the algorithm, one must first understand the algorithm presented by Ardila et al. in [3] which we present now along with an example. Recall that we consider a Le-graph together with an embedding in a Le-diagram from which it was constructed. On the other hand, for every Le-diagram we can easily construct a Le-graph. Thus, we will abuse the nomenclature and speak about Le-graphs and Le-diagrams as the same object.

Lemma 2. *The following algorithm gives a bijection α between the Le-diagrams of type (d, n) and the decorated permutations on n letters with d weak excedances.*

1. *Replace each dot in the Le-diagram D with an uncross $\diagdown\!\!\!\diagup$, and each empty box in D with a cross*
2. *The south and east border of Y_λ gives rise to a path of length n from the northeast corner to the southwest corner of the $d \times (n - d)$ rectangle. Label the edges of this path with the numbers 1 through n.*
3. *Label the edges of the north and west border of Y_λ so that opposite horizontal edges and opposite vertical edges have the same label.*
4. *By following the "pipes" from the northwest border to the southeast border of the Young diagram in the resulting "pipe dream" we get a permutation σ. If the pipe originating at i ends at j, we define $\sigma(i) = j$. (No right angles turns are possible. A turn is only possible with an uncross.)*
5. *If $\sigma(j) = j$ and j labels two horizontal (respectively vertical) edges of Y_λ, then $\sigma(j) = \underline{j}$ (respectively $\sigma(j) = \bar{j}$).*

The following is an example that illustrates this procedure for a Le-diagram D.

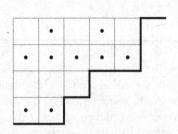
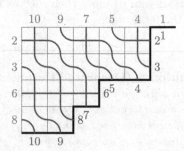

Fig. 3. A Le-diagram D (on the left) with the pipe dream (on the right) constructed when computing $\alpha(D)$ according to Lemma 2. In this case $\alpha = (D) = (\underline{1}, 7, 9, 3, 2, \bar{6}, 5, 10, 4, 8)$.

5   From Decorated Permutation to Le-Diagram

5 As we have already mentioned, the aim of In this section we to propose a polynomial-time computation of α^{-1}, i.e., the inverse function of the mapping defined by the algorithm in Lemma 2. The Le-graph-construction algorithm takes a decorated permutation σ as an input and outputs the Le-diagram corresponding to σ. It is evident that the first step of the algorithm can be inverted in linear time with respect to the size of the Le-diagram. For this reason we will identify Le-diagrams and pipe dreams, according to the construction in the first step of the algorithm in Lemma 2. Allow us to formalize the concept of pipe dream. A *pipe dream* D consists of a *shape*, a *labelling* and a *filling*. The shape of D is a Young diagram Y contained in a $d \times (n - d)$ rectangle. The labelling of D is a pair of labelling, one of the southeast border SE of Y and the second of the northwest border NW of Y. The filling F of D is a function from the boxes of D to the set $\{\diagdown, +\}$; the first element is an uncross and the second is a cross. An *empty pipe dream* ED consists of the shape together with a labelling as defined above, but with no filling. So every pipe dream can be described as an ordered pair (ED, F). In the rest of this work we stick to the labelling used in Lemma 2. In other words, if we have an empty pipe dream ED, then we label the steps of the southeast borders from 1 to n (from northeast to southwest) and label the steps of the north and west border by the same label of the step it is facing. We call this the AL labelling. Notice that every pipe dream D defines a bijection $f_D : NW \to SE$ by following the pipes from the northwest border to the southeast border. We call this function the *bijection defined* by D.

We start by presenting our algorithm along with an example and then prove it in Theorem 1. Let $\sigma : NW \to SE$ be a decorated permutation on the set $[n]$ of a positroid $\mathcal{P}$ whose pipe dream $D = (ED, F)$ we want to find. First, we have to determine the shape of the empty pipe dream ED. For that purpose it suffices to determine which steps are the vertical steps of the southeast border. Notice that, starting from the northwest border, a cross or an uncross can only take a path downwards or to the right. When using the AL labelling of ED, for any step with label i on the north border, we must have that $\sigma(i) \leq i$ and for any step with label j on the west border, we must have $\sigma(j) \geq j$. Thus, the set B of all vertical edges is the set of weak 1-excedances of the decorated permutation σ. Observe that the color of the fixed points on decorated permutation are important to determine the vertical steps of the southeast border. The empty pipe dream ED we get from σ in this way is called the *empty pipe dream defined* by σ and we denote it by $ED(\sigma)$. E.g. consider the decorated permutation σ from the previous example. The set of weak 1-excedances is $B = \{2, 3, 6, 8\}$, thus the second, third, sixth and eighth steps are vertical. Having determined the shape of the southeast border of ED, we add boxes to the left of the path in the Ferrers shape and give the borders the AL labelling. Now we still have to find a filling F such that σ is the bijection defined by $D = (ED, F)$.

We present now a recursive algorithm that generates the filling F of D that satisfies the ⅃-property, and later prove that such a filling always exists. The algorithm consists of three steps which we explain separately in the following subsections.

5.1 Removing Fixed Points

After constructing the shape of ED and labelling the borders with the AL labelling, we start the filling process. Thepaths of fixed points in the decorated permutation are easy to fill. For instance, in the example of Lemma 2, there is only one way to link 6 to 6 and that's with a series of crossings from left to right. Similarly, with "clockwise" fixed point, we draw a series of crossings from top to bottom. Therefore, for the simplicity of the algorithm we remove fixed points, since they can be added later on after completing the filling of ED. Removing fixed points means removing the entire row or column of the fixed point from ED. Since these rows and columns are filled with crosses, removing them will not affect any other paths. We now have a new empty pipe dream ED' that we relabel again with the AL labelling. The decorated permutation will change as well and it will be computed using the function l.

Let $X = \{x_1, x_2, \ldots, x_t\}$ be the set of fixed points in the decorated permutation σ such that $x_1 < x_2 < \cdots < x_t$. First we remove X from σ and then define l as follows:

$$l : [n] - X \longrightarrow [n - t]$$

$$e \longmapsto \begin{cases} e - i & \text{if } x_i < e < x_{i+1} \\ e - t & \text{if } e > x_t \\ e & \text{otherwise.} \end{cases}$$

Note that e is a non-fixed element of the permutation σ i.e., $e = \sigma(i)$ and $\sigma(i) \neq i$. For example let us consider the decorated permutation $\sigma = (2, 3, 5, \underline{4}, 1, \overline{6}, 8, 7, \underline{9}, 11, 10, \overline{12}, 14, 13)$. Here σ has 4 fixed points $x_1 = 4$, $x_2 = 6$, $x_3 = 9$ and $x_4 = 12$. Every element $e > 12$ in σ will be replaced by $e - 4$ and every element $e < 4$ is invariant. Moreover, 5 is replaced by $5 - 1 = 4$ (since $x_1 < 5 < x_2$), 8 by $8 - 2 = 6$ and 7 by $7 - 2 = 5$, 11 by $11 - 3 = 8$ and 10 by $10 - 3 = 7$. The new permutation we get after removing all fixed points is $\sigma' = (2, 3, 4, 1, 6, 5, 8, 7, 10, 9)$.

5.2 Filling the Bottom Row

After removing rows and columns corresponding to fixed points from ED, a new empty pipe dream ED' is obtained. The algorithm will now fill the bottom row of ED', from left to right.

Proposition 3. *After removing rows and columns corresponding to fixed points from an empty pipe dream ED, the bottom row of the new empty pipe dream ED', is either filled with uncrosses or with a series of crosses followed by a series of uncrosses.*

Proof. Recall that the Le-diagram of $\mathcal{P}$ fulfills the J-property. This implies that a filling F of a pipe dream satisfies the J-property if there is no box with a cross which has an uncross above it in the same column and to the left of it in the same row. So, by identifying empty boxes with crosses and dots with uncrosses, we have a bijection between pipe dreams that satisfy the J-property and Le-diagrams that satisfy the same property. In order to fill the bottom row, we look at the lowest vertical i of the west border and its image in the new decorated permutation σ'. Since i is a vertical vertex, we have that $\sigma'(i) > i$, which means that $\sigma'(i)$ is on the south border below i. Thus, there is always only one way to route this vertical to its image. That is by one uncross or a series of crosses that ends with an uncross. After we complete the path from i to $\sigma'(i)$, if there is still empty boxes to the right of the box with an uncross, they should be filled with uncrosses, otherwise we get a fixed point, since there cannot be an uncross above it by the J-property. □

Let us consider the permutation from Fig. 3: $\sigma = (\underline{1}, 7, 9, 3, 2, \overline{6}, 5, 10, 4, 8)$ as an example, and let ED be the empty pipe dream constructed from σ. After using the function l to remove fixed points we get a new empty pipe dream ED' and the permutation $\sigma' = (5, 7, 2, 1, 4, 8, 3, 6)$. We have $\sigma'(6) = 8$ and there is one way to route this, namely with an uncross. The box to the right should also contain an uncross, since otherwise we either get a fixed point or the J-property is no longer fulfilled as we have already shown in the proof of Proposition 3. The following is ED' and the filling of the last row.

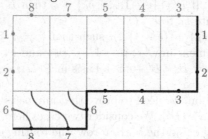

5.3 Contracting an Element

Once the last row of ED' is filled, the labels from the last row can now be shifted to the upper row following the partial paths formed from the filling of the last row. By doing so, the last row is in fact no longer needed. We can remove it from ED', and the algorithm will now work on a new empty pipe dream ED'' that has one element less than ED'. As a matter of fact, the positroid $\mathcal{P}(D'')$ where D'' is the pipe dream we get after removing the last row from a pipe dream D' correspond to a minor of the positroid $\mathcal{P}(D')$. We prove this in the next proposition.

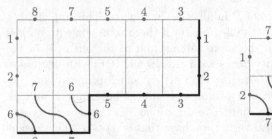
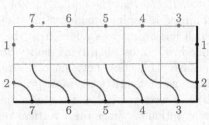

As shown in the figures above, after filling the last row and shifting the labels following the partial paths, we can now delete the last row, relabel the borders using the AL labelling and work on a new empty pipe dream. Certainly, the decorated permutation changes as well since now we have one less element, which is 8 in this example.

Proposition 4. *Let $\mathcal{P}(D)$ be the positroid coming from a Le-diagram D, and a decorated permutation σ, and let v_k be the bottom vertical in D. Removing the bottom row from D, is in fact contracting $e = \sigma(v_k)$ in $\mathcal{P}(D)$.*

Proof. Let D' be the Le-diagram obtained after removing the bottom row from D. First, we have to prove that any basis B' of the positroid $\mathcal{P}(D')$ is equal to $B - e$ for a basis B in $\mathcal{P}(D)$ such that $e \in B$. Since $e = \sigma(v_k)$ and v_k is a vertical, we have $e \geq v_k$. Let $e - 1, e - 2, \ldots, v_k$ denote the elements before e including the bottom vertical v_k. If $B' \cap \{e - 1, \ldots, v_k\} = \phi$ then $B = B' + e$ is a base in $\mathcal{P}(D)$ since we can route v_k to e.

If $B' \cap \{e - 1, \ldots, v_k\} = \{e - j\}_{j \in J}$ such that $\emptyset \neq J \subseteq \{1, \ldots, e - v_k\}$, let $e - i$ be the largest element in $B' \cap \{e - 1, \ldots, v_k\}$ and let $e - i$ be routed to a vertical v^i in the D. Now $B = B' + e$ is a basis in $\mathcal{P}(D)$, since we can route e to v^i, and reroute $e - i$ to v_k.

For the other direction, we need to prove that any basis B in $\mathcal{P}(D)$ such that $e \in B$, $B - e$ is a basis in $\mathcal{P}(D')$. We consider two cases here. If e was routed to v_k in B, then by removing e and the bottom row, the routing of the other elements stays the same in $B - e$, and $B - e$ is a basis in $\mathcal{P}(D')$. However, if e was not routed to v_k but to a vertical v^e, this means that one element from $\{e - 1, \ldots, v_k - 1\}$ was routed to v_k or $v_k \in B$. Now let $B \cap \{e - 1, \ldots, v_k\} = \{e - j\}_{j \in J}$ such that $J = \{j_1, j_2, \ldots, j_p\} \neq \emptyset$ and J is arranged in increasing order. Now $B - e$ is a basis in D' since we can route $e - j_1$ to v^e, $e - j_\ell$ to $v^{e - j_{\ell - 1}}$ (that is the vertical that $e - j_1$ was routed to in B) for $\ell = 2, \ldots, p$. $\qquad \square$

In order to compute the new permutation after contracting an element e, we use the function l_e which we define as follows:

$$l_e : [n] - e \longrightarrow [n - 1]$$

$$x \longmapsto \begin{cases} x - 1 & \text{if} \quad x > e \\ x & \text{otherwise.} \end{cases}$$

Note that x is an element of the permutation σ' different from e i.e., $x = \sigma'(i)$ implies $\sigma'(i) \neq e$. Positroids are closed under minors, and therefore after contracting an element and computing the new permutation, by induction we can now repeat the same process. We remove fixed points, fill the last row and then contract the image of the last vertical. We repeat until the pipe dream is completely filled.

For the reader's sake we continue with the example. After contracting 8, the new permutation is now $(5, 7, 2, 1, 4, 3, 6)$. We see that there are no fixed points, so we can skip the first step and start by filling the last row. The image of 2 (the last vertical) is 7 so the first box to the left of the last row contains an uncross and now the other boxes in that row should also contain uncrosses. Now we contract 7, that means we remove the last row and shift all the labels from the last row following the partial paths, and we compute the new permutation using the function l_7. The new permutation is now $(5, \underline{2}, 1, \underline{4}, 3, \underline{6})$. We got now three new fixed points. We remove them using the function l and we get the permutation $(3, 1, 2)$. Now we only have one row which we fill with uncrosses since the image of 1 is 3, and we get the following pipe dream.

All the boxes of ED were filled with crosses and uncrosses during the process. The final output of the algorithm, which is the complete filling F of ED satisfying the ⅃-property, can be computed in two ways.

1. The algorithm enumerates each box of the initial empty diagram ED and saves the filling of each box during the process. The output is then an array containing the filling of all boxes.

2. The algorithm does not store the filling of each box during the process but keeps track of the fixed points deleted, and the elements contracted, at each step. Then, after the process is completed and we are down to one row, we start to reconstruct the original pipe dream by adding deleted fixed points using l^{-1} and contracted elements using l_e^{-1}.

Theorem 1. *Let σ be a decorated permutation and ED the empty pipe dream defined by σ and has the AL labelling. There exists a unique filling F such that σ equals the bijection defined by (ED, F) and F satisfies the ⅃-property.*

Proof. We proceed by strong induction over the number of boxes in ED. The base case is when there are no boxes in ED so the southeast border equals the northwest border and all elements in σ are fixed points. In this case the empty filling $\emptyset$ obviously satisfies the ⅃-property. For the inductive case we consider two cases.

Case 1: σ has a fixed point i.

Let i be the label of a north step such that $\sigma(i) = i$ (if i is the label of a west step, the argument is similar to the following one). Consider the empty pipe dream ED' obtained by removing the column C determined by the step with label i, and let $\sigma' : [n] - i \rightarrow [n] - \sigma(i)$ be the restriction of σ to the set of labels $[n] - i$. Suppose F' is a filling of ED' such that (ED, F') defines the bijection σ'. Consider the extension F of F' where every box of C is filled with a cross. It is not hard to notice that (ED, F) defines the function σ. Moreover, since every box with a cross in F that does not belong to ED' has only crosses on top, F satisfies the J-property whenever F' does.

Case 2: σ has no fixed point.

Let R be the bottom row of ED and i the label of the lowest vertical of the west border. Since σ has no fixed points and i is the label of a vertical step, this means that $\sigma(i)$ is the label of a step on the south border below i. Let B be the box whose bottom edge is $\sigma(i)$. Consider the partial filling F_0 defined for the boxes of R as follows: every box to the left of B is filled with a cross, and every box to the right of B including B is filled with an uncross. Notice that i reaches $\sigma(i)$ by following the pipe dream of F_0. Let ED' be the empty pipe dream obtained from ED after removing R. Preserve the labelling of all steps not in R and shift all labels of the steps in R along the pipes of the partial filling F_0. Notice that the labels of the northwest border of ED' are in $[n] - i$ and the labels of the southeast border are in $[n] - \sigma(i)$. So let $\sigma' : [n] - i \rightarrow [n] - \sigma(i)$ be the restriction of σ to $[n] - i$. Again, by extending the filling F' obtained by our induction hypothesis to the filling $F' \cup F_0$, we conclude that the function defined by $(ED, F' \cup F_0)$ equals σ. Finally the fact that $F' \cup F_0$ satisfies the J-property, follows from the induction hypothesis and the fact that every cross in R has only crosses to the left.   □

We highlight that, given an empty pipe dream ED defined by a decorated permutation $\sigma : [n] \rightarrow [n]$, the inductive proof of Theorem 1 yields an $\mathcal{O}(n^2)$-time construction of the filling F such that σ is the bijection defined by (ED, F).

Lemma 3. *For any decorated permutation $\sigma : [n] \rightarrow [n]$, the Le-diagram $\alpha^{-1}(\sigma)$ can be constructed in quadratic time with respect to n.*

Proof. First, notice that we can construct an empty pipe dream $ED(\sigma)$ in $\mathcal{O}(n^2)$-time. By Theorem 1, there is a filling F such that the bijection defined by $(ED(\sigma), F)$ is σ. Clearly, this function coincides with the permutation in Lemma 4, so if $D = (ED(\sigma), F)$ then $\alpha(D) = \sigma$. Finally, since F can be constructed in $\mathcal{O}(n^2)$-time then the Le-diagram D can also be constructed $\mathcal{O}(n^2)$-time.   □

By composing the construction in the proof of Lemma 3 and Corollary 1 we obtain a quadratic time algorithm that calculates the rank of a given set of a positroid represented by a decorated permutation.

Corollary 2. *The rank of a given set of a positroid represented by a decorated permutation can be computed in polynomial time from the decorated permutation.*

6 Discussion

This paper was inspired by the work of Oh and Mcalmon in [5]. We noticed that the rank of an arbitrary subset of a positroid can easily be computed from the Le-diagram instead of the decorated permutation. Later, we showed that given a decorated permutation σ we can efficiently compute the Le-diagram that represents the same positroid as σ. It is worth mentioning that an equivalent way of doing this would be to compute the Grassmann Necklace of a given decorated permutation and then use the algorithm of Agarwala and Fryer in [1] that constructs the Le-diagram associated to the Grassmann Necklace.

To conclude this brief work we stand out the following. It is important to notice that the all along the wall routings defined in Sect. 3 stem from the linear-time algorithm of Ulrick Brandes and Dorothea Wagner in [4] for the arc disjoint Menger problem in planar directed graphs. The problem of finding a maximum number of internally vertex-disjoint s-t paths can be reduced to the arc-disjoint case by replacing each vertex $v \neq s, t$ by two vertices v', v'', while each arc with head v is redirected to v' and each arc with tail v is redirected from v''; moreover, an arc (v', v'') is added ([8] page 137). In general, this reduction does not preserve planarity, but in the case of Le-graphs planarity is preserved. Nonetheless, when adding a source and a sink to the Le-graph, planarity might be destroyed. Thus, we could not simply reduce our problem to the one considered in [4], but a similar technique to the one used in [4] also worked for us.

References

1. Agarwala, S., Fryer, S.: An algorithm to construct the Le diagram associated to a Grassmann necklace. Glasg. Math. J. **62**(1), 85–91 (2020)
2. Albrecht, I.: Contributions to the problems of recognizing and coloring gammoids. FernUniversität in Hagen (2018). https://doi.org/10.18445/20180820-090543-4
3. Ardila, F., Rincón, F., Williams, L.: Positroids and non-crossing partitions. Trans. Amer. Math. Soc. **368**(1), 337–363 (2016)
4. Brandes, U., Wagner, D.: A linear time algorithm for the arc disjoint Menger problem in planar directed graphs. In: Burkard, R., Woeginger, G. (eds.) ESA 1997. LNCS, vol. 1284, pp. 64–77. Springer, Heidelberg (1997). https://doi.org/10.1007/3-540-63397-9_6
5. Mcalmon, R., Oh, S.: The rank function of a positroid and non-crossing partitions. Electron. J. Combin., 27(1) (2020). Paper No. 1.11, 13
6. Oxley, J.: Matroid theory, second edition. In: Oxford Graduate Texts in Mathematics, vol. 12. Oxford University Press, Oxford (2011)
7. Postnikov, A.: Total positivity, grassmannians, and networks (2006)
8. Schrijver, A.: Combinatorial Optimization: Polyhedra and Efficiency, volume B (2003)

Minimizing Query Frequency to Bound Congestion Potential for Moving Entities at a Fixed Target Time

William Evans[(✉)] [iD] and David Kirkpatrick [iD]

Computer Science, University of British Columbia, Vancouver, Canada
{will,kirk}@cs.ubc.ca

Abstract. Consider a collection of entities moving continuously with bounded speed, but otherwise unpredictably, in some low-dimensional space. Two such entities encroach upon one another at a fixed time if their separation is less than some specified threshold. Encroachment, of concern in many settings such as collision avoidance, may be unavoidable. However, the associated difficulties are compounded if there is uncertainty about the precise location of entities, giving rise to potential encroachment and, more generally, potential congestion within the full collection.

We consider a model in which entities can be queried for their current location (at some cost) and the uncertainty region associated with an entity grows in proportion to the time since that entity was last queried. The goal is to maintain low potential congestion, measured in terms of the (dynamic) intersection graph of uncertainty regions, at specified (possibly *all*) times, using the lowest possible query cost. Previous work [SoCG'13, EuroCG'14, SICOMP'16, SODA'19], in the same uncertainty model, addressed the problem of minimizing the congestion potential *of point entities* using location queries of some bounded frequency. It was shown that it is possible to design query schemes that are $O(1)$-competitive, in terms of worst-case congestion potential, with other, even clairvoyant query schemes (that exploit knowledge of the trajectories of all entities), subject to the same bound on query frequency.

In this paper we initiate the treatment of a more general problem with the dual optimization objective: minimizing the *query frequency*, measured as the reciprocal of the minimum time between queries (granularity), while guaranteeing a fixed bound on congestion potential *of entities with positive extent* at one specified target time. This complementary objective necessitates quite different schemes and analyses. Nevertheless, our results parallel those of the earlier papers, specifically tight competitive bounds on required query frequency.

Keywords: data in motion · uncertain inputs · collision avoidance · online algorithms · competitive analysis

This work was funded in part by Discovery Grants from the Natural Sciences and Engineering Research Council of Canada.

H. Fernau and K. Jansen (Eds.): FCT 2023, LNCS 14292, pp. 162–175, 2023.
https://doi.org/10.1007/978-3-031-43587-4_12

1   Introduction

This paper addresses a fundamental issue in algorithm design, of both theoretical and practical interest: how to cope with unavoidable imprecision in data. We focus on a class of problems associated with location uncertainty arising from the motion of independent entities when location queries to reduce uncertainty are expensive. For concreteness, imagine a collection of robots following unpredictable trajectories with bounded speed. If an individual robot is not monitored continuously there is uncertainty, growing with the duration of unmonitored activity, concerning its precise location. This portends some risk of collision with neighbouring robots, necessitating some perhaps costly collision avoidance protocol. Nevertheless, robots that are known to be well-separated at some point in time will remain free of collision for the near future. How then should a limited query budget be allocated over time so as to minimize the risk of collisions or, more realistically, help focus collision avoidance measures on robot pairs that are in serious risk of collision?

We adopt a general framework for addressing such problems, essentially the same as the one studied by Evans et al. [10] (which unifies and improves [8,9]) and by Busto et al. [5]. In this model, an entity may be queried at any time in order to reveal its true location but between queries, location uncertainty, represented by a region surrounding the last known location, grows. Our goal is to understand with what frequency such queries need to be performed, and which entities should be queried, in order to maintain a particular measure of the congestion potential of the entities, formulated in terms of the overlap of their uncertainty regions. We describe query schemes that ensure a specified bound on two measures of congestion potential at a specified target time. Our schemes are shown to be competitive, in terms of query granularity, with any other schemes that ensure the same bound.

While the problem of guaranteeing low congestion potential at a fixed target time is of interest in its own right, it also serves to set the stage for the more ambitious task of guaranteeing low congestion potential continuously (i.e. at all times). This task is taken up in an expanded version of this paper (see [7]) where we present query schemes to maintain several measures of congestion potential that, over every modest-sized time interval, are competitive in terms of the frequency of their queries, with any scheme that maintains the same measure over that interval alone.

1.1   The Query Model

To facilitate comparisons with earlier results, we adopt much of the notation used by Evans et al. [10] and Busto et al. [5]. Let $\mathcal{E}$ be a set $\{e_1, e_2, \ldots, e_n\}$ of (mobile) entities. Each entity e_i is modelled as a d-dimensional closed ball with fixed extent and bounded speed, whose position (centre location) at any time is specified by the (unknown) continuous function ζ_i from $[0, \infty)$ (time) to $\mathbb{R}^d$. We take the entity radius to be our *unit of distance*, and take the time for an entity moving at maximum speed to move a unit distance to be our *unit of time*.

The n-tuple $(\zeta_1(t), \zeta_2(t), \ldots, \zeta_n(t))$ is called the $\mathcal{E}$-*configuration* at time t. Entities e_i and e_j are said to *encroach* upon one another at time t if the distance $\|\zeta_i(t) - \zeta_j(t)\|$ between their centres is less than some fixed *encroachment threshold* Ξ. For simplicity we assume to start that the distance between entity centres is always at least 2, i.e. the *separation* $\|\zeta_i(t) - \zeta_j(t)\| - 2$ between entities e_i and e_j at time t, is always at least zero—so entities never properly intersect, and that the encroachment threshold is exactly 2 (i.e. we are only concerned with avoiding entity contact). The concluding section considers a relaxation (and decoupling) of these assumptions in which $\|\zeta_i(t) - \zeta_j(t)\|$ is always at least some positive

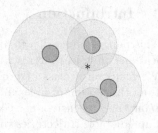

Fig. 1. Uncertainty regions (light grey) of four unit-radius entities (dark grey). (Color figure online)

constant ρ_0 (possibly less than 2), and the encroachment threshold Ξ is some constant at least ρ_0.

We wish to maintain knowledge of the positions of the entities over time by making location queries to individual entities, each of which returns the exact position of the entity at the time of the query. A *(query) scheme* $\mathbb{S}$ is just an assignment of location queries to time instances. We measure the performance of a scheme over a specified time interval T as the minimum query *granularity* (the time between consecutive queries) over T.

At any time $t \geq 0$, let $p_i^{\mathbb{S}}(t)$ denote the time, prior to t, that entity e_i was last queried; we define $p_i^{\mathbb{S}}(0) = -\infty$. The *uncertainty region* of e_i at time t, denoted $u_i^{\mathbb{S}}(t)$, is defined as the ball with centre $\zeta_i(p_i^{\mathbb{S}}(t))$ and radius $1 + t - p_i^{\mathbb{S}}(t)$; note that $u_i^{\mathbb{S}}(0)$ is unbounded. (We omit $\mathbb{S}$ when it is understood and the dependence on t when t is fixed.) Fig. 1 illustrates the uncertainty regions of four unit-radius entities shown at their most recently known locations, four, three, two and one time unit in the past, respectively.

The set $U(t) = \{u_1(t), \ldots, u_n(t)\}$ is called the *(uncertainty) configuration* at time t. Entity e_i is said to *potentially encroach* upon entity e_j in configuration $U(t)$ if $u_i(t) \cap u_j(t) \neq \emptyset$ (that is, there are potential locations for e_i and e_j at time t such that $e_i \cap e_j \neq \emptyset$).

In this way any configuration U gives rise to an associated (symmetric) *potential encroachment graph* PE^U on the set $\mathcal{E}$. Note that, by our assumptions above, the potential encroachment graph associated with the initial uncertainty configuration $U(0)$ is complete.

We define the following notions of *congestion potential* (called *interference potential* in [5]) in terms of configuration U and the graph PE^U.

- The **(uncertainty) max-degree** (hereafter **degree**) of the configuration U is given by $\delta^U = \max_i\{\delta_i^U\}$ where δ_i^U is defined as the degree of entity e_i in PE^U (the maximum number of entities e_j, *including* e_i, that potentially encroach upon e_i in configuration U).
- The **(uncertainty) ply** ω^U of configuration U is the maximum number of uncertainty regions in U that intersect at a single point. This is the largest

number of entities in configuration U whose mutual potential encroachment is witnessed by a single point.

- The (**uncertainty**) **thickness** χ^U of configuration U is the chromatic number of PE^U. This is the size of the smallest decomposition of $\mathcal{E}$ into independent sets (sets with no potential for encroachment) in configuration U.

The configuration illustrated in Fig. 1 has uncertainty degree four and uncertainty ply three (witnessed by point $*$).

Note that $\omega^U \leq \chi^U \leq \delta^U$, so upper bounds on uncertainty degree, and lower bounds on uncertainty ply, apply to all three measures. As we will see, even when we seek to minimize congestion at one fixed target time, the query frequency required to guarantee that uncertainty degree does not exceed some fixed value x can exceed the query frequency required to guarantee that uncertainty ply does not exceed x, by a factor of $\Theta(x)$.

The assumption that entities never properly intersect is helpful since it means that if uncertainty regions are kept sufficiently small, uncertainty ply can be kept to at most two. Similarly, for x larger than some dimension-dependent sphere-packing constant, it is always possible to maintain uncertainty degree at most x, using sufficiently high query frequency.

1.2   Related Work

One of the most widely-studied approaches to computing functions of moving entities uses the *kinetic data structure* model which assumes precise information about the future trajectories of the moving entities and relies on elementary geometric relations among their locations along those trajectories to certify that a combinatorial structure of interest, such as their convex hull, remains essentially the same. The algorithm can anticipate when a relation will fail, or is informed if a trajectory changes, and the goal is to update the structure efficiently in response to these events [1,11,12]. Another less common model assumes that the precise location of *every* entity is given to the algorithm periodically. The goal again is to update the structure efficiently when this occurs [2–4].

More similar to ours is the framework introduced by Kahan [13,14] for studying data in motion problems that require repeated computation of a function (geometric structure) of data that is moving continuously in space where data acquisition via queries is costly. There, location queries occur simultaneously in batches, triggered by requests (Kahan refers to these requests as "queries") to compute some function at some time rather than by a requirement to maintain a structure or property at all times. Performance is compared to a "lucky" algorithm that queries the minimum amount to calculate the function. Kahan's model and use of competitive evaluation is common to much of the work on query algorithms for uncertain inputs (see Erlebach and Hoffmann's survey [6]).

As mentioned, our model is essentially the same as the one studied by Evans et al. [10] and by Busto et al. [5], both of which focus on point entities. Like the current paper, paper [10] contains strategies whose goal is to guarantee competitively low congestion potential, compared to any other (even clairvoyant) scheme, at one specified target time. It provides precise descriptions of the impact on this guarantee using several measures of initial location knowledge and available lead time before the target. The other paper [5] contains a scheme for guaranteeing competitively low congestion potential at *all* times. For this more challenging task the scheme maintains congestion potential over time that is within a constant factor of that maintained by any other scheme over modest-sized time intervals. All of these results dealt with *optimizing congestion potential measures subject to fixed query frequency*.

In this paper, we consider the dual problem: *optimizing query frequency required to guarantee fixed bounds on congestion*. These two problems are fundamentally different: being able to optimize congestion using fixed query frequency provides little insight into how to optimize query frequency to maintain a fixed bound on congestion. In particular, even for stationary entities, a small change in the congestion bound can lead to an arbitrarily large change in the required query frequency. Our frequency optimization involves maximizing the minimum query granularity which requires more than just minimizing the number of queries made over a specified interval.

1.3   Our Results

The overarching goal of this line of research is to formulate efficient query schemes that, for all possible collections of moving entities, maintain fixed bounds on congestion potential measures continuously (i.e. at all times). Naturally for many such collections the required query frequency changes over time as entities cluster and spread, so efficient query schemes need to adapt to changes in the configuration of entities. While such changes are continuous and bounded in rate, they are only discernible through queries to individual entities, so entity configurations are *never* known precisely; future configurations are of course entirely hidden. In this latter respect our schemes and the competitive analysis of their efficiency, using as a benchmark a *clairvoyant scheme* that bases its queries on full knowledge of all entity trajectories (and hence all future configurations), resembles familiar scenarios that arise in the design and analysis of on-line algorithms.

Our goal in this paper is to show how to optimize query frequency to guarantee low congestion potential at one fixed target time, say time τ in the future, starting from a state of complete uncertainty of entity locations. In an expanded version of this paper [7] we turn our attention to the optimization of query frequency to guarantee low congestion potential continuously. Motivation for restricting attention to a fixed target time comes in part from the desire to prepare for a computation, at some known time in the future, whose efficiency depends on this low congestion potential (see [15] for example). As we will see, it also plays an important role in the efficient initialization of query schemes that optimize queries to guarantee low congestion potential continuously, from some point in time onward [7], and provides an informative contrast to those schemes.

We begin by describing a query scheme to achieve uncertainty ply at most x at one fixed target time in the future, reminiscent of the objective in [10]. The detailed description and analysis of our fixed target time scheme shows that uncertainty ply at most x can be achieved using query granularity that is at most a factor $\Theta(x)$ smaller than that used by any, even clairvoyant, query scheme to achieve the same goal. A similar, but more intricate scheme and analysis establishes the same result for uncertainty degree. An example shows that the competitive factor for both congestion potential measures is asymptotically optimal in the worst case. Nevertheless, if we relax our objective, allowing instead uncertainty degree at most $x+\Delta$, where $1 \leq \Delta \leq x$, the competitive factor $\Theta(x)$ drops to $\Theta(\frac{x}{1+\Delta})$. Again, this competitive factor is shown to be asymptotically optimal in the worst case. This analysis of a query scheme that solves a slightly relaxed optimization, relative to a clairvoyant scheme that solves the un-relaxed optimization, foreshadows similar analyses of our schemes for continuous-time query optimization [7].

In the concluding discussion, we describe modifications to our model that make our query optimization framework even more broadly applicable.

2   Geometric Preliminaries

In any $\mathcal{E}$-configuration $Z = (z_1, z_2, \ldots, z_n)$ and for any positive integer x, we call the separation between e_i and its xth closest neighbour (not including e_i) its x-separation, and denote it by $\sigma_i^Z(x)$. We call the closed ball with radius (called the x-radius of e_i) $r_i^Z(x) = \sigma_i^Z(x)+1$ and centre z_i, the x-ball of e_i, and denote it by $B_i^Z(x)$ (cf. Figure 2). We will omit Z when the configuration is understood. Note that, for all entities e_i and e_j,

$$\sigma_j^Z(x) \leq \|z_j - z_i\| + \sigma_i^Z(x), \qquad (1)$$

since the ball with radius $\|z_j - z_i\| + \sigma_i^Z(x)$ centred at z_j contains the ball with radius $\sigma_i^Z(x)$ centred at z_i (by the triangle inequality).

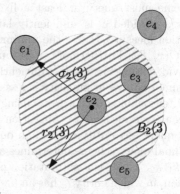

Fig. 2. A configuration of five unit radius entities. The 3-ball $B_2(3)$ of entity e_2 is shown shaded.

We have assumed that entities do not properly intersect. Define $c_{d,x}$ to be the smallest constant such that a unit-radius d-dimensional ball B can have x disjoint unit-radius d-dimensional balls (not including itself) with separation from B at most $c_{d,x}$. Thus, $\sigma_i^Z(x) \geq c_{d,x}$ and hence

$$\frac{1 - \lambda_{d,x}}{\lambda_{d,x}} \sigma_i^Z(x) \geq 1 \text{ and } r_i^Z(x) \leq \sigma_i^Z(x)/\lambda_{d,x} \qquad (2)$$

where $\lambda_{d,x} = \frac{c_{d,x}}{1+c_{d,x}}$. Observe that, for any $\xi \geq 0$, $c_{d,x} \geq \xi$ provided $x \geq (3+\xi)^d$, since unit-radius balls with separation at most ξ from B must all fit within a ball of radius $3 + \xi$ concentric with B. Thus $1/2 \leq \lambda_{d,x} < 1$ if $x \geq 4^d$.

Let X_d be the largest value of x for which $c_{d,x} = 0$ (e.g., $X_2 = 6$). Clearly, if $x \leq X_d$, there are entity configurations Z with $\sigma_i^Z(x) = 0$. Thus, for such x, maintaining uncertainty degree at most x, even at one specified target time, might be impossible in general, for any query scheme.

Remark. Hereafter *we will assume that x, our bound on congestion potential, is greater than X_d.* The constants X_d and $\lambda_{d,x}$ will factor in both the formulation and analysis of our query schemes in (arbitrary, but fixed) dimension d.

Since we assume $x > X_d$, $0 < \lambda_{d,x} < 1$. If the reader prefers to focus on dimension 2, then it will be safe to assume hereafter that $x \geq 16$ (and so $1/2 \leq \lambda_{d,x} < 1$).

3   Query Optimization at a Fixed Target Time

Suppose our goal, for a given entity set $\mathcal{E}$, is to optimize queries to guarantee low congestion potential at some fixed target time τ in the future, starting from a state of complete uncertainty of entity locations.

Since we assume unbounded uncertainty regions at the start, any query scheme must query at least a fixed fraction of the entities over the interval $[0, \tau]$, provided $\mathcal{E}$ is sufficiently large compared to the allowed potential congestion measure. Hence the minimum query granularity over this interval must be $O(\tau/|\mathcal{E}|)$. Furthermore, if granularity is not an issue, $O(|\mathcal{E}|)$ queries suffice, provided they are made sufficiently close to the target time. Maximizing the minimum query granularity is less straightforward. Nevertheless, it is clear that any reasonable query scheme using minimum query granularity γ, that guarantees a given measure of congestion potential at most x at the target time τ, will not query any entity more than once within the final n queries. Thus any such optimal query scheme determines a radius $1 + k_i\gamma$ for each entity $e_i \in \mathcal{E}$, where (i) $k_1, k_2, \ldots, k_n$ is a permutation of $1, 2, \ldots, n$, and (ii) the uncertainty configuration, in which entity e_i has an uncertainty region u_i with centre $\zeta_i(\tau - k_i\gamma)$ and radius $1 + k_i\gamma$, has the given congestion measure at most x. For any measure, we associate with $\mathcal{E}$ an *intrinsic fixed-target granularity*, defined to be the largest γ for which these conditions are satisfiable.

It is not hard to see that, by projecting the current uncertainty regions to the target time (assuming no further queries), some entities can be declared "safe" (meaning their projected uncertainty regions cannot possibly contribute to making a congestion measure, for itself or any other entity, exceed x at the target time). This idea is exploited in query schemes that query entities in rounds of geometrically decreasing duration, following each of which a subset of such "safe" entities are set aside with no further attention, until no "unsafe" entities remain.

The Fixed-Target-Time-ply (FTT-ply) query scheme. This query scheme shows that, for any Δ, $0 \le \Delta \le x$, uncertainty ply at most $x+\Delta$ can be guaranteed at a fixed target time using minimum query granularity that is at most $\Theta(\frac{x}{1+\Delta})$ smaller than that used by any (even clairvoyant) query scheme that guarantees uncertainty ply at most x. Since the uncertainty regions of all entities are unbounded at time 0, none of the entities are $(x+\Delta)$-ply-safe to start (assuming $x+\Delta < n$). Thus any scheme, including a clairvoyant scheme, must query all but x of the entities at least once in order to avoid ply greater than x at the target time. The FTT-ply$[x+\Delta]$ scheme starts by querying all entities in a single round using query granularity $\frac{\tau}{2n}$, which is $O(1)$-competitive, assuming $n - (x+\Delta) = \Omega(x+\Delta)$, with what must be done by any other scheme.

At this point, the FTT-ply$[x+\Delta]$ scheme identifies the set of n_1 entities that are not yet $(x+\Delta)$-ply-safe (the *unsafe survivors*). All other entities are set aside and attract no further queries.

The scheme then queries, in a second round, all n_1 survivors using query granularity $\frac{\tau}{4n_1}$. In general, after the rth round, the scheme identifies n_r unsafe survivors which, assuming $n_r > 0$, continue into an $(r + 1)$st round using granularity $\frac{\tau}{2^{r+1}n_r}$. The rth round completes at time $\tau - \tau/2^r$. Furthermore, all entities that have not been set aside have a projected uncertainty region whose radius is in the range $(1 + \tau/2^r, 1 + \tau/2^{r-1}]$.

Theorem 1. *For any Δ, $0 \le \Delta \le x$, the FTT-ply$[x+\Delta]$ query scheme guarantees uncertainty ply at most $x+\Delta$ at target time τ and uses minimum query granularity over the interval $[0, \tau]$ that is at most a factor $\Theta(\frac{x}{1+\Delta})$ smaller than the intrinsic fixed-target granularity for guaranteeing uncertainty ply at most x.*

Proof. We claim that any query scheme $\mathbb{S}$ that guarantees uncertainty ply at most x at time τ must use at least $\Theta(\frac{n_r(1+\Delta)}{x+\Delta})$ queries after the start of the rth query round of the FTT-ply$[x+\Delta]$ query scheme; any fewer queries would result in one or more entities having ply greater than x at the target time.

To see this observe first that each of the n_r unsafe survivors is either queried by $\mathbb{S}$ after the start of the rth query round or has its projected uncertainty ply reduced to at most x by at least $1+\Delta$ queries to its projected uncertainty neighbours after the start of the rth query round. Assuming that fewer than $n_r/2$ unsafe survivors are queried by $\mathbb{S}$ after the start of the rth query round, we argue that at least $\frac{n_r(1+\Delta)}{2 \cdot 4^d(x+\Delta)}$ queries must be made after the start of the rth query round to reduce the projected uncertainty ply of the remaining unsafe survivors to some value at most x.

Note that any query after the start of the rth round to an entity set aside in an earlier round cannot serve to lower the projected uncertainty ply of any of the n_r unsafe survivors. Furthermore, any query to one of the survivors of the $(r - 1)$st round can serve to decrease by one the projected uncertainty ply of at most $4^d(x+\Delta)$ of the unsafe survivors whose projected uncertainty ply is at most $x+\Delta$. (This follows because (i) the projected uncertainty regions of all survivors are within a factor of 2 in size, and (ii) any collection of $4^d\hat{x}$ unit radius balls that are all contained in a ball of radius 4, must have ply at least $\hat{x}$.) Thus any scheme that guarantees uncertainty ply at most x at time τ must make at least $\frac{n_r(1+\Delta)}{2\cdot4^d(x+\Delta)}$ queries after the start of the rth query round.

Since query scheme $\mathbb{S}$ must use at least $\frac{n_r(1+\Delta)}{2\cdot4^d(x+\Delta)} \geq \frac{n_r}{4^{d+1}}\frac{1+\Delta}{x}$ queries over the interval $[\tau - \tau/2^{r-1}, \tau]$, it follows that our query scheme is $\Theta(\frac{x}{1+\Delta})$-competitive, in terms of minimum query granularity, with any, even clairvoyant, query scheme that guarantees uncertainty ply at most x at the target time. □

The Fixed-Target-Time-degree (FTT-degree) query scheme. This somewhat more involved query scheme shows that for any Δ, $0 \leq \Delta \leq x$, uncertainty degree at most $x+\Delta$ can be guaranteed at a fixed target time using minimum query granularity that is at most $\Theta(\frac{x}{1+\Delta})$ smaller than that used by any query scheme that guarantees uncertainty degree at most x. As before, since the projected uncertainty regions of all entities are unbounded at time 0, any scheme, including a clairvoyant scheme, must query all but x of the entities at least once in order to avoid degree (and also ply) greater than x at the target time. The FTT-degree$[x+\Delta]$ scheme starts by querying all entities in a single round using query granularity $\frac{\tau}{2n}$, which is $O(1)$-competitive, assuming $n-(x+\Delta) = \Omega(x+\Delta)$, with what must be done by any other scheme.

At this point, the FTT-degree$[x+\Delta]$ scheme identifies two sets of entities (i) the n_1 entities that are not yet $(x+\Delta)$-degree-safe (the *unsafe survivors*), and (ii) the m_1 entities that are $(x+\Delta)$-degree-safe and whose projected uncertainty region intersects the projected uncertainty region of one or more of the unsafe survivors (the *safe survivors*). All other entities are set aside and attract no further queries.

The scheme then queries, in a second round, all n_1+m_1 survivors using query granularity $\frac{\tau}{4(n_1+m_1)}$. In general, after the rth round, the scheme identifies n_r unsafe survivors and m_r safe survivors, which, assuming $n_r + m_r > 0$, continue into an $(r+1)$st round using granularity $\frac{\tau}{2^{r+1}(n_r+m_r)}$. The rth round completes at time $\tau - \tau/2^r$. Furthermore, all entities that have not been set aside have a projected uncertainty region whose radius is in the range $(1+\tau/2^r, 1+\tau/2^{r-1}]$.

Theorem 2. *For any Δ, $0 \leq \Delta \leq x$, the FTT-degree$[x+\Delta]$ query scheme guarantees uncertainty degree at most $x+\Delta$ at target time τ and uses minimum query granularity over the interval $[0, \tau]$ that is at most a factor $\Theta(\frac{x}{1+\Delta})$ smaller than the intrinsic fixed-target granularity for guaranteeing uncertainty degree at most x.*

Proof. We claim that any query scheme $\mathbb{S}$ that guarantees uncertainty degree at most x at time τ must use at least $\Theta(\frac{(n_r+m_r)(1+\Delta)}{x+\Delta})$ queries after the start of the rth query round of the FTT-degree$[x+\Delta]$ query scheme; any fewer queries would result in one or more entities having degree greater than x at the target time.

To see this observe first that each of the n_r unsafe survivors must be *satisfied*, meaning, is either queried by $\mathbb{S}$ after the start of the rth query round or has its projected uncertainty degree reduced below x by at least $1+\Delta$ queries to its projected uncertainty neighbours after the start of the rth query round. Assuming that fewer than $n_r/2$ unsafe survivors are queried by $\mathbb{S}$ after the start of the rth query round, we argue that at least $\frac{n_r(1+\Delta)}{2 \cdot 4^d(x+\Delta)}$ queries must be made after the start of the rth query round to reduce below x the projected uncertainty degree of the remaining unsafe survivors.

Note that any query after the start of the rth round to an entity set aside in an earlier round cannot serve to lower the projected uncertainty degree of any of the n_r unsafe survivors. As in the proof of Theorem 1, any query to one of the survivors of the $(r-1)$st round can serve to decrease by one the projected uncertainty degree of at most $4^d(x+\Delta)$ of the unsafe survivors whose uncertainty degree is at most $x+\Delta$. Thus any scheme that guarantees uncertainty degree at most x at time τ must make at least $\frac{n_r(1+\Delta)}{2 \cdot 4^d(x+\Delta)}$ queries after the start of the rth query round.

Similarly, observe that each of the m_r safe survivors must have each of its unsafe neighbours satisfied in the sense described above. But, since the projected uncertainty regions of all survivors are within a factor of 2 in size, each query that serves to lower the projected uncertainty degree of an unsafe neighbour of some safe survivor e_i must be to an entity e_j that has the projected uncertainty region of e_i in its projected uncertainty *near-neighbourhood* (the ball centred at z_j, whose radius is 9 times the projected uncertainty radius of e_j). But e_j has at most $18^d(x+\Delta)$ such safe near-neighbours, since any collection of $18^d\hat{x}$ unit radius balls that are all contained in a ball of radius 18, must have degree (and also ply) at least $\hat{x}$.

It follows that, even if a query to e_j lowers the projected uncertainty degree of all of the unsafe neighbours of e_i, a total of at least $\frac{m_r(1+\Delta)}{18^d(x+\Delta)}$ queries must be made after the start of the rth query round by any scheme that guarantees uncertainty degree at most x at time τ.

Thus, query scheme $\mathbb{S}$ must use at least

$$\max\left\{ \frac{n_r(1+\Delta)}{2 \cdot 4^d(x+\Delta)}, \frac{m_r(1+\Delta)}{18^d(x+\Delta)} \right\} \geq \frac{n_r + m_r}{2 \cdot (18)^d} \frac{1+\Delta}{x+\Delta} \geq \frac{n_r + m_r}{4 \cdot (18)^d} \frac{1+\Delta}{x}$$

queries over the interval $[\tau - \tau/2^{r-1}, \tau]$. It follows that our query scheme is $\Theta(\frac{x}{1+\Delta})$-competitive, in terms of minimum query granularity, with any, even clairvoyant, query scheme that guarantees uncertainty degree at most x at the target time.   □

The competitive factor in both of the preceding theorems is worst-case optimal. Specifically, the following example demonstrates that, for $0 \leq \Delta < x$ degree at most x can be guaranteed at a fixed target time by a clairvoyant scheme that uses query granularity one, yet any non-clairvoyant scheme that guarantees ply at most $x+\Delta$ at the target time must use query granularity that is $O(\frac{x}{1+\Delta})$.

Example 1. Imagine a configuration involving two collections A and B each with $(x+1+\Delta)/2$ point entities located in $\mathbb{R}^1$, on opposite sides of a point O. At time 0 all of the entities are at distance $x+3+3\Delta$ from O, but have unbounded uncertainty regions. All entities begin by moving towards O at unit speed, but at time $x+1+\Delta$ a subset of $1+\Delta$ entities in both A and B (the *special* entities) change direction and move away from O at unit speed, while all of the others carry on until the target time $x+3+3\Delta$ when they simultaneously reach O and stop.

To avoid uncertainty degree greater than x at the target time a clairvoyant scheme needs only to (i) query all entities (in arbitrary order) up to time $x+1+\Delta$, and then (ii) query just the special entities (in arbitrary order) in the next $2(1+\Delta)$ time prior to the target, using query granularity 1, since doing so will leave the uncertainty regions of the $(x+1+\Delta)/2$ entities in A disjoint from the uncertainty regions of the $1+\Delta$ special entities in B, and vice versa.

On the other hand, to avoid ply $x+1+\Delta$ at the target time any non-clairvoyant scheme must query at least one of the special entities (in either A or B) in the last $2(1+\Delta)$ time before the target. Since special and non-special entities are indistinguishable before this time interval, at least $x/2+1$ entities in at least one of A or B must be queried in the last $2(1+\Delta)$ time before the target in order to be sure that at least one special entity is queried late enough to confirm that its uncertainty region will not contain O at the target time. This requires query granularity at most $\frac{4(1+\Delta)}{x+2}$. Thus in the worst case every scheme that achieves uncertainty ply at most $x+\Delta$ at the target time needs to use at least a factor $\Theta(\frac{x}{1+\Delta})$ smaller query granularity on some instances than the best query scheme for achieving uncertainty degree at most x at the target time on those same instances. □

Theorems 1 and 2 speak to the query frequency requirements for bounding congestion at a fixed target time, measured in terms of ply or degree individually. This leaves open the question of how these measures relate to one another. The following example demonstrates that in some cases the granularity required to bound congestion degree by $x+\Delta$ can be a factor $\Theta(\frac{x}{1+\Delta})$ smaller than that required to bound congestion ply by x.

Example 2. The example involves two clusters A and B of $(x+1+\Delta)/2$ point entities separated by distance $4(1+\Delta)$. To maintain uncertainty ply at most x it suffices to query $1+\Delta$ entities in both clusters once every $2(1+\Delta)$ steps, which can be achieved with query frequency one. Since the uncertainty regions associated with queried entities in cluster A never intersect the uncertainty regions associated with queried entities in cluster B, the largest possible ply involves entities in one cluster (say A) together with unqueried entities in the other cluster (B), for a total of x.

On the other hand, to maintain degree at most $x+\Delta$ no uncertainty region can be allowed to have radius $4(1+\Delta)$. Thus all $x+1+\Delta$ entities need to be queried with frequency at least $1/(4(1+\Delta))$, giving a total query demand of $x+1+\Delta$ over any time interval of length $4(1+\Delta)$. □

Nevertheless, bounding congestion degree at a fixed target time cannot be too much worse than bounding congestion ply.

Theorem 3. *The FTT-degree[$x+\Delta$] scheme uses a query granularity that is at most a factor $\frac{x^2}{1+\Delta}$ smaller than the best, even clairvoyant, scheme that guarantees ply at most x.*

Proof. (Sketch) The idea is that, in FTT-degree[$x+\Delta$], m_r (the number of safe survivors in round r) is $O(n_r(x+\Delta))$. Thus the "extra" queries (to handle the safe survivors) are at most a factor $x+\Delta$ more numerous than the queries to handle the unsafe survivors. If FTT-ply[$x+\Delta$] is modified so that the queries in round r occur with granularity $\frac{\tau}{2^{r+1}n_r}$ (i.e. half of their previous granularity), completing at the midpoint of the round, and FTT-degree[$x+\Delta$] is modified so that the queries in round r occur with granularity $\frac{\tau}{2^{r+1}(n_r+m_r)}$ (i.e. half of their previous granularity), starting at the midpoint of the round. Since the queries of FTT-degree[$x+\Delta$] in round r now occur after the corresponding queries in FTT-ply[$x+\Delta$], it is straightforward to see that the unsafe survivors in round r of FTT-degree[$x+\Delta$] are no more numerous than the unsafe survivors in round r of FTT-ply[$x+\Delta$]. It follows that the granularity of queries in FTT-degree[$x+\Delta$] is no more than a factor $\Theta(x+\Delta)$ smaller than that of FTT-ply[$x+\Delta$]. □

4   Towards Continuous Query Optimization

If the configuration of n entities $\mathcal{E}$ does not change over time, uncertainty degree at most x can be maintained on a continuous basis using granularity at least $\Theta(\frac{\gamma_{\mathcal{E},x}}{\ln n})$, where $\gamma_{\mathcal{E},x}$ denotes the intrinsic fixed-target granularity required to achieve uncertainty degree at most x at any fixed time. Furthermore, a simple example shows that this $\Theta(\frac{1}{\ln n})$ gap is unavoidable in the worst case. See [7, Appendix C] for details.

While the case of stationary entities exhibits some of the difficulties in maintaining uncertainty regions with low congestion, mobile entities add an additional level of complexity. Since an $\mathcal{E}$-configuration may now change over time, we add a parameter t to our stationary definitions, and refer to $B_i(x,t)$, $\sigma_i(x,t)$, and $r_i(x,t)$ in place of their stationary counterparts at time t, where it is understood that the configuration in question is just the $\mathcal{E}$-configuration at time t.

Perception Versus Reality. For any query scheme, the true location of a moving entity e_i at time t, $\zeta_i(t)$, may differ from its *perceived location*, $\zeta_i(p_i(t))$, its location at the time of its most recent query. Let $N_i(x,t)$ be e_i plus the set of x entities whose perceived locations at time t are closest to the perceived location of e_i at time t. The *perceived x-separation* of e_i at time t, denoted $\widetilde{\sigma}_i(x,t)$, is the separation between e_i and its perceived xth-nearest-neighbour at time t, i.e., $\widetilde{\sigma}_i(x,t) = \max_{e_j \in N_i(x,t)} \|\zeta_i(p_i(t)) - \zeta_j(p_j(t))\| - 2$. The *perceived x-radius* of e_i at time t, denoted $\widetilde{r}_i(x,t)$, is just $1 + \widetilde{\sigma}_i(x,t)$.

Since a scheme only knows the perceived locations of the entities, it is important that each entity e_i be probed sufficiently often that its perceived x-separation $\widetilde{\sigma}_i(x,t)$ closely approximates its true x-separation $\sigma_i(x,t)$ at all times t. It turns out that once a close relationship between perception and reality has been established, it can be sustained by ensuring that the time between queries to an entity is bounded by some small fraction of its perceived x-separation. See [7, Lemma 8] for details.

Prior to performing any queries, our perception of the x-separation between entities is far from reality. Fortunately, this close relationship can be initialized at some time t_0 by using a modified version of the FTT-degree$[x + \Delta]$ scheme of Sect. 3, using higher query frequency and a more restrictive criterion than $(x + \Delta)$-degree-safety. See [7, Lemma 9] for details.

5   Discussion

To this point, we have assumed that the distance between entity centres is always at least 2 (i.e. entities never properly intersect), and that the encroachment threshold is exactly 2 (i.e. we are only concerned with avoiding entity contact). However, without changing the units of distance and time, we can model a collection of unit-radius entities, any pair of which possibly intersect but whose centres always maintain distance at least some positive constant $\rho_0 < 2$, by simply scaling the constant $c_{d,x}$ by $\rho_0/2$ and the constant $\lambda_{d,x}$ accordingly. Similarly (and simultaneously), we can model a collection of unit-radius entities with encroachment threshold $\Xi > 2$ by (i) changing the *basic uncertainty radius* (the radius of the uncertainty region of an entity immediately after it has been queried) to $\Xi/2$ thereby ensuring that entities with disjoint uncertainty regions do not encroach on one another, and (ii) changing X_d to be the largest x such that $c_{d,x} \geq \Xi - 2$ since for x exceeding this changed X_d there can be at most $x - 1$ entities that are within the encroachment threshold of any fixed entity.

This flexibility makes it possible to relax our assumption that location queries are answered exactly, since potential error in the response to location queries can be modelled by an increase in the basic uncertainty radius. Furthermore, it increases significantly the scope of applications of our results.

Recall the problem concerning collision avoidance mentioned in the introduction, where entity encroachment might reasonably be held to hold well before contact. It follows from our results that, by achieving uncertainty degree at most x at time τ, we obtain for each entity e_i a certificate identifying the, at most $x - 1$, other entities that could potentially encroach upon e_i at that time (those warranting more careful local monitoring). An additional application, considered in [5], concerns entities that are mobile transmission sources with associated broadcast ranges, that one would expect might sometimes properly intersect, where the goal is to minimize the number of broadcast channels at time τ so as to eliminate potential transmission interference. In this case, achieving uncertainty thickness at most x using minimum query frequency serves to obtain a fixed bound on the number of broadcast channels required at time τ, an objective that seems to be at least as well-motivated as optimizing the number of channels for a fixed query frequency (the objective in [5]).

References

1. Basch, J., Guibas, L.J., Hershberger, J.: Data structures for mobile data. J. Algorithms **31**(1), 1–28 (1999)
2. de Berg, M., Roeloffzen, M., Speckmann, B.: Kinetic compressed quadtrees in the black-box model with applications to collision detection for low-density scenes. In: Epstein, L., Ferragina, P. (eds.) ESA 2012. LNCS, vol. 7501, pp. 383–394. Springer, Heidelberg (2012). https://doi.org/10.1007/978-3-642-33090-2_34
3. de Berg, M., Roeloffzen, M., Speckmann, B.: Kinetic convex hulls, Delaunay triangulations and connectivity structures in the black-box model. J. Comput. Geom. **3**(1), 222–249 (2012)
4. de Berg, M., Roeloffzen, M., Speckmann, B.: Kinetic 2-centers in the black-box model. In: Symposium on Computational Geometry, pp. 145–154 (2013)
5. Busto, D., Evans, W., Kirkpatrick, D.: Minimizing interference potential among moving entities. In: Proceedings of the ACM-SIAM Symposium on Discrete Algorithms (SODA), pp. 2400–2418 (2019)
6. Erlebach, T., Hoffmann, M.: Query-competitive algorithms for computing with uncertainty. Bull. Eur. Assoc. Theor. Comput. Sci. **2**(116) (2015)
7. Evans, W., Kirkpatrick, D.: Frequency-competitive query strategies to maintain low congestion potential among moving entities (2023). arXiv:2205.09243
8. Evans, W., Kirkpatrick, D., Löffler, M., Staals, F.: Competitive query strategies for minimising the ply of the potential locations of moving points. In: Symposium on Computational Geometry, pp. 155–164 (2013)
9. Evans, W., Kirkpatrick, D., Löffler, M., Staals, F.: Query strategies for minimizing the ply of the potential locations of entities moving with different speeds. In: Abstr. 30th European Workshop on Computational Geometry (EuroCG) (2014)
10. Evans, W., Kirkpatrick, D., Löffler, M., Staals, F.: Minimizing co-location potential of moving entities. SIAM J. Comput. **45**(5), 1870–1893 (2016)
11. Guibas, L.J.: Kinetic data structures: a state of the art report. In: Proceedings of the Third Workshop on the Algorithmic Foundations of Robotics on Robotics: The Algorithmic Perspective, pp. 191–209. WAFR '98, A. K. Peters Ltd, USA (1998)
12. Guibas, L.J., Roeloffzen, M.: Modeling motion. In: Toth, C.D., O'Rourke, J., Goodman, J.E. (eds.) Handbook of Discrete and Computational Geometry, chap. 53, pp. 1401–1420. CRC Press (2017)
13. Kahan, S.: A model for data in motion. In: Twenty-third Annual ACM Symposium on Theory of Computing, pp. 265–277. STOC '91 (1991)
14. Kahan, S.: Real-Time Processing of Moving Data. Ph.D. thesis, University of Washington (1991)
15. Löffler, M., Snoeyink, J.: Delaunay triangulation of imprecise points in linear time after preprocessing. Comput. Geom.: Theory Appl. **43**(3), 234–242 (2010)

Complexity of Conformant Election Manipulation

Zack Fitzsimmons[1]([⊠]) and Edith Hemaspaandra[2]

[1] College of the Holy Cross, Worcester, MA 01610, USA
zfitzsim@holycross.edu
[2] Rochester Institute of Technology, Rochester, NY 14623, USA

Abstract. It is important to study how strategic agents can affect the outcome of an election. There has been a long line of research in the computational study of elections on the complexity of manipulative actions such as manipulation and bribery. These problems model scenarios such as voters casting strategic votes and agents campaigning for voters to change their votes to make a desired candidate win. A common assumption is that the preferences of the voters follow the structure of a domain restriction such as single peakedness, and so manipulators only consider votes that also satisfy this restriction. We introduce the model where the preferences of the voters define their own restriction and strategic actions must "conform" by using only these votes. In this model, the election after manipulation will retain common domain restrictions. We explore the computational complexity of conformant manipulative actions and we discuss how conformant manipulative actions relate to other manipulative actions.

1 Introduction

The computational study of election problems is motivated by the utility of elections to aggregate preferences in multiagent systems and to better understand the computational tradeoffs between different rules. A major direction in this area has been to study the computational complexity of manipulative actions on elections (see, e.g., Faliszewski and Rothe [15]).

The problems of manipulation [1] and bribery [14] in elections represent two important ways that agent(s) can strategically affect the outcome of an election. Manipulation models the actions of a collection of strategic voters who seek to ensure that their preferred candidate wins by casting strategic votes. Bribery models the actions of an agent, often referred to as the briber, who sets the votes of a subcollection of the voters to ensure that the briber's preferred candidate wins. These problems each relate nicely to real-world scenarios such as how voters may attempt to work together to strategically vote, or the actions of a campaign manager looking to influence the preferences of a group of voters to ensure their candidate wins.

In the manipulation problem each manipulator can cast any strategic vote, and similarly for the bribery problem the votes can be set to any collection

H. Fernau and K. Jansen (Eds.): FCT 2023, LNCS 14292, pp. 176–189, 2023.
https://doi.org/10.1007/978-3-031-43587-4_13

of preferences. However, this is not always a reasonable assumption to make. Voters may have preferences that satisfy a domain restriction such as single-peaked preferences [6] or single-crossing preferences [27] where the manipulator or the briber are restricted to using votes that also satisfy the restriction. We introduce new models of manipulation and bribery where the votes cast by the manipulators or set by the briber must have already been stated in the election, i.e., the votes must *conform* to the views of the electorate. In these conformant models of manipulation and bribery the election after the given manipulative action will retain common domain restrictions such as being single-peaked or single-crossing.

We consider how the computational complexity of our conformant models of manipulation and bribery compare to the standard models. Specifically, we show that there are settings where manipulation and bribery are easy in the standard model, but computationally difficult in the conformant model, and vice versa. This shows that there is no reduction in either direction between the standard and conformant cases (unless P = NP).

Conformant manipulation and bribery are also each related to electoral control. The study of electoral control was introduced by Bartholdi, Tovey, and Trick [2], and it models the actions of an agent who can modify the structure of an election to ensure a preferred candidate wins (e.g., by adding or deleting voters). We explore the connection between conformant manipulation and the exact variant of voter control as well as conformant bribery and the model of control by replacing voters introduced by Loreggia et al. [26]. This includes showing reductions between these problems as well as showing when such reductions cannot exist. Exact versions of electoral control problems can model scenarios where the election chair seeks to ensure their preferred outcome by adding exactly the number of voters required to meet the quorum for a vote. This is in line with the standard motivation for control by replacing voters which includes settings such as voting in a parliament where the chair may replace some of the voters, but makes sure to leave the total number the same to avoid detection [26].

Our main contributions are as follows.

- We introduce the problems of conformant manipulation and conformant bribery, which model natural settings for manipulative attacks on elections.
- We show that there is no reduction in either direction between the standard and conformant versions of manipulation (Theorems 1 and 2) and between the standard and conformant versions of bribery (Theorem 8 and Corollary 2) (unless P = NP).
- We show conformant manipulation reduces to exact control by adding voters (Theorem 3), conformant bribery reduces to control by replacing voters (Theorem 11), and that reductions do not exist in the other direction (Theorems 4 and 12) (unless P = NP).
- We obtain a trichotomy theorem for the complexity of exact control by adding voters problem (Theorem 5) for the important class of pure scoring rules.

Due to space constraints some of our proofs have been deferred to the full version [17].

2   Related Models

Since the conformant model of manipulation uses only the votes stated in the initial election, it is related to the possible winner problem with uncertain weights introduced by Baumeister et al. [3], which was recently extended by Neveling et al. [29]. In this problem weights of voters are initially unset and it asks if there exists a way to set the weights such that a given candidate is a winner. As Baumeister et al. [3] mention, this generalizes control by adding/deleting voters (rather than manipulation or bribery).

There are other models for manipulative actions that have a similar motivation to our conformant models, i.e., to have the manipulative action not stand out. Examples are bribery restrictions where the briber cannot put the preferred candidate first in the bribed votes (negative bribery [14]), restrictions on how much the votes of voters are changed [34], and restrictions on which voters càn be bribed [10]. However, it is easy to see that in these models all sorts of new votes can be used by the briber, not just votes appearing in the initial election.

Another model in which the votes of the strategic agents are restricted is that of manipulative actions on restricted domains such as single-peaked [6] and single-crossing preferences [27]. For example, for manipulation of a single-peaked election the manipulators must all cast votes that are single-peaked with respect to the rest of the electorate [32]. Notice that in the conformant models we typically keep the domain restriction, but manipulative actions for domain restrictions are quite different since in those settings the manipulators can cast a vote not stated by any of the nonmanipulators as long as it satisfies the given restriction.

3   Preliminaries

An election (C, V) consists of a set of candidates C and a collection of voters V. Each voter $v \in V$ has a corresponding vote, which is a strict total order preference over the set of candidates.

A voting rule, $\mathcal{E}$, is a mapping from an election to a subset of the candidate set referred to as the winner(s).

3.1   Scoring Rules

Our results focus on (polynomial-time uniform) pure scoring rules. A scoring rule is a voting rule defined by a family of scoring vectors of the form $\langle \alpha_1, \alpha_2, \ldots, \alpha_m \rangle$ with $\alpha_i \geq \alpha_{i+1}$ such that for a given election with m candidates the m-length scoring vector is used and each candidate receives α_i points for each vote where they are ranked ith. The candidate(s) with the highest score win. We use the notation score(a) to denote the score of a candidate a in a given election. Important examples of scoring rules are

- k-Approval, $\langle \underbrace{1, \ldots, 1}_{k}, 0, \ldots, 0 \rangle$

- k-Veto, $\langle 0, \ldots, 0, \underbrace{-1, \ldots, -1}_{k} \rangle$
- Borda, $\langle m - 1, m - 2, \ldots, 1, 0 \rangle$
- First-Last, $\langle 1, 0, \ldots, 0, -1 \rangle$[1]

Note that there are uncountably many scoring rules, that scoring vectors may not be computable, and that the definition of scoring rule does not require any relationship between the scoring vectors for different numbers of candidates. To formalize the notion of a natural scoring rule, we use the notion of *(polynomial-time uniform) pure scoring rules* [4]. These are families of scoring rules where the scoring vector for $m + 1$ candidates can be obtained from the scoring vector for m candidates by adding one coefficient and for which there is a polynomial-time computable function that outputs, on input 0^m, the scoring vector for m. Note that the election rules above are all pure scoring rules. Also note that manipulative action problems for pure scoring rules are in NP.

3.2 Manipulative Actions

Two of the most commonly-studied manipulative actions on elections are manipulation [1] and bribery [14]. We consider conformant variants of these standard problems by requiring that the strategic votes cast by the manipulators or set by the briber must have appeared in the initial election. We define these problems formally below.

Name: $\mathcal{E}$-Conformant Manipulation
Given: An election (C, V), a collection of manipulative voters W, and a preferred candidate p.
Question: Does there exist a way to set the votes of the manipulators in W using only the votes that occur in V such that p is an $\mathcal{E}$-winner of the election $(C, V \cup W)$?

Name: $\mathcal{E}$-Conformant Bribery
Given: An election (C, V), a bribe limit k, and a preferred candidate p.
Question: Does there exist a way to set the preferences of a subcollection of at most k voters in V to preferences in V such that p is an $\mathcal{E}$-winner?

3.3 Computational Complexity

We assume the reader is familiar with the complexity classes P and NP, polynomial-time many-one reductions, and what it means to be complete for a given class. Our NP-completeness proofs utilize reductions from the well-known NP-complete problem 3-Dimensional Matching [23].

Name: 3-Dimensional Matching (3DM)

---

[1] We will see that this rule exhibits very unusual complexity behavior. This rule has also been referred to as "best-worst" in social choice (see, e.g., [24]).

Given: Pairwise disjoint sets X, Y, and Z such that $\|X\| = \|Y\| = \|Z\| = k > 0$ and set $\mathcal{M} \subseteq X \times Y \times Z$.

Question: Does there exist a $\mathcal{M}' \subseteq \mathcal{M}$ of size k such that each $a \in X \cup Y \cup Z$ appears exactly once in $\mathcal{M}'$.

3-Dimensional Matching remains NP-complete when each element $a \in X \cup Y \cup Z$ appears in exactly three triples (Garey and Johnson [19] show this for at most three triples, which can be adapted to exactly three by the approach from Papadimitriou and Yannakakis [30]). Note that in that case $\|\mathcal{M}\| = 3k$.

For our polynomial-time algorithms, we will reduce to polynomial-time computable (edge) matching problems. We define the most general version we use, Max-Weight b-Matching for Multigraphs, below. The version of this problem for simple graphs was shown to be in P by Edmonds and Johnson [12], and as explained in [20, Section 7], it is easy to reduce such problems to Max-Weight Matching, which is well-known to be in P [11], using the construction from [31]. (Note that we can assume that the b-values are bound by the number of edges in the graph.)

Name: Max-Weight b-Matching for Multigraphs

Given: An edge-weighted multigraph $G = (V, E)$, a function $b : V \to \mathbb{N}$, and integer $k \geq 0$.

Question: Does there exist an $E' \subseteq E$ of weight at least k such that each vertex $v \in V$ is incident to at most $b(v)$ edges in E'?

In addition to NP-hardness and polynomial-time results, we have results that link the complexity of voting problems to the complexity of Exact Perfect Bipartite Matching [30].

Name: Exact Perfect Bipartite Matching

Given: A bipartite graph $G = (V, E)$, a set of red edges $E' \subseteq E$, and integer $k \geq 0$.

Question: Does G contain a perfect matching that contains exactly k edges from E'?

This problem was shown to be in RP by Mulmuley, Vazirani, and Vazirani [28], but it is a 40-year-old open problem whether it is in P.

4   Conformant Manipulation

The problem of manipulation asks if it is possible for a given collection of manipulative voters to set their votes so that their preferred candidate wins. This problem was first studied computationally by Bartholdi, Tovey, and Trick [1] for the case of one manipulator and generalized by Conitzer, Sandholm, and Lang [8] for the case of a coalition of manipulators.

In our model of conformant manipulation the manipulators can only cast votes that at least one nonmanipulator has stated. As mentioned in the introduction, this is so that the manipulators vote realistic preferences for the given election by conforming to the preferences already stated. Since this modifies the standard model of manipulation, we will consider how the complexity of these problems relate to one another.

Manipulation is typically easy for scoring rules. Hemaspaandra and Schnoor [22] showed that manipulation is in P for every pure scoring rule with a constant number of different coefficients. However, our model of conformant manipulation is NP-complete for even the simple rule 4-approval. To show hardness, we use a construction similar to the construction that shows hardness for control by adding voters from Lin [25].

Theorem 1. *4-Approval Conformant Manipulation is* NP-*complete.*

Proof. Let the pairwise disjoint sets X, Y, and Z such that $\|X\| = \|Y\| = \|Z\| = k > 0$ and $\mathcal{M} \subseteq X \times Y \times Z$ be an instance of 3DM where each $a \in X \cup Y \cup Z$ appears in exactly three triples. Note that $\|\mathcal{M}\| = 3k$. We construct an instance of conformant manipulation as follows.

Let the candidate set C consist of preferred candidate p, and for each a in $X \cup Y \cup Z$, we have candidate a and three padding candidates a_1, a_2, and a_3. We now construct the collection of nonmanipulators.

- For each (x, y, z) in $\mathcal{M}$, we have a voter voting $p > x > y > z > \cdots$.
- For each a in $X \cup Y \cup Z$, we have $4k - 4$ voters voting $a > a_1 > a_2 > a_3 > \cdots$.

We have k manipulators.

Note that we have the following scores from the nonmanipulators. $\text{score}(p) = 3k$ and for $a \in X \cup Y \cup Z$, $\text{score}(a) = 4k - 4 + 3 = 4k - 1$ and $\text{score}(a_1) = \text{score}(a_2) = \text{score}(a_3) = 4k - 4$.

If there is a matching, let the k manipulators vote corresponding to the matching. Then p's score increases by k, for a total of $4k$, and for each $a \in X \cup Y \cup Z$, $\text{score}(a)$ increases by 1 for a total of $4k$. The scores of the dummy candidates remain unchanged. Thus, p is a winner.

For the converse, suppose the manipulators vote such that p is a winner. Since $k > 0$, after manipulation there is a candidate a in $X \cup Y \cup Z$ with score at least $4k$. The highest possible score for p after manipulation is $4k$, and this happens only if p is approved by every manipulator. It follows that every manipulator approves p and for every a in $X \cup Y \cup Z$, a is approved by at most one manipulator. This implies that the votes of the manipulators correspond to a cover. □

We just saw a case where the complexity of conformant manipulation is harder than the standard problem (unless P = NP). This is not always the case. One setting where it is clear to see how to determine if conformant manipulation is possible or not is when there are only a fixed number of manipulators. We have only a polynomial number of votes to choose from (the votes of the nonmanipulators) and so a fixed number of manipulators can brute force these choices in polynomial time as long as determining the winner can be done in polynomial time.

Theorem 2. *Conformant Manipulation is in* P *for every voting rule with a polynomial-time winner problem when there are a fixed number of manipulators.*

This behavior is in contrast to what can occur for the standard model of manipulation. One well-known example is that manipulation for the Borda rule is NP-complete even when there are only two manipulators [5,9]. Intuitively the hardness of manipulation in this case is realized by the choice of different votes that the manipulator(s) have.

Corollary 1. *For Borda, Manipulation with two manipulators is* NP-*complete, but Conformant Manipulation with two manipulators is in* P.

In some ways conformant manipulation acts more like the problem of electoral control introduced by Bartholdi, Tovey, and Trick [2], specifically *Control by Adding Voters*, which asks when given an election, a collection of unregistered voters, add limit k, and preferred candidate p, if p is a winner of the election after adding at most k of the unregistered voters. In conformant manipulation we can think of the nonmanipulative voters as describing the different votes to choose from for the manipulators.

At first glance it may appear that there is a straightforward reduction from conformant manipulation to control by adding voters, but in conformant manipulation *all* k of the manipulators must cast a vote, while in control by adding voters *at most* k votes are added. In this way conformant manipulation is closer to the "exact" variant of control by adding voters where *exactly* k unregistered voters must be added. The following is immediate.

Theorem 3. *Conformant Manipulation polynomial-time many-one reduces to Exact Control by Adding Voters.*

Below we show that conformant manipulation is in P for the voting rule 3-approval, but exact control by adding voters is NP-complete. And so there is no reduction from exact control by adding voters to conformant manipulation (unless P = NP).

Theorem 4. *For 3-Approval, Exact Control by Adding Voters is* NP-*complete, but Conformant Manipulation is in* P.

Proof. Given an election (C, V), k manipulators, and a preferred candidate p, we can determine if conformant manipulation is possible in the following way.

If there is no nonmanipulator that approves p then p is a winner if and only if there are no voters.

If there is at least one nonmanipulator that approves p, the manipulators will all cast votes that approve of p, and so we know the final score of p after manipulation: $fs_p = \text{score}(p) + k$. However, these manipulator votes will each also approve of two other candidates. To handle this we can adapt the approach used by Lin [25] to show control by adding voters is in P for 3-approval elections, which constructs a reduction to Max-Cardinality b-Matching for Multigraphs.

For each candidate $a \neq p$, let $b(a) = fs_p - \text{score}(a)$, i.e., the maximum number of approvals that a can receive from the manipulators without beating p. If $b(a)$ is negative for any candidate $a \neq p$ then conformant manipulation is not possible. For each distinct nonmanipulative vote of the form $\{p, a, b\} > \ldots$, add k edges between a and b. Conformant manipulation is possible if and only if there is a b-edge matching of size at least k.

We now consider the complexity of exact control by adding voters. Given an instance of 3-Dimensional Matching: pairwise disjoint sets X, Y, and Z such that $\|X\| = \|Y\| = \|Z\| = k$, and $\mathcal{M} \subseteq X \times Y \times Z$ with $\mathcal{M} = \{M_1, \ldots, M_{3k}\}$ we construct the following instance of exact control by adding voters.

Let the candidate set $C = \{p, d_1, d_2\} \cup X \cup Y \cup Z$, the preferred candidate be p, and the add limit be k.

Let there be one registered voter voting $p > d_1 > d_2 > \ldots$, and let the set of unregistered voters consist of one voter voting $x > y > z > \ldots$ for each $M_i = (x, y, z)$.

It is easy to see that p can be made a winner by adding exactly k unregistered voters if and only if there is a matching of size k. $\qquad\square$

The standard case of control by adding voters for 3-approval is in P [25], but as shown above the exact case is NP-complete. Related work that mentions exact control has results only where the exact variant is also easy [13,16]. Note that, as observed in [16], control polynomial-time reduces to exact control, since, for example, p can be made a winner by adding at most k voters if and only if p can be made a winner by adding 0 voters or 1 voter or 2 voters or $\ldots$), and so if exact control is easy, the standard case will be as well, and if the standard case is hard, then the exact case will be hard. Note that we are using a somewhat more flexible notion of reducibility than many-one reducibility here, since we are allowing the disjunction of multiple queries to the exact control problem. Such a reduction is called a *disjunctive truth-table (dtt) reduction*. This type of reduction is still much less flexible than a Turing reduction.

Is 3-approval special? For the infinite and natural class of pure scoring rules, Table 1 completely classifies the complexity of exact control by adding voters and compares this behavior to the complexity of control by adding voters [21] and control by deleting voters [22]. In particular, and in contrast to earlier results, we obtain a *trichotomy* theorem for exact control by adding voters.[2]

Theorem 5. *For every pure scoring rule f,*

1. *If f is ultimately (i.e., for all but a finite number of candidates) equivalent to 0-approval, 1-approval, 2-approval, 1-veto, or 2-veto, exact control by adding voters is in P.*
2. *If f is ultimately equivalent to first-last, then exact control by adding voters is (dtt) equivalent to the problem Exact Perfect Bipartite Matching [30].*

---

[2] Exact Perfect Bipartite Matching [30] is defined in Sect. 3.3. As mentioned there the complexity of this problem is still open. And so Theorem 5 is a trichotomy theorem unless we solve a 40-year-old open problem.

Table 1. This table classifies the complexity of all pure scoring rules for the specified control action. A scoring rule entry represents all pure scoring rules that are ultimately equivalent to that scoring rule. The dichotomy for control by adding voters is due to [21], the dichotomy for control by deleting voters to [22], and the result for exact control by adding voters for first-last is due to [16]. EPBM stands for the Exact Perfect Bipartite Matching, which is defined in Sect. 3.3.

	P	eq. to EPBM	NP-complete
Exact Control by Adding Voters	0/1/2-approval, 1/2-veto	first-last	all other cases
Control by Adding Voters	0/1/2/3-approval, 1/2-veto, first-last, $\langle \alpha, \beta, 0, \ldots, 0 \rangle$		all other cases
Control by Deleting Voters	0/1/2-approval, 1/2/3-veto, first-last, $\langle 0, \ldots, 0, -\beta, -\alpha \rangle$		all other cases

3. *In all other cases, exact control by adding voters is NP-complete (under dtt reductions).*

Proof.

1. The case for 0-approval is trivial, since all candidates are always winners. The 1-approval and 1-veto cases follow by straightforward greedy algorithms. The case for 2-approval can be found in [16] and the case for 2-veto is similar.
2. The case for first-last can be found in [16].
3. Note that the remaining cases are hard for control by adding voters or for control by deleting voters. We have already explained how we can dtt reduce control by adding voters to exact control by adding voters. Similarly, we can dtt reduce control by deleting voters to exact control by adding voters, since p can be made a winner by deleting at most k voters if and only p can be made a winner by adding to the empty set $n - k$ voters or $n - k + 1$ voters or $n - k + 2$ voters or … or n voters, where n is the total number of voters. It follows that all these cases are NP-complete (under dtt reductions).   □

In this section, we compared conformant manipulation with manipulation and conformant manipulation with exact control by adding voters. We can also look at conformant manipulation versus control by adding voters. Here we also find voting rules where the manipulative actions differ. In particular, control by adding voters for first-last is in P [21], but we show in the full version that conformant manipulation for first-last is equivalent to exact perfect bipartite matching.

Theorem 6. *First-Last Conformant Manipulation is equivalent to Exact Perfect Bipartite Matching (under dtt reductions).*

We also show in the full version that there exists a voting rule where control by adding voters is hard and conformant manipulation is easy.

Theorem 7. *There exists a voting rule where Control by Adding Voters is NP-complete, but Conformant Manipulation is in P.*

This shows that a reduction from control by adding voters to conformant manipulation does not exist (unless P = NP).

5   Conformant Bribery

We now turn to our model of conformant bribery. The standard bribery problem introduced by Faliszewski, Hemaspaandra, and Hemaspaandra [14] asks when given an election, a bribe limit k, and a preferred candidate p, if there exists a subcollection of at most k voters whose votes can be changed such that p is a winner. In our model of conformant bribery the votes can only be changed to votes that appear in the initial election. As with conformant manipulation, this is so that the votes changed to preferences that are still realistic with respect to the preferences already stated. Notice how this also nicely models how a voter can convince another to vote their same vote. In the same way as we did with manipulation in the previous section, we can compare the complexity behavior of our conformant model with respect to the standard model.

Bribery is in P for the voting rule 3-veto [25], but we show below that our model of conformant bribery is NP-complete for this rule.

Theorem 8. *3-Veto Conformant Bribery is NP-complete.*

Proof. Let X, Y, and Z be pairwise disjoint sets such that $\|X\| = \|Y\| = \|Z\| = k$, and $\mathcal{M} \subseteq X \times Y \times Z$ with $\mathcal{M} = \{M_1, \ldots, M_{3k}\}$ be an instance of 3DM where each element $a \in X \cup Y \cup Z$ appears in exactly three triples. We construct an instance of conformant bribery as follows.

Let the candidate set $C = \{p\} \cup X \cup Y \cup Z \cup \{p_1, p_2\}$. Let p be the preferred candidate and let k be the bribe limit. Let there be the following voters.

- For each $M_i = (x, y, z)$,
    • One voter voting $\cdots > x > y > z$
- $k + 4$ voters voting $\cdots > p > p_1 > p_2$

We view the corresponding scoring vector for 3-veto as $\langle 0, \ldots, 0, -1, -1, -1 \rangle$ to make our argument more straightforward. And so, before bribery score$(p) =$ score$(p_1) =$ score$(p_2) = -k - 4$ and for each $a \in X \cup Y \cup Z$, score$(a) = -3$.

If there exists a matching $\mathcal{M}' \subseteq \mathcal{M}$ of size k, for each $M_i \in \mathcal{M}'$ such that $M_i = (x, y, z)$ we can bribe one of the voters voting $\cdots > p > p_1 > p_2$ to vote $\cdots > x > y > z$. Since $\mathcal{M}'$ is a matching the score of each candidate $a \in X \cup Y \cup Z$ decreases by 1 to be -4, and since k of the voters vetoing p are bribed the score of p increases by k to -4 and p is a winner.

For the converse, suppose there is a successful conformant bribery. Only the voters vetoing p should be bribed, and so the score of p after bribery is -4. The score of each candidate $a \in X \cup Y \cup Z$ must decrease by at least 1, and so it is easy to see that a successful conformant bribery of at most k voters must correspond to a perfect matching.   □

We now consider the case where bribery is hard for the standard model, but easy in our conformant model. Since the briber is restricted to use only votes that appear in the initial election we have the same behavior as stated in Theorem 2 for conformant manipulation.

Theorem 9. *Conformant Bribery is in* P *for every voting rule with a polynomial-time winner problem when there is a fixed bribe limit.*

There are generally fewer results looking at a fixed bribe limit than there are looking at a fixed number of manipulators. One example is that for Single Transferable Vote (STV), bribery is NP-complete even when the bribe limit is 1 [33], but our focus is on scoring rules. Brelsford at al. [7] show bribery is NP-complete for Borda, but do not consider a fixed bribe limit. However, it is easy to adapt the NP-hardness proof for Borda manipulation with two manipulators from Davies et al. [9]. The main idea is to add two voters that are so bad for the preferred candidate that they have to be bribed.

Theorem 10. *Borda Bribery with a bribe limit of 2 is NP-complete.*

Proof. We need to following properties of the instance of Borda Manipulation with two manipulators constructed in Davies et al. [9]. (For this proof we use the notation from Davies et al. [9].) The constructed election has a collection of (nonmanipulative) voters V and $q + 3$ candidates. Preferred candidate p scores C and candidate a_{q+1} scores $2(q+2) + C$. This implies that in order for p to be a winner, the two manipulators must vote p first and a_{q+1} last. We add two voters voting $a_{q+1} > \cdots > p$. Note that such votes are very bad for p and that we have to bribe two voters voting $a_{q+1} > \cdots > p$. In order to ensure that we bribe exactly the two added voters, it suffices to observe that we can ensure in the construction from Davies et al. [9] that p is never last in any vote in V. So, p can be made a winner by bribing two voters in $V \cup \{a_{q+1} > \cdots > p, a_{q+1} > \cdots > p\}$ if and only if p can be made a winner by two manipulators in the election constructed by Davies et al. [9]. □

Corollary 2. *For Borda, Bribery is NP-complete with a bribe limit of 2, but Conformant Bribery is in* P *with a bribe limit of 2.*

Bribery can be thought of as control by deleting voters followed by manipulation. For conformant bribery we can see that the same will hold, just with conformant manipulation. However, we also have a correspondence to the problem of control by replacing voters introduced by Loreggia et al. [26]. Control by replacing voters asks when given an election, a collection of unregistered voters, parameter k, and preferred candidate p, if p can be made a winner by replacing at most k voters in the given election with a subcollection of the unregistered voters of the same size. It is straightforward to reduce conformant bribery to control by replacing voters (for each original voter v, we have a registered copy of v and k unregistered copies of v), and so we inherit polynomial-time results from this setting.

Theorem 11. *Conformant Bribery polynomial-time many-one reduces to Control by Replacing Voters.*

It's natural to ask if there is a setting where conformant bribery is easy, but control by replacing voters is hard, and we show in the full version that this is in fact the case.

Theorem 12. *There exists a voting rule where Control by Replacing Voters is NP-complete, but Conformant Bribery is in* P.

In the related work on control by replacing voters, only the complexity for 2-approval remained open (see Erdélyi et al. [13]). This was recently shown to be in P by Fitzsimmons and Hemaspaandra [18]. This result immediately implies that conformant bribery for 2-approval is also in P.

Theorem 13. *2-Approval Conformant Bribery is in* P.

2-approval appears right at the edge of what is easy. For 3-approval, control by deleting voters and bribery are hard [25], control by replacing voters is hard [13], and we show in the full version that conformant bribery is hard as well (recall that for 3-approval, conformant manipulation (Theorem 4) and control by adding voters [25] are easy, essentially because all we are doing is "adding" votes that approve p).

Theorem 14. *3-Approval Conformant Bribery is NP-complete.*

As a final note, we mention that for first-last, conformant bribery, like conformant manipulation, is equivalent to exact perfect bipartite matching again showing the unusual complexity behavior of this rule.

Theorem 15. *First-Last Conformant Bribery is equivalent to Exact Perfect Bipartite Matching (under dtt reductions).*

The proof of the above theorem can be found in the full version.

6   Conclusion

The conformant models of manipulation and bribery capture a natural setting for election manipulation. We found that there is no reduction between the standard and conformant models in either direction (unless $P = NP$), and further explored the connection between these models and types of electoral control.

We found the first trichotomy theorem for scoring rules. This theorem concerns the problem of exact control by adding voters and highlights the unusual complexity behavior of the scoring rule first-last. We show that this unusual complexity behavior also occurs for our conformant models.

We also observed interesting behavior for exact variants of control, including a nontrivial case where the complexity of a problem increases when going from the standard to the exact case.

We see several interesting directions for future work. For example, we could look at conformant versions for other bribery problems (e.g., priced bribery) or for restricted domains such as single-peakedness. We are also interested in further exploring the complexity landscape of problems for the scoring rule first-last.

Acknowledgements. This work was supported in part by grant NSF-DUE-1819546. We thank the anonymous reviewers for their helpful comments and suggestions.

References

1. Bartholdi, J., Tovey, C., Trick, M.: The computational difficulty of manipulating an election. Soc. Choice Welfare **6**(3), 227–241 (1989)
2. Bartholdi, J., Tovey, C., Trick, M.: How hard is it to control an election? Math. Comput. Model. **16**(8/9), 27–40 (1992)
3. Baumeister, D., Roos, M., Rothe, J., Schend, L., Xia, L.: The possible winner problem with uncertain weights. In: Proceedings of the 20th European Conference on Artificial Intelligence, pp. 133–138 (2012)
4. Betzler, N., Dorn, B.: Towards a dichotomy of finding possible winners in elections based on scoring rules. J. Comput. Syst. Sci. **76**(8), 812–836 (2010)
5. Betzler, N., Niedermeier, R., Woeginger, G.: Unweighted coalitional manipulation under the Borda rule is NP-hard. In: Proceedings of the 22nd International Joint Conference on Artificial Intelligence, pp. 55–60 (2011)
6. Black, D.: On the rationale of group decision-making. J. Polit. Econ. **56**(1), 23–34 (1948)
7. Brelsford, E., Faliszewski, P., Hemaspaandra, E., Schnoor, H., Schnoor, I.: Approximability of manipulating elections. In: Proceedings of the 23rd National Conference on Artificial Intelligence, pp. 44–49 (2008)
8. Conitzer, V., Sandholm, T., Lang, J.: When are elections with few candidates hard to manipulate? J. ACM **54**(3), 1–33 (2007)
9. Davies, J., Katsirelos, G., Narodytska, N., Walsh, T., Xia, L.: Complexity of and algorithms for the manipulation of Borda, Nanson's and Baldwin's voting rules. Artif. Intell. **217**, 20–42 (2014)
10. Dey, P., Misra, N., Narahari, Y.: Frugal bribery in voting. Theor. Comput. Sci. **676**, 15–32 (2017)
11. Edmonds, J.: Maximum matching and a polyhedron with 0,1-vertices. J. Res. National Bureau Standards–B. Math. Math. Phys. **69B**(1/2), 125–130 (1965)
12. Edmonds, J., Johnson, E.: Matching: a well-solved class of integer linear programs. In: Combinatorial Structures and Their Applications (Gordon and Breach), pp. 89–92 (1970)
13. Erdélyi, G., Neveling, M., Reger, C., Rothe, J., Yang, Y., Zorn, R.: Towards completing the puzzle: complexity of control by replacing, adding, and deleting candidates or voters. Auton. Agent. Multi-Agent Syst. **35**(41), 1–48 (2021)
14. Faliszewski, P., Hemaspaandra, E., Hemaspaandra, L.: How hard is bribery in elections? J. Artif. Intell. Res. **35**, 485–532 (2009)
15. Faliszewski, P., Rothe, J.: Control and bribery in voting. In: Handbook of Computational Social Choice, pp. 146–168. Cambridge University Press (2016)
16. Fitzsimmons, Z., Hemaspaandra, E.: Insight into voting problem complexity using randomized classes. In: Proceedings of the 31st International Joint Conference on Artificial Intelligence, pp. 293–299 (2022)

17. Fitzsimmons, Z., Hemaspaandra, E.: Complexity of conformant election manipulation. Tech. Rep. arXiv:2307.11689 [cs.GT], arXiv.org (2023)
18. Fitzsimmons, Z., Hemaspaandra, E.: Using weighted matching to solve 2-approval/veto control and bribery. In: Proceedings of the 26th European Conference on Artificial Intelligence (2023), to appear
19. Garey, M., Johnson, D.: Computers and Intractability: a guide to the theory of NP-completeness. W. H, Freeman and Company (1979)
20. Gerards, A.: Matching. In: M.B. et al., (ed.) Handbooks in OR and MS Vol. 7, chap. 3, pp. 135–224. Cambridge University Press (1995)
21. Hemaspaandra, E., Hemaspaandra, L., Schnoor, H.: A control dichotomy for pure scoring rules. In: Proceedings of the 28th AAAI Conference on Artificial Intelligence, pp. 712–720 (2014)
22. Hemaspaandra, E., Schnoor, H.: Dichotomy for pure scoring rules under manipulative electoral actions. In: Proceedings of the 22nd European Conference on Artificial Intelligence, pp. 1071–1079 (2016)
23. Karp, R.: Reducibility among combinatorial problems. In: Proceedings of Symposium on Complexity of Computer Computations, pp. 85–103 (1972)
24. Kurihara, T.: Axiomatic characterisations of the basic best-worst rule. Econ. Lett. **172**, 19–22 (2018)
25. Lin, A.: The complexity of manipulating k-approval elections. In: Proceedings of the 3rd International Conference on Agents and Artificial Intelligence, pp. 212–218 (2011)
26. Loreggia, A., Narodytska, N., Rossi, F., Venable, K., Walsh, T.: Controlling elections by replacing candidates or votes. In: Proceedings of the 14th International Conference on Autonomous Agents and Multiagent Systems, pp. 1737–1738 (2015)
27. Mirrlees, J.: An exploration in the theory of optimum income taxation. Rev. Econ. Stud. **38**(2), 175–208 (1971)
28. Mulmuley, K., Vazirani, U., Vazirani, V.: Matching is as easy as matrix inversion. Combinatorica **7**(1), 105–113 (1987)
29. Neveling, M., Rothe, J., Weishaupt, R.: The possible winner problem with uncertain weights revisited. In: Proceedings of the 23rd International Symposium on Fundamentals of Computation Theory, pp. 399–412 (2021)
30. Papadimitriou, C., Yannakakis, M.: The complexity of restricted spanning tree problems. J. ACM **29**(2), 285–309 (1982)
31. Tutte, W.: A short proof of the factor theorem for finite graphs. Can. J. Math. **6**, 347–352 (1954)
32. Walsh, T.: Uncertainty in preference elicitation and aggregation. In: Proceedings of the 22nd National Conference on Artificial Intelligence, pp. 3–8 (2007)
33. Xia, L.: Computing the margin of victory for various voting rules. In: Proceedings of the 12th ACM Conference on Electronic Commerce, pp. 982–999 (2012)
34. Yang, Y., Shrestha, Y., Guo, J.: On the complexity of bribery with distance restrictions. Theor. Comput. Sci. **760**, 55–71 (2019)

α-β-Factorization and the Binary Case of Simon's Congruence

Pamela Fleischmann[1](✉), Jonas Höfer[2], Annika Huch[1], and Dirk Nowotka[1]

[1] Kiel University, Kiel, Germany
{fpa,dn}@informatik.uni-kiel.de, stu216885@mail.uni-kiel.de
[2] University of Gothenburg, Gothenburg, Sweden
jonas.hofer@gu.se

Abstract. In 1991 Hébrard introduced a factorization of words that turned out to be a powerful tool for the investigation of a word's scattered factors (also known as (scattered) subwords or subsequences). Based on this, first Karandikar and Schnoebelen introduced the notion of k-richness and later on Barker et al. the notion of k-universality. In 2022 Fleischmann et al. presented at DCFS a generalization of the arch factorization by intersecting the arch factorization of a word and its reverse. While the authors merely used this factorization for the investigation of shortest absent scattered factors, in this work we investigate this new α-β-factorization as such. We characterize the famous Simon congruence of k-universal words in terms of 1-universal words. Moreover, we apply these results to binary words. In this special case, we obtain a full characterization of the classes and calculate the index of the congruence. Lastly, we start investigating the ternary case, present a full list of possibilities for $\alpha\beta\alpha$-factors, and characterize their congruence.

1 Introduction

A *scattered factor*, *subsequence*, *subword* or *scattered subword* of a word w is a word that is obtained by deleting any number of letters from w while preserving the order of the remaining letters. For example, oiaoi and cmbntrcs are both scattered factors of combinatorics. In contrast to a factor, like combinat, a scattered factor is not necessarily contiguous. Note that a scattered factor v can occur in different ways inside a word w, for example, ab occurs in aab as a̱a̱ḇ and a̱a̱ḇ as marked by the lines below the letters. The relation of u being a scattered factor of v is a partial order on words.

In this paper, we focus on the congruence relation $\sim_k$ for $k \in \mathbb{N}_0$ which is known as Simon's congruence [22]. For two words, we have $u \sim_k v$ iff they share all scattered factors up to length k. Unions of the congruence classes of this relation are used to form the *piecewise testable languages* (first studied by Simon [22]), which are a subclass of the regular languages (they are even subregular).

A long-standing open question, posed by Sakarovitch and Simon [21], is the exact structure of the congruence classes of $\sim_k$ and the index of the congruence relation itself. Two existing results include a characterization of the congruence in terms of a special common upper bound of two words [23, Lemma 6], as well as a characterization of the (not unique) shortest elements of the congruence classes [21, Theorem 6.2.9] and [1,4,22]. The index of the relation is described asymptotically by Karandikar et al. [12]. Currently, no exact formula is known. One approach for studying scattered factors in words is based on the notion of *scattered factor universality*. A word w is called ℓ-*universal* if it contains all words of length ℓ over a given alphabet as scattered factors. For instance, the word $\texttt{alfalfa}^1$ is 2-universal since it contains all words of length two over the alphabet $\{\texttt{a},\texttt{l},\texttt{f}\}$ as scattered factors. Barker et al. and Fleischmann et al. [1,5] study the universality of words, as well as how the universality of a word changes when considering repetitions of a word. Fleischmann et al. [6] investigate the classes of Simon's congruence separated by the number of shortest absent scattered factors, characterize the classes for arbitrary alphabets for some fixed numbers of shortest absent scattered factors and give explicit formulas for these subsets. The shortest absent scattered factors of $\texttt{alfalfa}$ are $\texttt{fff}$, $\texttt{ffl}$ $\texttt{lll}$, and $\texttt{fll}$. A main tool in this line of research is a newly introduced factorization, known as the α-β-*factorization* [6] which is based on the arch factorization by Hébrard [9]. The arch factorization factorizes a word into factors of minimal length containing the complete alphabet. The α-β-factorization takes also the arch factorization of the reversed word into account. Kosche et al. [16] implicitly used this factorization to determine shortest absent scattered factors in words. In this paper, we study this factorization from a purely combinatorial point of view. The most common algorithmic problems regarding Simon's congruence are SIMK (testing whether two words u, v are congruent for a fixed k) and MAXSIMK (the optimization problem of finding the largest k such that they are congruent). The former was approached by finding the (lexicographical least element of the) minimal elements of the congruence classes of u and v. Results regarding normal forms and the equation $pwq \sim_k r$ for given words p, q, r can be found in [17,20]. The computation of the normal form was improved first by Fleischer et al. [4] and later by Barker et al. [1]. The latter was approached in the binary case by Hébrard [9], and was solved in linear time using a new approach by Gawrychowski et al. [8]. A new perspective on $\sim_k$ was recently given by Sungmin Kim et al. [14,15] when investigating the congruence's closure and pattern matching w.r.t. $\sim_k$.

Our Contribution. We investigate the α-β-factorization as an object of independent interest and give necessary and sufficient conditions for the congruence of words in terms of their factors. We characterize $\sim_k$ in terms of 1-universal words through their $\alpha\beta\alpha$-factors. We use these results to characterize the congruence classes of binary words and their cardinality, as well as to calculate the index in this special case. Moreover, we give a short and conceptually straightforward algorithm for MAXSIMK for binary words. Lastly, we start to transfer the previous results to the ternary alphabet.

Structure of the Work. First, in Sect. 2 we establish basic definitions and notation. In Sect. 3, we give our results regarding the α-β-factorization for arbitrary

---

[1] Alfalfa (Medicago sativa) is a plant whose name means *horse food* in Old Persian.

alphabets, including the characterization of the congruence of words w.r.t. $\sim_k$ in terms of their $\alpha\beta\alpha$-factors. Second, in Sect. 4, we present our results regarding binary words. We characterize the congruence classes of binary words in terms of their α- and β-factors, and apply them to calculate the index of $\sim_k$ in this special case. Third, in Sect. 5, we consider a ternary alphabet and investigate the cases for the β-factors. Last, in Sect. 6, we conclude and give ideas for further research.

2 Preliminaries

We set $\mathbb{N} := \{1, 2, 3, \dots\}$ and $\mathbb{N}_0 := \{0\} \cup \mathbb{N}$ as well as $[m] := \{1, \dots, m\}$ and $[m]_0 := \{0\} \cup [m]$. We denote disjoint unions by $\sqcup$. If there exists a bijection between two sets A, B, then we write $A \cong B$. An *alphabet* is a finite set Σ whose elements are called *letters*. An alphabet of cardinality $i \in \mathbb{N}$ is abbreviated by Σ_i. A *word* w is a finite sequence of letters from Σ where $w[i]$ denotes the ith letter of w. The set of all words over the alphabet Σ is denoted by Σ^* and the *empty word* by ε. Set $\Sigma^+ := \Sigma^* \setminus \{\varepsilon\}$. The *length* $|w|$ of w is the number of letters in w, i.e., $|\varepsilon| = 0$. We denote the set of all words of length $k \in \mathbb{N}_0$ by Σ^k and set $\Sigma^{\leq k} := \{w \in \Sigma^* \mid |w| \leq k\}$. Set $\text{alph}(w) := \{w[i] \in \Sigma \mid i \in [|w|]\}$. Set $|w|_a := |\{i \in [|w|] \mid w[i] = a\}|$ for all $a \in \Sigma$. A word $u \in \Sigma^*$ is called *factor* of $w \in \Sigma^*$ if there exist $x, y \in \Sigma^*$ with $w = xuy$. In the case that $x = \varepsilon$, u is called *prefix* of w and *suffix* if $y = \varepsilon$. The factor of w from its ith letter to its jth letter is denoted by $w[i..j]$ for $1 \leq i \leq j \leq |w|$. For $j < i$ we define $w[i..j] := \varepsilon$. If $w = xy$ we write $x^{-1}w$ for y and wy^{-1} for x. For $u \in \Sigma^*$ we set $u^0 := \varepsilon$ and inductively $u^\ell := uu^{\ell-1}$ for all $\ell \in \mathbb{N}$. For $w \in \Sigma^*$ define w^R as $w[|w|] \cdots w[1]$. For more background information on *combinatorics on words* see [18].

Now, we introduce the main notion of our work, the scattered factors also known as (scattered) subwords or subsequence (also cf. [21]).

Definition 1. *A word* $u \in \Sigma^*$ *of length* $n \in \mathbb{N}_0$ *is called a* scattered factor *of* $w \in \Sigma^*$ *if there exist* $v_0, \dots, v_n \in \Sigma^*$ *with* $w = v_0 u[1] v_1 \cdots v_{n-1} u[n] v_n$, *denoted by* $u \preceq w$. *Let* $\text{ScatFact}(w) := \{v \in \Sigma^* \mid v \preceq w\}$ *as well as* $\text{ScatFact}_k(w) := \text{ScatFact}(w) \cap \Sigma^k$ *and* $\text{ScatFact}_{\leq k}(w) := \text{ScatFact}(w) \cap \Sigma^{\leq k}$.

For instance, we have **and** $\preceq$ **agenda** but **nada** $\npreceq$ **agenda**. For comparing words w.r.t. their scattered factors, Simon introduced a congruence relation nowadays known as *Simon's congruence* [22]. Two words are called Simon k-congruent, if they have the same set of scattered factors up to length k. We refer to this k as the *level* of the congruence. This set is the *full k-spectrum* of a word, whereas the *k-spectrum* only contains all scattered factors of exactly length k.

Definition 2. *Let* $k \in \mathbb{N}$. *Two words* $u, v \in \Sigma^*$ *are called* Simon k-congruent $(u \sim_k v)$ *iff* $\text{ScatFact}_{\leq k}(u) = \text{ScatFact}_{\leq k}(v)$. *Let* $[u]_{\sim_k}$ *denote the congruence class of* u *w.r.t.* $\sim_k$.

For instance, over $\Sigma = \{a, b\}$, the words **abaaba** and **baab** are Simon 2-congruent since both contain each all words up to length 2 as scattered factors.

On the other hand, they are not Simon 3-congruent since we have aaa $\preceq$ abaaba but aaa $\not\preceq$ baab.

Starting in [11–13] special k-spectra were investigated in the context of piecewise testable languages: the *rich* resp. k-*rich* words. This work was pursued from the perspective of the universality problem for languages in [1,3,5,8] with the new notion of k-*universal* words.

Definition 3. *A word $w \in \Sigma^*$ is called k-universal w.r.t. Σ if $\mathrm{ScatFact}_k(w) = \Sigma^k$. The maximal k such that w is k-universal is denoted by $\iota_\Sigma(w)$ and called w's universality index.*

Remark 4. If we investigate a single word $w \in \Sigma^*$, we assume $\Sigma = \mathrm{alph}(w)$ implicitly and omit the Σ as index of ι.

In [6] the notion of universality was extended to m-nearly k-universal words, which are words where exactly m scattered factors of length k are absent, i.e., $|\mathrm{ScatFact}_k(w)| = |\Sigma|^k - m$. In the last section of their paper the authors introduce a factorization of words based on the arch factorization (cf. [9]) in order to characterize the 1-nearly k-universal words with $\iota(w) = k-1$. This work is closely related to the algorithmic investigation of shortest absent scattered factors [8]. Therefore, we introduce first the arch factorization and based on this the α-β-factorization from [6]. An arch is a factor of minimal length (when read from left to right) containing the whole alphabet. Consider the word $w =$ abaccaabca. This leads to the arch factorization (abac) $\cdot$ (caab) $\cdot$ ca where the arches are visualized by the brackets.

Definition 5. *For a word $w \in \Sigma^*$ the arch factorization is given by $w =:$ $\mathrm{ar}_1(w) \cdots \mathrm{ar}_k(w)\,\mathrm{re}(w)$ for $k \in \mathbb{N}_0$ with $\mathrm{alph}(\mathrm{ar}_i(w)) = \Sigma$ for all $i \in [k]$, the last letter of $\mathrm{ar}_i(w)$ occurs exactly once in $\mathrm{ar}_i(w)$ for all $i \in [k]$, and $\mathrm{alph}(\mathrm{re}(w)) \subset \Sigma$. The words $\mathrm{ar}_i(w)$ are called arches of w and $\mathrm{re}(w)$ is the rest of w. Define the modus of w as $\mathrm{m}(w) := \mathrm{ar}_1(w)[|\,\mathrm{ar}_1(w)|] \cdots \mathrm{ar}_k(w)[|\,\mathrm{ar}_k(w)|] \in \Sigma^k$. For abbreviation let $\mathrm{ar}_{i..j}(w)$ denote the concatenation from the ith arch to the jth arch.*

The following remark is a direct consequence of the combination of the k-universality and the arch factorization.

Remark 6. Let $w, w' \in \Sigma^*$ such that $w \sim_k w'$ for some $k \in \mathbb{N}_0$. Then either both w, w' have k or more arches or they both have less than k and the same number of arches. Moreover, we have $\iota(w) = k$ iff w has exactly k arches.

A generalization of the arch factorization was introduced in [6] inspired by [16]. In this factorization not only the arch factorization of a word w but also the one of w^R is taken into consideration. If both arch factorisations, i.e., the one of w and the one of w^R are considered simultaneously, we get overlaps of the arches and special parts which start at a modus letter of a *reverse* arch and end in a modus letter of an arch. For better readability, we use a specific notation for the arch factorisation of w^R where we read the parts from left to right: let $\bar{\mathrm{ar}}_i(w) := (\mathrm{ar}_{\iota(w)-i+1}(w^R))^R$ the ith *reverse arch*, let $\bar{\mathrm{re}}(w) := (\mathrm{re}(w^R))^R$ the *reverse rest*, and define the *reverse modus* $\bar{\mathrm{m}}(w)$ as $\mathrm{m}(w^R)^R$.

Definition 7. *For $w \in \Sigma^*$ define w's α-β-factorization (cf. Figure 1) by $w =:$ $\alpha_0 \beta_1 \alpha_1 \cdots \alpha_{\iota(w)-1} \beta_{\iota(w)} \alpha_{\iota(w)}$ with $\mathrm{ar}_i(w) = \alpha_{i-1}\beta_i$ and $\bar{\mathrm{ar}}_i(w) = \beta_i \alpha_i$ for all $i \in [\iota(w)]$, $\bar{\mathrm{re}}(w) = \alpha_0$, as well as $\mathrm{re}(w) = \alpha_{\iota(w)}$. Define $\mathrm{core}_i := \varepsilon$ if $|\beta_i| \in \{1,2\}$ and $\mathrm{core}_i = \beta_i[2..|\beta_i|-1]$ otherwise, i.e., as the β_i without the associated letters of the modus and reverse modus.*

For example, consider $w = \mathsf{bakebananacake} \in \{\mathsf{a,b,c,k,e}\}^*$. We get $\mathrm{ar}_1(w) = \mathsf{bakebananac}$, $\mathrm{re}(w) = \mathsf{ake}$ and $\bar{\mathrm{ar}}_1(w) = \mathsf{bananacake}$, $\bar{\mathrm{re}}(w) = \mathsf{bake}$. Thus, we have $\alpha_0 = \mathsf{bake}$, $\beta_1 = \mathsf{bananac}$ and $\alpha_1 = \mathsf{ake}$. Moreover, we have $\mathrm{m}(w) = \mathsf{c}$ and $\bar{\mathrm{m}}(w) = \mathsf{b}$. This leads to $\mathrm{core}_1 = \mathsf{anana}$.

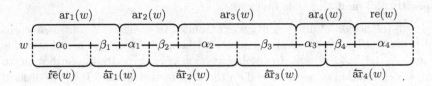

Fig. 1. α-β-Factorization of a word w with 4 arches.

Remark 8. In contrast to the arch factorization, the α-β-factorization is left-right-symmetric. Note that the ith reverse arch always starts inside the ith arch since otherwise an arch or the rest would contain at least two reverse arches or a complete arch and thus the arch would contain the complete alphabet more than once or once.

For better readability, we do not parametrize the α_i and β_i by w. Instead, we denote the factors according to the word's name, i.e. $\tilde{\alpha}_i \tilde{\beta}_{i+1}$ is an arch of $\tilde{w}$.

Remark 9. Since $|\mathrm{alph}(\alpha_i)| \leq |\Sigma| - 1$ we can build the arch factorization of α_i w.r.t. some Ω with $\mathrm{alph}(\alpha_i) \subseteq \Omega \in \binom{\Sigma}{|\Sigma|-1}$. This yields the same factorization for all Ω because either $\mathrm{alph}(\alpha_i) = \Omega$ or $\mathrm{alph}(\alpha_i) \subset \Omega$ and thus $\mathrm{re}(\alpha_i) = \alpha_i$.

Last, we recall three lemmata regarding Simon's congruence which we need for our results. The first lemma shows that if we prepend or append a sufficiently universal word to two congruent words each, we obtain congruent words with an increased level of congruence.

Lemma 10 ([12, **Lemma 4.1**][13, **Lemma 3.5**]). *Let $w, \tilde{w} \in \Sigma^*$ such that $w \sim_k \tilde{w}$, then for all $u, v \in \Sigma^*$ we have $uwv \sim_{\iota(u)+k+\iota(v)} u\tilde{w}v$.*

The next lemma characterizes the omittance of suffixes when considering words up to $\sim_k$.

Lemma 11 ([23, **Lemma 3**]). *Let $u, v \in \Sigma^+$ and $\mathsf{x} \in \Sigma$. Then, $uv \sim_k u$ iff there exists a factorization $u = u_1 u_2 \cdots u_k$ such that $\mathrm{alph}(u_1) \supseteq \mathrm{alph}(u_2) \supseteq \ldots \supseteq \mathrm{alph}(u_k) \supseteq \mathrm{alph}(v)$.*

The last lemma characterizes letters which can be omitted when we consider words up to $\sim_k$. The last two of its conditions follow from the previous lemma.

Lemma 12 ([23, **Lemma 4**]). *Let $u, v \in \Sigma^*$ and $\mathrm{x} \in \Sigma$. Then, $uv \sim_k u\mathrm{x}v$ iff there exist $p, p' \in \mathbb{N}_0$ with $p + p' \geq k$ and $u\mathrm{x} \sim_p u$ and $\mathrm{x}v \sim_{p'} v$.*

3 α-β-Factorization

In this section, we investigate the α-β-factorization based on results of [12] in the relatively new light of factorizing an arch into an α and a β part. The main result states that it suffices to look at 1-universal words in order to gain the information about the congruence classes of $\sim_k$.

Remark 13. By the left-right symmetry of the α-β-factorisation, it suffices to prove most of the claims only for one direction (reading the word from left to right) and the other direction (reading the word from right to left) follows immediately. Thus, these claims are only given for one direction and it is not always mentioned explicitly that the analogous claim holds for the other direction.

Our first lemma shows that *cutting of ℓ arches* from two k-congruent words each, leads to $(k-\ell)$-congruence. Here, we use the α-β-factorization's symmetry.

Lemma 14. *Let $w, \tilde{w} \in \Sigma^*$ with $w \sim_k \tilde{w}$ and $\iota(w) = \iota(\tilde{w}) < k$. Then we have $\alpha_i \beta_{i+1} \alpha_{i+1} \cdots \alpha_j \sim_{k-\iota(w)+j-i} \tilde{\alpha}_i \tilde{\beta}_{i+1} \tilde{\alpha}_{i+1} \cdots \tilde{\alpha}_j$ for all $0 \leq i \leq j \leq \iota(w)$.*

The following proposition shows that two words having exactly the same β-factors are k-congruent iff the corresponding α-factors are congruent at a smaller level. The proof uses a similar idea to the one presented by Karandikar et al. [12, Lemma 4.2].

Proposition 15. *For all $w, \tilde{w} \in \Sigma^*$ with $m := \iota(w) = \iota(\tilde{w}) < k$ such that $\beta_i = \tilde{\beta}_i$ for all $i \in [m]$, we have $w \sim_k \tilde{w}$ iff $\alpha_i \sim_{k-m} \tilde{\alpha}_i$ for all $i \in [m]_0$.*

As an immediate corollary, we obtain the following statement which allows us to normalize the α-factors when proving congruence of words.

Corollary 16. *Let $w, \tilde{w} \in \Sigma^*$ with $m := \iota(w) = \iota(\tilde{w}) < k$, then $w \sim_k \tilde{w}$ iff $\alpha_i \sim_{k-m} \tilde{\alpha}_i$ for all $i \in [m]_0$ and for $w' := \alpha_0 \tilde{\beta}_1 \alpha_1 \cdots \tilde{\beta}_m \alpha_m$ we have $w \sim_k w'$.*

Next, we show the central result for this section. We can characterize the congruence of words by the congruence of their $\alpha\beta\alpha$-factors. Therefore, it suffices to consider 1-universal words in general. Again, the proof uses Lemma 10 and is inspired by Karandikar et al. [12, Lemma 4.2] and repeatedly exchanges factors up to k-Simon congruence.

Theorem 17. *Let $w, \tilde{w} \in \Sigma^*$ with $m := \iota(w) = \iota(\tilde{w}) < k$. Then, $w \sim_k \tilde{w}$ iff $\alpha_{i-1} \beta_i \alpha_i \sim_{k-m+1} \tilde{\alpha}_{i-1} \tilde{\beta}_i \tilde{\alpha}_i$ for all $i \in [m]$.*

Proof. Assume $w \sim_k \tilde{w}$, then the congruences follow directly by Lemma 14 for $i, j \in \mathbb{N}_0$ with $|j - i| = 1$.

Assume $\alpha_{i-1}\beta_i\alpha_i \sim_{k-m+1} \tilde{\alpha}_{i-1}\tilde{\beta}_i\tilde{\alpha}_i$ for all $i \in [m]$. By Lemma 14, we obtain that $\alpha_i \sim_{k-m} \tilde{\alpha}_i$ for all $i \in [m]_0$. By Corollary 16, we have $\alpha_{i-1}\beta_i\alpha_i \sim_{k-m+1} \alpha_{i-1}\tilde{\beta}_i\alpha_i$ for all $i \in [m]$, and it suffices to show that $w \sim_k \alpha_0\tilde{\beta}_1\alpha_1 \cdots \beta_m\alpha_m$. Now, we have by repeated applications of Lemma 10 that

$$\alpha_0\beta_1\alpha_1 \cdot \beta_2\alpha_2 \cdots \beta_m\alpha_m \sim_k \alpha_0\tilde{\beta}_1\alpha_1 \cdot \beta_2\alpha_2 \cdots \beta_m\alpha_m \sim_k \ldots \sim_k \tilde{w}. \qquad \square$$

In the light of Theorem 17, in the following, we consider some special cases of these triples w.r.t. the alphabet of both involved α. Hence, let $w, \tilde{w} \in \Sigma^*$ with $1 = \iota(w) = \iota(\tilde{w})$.

Proposition 18. *Let* $\alpha_0 = \alpha_1 = \tilde{\alpha}_0 = \tilde{\alpha}_1 = \varepsilon$. *Then* $w \sim_k \tilde{w}$ *iff* $k = 1$ *or* $k \geq 2$, $\mathrm{m}(w) = \mathrm{m}(\tilde{w})$, $\hat{\mathrm{m}}(w) = \hat{\mathrm{m}}(\tilde{w})$, *and* $\mathrm{core}_1 \sim_k \widetilde{\mathrm{core}}_1$.

Proposition 19. *Let* $\mathrm{alph}(\alpha_0) = \mathrm{alph}(\alpha_1) = \mathrm{alph}(\tilde{\alpha}_0) = \mathrm{alph}(\tilde{\alpha}_1) \in \binom{\Sigma}{|\Sigma|-1}$. *We have* $w \sim_k \tilde{w}$ *iff* $\alpha_i \sim_{k-1} \tilde{\alpha}_i$ *for all* $i \in [1]_0$.

In the last two propositions, we considered special cases of congruence classes, where all words in such a congruence class have not only the same modus but also the same reverse modus. This is not necessarily always the case witnessed by $w = \mathtt{ababeabab} \cdot \mathtt{abecd} \cdot \mathtt{cdcdcd} \sim_4 \mathtt{ababeabab} \cdot \mathtt{baedc} \cdot \mathtt{cdcdcd} = \tilde{w}$ with $\mathrm{m}(w) = \mathtt{d} \neq \mathtt{c} = \mathrm{m}(\tilde{w})$ and $\hat{\mathrm{m}}(w) = \mathtt{a} \neq \mathtt{b} = \hat{\mathrm{m}}(\tilde{w})$. This case occurs if one of the α satisfies $\alpha_0\mathtt{x} \sim_{k-1} \alpha_0$ and the alphabet of α_1 factor is missing at least $\mathtt{x}$ for all $\mathtt{x} \in \{\hat{\mathrm{m}}(\tilde{w}) \mid \tilde{w} \in [w]_{\sim_k}\}$. The conditions for $\mathrm{m}(w)$ are analogous. In the last proposition of this section, we show a necessary condition for the α-factors of words which are congruent to words with a different modus. The proof uses the same factorization as the proof of Lemma 11 (cf. [23, Lemma 3]). By identifying permutable factors, similar ideas also appear when characterizing the shortest elements in congruence classes (cf. [21, Theorem 6.2.9][4, Proposition 6]).

Proposition 20. *Let* $w \in \Sigma^*$ *with* $\iota(w) = 1$, $k \in \mathbb{N}$, *and* $\hat{\mathrm{M}} := \{\hat{\mathrm{m}}(\tilde{w})[1] \mid \tilde{w} \in [w]_{\sim_k}\}$, *i.e., we capture all modus letters of words which are k-congruent to w. If* $|\hat{\mathrm{M}}| \geq 2$ *then there exists a factorization* $\alpha_0 =: u_1 \cdots u_{k-1}$ *with* $\mathrm{alph}(u_1) \supseteq \ldots \supseteq \mathrm{alph}(u_{k-1}) \supseteq \hat{\mathrm{M}}$.

4   The Binary Case of Simon's Congruence

In this section, we apply our previous results to the special case of the binary alphabet. Here, for $x \in \Sigma$, let $\bar{x}$ be the well defined other letter of Σ. First, we characterize the congruence of binary words in terms of α- and β-factors. We show that in this scenario in each congruence class of a word w with at most k arches, we have $|\{\hat{\mathrm{m}}(\tilde{w}) \mid \tilde{w} \in [w]_{\sim_k}\}| = 1$ (cf. Proposition 20). We present results such that a full characterization of the structure of the classes in the binary case is given, implying as a byproduct a simple algorithm for MAXSIMK in this special case (cf. [9]). Moreover, we can calculate $|\Sigma_2^*/\sim_k|$.

Proposition 21. *For all $w \in \Sigma_2^*$, we have for all $i \in [\iota(w)]$*
1. $\beta_i \in \{a, b, ab, ba\}$,
2. if $\beta_i = x$, then $\alpha_{i-1}, \alpha_i \in \overline{x}^+$ with $x \in \Sigma_2$,
3. if $\beta_i = x\overline{x}$, then $\alpha_{i-1} \in x^*$ and $\alpha_i \in \overline{x}^*$ with $x \in \Sigma_2$.

Thus, we get immediately that the $\alpha\beta\alpha$-factors are of the following forms: $a^{\ell_1+1}ba^{\ell_2+1}$, $b^{\ell_1+1}ab^{\ell_2+1}$, $b^{\ell_3}baa^{\ell_4}$, or $a^{\ell_3}abb^{\ell_4}$ for some $\ell_1, \ell_2, \ell_3, \ell_4 \in \mathbb{N}_0$. The following lemma shows that in the binary case the k-congruence of two words with identical universality less than k leads to the same modi and same β.

Lemma 22. *Let $w, w' \in \Sigma_2^*$ with $w \sim_k w'$ and $m := \iota(w) = \iota(w') < k$, then $m(w) = m(w')$ and thus, $\beta_i = \beta_i'$ for all $i \in [m]$.*

Combining the Lemmata 22, 14 and Proposition 15 yields the following characterization of $\sim_k$ for binary words in terms of unary words and factors.

Theorem 23. *Let $w, w' \in \Sigma_2^*$ such that $m := \iota(w) = \iota(w') < k$, then $w \sim_k w'$ iff $\beta_i = \beta_i'$ for all $i \in [m]$ and $\alpha_i \sim_{k-m} \alpha_i'$ for all $i \in [m]_0$.*

Using the characterization, we can also give an $\mathcal{O}(|u|+|v|)$-time algorithm for finding the largest k with $u \sim_k v$ for $u, v \in \Sigma_2^*$. This special case was originally solved by Hébrard [9] just considering arches. Recently, a linear time algorithm for arbitrary alphabets was presented by Gawrychowski et al. [8]. Nonetheless, we give Algorithm 1, as it is a conceptually simple algorithm exploiting that α_i factors can be treated similar to $re(w)$ in the arch factorization.

Algorithm 1: MaxSimK for binary words

Input: $u, \tilde{u} \in \Sigma_2^*$
Result: if $u = u'$ then ∞ and otherwise the maximum k such that $u \sim_k \tilde{u}$
1   $(\alpha_0, \beta_1, \ldots, \alpha_{\iota(u)}) := \alpha\text{-}\beta\text{-}\mathrm{FACT}(u);$ // w.r.t. Σ_2
2   $(\tilde{\alpha}_0, \tilde{\beta}_1, \ldots, \tilde{\alpha}_{\iota(\tilde{u})}) := \alpha\text{-}\beta\text{-}\mathrm{FACT}(\tilde{u});$
3   **if** $\iota(u) \neq \iota(\tilde{u}) \lor \mathrm{alph}(u) \neq \mathrm{alph}(\tilde{u})$ **then**   // 2nd condition for $u = x^i, \tilde{u} = \overline{x}^j$
4 | **return** $\min(\iota(u), \iota(\tilde{u}));$
5   **else if** $\beta_1 = \tilde{\beta}_1 \land \cdots \land \beta_{\iota(u)} = \tilde{\beta}_{\iota(\tilde{u})}$ **then**
6 | **for** $i \in [\iota(u)]_0$ **do**   // solve MaxSimK for unary α pairs
7 | $\lfloor$ $e_i := $ **if** $|\alpha_i| = |\tilde{\alpha}_i|$ **then** ∞ **else** $\min(|\alpha_i|, |\tilde{\alpha}_i|)$
8 | **return** $\iota(u) + \min\{e_i \mid i \in [\iota(u)]_0\};$
9   **else**
10 | **return** $\iota(u).$

We can use Theorem 23 to answer a number of questions regarding the structure of the congruence classes of $\Sigma_2^*/\sim_k$. For instance, for each w with $|[w]_{\sim_k}| = \infty$, we have $x^k \preceq w$ for some $x \in \Sigma$ by the pigeonhole-principle. The contrary is not true in general witnessed by the word $v = bbabb$ with respect to $\sim_4$. Its scattered factors of length four are $bbab, babb$ and $bbbb$. Therefore,

Table 1. Index of $\sim_k$ restricted to binary words with a fixed number of arches

		Number of Arches								
		0	1	2	3	4	5	6	7	m
Scat Fact Length	1	3	1							
	2	5	10	1						
	3	7	26	34	1					
	4	9	50	136	116	1				
	5	11	82	358	712	396	1			
	6	13	122	748	2 564	3 728	1 352	1		
	7	15	170	1 354	6 824	18 364	19 520	4 616	1	
	k	$2k+1$								

each word in its class contains exactly one a (aa $\not\preceq v$ but a $\preceq v$ is), at least two b succeeding and preceding the a (bba, abb $\preceq v$) but not more than two b (bbba, abbb $\not\preceq v$). Therefore, bbabb is the only word in this class, but it contains b^4. By a famous result of Simon [21, Corollary 6.2.8], all congruence classes of $\sim_k$ are either infinite or singletons. In the binary case, we can give a straightforward characterization of the finite/singleton and infinite classes.

Theorem 24. *Let $w \in \Sigma_2^*$, then $|[w]_{\sim_k}| < \infty$. In particular, we have $|[w]_{\sim_k}| = 1$ iff $\iota(w) < k$ and $|\alpha_i| < k - \iota(w)$ for all $i \in [\iota(w)]_0$.*

In the following, we will use Theorem 23 to derive a formula for the precise value of $|\Sigma_2^*/\sim_k|$. Note that in the unary case, we have $|\Sigma_1^*/\sim_k| = k + 1$ because the empty word has its own class. By Remark 6, we know that there exists exactly one class w.r.t. $\sim_k$ of words with k arches. We can consider the other classes by the common number of arches of their elements. By Theorem 23, we can count classes based on the valid combinations of β-factors and number of classes for each α-factors. Because the α are unary, we already know their number of classes. These valid combinations are exactly given by Proposition 21. The first values for the number of classes separated by the number of arches are given in Table 1.

Theorem 25. *The number of congruence classes of $\Sigma_2^*/\sim_k$ of words with $m < k$ arches (the entries of Table 1) is given by*

$$\left\| \begin{pmatrix} k-m & k-m & k-m \\ 1 & 2 & 1 \\ k-m & k-m & k-m \end{pmatrix}^m \cdot \begin{pmatrix} k-m \\ 1 \\ k-m \end{pmatrix} \right\|_1 = c_k^m$$

where $c_k^{-1} := 1$, $c_k^0 := 2k+1$, and $c_k^m := 2 \cdot (k-m+1) \cdot c_{k-1}^{m-1} - 2 \cdot (k-m) \cdot c_{k-2}^{m-2}$ where $\|\cdot\|_1$ denotes the 1-norm.

Proof. First, we show that the matrix representation produces the correct values, then we show the characterization as recurrence. Note that $k - m$ is fixed on the diagonals of Table 1. Therefore, increasing both, increases just the exponent of the matrix. We show that the first column is correct and then proceed by induction along the diagonals. Denote the above matrix by $D_{k,m}$.

Let $k \in \mathbb{N}_0$ and $w \in \Sigma^*$ with $m := \iota(w) < k$. For $i \in [m]_0$, all elements $v \in [w]_{\sim_k}$ have $k-m$ congruent α_i by Theorem 23. By definition, their alphabets are proper subsets of Σ_2. Therefore, they are either empty or non-empty unary words consisting of just a or b. We separate the choice of α_i into these three cases. Let $M_\varepsilon^\ell := \{[w] \in \Sigma_2^*/\!\sim_{(k-m)+\ell} \ | \ \iota(w) = \ell, \alpha_0 \sim_{k-m} \varepsilon\}$ and $M_\mathsf{x}^\ell := \{[w] \in \Sigma_2^*/\!\sim_{(k-m)+\ell} \ | \ \iota(w) = \ell, \alpha_0 \sim_{k-m} \mathsf{x}^r, r \in \mathbb{N}\}$ for $\mathsf{x} \in \Sigma_2$ be sets of $\ell + (k-m)$ congruence classes of words with ℓ arches, separated by the alphabet of α_0. Denote by $e_{k,m} := (|M_\mathsf{a}^0|, |M_\varepsilon^0|, |M_\mathsf{b}^0|)^\intercal = (k-m, 1, k-m)^\intercal$ the number of classes for zero arches. We show $\|D_{k,m}^\ell \cdot e_{k,m}\|_1 = (|M_\mathsf{a}^\ell|, |M_\varepsilon^\ell|, |M_\mathsf{b}^\ell|)^\intercal$. There are four choices for β_i which are given by Proposition 21. Each choice of β_{i+1} depends on the preceding α_i and limits the choices for the succeeding α_{i+1}. These are given by Proposition 21, and correspond to the entries of the matrix because for $\ell \geq 1$ we have

$$M_\varepsilon^\ell = \{[w]_{\sim_{(k-m)+\ell}} \in M_\varepsilon^\ell \mid \mathsf{x} \in \Sigma_2, \beta_1(w) = \overline{\mathsf{x}}\mathsf{x}, \alpha_1(w) \sim_{k-m} \varepsilon\}$$
$$\sqcup \{[w]_{\sim_{(k-m)+\ell}} \in M_\varepsilon^\ell \mid \mathsf{x} \in \Sigma_2, \beta_1(w) = \overline{\mathsf{x}}\mathsf{x}, \alpha_1(w) \sim_{k-m} \mathsf{x}^r, r \in \mathbb{N}\}$$
$$\cong \{\mathsf{ab}, \mathsf{ba}\} \times M_\varepsilon^{\ell-1} \sqcup \{\mathsf{ab}\} \times M_\mathsf{b}^{\ell-1} \sqcup \{\mathsf{ba}\} \times M_\mathsf{a}^{\ell-1}$$
$$M_\mathsf{x}^\ell = \{[w]_{\sim_{(k-m)+\ell}} \in M_\mathsf{x}^\ell \mid \beta_1(w) = \overline{\mathsf{x}}\}$$
$$\sqcup \{[w]_{\sim_{(k-m)+\ell}} \in M_\mathsf{x}^\ell \mid \beta_1(w) = \mathsf{x}\overline{\mathsf{x}}, \alpha_1(w) \sim_{k-m} \overline{\mathsf{x}}^r, r \in \mathbb{N}\}$$
$$\sqcup \{[w]_{\sim_{(k-m)+\ell}} \in M_\mathsf{x}^\ell \mid \beta_1(w) = \mathsf{x}\overline{\mathsf{x}}, \alpha_1(w) \sim_{k-m} \varepsilon\}$$
$$\cong [k-m] \times (\{\overline{\mathsf{x}}\} \times M_\mathsf{x}^{\ell-1} \sqcup \{\mathsf{x}\overline{\mathsf{x}}\} \times M_{\overline{\mathsf{x}}}^{\ell-1} \sqcup \{\mathsf{x}\overline{\mathsf{x}}\} \times M_\varepsilon^{\ell-1}).$$

Therefore, each multiplication with the matrix increases the number ℓ of arches by one. Thus, for $m = \ell$ we have the desired value as M_ε^m and M_x^m are sets of k congruence classes with m arches. Therefore, $\|D_{k,m}^m \cdot e_{k,m}\|_1$ corresponds to the number of classes with respect to $\sim_k$ of words with m arches.

The equivalence of the two formulas is left to show. The characteristic polynomial of $D_{k,m}$ is given by $\chi_{D_{k,m}} = \det(D_{k,m} - \lambda I) = -\lambda^3 + 2\lambda^2 + 2(k - m)\lambda^2 - 2(k-m)\lambda$. By the Cayley-Hamilton theorem [7], $D_{k,m}$ is a root of its characteristic polynomial and thus satisfies the recurrence

$$D_{k,m}^{\ell+2} = 2 \cdot D_{k,m}^{\ell+1} + 2 \cdot (k-m) \cdot D_{k,m}^{\ell+1} - 2 \cdot (k-m) \cdot D_{k,m}^\ell$$
$$= 2 \cdot (k-m+1) \cdot D_{k,m}^{\ell+1} - 2 \cdot (k-m) \cdot D_{k,m}^\ell$$

for $\ell \in \mathbb{N}$. Note that $e_{k,m} = e_{k+\ell,m+\ell}$ for all $\ell \in \mathbb{N}_0$. Now we conclude by induction that

$$\|D_{k+2,m+2}^{m+2} \cdot e_{k+2,m+2}\|_1 = \|D_{k,m}^{m+2} \cdot e_{k,m}\|_1$$
$$= \|(2 \cdot (k-m+1) \cdot D_{k,m}^{m+1} - 2 \cdot (k-m) \cdot D_{k,m}^m) \cdot e_{k,m}\|_1$$
$$= 2 \cdot (k-m+1) \cdot \|D_{k,m}^{m+1} \cdot e_{k,m}\|_1 - 2 \cdot (k-m) \cdot \|D_{k,m}^m \cdot e_{k,m}\|_1$$
$$= 2 \cdot (k-m+1) \cdot c_{k+1}^{m+1} - 2 \cdot (k-m) \cdot c_k^m = c_{k+2}^{m+2},$$

Table 2. Number of classes of perfect universal binary words restricted to a fixed number of arches

		Number of Arches								
		0	1	2	3	4	5	6	7	8	m
Scat. Fact Length	2	1	4	1							
	3	1	6	14	1						
	4	1	8	32	48	1					
	5	1	10	58	168	164	1				
	6	1	12	92	416	880	560	1			
	7	1	14	134	840	2 980	4 608	1 912	1		
	8	1	16	184	1 488	7 664	21 344	24 344	6 528	1	
	k	1	$2k$								

because $\|u \pm v\|_1 = \|u\|_1 \pm \|v\|_1$ for all $u = (u_i), v = (v_i) \in \mathbb{R}^n$ for which $u_j v_j \geq 0$ for all $j \in [n]$.   □

Remark 26. Note that by setting $\Delta := k - m$, the family of recurrences depends only on one variable Δ, because $k - m = (k - \ell) - (m - \ell)$ holds for all $\ell \in \mathbb{N}$.

Remark 27. Some sequences in Table 1 are known sequences. The first and second diagonal are A007052 and A018903 resp. in [19]. Both sequences are investigated in the work of Janjic [10]. There, the two sequences appear as the number of compositions of $n \in \mathbb{N}$, considering three (resp. five) differently colored 1s. Furthermore, the sequences c_k^m seem to be equivalent to the family of sequences (s_n) where $s_0 = 1$ and s_1 is fixed and s_{n+2} is the smallest number such that $\frac{s_{n+2}}{s_{n+1}} > \frac{s_{n+1}}{s_n}$. These sequences where studied by Boyd [2].

By Remark 6, we can count the number of classes separated by the universality of words with less than k arches. This leads to the following immediate corollary which allows us to efficently calculate $|\Sigma_2^* / {\sim_k}|$.

Corollary 28. *Let $k \in \mathbb{N}_0$. Over a binary alphabet, the number of congruence classes of $\sim_k$ is given by $|\Sigma_2^* / {\sim_k}| = 1 + \sum_{m=0}^{k-1} c_k^m$.*

The first values of the sequence, some of which are already given in [12], are

$$1, 4, 16, 68, 312, 1560, 8528, 50864, 329248, 2298592, 17203264, 137289920,$$
$$1162805376, 10409679744, 98146601216, 971532333824, 10068845515264, \ldots$$

We can use the idea of Theorem 25 to count the number of perfect k-universal words, i.e., k-universal words with an empty rest (cf. [5]). We can count them by replacing the vector from Theorem 25 with the initial distribution of α_i values with $(0, 1, 0)^\mathsf{T}$. Thus, the formula counts words starting or ending with an empty α. Because the matrix does not change, we obtain the same recurrence with different initial values. The þk diagonal, shifted by one, is now given by the Lucas sequence of the first kind $U(2 \cdot k + 2, 2 \cdot k)$, where $U_n(P, Q)$ is given by

$U_0(P,Q) = 0$, $U_1(P,Q) = 1$, $U_n(P,Q) = P \cdot U_{n-1}(P,Q) - Q \cdot U_{n-2}(P,Q)$. The first calculated values are given in Table 2. The first three diagonals of the table are the known integer sequences A007070, A084326, and A190978 in [19].

5   Towards the Ternary Case of Simon's Congruence

In the following, we will consider cases for the ternary alphabet based on the alphabets of the $\dot{\alpha}$-factors with the goal of proving similar results to Proposition 21 and Theorem 23 for ternary words, leading to Theorem 30. By Theorem 17, it suffices to consider $\alpha\beta\alpha$-factors for characterizing congruence classes. In Sect. 3 we already considered some cases for $\alpha\beta\alpha$-factors for arbitrary alphabets. Note that if $m_1(w) = \tilde{m}_1(w)$ then $core_1 = \varepsilon$. Otherwise, if $m_1(w) \neq \tilde{m}_1(w)$, then $core_1 \in (\Sigma \setminus \{m_1(w), \tilde{m}_1(w)\})^*$. Thus, cores of ternary words are unary, and we denote the well-defined letter of the core by $y \in \Sigma_3$.

We use a variant of the *Kronecker-δ* for a boolean predicate P as $\delta_{P(x)} = 1$ if $P(x)$ is true and 0 otherwise to express a condition on the alphabet of the rest of a binary α-factor (cf. Figure 2). If an α_0's rest contains the letter y different from the reverse modus $x := \tilde{m}(w)$, then $re(\alpha_0)\,\tilde{m}(w)$ builds another arch ending before the core (left). This lowers the level of congruence, up to which we can determine the core, by one. If $y \npreceq re(\alpha_0)$ the next y is in the core (right).

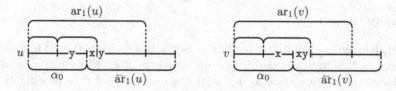

Fig. 2. Factorization of α in the ternary case assuming core $\in y^+$.

We always assume that $k \geq 2$ because we characterize the congruence of 1-universal words. Moreover, let $w, \tilde{w} \in \Sigma_3^*$ with $1 = \iota(w) = \iota(\tilde{w})$.

First, we prove a useful lemma which characterizes the congruence of two ternary words with the same modus and reverse modus. Together with Proposition 20, this immediately implies several cases.

Lemma 29. *Let* $m(w) = m(\tilde{w})$ *and* $\tilde{m}(w) = \tilde{m}(\tilde{w})$, *we have* $w \sim_k \tilde{w}$ *iff* $\alpha_i \sim_{k-1} \tilde{\alpha}_i$ *for all* $i \in [1]_0$ *and* $core_1 \sim_{k-c} \widetilde{core}_1 \in y^*$ *where* $c := \iota(\alpha_0) + \delta_{y \preceq re(\alpha_0)} + \iota(\alpha_1) + \delta_{y \preceq \tilde{re}(\alpha_1)}$.

Since in the ternary case, there are congruent words having different modi or reverse modi, Lemma 29 does not imply a full characterization. This leads to two cases in the following classification (case 3 and 5 out of the 9 cases in Table 3). These two cases correspond to the first case in the following theorem.

Table 3. The possibilities for the β-factor of $w = \alpha_0\beta\alpha_1$, assuming $\mathsf{a}, \mathsf{b}, \mathsf{c} \in \Sigma_3$ different. Note that in the cases $(1,1)$, $(1,0)$, $(0,0)$ the letters not fixed by the α-factors can be chosen arbitrarily but differently from Σ_3.

$|\mathrm{alph}(\alpha_0)|,|\mathrm{alph}(\alpha_1)|$	$\mathrm{alph}(\alpha_0)$	$\mathrm{alph}(\alpha_1)$	β RegExp	Stated In
2,2	$\{\mathsf{a},\mathsf{b}\}$	$\{\mathsf{a},\mathsf{c}\}$	$\mathsf{ba^*c}$	Proposition 19
	$\{\mathsf{a},\mathsf{b}\}$	$\{\mathsf{a},\mathsf{b}\}$	c	
2,1	$\{\mathsf{a},\mathsf{b}\}$	$\{\mathsf{c}\}$	$(\mathsf{ab^+}\mid\mathsf{ba^+})\mathsf{c}$	
	$\{\mathsf{a},\mathsf{b}\}$	$\{\mathsf{a}\}$	$\mathsf{ba^*c}$	
2,0	$\{\mathsf{a},\mathsf{b}\}$	$\emptyset$	$(\mathsf{ab^+}\mid\mathsf{b^+a})\mathsf{c}$	
1,1	$\{\mathsf{a}\}$	$\{\mathsf{b}\}$	$\mathsf{ab^+c}\mid\mathsf{ac^+b}\mid\mathsf{ca^+b}$	
	$\{\mathsf{a}\}$	$\{\mathsf{a}\}$	$\mathsf{ba^*c}$	
1,0	$\{\mathsf{a}\}$	$\emptyset$	$\mathsf{ba^*c}\mid\mathsf{ab^+c}$	
0,0	$\emptyset$	$\emptyset$	$\mathsf{ab^+c}$	Proposition 18

Theorem 30. *For $w, \tilde{w} \in \Sigma_3^*$ we have $w \sim_k \tilde{w}$ iff $\alpha_i \sim_{k-1} \tilde{\alpha}_i$ for all $i \in [1]_0$, and one of the following*
1. $|\mathrm{alph}(\alpha_i)| = 2$, $\mathrm{alph}(\alpha_{1-i}) \cap \mathrm{alph}(\alpha_i) = \emptyset$, and $\iota(\alpha_i) \geq k-1$ for some $i \in [1]_0$,
2. $\mathrm{m}(w) = \mathrm{m}(\tilde{w})$, $\bar{\mathrm{m}}(w) = \bar{\mathrm{m}}(\tilde{w})$, core $\sim_{k-c} \widetilde{\text{core}}$ with $c := \iota(\alpha_0) + \delta_{\mathsf{y} \preceq \alpha_0} + \iota(\alpha_1) + \delta_{\mathsf{y} \in \alpha_1}$.
For all possibilities distinguishing the β-factors, see Table 3.

6   Conclusion

In 2021, Kosche et al. [16] first implicitly used a new factorization to find absent scattered factors in words algorithmically. Later, in 2022, Fleischmann et al. [6] introduced this factorization as α-β-factorization and used it to investigate the classes of Simon's congruence separated by the number of shortest absent scattered factors, to characterize the classes for arbitrary alphabets for some fixed numbers of shortest absent scattered factors and to give explicit formulas for these subsets. In this paper, we investigated the α-β-factorization as an object of intrinsic interest. This leads to a result characterizing k-congruence of m-universal words in terms of their 1-universal $\alpha\beta\alpha$-factors. In the case of the binary and ternary alphabet, we fully characterized the congruence of words in terms of their single factors. Moreover, using this characterization, we gave a formula for the number of classes of binary words for each k, characterized the finite classes, and gave a conceptually simple linear time algorithm for testing $\mathrm{MaxSimK}$ for binary words.

The modus of the layered arch factorizations used in the proof of Propostion 20 and throughout the literature [4, 21, 23], can be regarded as the optimal word to jump to certain letters in certain parts of the word. The α-β-factorization encapsulates the first layer (arches w.r.t. Σ) of these factorizations for all indicies. For small alphabets (this paper) and shortest absent scattered factors (c.f. [6]) this allows the characterization and enumeration of classes. Extending this idea to lower layers (arches w.r.t. some $\Omega \subset \Sigma$), is left as future work.

References

1. Barker, L., Fleischmann, P., Harwardt, K., Manea, F., Nowotka, D.: Scattered factor-universality of words. In: DLT (2020)
2. Boyd, D.W.: Linear recurrence relations for some generalized Pisot sequences. In: CNTA, pp. 333–340 (1993)
3. Day, J., Fleischmann, P., Kosche, M., Koß, T., Manea, F., Siemer, S.: The edit distance to k-subsequence universality. In: STACS, vol. 187, pp. 25:1–25:19 (2021)
4. Fleischer, L., Kufleitner, M.: Testing Simon's congruence. In: Potapov, I., Spirakis, P.G., Worrell, J. (eds.) MFCS, vol. 117, pp. 62:1–62:13 (2018)
5. Fleischmann, P., Germann, S., Nowotka, D.: Scattered factor universality - the power of the remainder. Preprint arXiv:2104.09063 (published at RuFiDim) (2021)
6. Fleischmann, P., Haschke, L., Huch, A., Mayrock, A., Nowotka, D.: Nearly k-universal words - investigating a part of Simon's congruence. In: DCFS. LNCS, vol. 13439, pp. 57–71. Springer (2022). https://doi.org/10.1007/978-3-031-13257-5_5
7. Frobenius, H.: Über lineare Substitutionen und bilineare Formen, vol. 1878. De Gruyter (1878)
8. Gawrychowski, P., Kosche, M., Koß, T., Manea, F., Siemer, S.: Efficiently testing Simon's congruence. In: STACS. LIPIcs, vol. 187, pp. 34:1–34:18 (2021)
9. Hébrard, J.: An algorithm for distinguishing efficiently bit-strings by their subsequences. TCS **82**(1), 35–49 (1991)
10. Janjic, M.: Generalized compositions with a fixed number of parts. arXiv:1012.3892 (2010)
11. Karandikar, P., Schnoebelen, P.: The height of piecewise-testable languages with applications in logical complexity. In: CSL (2016)
12. Karandikar, P., Kufleitner, M., Schnoebelen, P.: On the index of Simon's congruence for piecewise testability. Inf. Process. Lett. **115**(4), 515–519 (2015)
13. Karandikar, P., Schnoebelen, P.: The height of piecewise-testable languages and the complexity of the logic of subwords. LMCS **15**(2) (2019)
14. Kim, S., Han, Y., Ko, S., Salomaa, K.: On simon's congruence closure of a string. In: Han, Y., Vaszil, G. (eds.) DCFS. LNCS, vol. 13439, pp. 127–141. Springer (2022). https://doi.org/10.1007/978-3-031-13257-5_10
15. Kim, S., Ko, S., Han, Y.: Simon's congruence pattern matching. In: Bae, S.W., Park, H. (eds.) ISAAC. LIPIcs, vol. 248, pp. 60:1–60:17. Schloss Dagstuhl - Leibniz-Zentrum für Informatik (2022)
16. Kosche, M., Koß, T., Manea, F., Siemer, S.: Absent subsequences in words. In: Bell, P.C., Totzke, P., Potapov, I. (eds.) RP 2021. LNCS, vol. 13035, pp. 115–131. Springer, Cham (2021). https://doi.org/10.1007/978-3-030-89716-1_8
17. Kátai-Urbán, K., Pach, P., Pluhár, G., Pongrácz, A., Szabó, C.: On the word problem for syntactic monoids of piecewise testable languages. Semigroup Forum **84**(2) (2012)
18. Lothaire, M.: Combinatorics on Words, 2nd edn. Cambridge Mathematical Library, Cambridge University Press, Cambridge (1997)
19. OEIS Foundation Inc.: The on-line encyclopedia of integer sequences (2022). http://oeis.org
20. Pach, P.: Normal forms under Simon's congruence. Semigroup Forum **97**(2) (2018)
21. Sakarovitch, J., Simon, I.: Subwords, chap. 6, pp. 105–144. In: Cambridge Mathematical Library [18], 2 edn. (1997)

22. Simon, I.: Hierarchies of Events with Dot-depth One. Ph.D. thesis, University of Waterloo, Department of Applied Analysis and Computer Science (1972)
23. Simon, I.: Piecewise testable events. In: Brakhage, H. (ed.) GI-Fachtagung 1975. LNCS, vol. 33, pp. 214–222. Springer, Heidelberg (1975). https://doi.org/10.1007/3-540-07407-4_23

Bounds for c-Ideal Hashing

Fabian Frei[✉] and David Wehner

ETH Zürich, Zürich, Switzerland
`{fabian.frei,david.wehner}@inf.ethz.ch`

Abstract. In this paper, we analyze hashing from a worst-case perspective. To this end, we study a new property of hash families that is strongly related to d-perfect hashing, namely c-ideality. On the one hand, this notion generalizes the definition of perfect hashing, which has been studied extensively; on the other hand, it provides a direct link to the notion of c-approximativity. We focus on the usually neglected case where the average load α is at least 1 and prove upper and lower parametrized bounds on the minimal size of c-ideal hash families.

As an aside, we show how c-ideality helps to analyze the *advice complexity* of hashing. The concept of advice, introduced a decade ago, lets us measure the information content of an online problem. We prove hashing's advice complexity to be linear in the hash table size.

1 Introduction

Hashing is one of the most popular tools in computing, both from a practical and a theoretical perspective. Hashing was invented as an efficient method to store and retrieve information. Its success began at latest in 1968 with the seminal paper "Scatter Storage Techniques" by Robert Morris [19], which was reprinted in 1983 with the following laudation [20]:

> From time to time there is a paper which summarizes an emerging research area, focuses the central issues, and brings them to prominence. Morris's paper is in this class. [...] It brought wide attention to hashing as a powerful method of storing random items in a data structure with low average access time.

This quote illustrates that the practical aspects were the focus of the early research, and rightly so. Nowadays, hashing has applications in many areas in computer science. Before we cite examples of how hashing can be applied, we state the general setting.

1.1 General Setting and Notation

We have a large set U, the *universe*, of all possible elements, the *keys*. In our example, this would be the set of all possible car plates, so all strings of length 8 over the alphabet $\{A, \ldots, Z, 0, \ldots, 9\}$. Then, there is a subset $S \subseteq U$ of keys from the universe. This set stands for the unknown elements that appear in our application. In our example, S corresponds to the set of all car plates that we see on this day. Then, there is a small set T, the *hash table*, whose m elements are called *cells* or *slots*. In our example, T corresponds to our notebook and we organize the entries in our notebook according to the last character on the car plate, so each cell corresponds to a single letter of the alphabet. Typically, the universe is huge in comparison to the hash table, for instance, $|U| = 2^{|T|}$. Every function from U to T is called a *hash function*; it *hashes* (i.e., maps) keys to cells. We have to choose a function h from U to T such that the previously unknown set S is distributed as evenly as possible among the cells of our hash table. For an introduction to hashing, one of the standard textbooks is the book by Mehlhorn [16]. We recommend the newer book by Mehlhorn and Sanders [17], and to German readers in particular its German translation by Dietzfelbinger, Mehlhorn and Sanders [5].

While U and T can be arbitrary finite sets, we choose to represent their elements by integers—which can always be achieved by a so-called pre-hash function—and let $U := \{1, \ldots, u\}$ and $T := \{1, \ldots, m\}$. For convenience, we abbreviate $\{1, \ldots, k\}$ by $[k]$ for any natural number k. We assume the size n of the subset to be at least as large as the size of the hash table, that is, $|S| = n \geq m$ and, to exclude the corner case of hashing almost the entire universe, also $|U| \geq n^2$.

1.2 Applications of Hashing

Among the numerous applications of hashing, two broad areas stand out. First, as we have seen, hashing can be used to store and retrieve information. With hashing, inserting new information records, deleting records, and searching records can all be done in expected constant time, that is, the number of steps needed to insert, find, or delete an information record does not depend on the size of U, S, or T. One example of this are dictionaries in Python, which are implemented using a hash table and thus allow for value retrieval by key, inserting new key-value-pairs, and deleting key-value-pairs in time independent of the dictionary size. Another example is the Rabin-Karp algorithm [12], which searches a pattern in a text and can be used for instance to detect plagiarism. The second main application area is cryptography. In cryptographic hashing, the hash function has to fulfill additional requirements, for example that it is computationally very hard to reconstruct a key x from its hash $h(x)$ or to find a colliding key, that is, a $y \in U$ with $h(y) = h(x)$. Typical use cases here are, for example, digital signatures and famous examples of cryptographic hash functions are the checksum algorithms such as MD5 or SHA, which are often used to check whether two files are equal. Moreover, and perhaps particularly interesting for computer science, hashing is a useful tool in the design of randomized algorithms and their derandomization. Further examples and more details can be found for example in the useful article by Luby and Wigderson [15].

1.3   Theory of Hashing

Soon after the early focus on the practical aspect, a rich theory on hashing evolved. In this theory, randomization plays a pivotal role. From a theory point of view, we aim for a hash function h that reduces collisions among the yet-to-be-revealed keys of S to a minimum. One possibility for selecting h is to choose the image for each key in U uniformly at random among the m cells of T. We can interpret this as picking a random h out of the family $\mathcal{H}_{all}$ of all hash functions.[1] Then the risk that two keys $x, y \in U$ collide is only $1/m$, that is, $\forall x, y \in U, x \neq y: \underset{h \in \mathcal{H}_{all}}{\mathbb{P}} (h(x) = h(y)) = \frac{1}{m}$. On the downside, this random process can result in computationally complex hash functions whose evaluation has space and time requirements in $\Theta(u \ln m)$. Efficiently computable functions are necessary in order to make applications feasible, however. Consequently, the assumption of such a simple uniform hashing remains in large part a theoretical one with little bearing on practical applications; it is invoked primarily to simplify theoretical manipulations.

The astonishing discovery of small hashing families that can take on the role of $\mathcal{H}_{all}$ addresses this problem. In their seminal paper in 1979, Larry Carter and Mark Wegman [3] introduced the concept of *universal hashing families* and showed that there exist small universal hashing families. A family $\mathcal{H}$ of hash functions is called *universal* if it can take the place of $\mathcal{H}_{all}$ without increasing the collision risk: $\mathcal{H}$ is universal $\iff \forall x, y \in U, x \neq y: \underset{h \in \mathcal{H}}{\mathbb{P}} (h(x) = h(y)) \leq 1/m$. In the following years, research has been successfully dedicated to revealing universal hashing families of comparably small size that exhibit the desired properties.

1.4   Determinism Versus Randomization

Deterministic algorithms cannot keep up with this incredible performance of randomized algorithms, that is, as soon as we are forced to choose a single hash function without knowing S, there is always a set S such that all keys are mapped to the same cell. Consequently, deterministic algorithms have not been of much interest. Using the framework of advice complexity, we measure how much additional information deterministic algorithms need in order to hold their ground or to even have the edge over randomized algorithms.

In order to analyze the advice complexity of hashing, we have to view hashing as an online problem. This is not too difficult; being forced to take irrevocable decisions without knowing the whole input is the essence of online algorithms, as we describe in the main introduction of this thesis. In the standard hashing setting, we are required to predetermine our complete strategy, that is, our hash function h, with no knowledge about S at all. In this sense, standard hashing is an ultimate online problem. We could relax this condition and require that $S = \{s_1, \ldots, s_n\}$ is given piecemeal and the algorithm has to decide to which cell the input s_i is mapped only upon its arrival. However, this way, an online algorithm could just distribute the input perfectly, which would lead, as in the case of $\mathcal{H}_{all}$, to computationally complex functions.

---

[1] Note that the number of hash functions, $|\mathcal{H}_{all}| = m^u$, is huge.

Therefore, instead of relaxing this condition, we look for the reason why there is such a gap between deterministic and randomized strategies. The reason is simple: Deterministic strategies are measured according to their performance in the worst case, whereas randomized strategies are measured according to their performance on average. However, if we measure the quality of randomized strategies from a worst-case perspective, the situation changes. In particular, while universal hashing schemes prove incredibly useful in everyday applications, they are far from optimal from a worst-case perspective, as we illustrate now.

We regard a hash function h as an algorithm operating on the input set of keys S. We assess the performance of h on S by the resulting *maximum cell load*: $\text{cost}(h, S) := \alpha_{\max} := \max\{\alpha_1, \ldots, \alpha_m\}$, where α_i is the *cell load* of cell i, that is, $\alpha_i := |h^{-1}(i) \cap S|$, the number of keys hashed to cell i. This is arguably the most natural cost measurement for a worst-case analysis. Another possibility, the total number of collisions, is closely related.[2]

The *average load* is no useful measurement option as it is always $\alpha := n/m$. The worst-case cost n occurs if all $n = |S|$ keys are assigned to a single cell. The optimal cost, on the other hand, is $\lceil n/m \rceil = \lceil \alpha \rceil$ and it is achieved by distributing the keys of S into the m cells as evenly as possible.

Consider now a randomized algorithm ALG that picks a hash function $h \in \mathcal{H}_{\text{all}}$ uniformly at random. A long-standing result [9], proven nicely by Raab and Steger [22], shows the expected cost $\mathbb{E}_{h \in \mathcal{H}_{\text{all}}}[\text{cost}(h, S)]$ to be in $\Omega\left(\frac{\ln m}{\ln \ln m}\right)$, as opposed to the optimum $\lceil \alpha \rceil$. The same holds true for smaller universal hashing schemes such as polynomial hashing, where we randomly choose a hash function h out of the family of all polynomials of degree $\mathcal{O}\left(\frac{\ln n}{\ln \ln n}\right)$. However, no matter which universal hash function family is chosen, it is impossible to rule out the absolute worst-case: For every chosen hash function h there is, due to $u \geq n^2$, a set S of n keys that are all mapped to the same cell.

1.5 Our Model and the Connection to Advice Complexity

We are, therefore, interested in an alternative hashing model that allows for meaningful algorithms better adapted to the worst case: What if we did not choose a hash function at random but could start with a family of hash functions and then always use the best function, for any set S? What is the trade-off between the size of such a family and the upper bounds on the maximum load they can guarantee? In particular, how large is a family that always guarantees an almost optimal load?

In our model, the task is to provide a set $\mathcal{H}$ of hash functions. This set should be small in size while also minimizing the term $\text{cost}(\mathcal{H}, S) := \max_{S \in \mathcal{S}} \min_{h \in \mathcal{H}} \text{cost}(h, S)$, which is the best cost bound that we can ensure across a given set S of inputs by using any hash function in $\mathcal{H}$. Hence, we look for a family $\mathcal{H}$ of hash functions that minimizes both $|\mathcal{H}|$ and $\text{cost}(\mathcal{H}, S)$. These two goals conflict, resulting in a trade-off, which we parameterize by $\text{cost}(\mathcal{H}, S)$, using the notion of *c-ideality*.

---

[2] This number can be expressed as $\sum_{k=1}^{m} \binom{\alpha_k}{2} \in \Theta\left(\alpha_1^2 + \ldots + \alpha_m^2\right) \subseteq \mathcal{O}\left(m \cdot \alpha_{\max}^2\right)$.

Definition 1. (c-Ideality). *Let $c \geq 1$. A function $h : U \to T$ is called c-ideal for a subset $S \subseteq U$ of keys if $\mathrm{cost}(h, S) = \alpha_{\max} \leq c\alpha$. In other words, a c-ideal hash function h assigns at most $c\alpha$ elements of S to each cell of the hash table T.*

Similarly, a family of hash functions $\mathcal{H}$ is called c-ideal for a family $\mathcal{S}$ of subsets of U if, for every $S \in \mathcal{S}$, there is a function $h \in \mathcal{H}$ such that h is c-ideal for S. This is equivalent to $\mathrm{cost}(\mathcal{H}, \mathcal{S}) \leq c\alpha$. If $\mathcal{H}$ is c-ideal for $\mathcal{S}_n := \{S \subseteq U; \ |S| = n\}$ (that is, all sets of n keys), we simply call $\mathcal{H}$ c-ideal.

We see that c-ideal families of hash functions constitute algorithms that guarantee an upper bound on $\mathrm{cost}(\mathcal{H}) := \mathrm{cost}(\mathcal{H}, \mathcal{S}_n)$, which fixes one of the two trade-off parameters. Now, we try to determine the other one and find, for every $c \geq 1$, the minimum size of a c-ideal family, which we denote by $H_c := \min\{|\mathcal{H}|; \ \mathrm{cost}(\mathcal{H}) \leq c\alpha\}$. Note that $c \geq m$ renders the condition of c-ideality void since $n = m\alpha$ is already the worst-case cost; we always have $\mathrm{cost}(\mathcal{H}) \leq n \leq c\alpha$ for $c \geq m$. Consequently, every function is m-ideal, every non-empty family of functions is m-ideal, and $H_m = 1$.

With the notion of c-ideality, we can now talk about c-competitiveness of hashing algorithms. Competitiveness is directly linked to c-ideality since, for each S, $\mathrm{cost}(\mathrm{ALG}(S)) := \alpha_{\max}$ and the cost of an optimal solution is always $\mathrm{cost}(\mathrm{OPT}(S)) = \lceil \alpha \rceil$. Therefore, a hash function that is c-ideal is c-competitive as well. Clearly, no single hash function is c-ideal; however, a hash function family $\mathcal{H}$ can be c-ideal and an algorithm that is allowed to choose, for each S, the best hash function among a c-ideal family $\mathcal{H}$ is thus c-competitive. Such a choice corresponds exactly to an online algorithm with advice that reads $\lceil \log(|\mathcal{H}|) \rceil$ advice bits, which is just enough to indicate the best function from a family of size $|\mathcal{H}|$. Since, by definition, there is a c-ideal hash function family of size H_c, there is an online algorithm with advice complexity $\lceil \log(H_c) \rceil$ as well. Moreover, there can be no c-competitive online algorithm ALG that reads less than $\lceil \log H_c \rceil$ advice bits: If there were such an algorithm, there would be a c-ideal hash function family of smaller size than H_c, which contradicts the definition of H_c. We will discuss the relation between c-ideal hashing and advice complexity in more depth in Sect. 6.

1.6   Organization

This paper is organized as follows. In Sect. 2, we give an overview of related work and our contribution. We present our general method of deriving bounds on the size of c-ideal families of hash functions in Sect. 3. Section 4 is dedicated to precisely calculating these general bounds. In Sect. 5, we give improved bounds for two edge cases. In Sect. 6, we analyze the advice complexity of hashing. We recapitulate and compare our results in Sect. 7.

2   Related Work and Contribution

There is a vast body of literature on hashing. Indeed, hashing still is a very active research area today and we cannot even touch upon the many aspects that have been considered. Instead, in this section, we focus on literature very closely connected to c-ideal hashing. For a coarse overview of hashing in general, we refer to the survey by Chi and Zhu [4] and the seminar report by Dietzfelbinger et al. [6].

The advice complexity of hashing has not yet been analyzed; however, c-ideality is a generalization of perfect k-hashing, sometimes also called k-perfect hashing. For $n \leq m$ (i.e., $\alpha \leq 1$) and $c = 1$, our definition of c-ideality allows no collisions and thus reduces to perfect hashing, a notion formally introduced by Mehlhorn in 1984 [16]. For this case, Fredman and Komlós [8] proved in 1984 the bounds

$$H_1 \in \Omega\left(\frac{m^{n-1}\log(u)(m-n+1)!}{m!\log(m-n+2)}\right) \text{ and } H_1 \in \mathcal{O}\left(\frac{-n\log(u)}{\log\left(1 - \frac{m!}{(m-n)!m^n}\right)}\right).$$

In 2000, Blackburn [2] improved their lower bound for u large compared to n. Recently, in 2022, Guruswami and Riazonov [10] improved their bound as well. None of these proofs generalize to $n > m$, that is, $\alpha > 1$.

Another notion that is similar to c-ideality emerged in 1995. Naor et al. [21] introduced (u, n, m)-splitters, which coincide with the notion of 1-ideal families for $\alpha \geq 1$. They proved a result that translates to

$$H_1 \in \mathcal{O}\left(\sqrt{2\pi\alpha}^m e^{\frac{m}{12\alpha}}\sqrt{n}\ln u\right). \tag{1}$$

Since the requirements for c-ideality are strongest for $c = 1$, this upper bound holds true for the general case of $c \geq 1$ as well.

We extend these three results to the general case of $n \geq m$ and $c \geq 1$. Moreover, we tighten the third result further for large c. Specifically, we prove the following new bounds:

$$H_c \geq (1-\varepsilon)\exp\left(\frac{m}{e^\alpha}(1-\varepsilon)\left(\frac{\alpha}{c\alpha+1}\right)^{c\alpha+1}\right) \qquad \text{(Theorem 3)} \qquad (2)$$

$$H_c \geq \frac{\ln u - \ln(c\alpha)}{\ln m} \qquad \text{(Theorem 5)} \qquad (3)$$

$$H_c \leq upperboundHc \qquad \text{(Theorem 4)} \qquad (4)$$

$$H_c \in \mathcal{O}\left(\frac{n\ln u}{\ln t}\right) \text{ for any } t \geq 1 \text{ and } c \in \omega\left(t\frac{\ln n}{\ln\ln n}\right) \qquad \text{(Theorem 6)} \qquad (5)$$

Note that (4) coincides with (1) for $c = 1$ and is only an improvement for α and c slightly larger than 1. Since $H_c \leq H_{c'}$ for $c \geq c'$, the bound still improves slightly upon (1) in general, depending on the constants hidden in the O-notation in (1). Interestingly, the size u of the universe does not appear in (2); a phenomenon discussed in [8]. Fredman and Komlós used information-theoretic results based on the Hansel Lemma [11] to obtain a bound that takes the universe size into account. Körner [14] expressed their approach in the language of graph entropy.

It is unclear, however, how these methods could be generalized to the case $\alpha > 1$ in any meaningful way.

The straightforward approach of proving a lower bound on H_1 is to use good Stirling estimates for the factorials in $H_1 \geq \binom{u}{n}/\binom{u/m}{\alpha}^m$; see Mehlhorn [16]. This yields a lower bound of roughly $\sqrt{2\pi\alpha}^{m-1}/\sqrt{m}$, which is better than (2) for $c = 1$, see Lemma 3 for details. Unfortunately, it turns out that we cannot obtain satisfying results for $c > 1$ with this approach. However, the results from Dubhashi and Ranjan [7] and the method of Poissonization enable us to circumnavigate this obstacle and derive Eq. (2).

We use our bounds to derive bounds for the advice complexity of hashing, which we will discuss in Sect. 6.

3   General Bounds on H_c

We present our bounds on H_c, which is the minimum size of c-ideal families of hash functions. First, we establish a general lower and upper bound.

We use a *volume bound* to lower-bound H_c. We need the following definition. Let M_c be the maximum number of sets $S \in \mathcal{S}_n$ that a single hash function can map c-ideally, that is,

$$M_c := \max_{h \in \mathcal{H}_{\mathrm{all}}} |\{S \in \mathcal{S}_n;\ \alpha_{\max} \leq c\alpha\}| = \max_{h \in \mathcal{H}_{\mathrm{all}}} |\{S \in \mathcal{S}_n;\ \forall i \in [m] : \alpha_i \leq c\alpha\}|.$$

Lemma 1 (Volume Bound). *The number of hash functions in a family of hash functions that is c-ideal is at least the number of sets in $\mathcal{S}_n$ divided by the number of sets for which a single hash function can be c-ideal, that is, $H_c \geq |\mathcal{S}_n|/M_c$.*

Proof. $|\mathcal{S}_n| = \binom{u}{n}$ is the number of subsets of size n in the universe U and M_c is the maximum number of such subsets for which a single function h can be c-ideal. A function family $\mathcal{H}$ is thus c-ideal for at most $|\mathcal{H}| \cdot M_c$ of these subsets. If $\mathcal{H}$ is supposed to be c-ideal for all subsets—that is, to contain a c-ideal hash function for every single one of them— we need $|\mathcal{H}| \cdot M_c \geq |\mathcal{S}_n|$.

To be able to estimate M_c, we consider hash functions that distribute the u keys of our universe as evenly as possible:

Definition 2 (Balanced Hash Function). *A balanced hash function h partitions the universe into m parts by allotting to each cell $\lceil u/m \rceil$ or $\lfloor u/m \rfloor$ elements. We denote the set of balanced hash functions by $\mathcal{H}_{\mathrm{eq}}$. We write h_{eq} to indicate that h is balanced.*[3]

Theorem 1, whose proof is omitted due to the space constraints, states that exactly all balanced hash functions attain the value $M_{\hat{c}}$. Therefore, we can limit ourselves to such functions.

---

[3] Note that, for any h_{eq}, there are exactly $(u \bmod m)$ cells of size $\lceil u/m \rceil$ since $\sum_{k=1}^{m} |h_{\mathrm{eq}}^{-1}(k)| = u$.

Theorem 1. (Balance Maximizes M_c). *A function is c-ideal for the maximal number of subsets if and only if it is balanced. In other words, the number of subsets that are hashed c-ideally by a hash function equals M_c if and only if h is balanced.*

Now fix a balanced hash function $h \in \mathcal{H}_{\mathrm{eq}}$. We switch to a randomization perspective: Draw an $S \in \mathcal{S}_n$ uniformly at random. The cell loads α_k are considered random variables that assume integer values based on the outcome $S \in \mathcal{S}_n$. The probability that our fixed h hashes the random S c-ideally is exactly

$$\mathbb{P}(\alpha_{\max} \leq c\alpha) = \frac{|\{S \in \mathcal{S}_n;\ h \text{ hashes } S \text{ } c\text{-ideally}\}|}{|\mathcal{S}_n|} = \frac{M_c}{|\mathcal{S}_n|}.$$

We suspend our analysis of the lower bound for the moment and switch to the upper bound to facilitate the comparison of the bounds. The proof of the following lemma is omitted to the space constraints.

Lemma 2 (Probability Bound). *We can bound the minimal size of a c-ideal family by $H_c \leq \left\lceil \frac{|\mathcal{S}_n|}{M_c} n \ln u \right\rceil$.*

We combine Lemma 1 and Lemma 2 and summarize our findings:

Corollary 1 (General Bounds on Family Size H_c). *The size of a c-ideal family of hash functions is bounded by $\frac{1}{\mathbb{P}(\alpha_{\max} \leq c\alpha)} \leq H_c \leq \frac{n \ln u}{\mathbb{P}(\alpha_{\max} \leq c\alpha)}$.*

Now, to lower-bound H_c, we first consider for $c = 1$ a straightforward application of Lemma 1 suggested by Mehlhorn [16] and then ponder whether we could extend this approach for c larger than 1.

Lemma 3. *The number of hash functions in a 1-ideal family of hash functions is bounded from below by approximately $\approx \frac{\sqrt{2\pi\alpha}^{m-1}}{\sqrt{m}}$.*

The natural extension of this approach yields $H_c \geq K(n, m, c\alpha)\binom{u}{n} / \left(\frac{u/m}{c\alpha}\right)^m$, where $K(n, m, c\alpha)$ denotes the number of compositions of n into m non-negative integers between 0 and $c\alpha$. This factor, which accounts for the number of possibilities to split the n keys into m different cells, equals 1 for $c = 1$. For general c, however, to the best of our knowledge, even the strongest approximations for $K(n, m, c\alpha)$ do not yield a meaningful lower bound. Therefore, we are forced to use a different strategy for general c and we estimate M_c via the probability $\mathbb{P}(\alpha_{\max} \leq c\alpha)$ in the following section.

4   Estimations for $\mathbb{P}(\alpha_{\max} \leq c\alpha)$

It remains to find good bounds on the probability $\mathbb{P}(\alpha_{\max} \leq c\alpha)$, which will immediately yield the desired bounds on H_c. We start by establishing an upper bound on $\mathbb{P}(\alpha_{\max} \leq c\alpha)$.

4.1 Upper Bound

Recall that we fixed a balanced hash function $h \in \mathcal{H}_{eq}$ and draw an $S \in \mathcal{S}_n$ uniformly at random. For every cell $k \in T$, we model its load α_k as a random variable. The joint probability distribution of the α_i follows the hypergeometric distribution, that is, $\mathbb{P}((\alpha_1, \ldots, \alpha_m) = (\ell_1, \ldots, \ell_m)) = \frac{\binom{|h^{-1}(1)|}{\ell_1} \cdots \binom{|h^{-1}(m)|}{\ell_m}}{\binom{u}{n}}$. There are two obstacles to overcome. First, calculating with probabilities without replacement is difficult. Second, the sum of the α_i is required to be n; in particular, the variables are not independent. The first obstacle can be overcome by considering drawing elements from U with replacement instead, that is, drawing multisets instead of sets. As the following lemma, whose proof is omitted due to the space constraints, shows, we do not lose much by this assumption.

Lemma 4 (Switching to Drawing With Replacement) *For any fixed $\varepsilon > 0$, there are u and n large enough such that $\frac{M_c}{|\mathcal{S}_n|} \leq (1 + \varepsilon)\mathbb{P}(T_i \leq c\alpha, i \in [m])$, where we use $T_1, \ldots T_m$ to model the cell loads as binomially distributed random variables with parameters n and $1/m$. In other words, for u and n large enough, we can consider drawing the elements of $S \in \mathcal{S}_n$ with replacement and only lose a negligible factor.*

If we consider drawing elements from U with replacement, we have a multinomial distribution, that is, $\mathbb{P}((T_1, \ldots, T_m) = (\ell_1, \ldots, \ell_m)) = \binom{n}{\ell_1, \ldots, \ell_m}/m^n$. This is easier to handle than the hypergeometric distribution. The T_i, $i \in [m]$, are still not independent, however. The strong methods by Dubhashi and Ranjan [7] provide us with a simple way to overcome this second obstacle:

Lemma 5 (Proposition 28 and Theorem 31 from [7]). *The joint distribution of the random variables α_i for $i \in [m]$ is upper-bounded by their product: $\mathbb{P}(\alpha_i \leq c\alpha, i \in [m]) \leq \prod_{i=1}^{m} \mathbb{P}(\alpha_i \leq c\alpha)$. The same holds if we consider the T_i, $i \in [m]$, instead of the α_i.*

We are ready to give an upper bound on the probability that a hash function family is c-ideal. The proof of the following theorem is omitted due to the space constraints.

Theorem 2 (Upper Bound on $\mathbb{P}(\alpha_{\max} \leq c\alpha)$). *For arbitrary $\varepsilon > 0$, $\mathbb{P}(\alpha_{\max} \leq c\alpha) \leq (1 + \varepsilon) \exp\left(-m e^{-\alpha}(1 - \varepsilon)\left(\frac{\alpha}{c\alpha + 1}\right)^{c\alpha + 1}\right)$, where n tends to infinity.*

Corollary 1 translates Theorem 2's upper bound into a lower bound on H_c.

Theorem 3 (Lower Bound on Family Size H_c). *For arbitrary $\varepsilon > 0$, the number of hash functions in a c-ideal family of hash functions is bounded from below by $H_c \geq \frac{1}{\mathbb{P}(\alpha_{\max} \leq c\alpha)} \geq (1 - \varepsilon) \exp\left(\frac{m}{e^\alpha}(1 - \varepsilon)\left(\frac{\alpha}{c\alpha + 1}\right)^{c\alpha + 1}\right)$.*

4.2   Lower Bound

Another way to overcome the obstacle that the variables are not independent would have been to apply a customized Poissonization technique. The main monograph presenting this technique is in the book by Barbour et al. [1]; the textbook by Mitzenmacher and Upfal [18] gives a good illustrating example. A series of arguments would have allowed us to bound the precision loss we incur by a constant factor of 2, leading to

$$\mathbb{P}(T_{\max} \leq c\alpha) \leq 2\mathbb{P}(Y \leq c\alpha)^m,$$

where we use $T_{\max}$ to denote $\max\{T_1, \ldots, T_m\}$ and where Y is a Poisson random variable with mean α. By using Lemma 5, we were able to abbreviate this approach. However, part of this Poissonization technique can be used for the lower bound on $\mathbb{P}(\alpha_{\max} \leq c\alpha)$.

Recall that a random variable X following the Poisson distribution—commonly written as $X \sim P_\lambda$ and referred to as a Poisson variable—takes on the value k with probability $\mathbb{P}(X = k) = \frac{1}{e^\lambda}\frac{\lambda^k}{k!}$. The following lemma, whose proof is omitted due to the space constraints, is the counterpart to Lemma 4.

Lemma 6. *We can bound $M_c/|\mathcal{S}_n| = \mathbb{P}(\alpha_{\max} \leq c\alpha)$ from below by $\mathbb{P}(T_{\max} \leq c\alpha)$.*

For the counterpart to Theorem 2, we use Poissonization to turn the binomial variables into Poisson variables.

The next lemma, whose proof is omitted due to the space constraints, shows that $(T_1, \ldots, T_m)$ has the same probability mass function as $(Y_1, \ldots, Y_n)$ under the condition $Y = n$.

Lemma 7 (Sum Conditioned Poisson Variables). *For any natural numbers $\ell_1, \ldots, \ell_m$ with $\ell_1 + \ldots + \ell_m = n$, we have that*

$$\mathbb{P}((Y_1, \ldots, Y_m) = (\ell_1, \ldots, \ell_m) \mid Y = n) = \mathbb{P}((T_1, \ldots, T_m) = (\ell_1, \ldots, \ell_m)).$$

The result of Lemma 7 immediately carries over to the conditioned expected value for any real function $f(\ell_1, \ldots, \ell_m)$, by the definition of expected values.

Corollary 2 (Conditioned Expected Value). *Let $f(\ell_1, \ldots, \ell_m)$ be any real function. We have $\mathbb{E}[f(Y_1, \ldots, Y_m) \mid Y = n] = \mathbb{E}[f(T_1, \ldots, T_m)]$.*

We are ready to state the counterpart to Theorem 2. The proof is omitted due to the space constraints.

Lemma 8 (Lower Bound on Non-Excess Probability). *Set $d = c\alpha$. We have $\mathbb{P}(T_{\max} \leq d) \geq \frac{\sqrt{2\pi n}}{(2\pi d)^{m/(2c)}} \frac{1}{c^n} \frac{1}{e^{\frac{m}{12cd}}} (\alpha + 1)^{m(1-\frac{1}{c})}.$*

Proof. Using Corollary 2, we can formulate this probability with Poisson variables:

$$\mathbb{P}(T_{\max} \le d) = \mathbb{E}[\chi_{T_{\max} \le d}]$$

$$\text{Corollary 2} \quad = \mathbb{E}[\chi_{\max\{Y_1,\ldots,Y_m\} \le d} \mid Y = n]$$

$$\text{Definition of conditional probability} \quad = \frac{\mathbb{E}[\chi_{\{Y=n\}} \chi_{\max\{Y_1,\ldots,Y_m\} \le d}]}{\mathbb{P}(Y = n)}.$$

We know that $Y \sim P_n$; hence, the denominator equals $n^n/(n!e^n)$. For the numerator, we use the fact that $Y_i \sim P_\alpha$ for all i to write:

$$\mathbb{E}[\chi_{\{Y=n\}} \chi_{\max\{Y_1,\ldots,Y_m\} \le d}] = \sum_{\substack{\ell_1,\ldots,\ell_m \in \mathbb{N} \\ \ell_1 + \ldots + \ell_m = n \\ \ell_1,\ldots,\ell_m \le d}} \prod_{i=1}^{m} \frac{1}{e^\alpha} \frac{\alpha^{\ell_i}}{\ell_i!}$$

We estimate the sum by finding a lower bound for both the number of summands and the products that constitute the summands. The product attains its minimum when the ℓ_i are distributed as asymmetrically as possible, that is, if almost all ℓ_i are set to either d or 0, with only one being $(n \mod d)$. This fact is an extension of the simple observation that for any natural number n, we have

$$n!n! = n(n-1)(n-2) \cdot \ldots \cdot 1 \cdot n! < 2n(2n-1)(2n-2) \cdot \ldots \cdot (n+1) \cdot n! = (2n)!.$$

To obtain a rigorous proof we extend the range of the expression to the real numbers and analyze the extremal values by observing where the projection of the gradient onto the hyperplane defined by the side condition vanishes. The details are omitted due to the space constraints.

The next step is to find a lower bound on the number of summands, that is, the number of integer compositions $\ell_1 + \ldots + \ell_m = n$. As we mentioned at the very end of Sect. 3, we are not aware of strong approximation for this number and the following estimation is very crude for many values of n, m, and c: If we let the first $m - \frac{m}{c}$ integers, $\ell_1, \ldots, \ell_{\frac{m}{c}}$, range freely between 0 and α, we can always ensure that the condition $\ell_1 + \ldots + \ell_m = n$ is satisfied by choosing the remaining $\frac{m}{c}$ summands $\ell_{\frac{m}{c}+1}, \ldots, \ell_m$ appropriately. Therefore, the number of summands is at least $(\alpha+1)^{m(1-\frac{1}{c})}$. Together with the previous calculation, this leads to

$$\mathbb{E}[\chi_{\{Y=n\}} \chi_{\max\{Y_1,\ldots,Y_m\} \le d}] \ge \frac{\alpha^n}{e^n} \frac{(\alpha+1)^{m(1-\frac{1}{c})}}{(d!)^{m/c}}.$$

Putting everything together and using the Stirling bounds [23] for the factorials, we obtain the desired result.

Combining the results in this section in the straightforward way—the execution is omitted due to the space constraints— yields the following upper bound on H_c.

Theorem 4 (Upper Bound on Minimal Family Size H_c). *The minimal number of hash functions in a c-ideal family of hash functions is bounded from above by $H_c \leq upperboundHc$.*

Theorem 4 states that for constant α, the number of functions such that for each set of keys of size n, there is a function that distributes this set among the hash table cells at least as good as c times an optimal solution is bounded from above by a number that grows exponentially in m, but only with the square root in n and only logarithmically with the universe size. However, our bounds do not match, hence the exact behavior of H_c within the given bounds remains obscure. It becomes easier if we analyze the advice complexity, using the connection between c-ideality and advice complexity described in the introduction of this paper. We first improve our bounds on H_c for some edge cases in the next section before we then use and interpret the results in Sect. 6.

5   Improvements for Edge Cases

As discussed in the introduction, the size u of the universe does not appear in the lower bound on H_c of Corollary 1. The following lower bound on H_c, which is a straightforward generalization of an argument presented in the classical textbook by Mehlhorn [16], features u in a meaningful way. The proof is omitted due to the space constraints.

Theorem 5. *We have that $H_c \geq \frac{\ln(u) - \ln(c\alpha)}{\ln(m)}$.*

This demonstrates that, while it is easy to find bounds that include the size of the universe, it seems to be very difficult to incorporate u into a general bounding technique that does not take it into account naturally, such as the first inequality of Corollary 1. However, it is not too difficult to obtain bounds that improve upon Corollary 1 for large—possibly less interesting—values of c. The proof for the following theorem is omitted due to the space constraints.

Theorem 6 (Yao Bound). *For every $c \in \omega\big(t \frac{\ln n}{\ln \ln n}\big)$, where $t \geq 1$, we have $H_c \in \mathcal{O}(\ln |\mathcal{S}_n| / \ln t)$. In particular, we obtain that $H_c \in \mathcal{O}(n \ln u)$ for $t \in \mathcal{O}(1)$.*

6   Advice Complexity of Hashing

With the conceptualization of hashing as an ultimate online problem mentioned in the introduction of this paper, we can use the bounds on H_c to provide bounds on the advice complexity of c-competitive algorithms. Theorem 5 immediately yields the following theorem.

Theorem 7. *Every ALG for hashing with less than $\ln(\ln(u) - \ln(c\alpha)) - \ln(\ln(m))$ advice bits cannot achieve a lower cost than $\mathrm{cost}(\mathrm{ALG}) = c\alpha$ and is thus not better than c-competitive. In other words, there exists an $S \subseteq U$ such that the output $h : U \to T$ of ALG maps at least $c\alpha$ elements of S to one cell.*

With a lower bound on the size of c-ideal families of hash functions, this bound can be improved significantly, as the following theorem shows.

Theorem 8. *Every* ALG *for hashing needs at least*

$$\log\left((1-\varepsilon) \exp\left(\frac{m}{e^\alpha}(1-\varepsilon) \left(\frac{\alpha}{c\alpha+1} \right)^{c\alpha+1} \right) \right)$$

advice bits in order to be c-competitive, for an arbitrary fixed $\varepsilon > 0$.

We want to determine the asymptotic behavior of this bound and hence analyze the term $(\alpha/(c\alpha+1))^{c\alpha+1}$. We are going to use the fact that $\lim_{n\to\infty}(1-1/n)^n = 1/e$.

$$\left(\frac{\alpha}{c\alpha+1} \right)^{c\alpha+1} = \left(\frac{\frac{1}{c}(c\alpha+1-1)}{c\alpha+1} \right)^{c\alpha+1}$$

$$= \left(\frac{1}{c}\left(1 - \frac{1}{c\alpha+1} \right) \right)^{c\alpha+1}$$

for any $\varepsilon' > 0$ and $c\alpha + 1$ large enough $\geq \left(\frac{1}{c} \right)^{c\alpha+1} (1-\varepsilon')\frac{1}{e}$

Therefore, the bound from Theorem 8 is in $\Omega\left(\frac{m}{e^\alpha}\left(\frac{1}{c} \right)^{c\alpha+1} \right)$. Before we interpret this result, we turn to the upper bounds. Theorem 4 yields the following result.

Theorem 9. *There is a c-competitive algorithm with advice that reads*

$$\log upperboundHc$$

$$\in upperboundHcOnotation$$

advice bits.

The factor after m is minimal for $c = \alpha = 1$; this minimal value is larger than 1.002. This upper bound is therefore always at least linear in m. For $c = \omega\left(\frac{\ln(n)}{\ln(\ln(n))} \right)$, we can improve on this and remove the last summand completely, based on Theorem 6.

Theorem 10. *For $c \in \omega\left(t\frac{\ln n}{\ln \ln n} \right)$, $t > 1$, there exists a c-competitive algorithm that reads $\mathcal{O}\left(\log\left(\frac{\ln |S_n|}{t}\ln t \right) \right)$ many advice bits. In particular, for $t \in \mathcal{O}(1)$, there exists an $\frac{\ln n}{\ln \ln n}$-competitive algorithm that reads $\mathcal{O}(\ln \ln u + \ln n)$ many advice bits.*

Theorem 8 and Theorem 9 reveal the advice complexity of hashing to be linear in the hash table size m. While the universe size u still appears in the upper bound, it functions merely as a summand and is mitigated by a double logarithm. Unless the key length $\log_2(u)$ is exponentially larger than the hash table size

m, the universe size cannot significantly affect this general behavior. The more immediate bounds for the edge cases do not reveal this dominance of the hash table over the universe. Moreover, changing the two parameters α and c has no discernible effect on the edge case bounds despite the exponential influence on the main bounds.

7   Conclusion

This paper analyzed hashing from an unusual angle by regarding hashing as an online problem and then studying its advice complexity. Online problems are usually studied with competitive analysis, which is a worst-case measurement. As outlined in the introduction, it is impossible to prevent a deterministic algorithm from incurring the worst-case cost by hashing all appearing keys into a single cell of the hash table. Therefore, randomized algorithms are key to the theory and application of hashing. In particular the surprising discovery of small universal hashing families gave rise to efficient algorithms with excellent expected cost behavior. However, from a worst-case perspective, the performance of randomized algorithms is lacking.

This motivated the conceptualization of c-ideal hashing families as a generalization of perfect k-hashing families to the case where $\alpha > 1$. Our goal was to analyze the trade-off between size and ideality of hashing families since this is directly linked to the competitiveness of online algorithms with advice. Our bounds generalize results by Fredman and Komlós [8] as well as Naor et al. [21] to the case $\alpha > 1$ and $c \geq 1$.

As a first step, we proved that balanced hash functions are suited best for hashing in the sense that they maximize the number of subsets that are hashed c-ideally. Building on this, we applied results by Dubhashi and Ranjan [7] to obtain our main lower bound of (2). Our second lower bound, (3), is a straightforward generalization of a direct approach for the special case $c = \alpha = 1$ by Mehlhorn [16]. We used two techniques to find complementing upper bounds. The first upper bound, (4), uses a Poissonization method combined with direct calculations and is mainly useful for $c \in o\left(\frac{\ln m}{\ln \ln m}\right)$. Our second upper bound, (5), relies on a Yao-inspired principle [13] and covers the case $c \in \omega\left(\frac{\ln m}{\ln \ln m}\right)$.

With these results on the size of c-ideal hash function families, we discovered that the advice complexity of hashing is linear in the hash table size m and only logarithmic in n and double logarithmic in u (see Schmidt and Siegel [24] for similar results for perfect hashing). Moreover, the influence of both α and c is exponential in the lower bound. In this sense, by relaxing the pursuit of perfection only slightly, the gain in the decrease of the size of a c-ideal hash function family can be exponential. Furthermore, only $O(\ln \ln u + \ln n)$ advice bits are necessary for deterministic algorithms to catch up with randomized algorithms.

Further research is necessary to close the gap between our upper and lower bounds. For the edge cases, that is, for $c \geq \log n / \log \log n$, the upper and lower bounds (5) and (3) differ by a factor of approximately $n \ln m$. The interesting case for us, however, is $c \leq \log n / \log \log n$, which is the observed worst-case cost for universal hashing. Contrasting (2) and (4), we note that the difference has two main reasons. First, there is a factor of $n \ln u$ that appears only in the upper bound; this factor stems from the general bounds in Corollary 1. Second, the probability $\mathbb{P}(\alpha_{\max} \leq c\alpha)$ is estimated from below in a more direct fashion than from above, leading to a difference between these bounds that increases with growing c. The reason for the more direct approach is the lack of a result similar to Lemma 5.

Moreover, it remains an open question whether it is possible to adapt the entropy-related methods in the spirit of Fredman-Komlós and Körner in such a way as to improve our general lower bound (2) by accounting for the universe size in a meaningful way.

References

1. Barbour, A.D., Holst, L., Janson, S.: Poisson Approximation. Clarendon Press, Oxford (1992)
2. Blackburn, S.R.: Perfect hash families: probabilistic methods and explicit constructions. J. Comb. Theory Ser. A **92**(1), 54–60 (2000)
3. Carter, L., Wegman, M.N.: Universal classes of hash functions. J. Comput. Syst. Sci. **18**(2), 143–154 (1979)
4. Chi, L., Zhu, X.: Hashing techniques: a survey and taxonomy. **50**(1) (2017)
5. Dietzfelbinger, M., Mehlhorn, K., Sanders, P.: Algorithmen und Datenstrukturen. Springer, Heidelberg (2014). https://doi.org/10.1007/978-3-642-05472-3
6. Dietzfelbinger, M., Mitzenmacher, M., Pagh, R., Woodruff, D.P., Aumüller, M.: Theory and applications of hashing (dagstuhl seminar 17181). Dagstuhl Rep. **7**(5), 1–21 (2017)
7. Dubhashi, D.P., Ranjan, D.: Balls and bins: a study in negative dependence. Random Struct. Algorithms **13**(2), 99–124 (1998)
8. Fredman, M.L., Komlós, J.: On the size of separating systems and families of perfect hash functions. SIAM J. Alg. Disc. Meth. **5**, 61–68 (1984)
9. Gonnet, G.H.: Expected length of the longest probe sequence in hash code searching. J. ACM **28**(2), 289–304 (1981)
10. Guruswami, V., Riazanov, A.: Beating Fredman-Komlós for perfect k-hashing. J. Comb. Theory, Ser. A **188**, 105580 (2022)
11. Hansel, G.: Nombre minimal de contacts de fermeture nécessaires pour réaliser une fonction booléenne symétrique de n variables. Comptes Rendus Hebdomadaires des Séances de l'Académie des Sciences, 258 (1964)
12. Karp, R.M., Rabin, M.O.: Efficient randomized pattern-matching algorithms. IBM J. Res. Dev. **31**(2), 249–260 (1987)
13. Komm, D.: An Introduction to Online Computation - Determinism, Randomization, Advice. Texts in Theoretical Computer Science. An EATCS Series. Springer (2016). https://doi.org/10.1007/978-3-319-42749-2
14. Körner, J.: Fredman-komlós bounds and information theory. SIAM J. Algebraic Discrete Methods **7**(4), 560–570 (1986)
15. Luby, M., Wigderson, A.: Pairwise independence and derandomization. Foundations and Trends in Theoretical Computer Science **1**(4) (2005)
16. Mehlhorn, K.: Data Structures and Algorithms 1: Sorting and Searching, volume 1 of EATCS Monographs on Theoretical Computer Science. Springer (1984). https://doi.org/10.1007/978-3-642-69672-5

17. Mehlhorn, K., Sanders, P.: Algorithms and Data Structures: The Basic Toolbox. Springer (2008). https://doi.org/10.1007/978-3-540-77978-0
18. Mitzenmacher, M., Upfal, E.: Probability and Computing. Cambridge University Press, Cambridge (2017)
19. Morris, R.H.: Scatter Storage Techniques. **11**(1) (1968)
20. Morris, R.H.: Scatter storage techniques (reprint). Commun. ACM **26**(1), 39–42 (1983)
21. Naor, M., Schulman, L.J., Srinivasan, A.: Splitters and near-optimal derandomization. In: 36th Annual Symposium on Foundations of Computer Science, Milwaukee, Wisconsin, USA, 23–25 October 1995, pp. 182–191 (1995)
22. Raab, M., Steger, A.: Balls into bins - a simple and tight analysis. In: Randomization and Approximation Techniques in Computer Science, Second International Workshop, RANDOM'98, Barcelona, Spain, October 8–10, 1998, Proceedings, pp. 159–170 (1998)
23. Robbins, H.E.: A remark on Stirling's formula. Am. Math. Mon. **62**, 26–29 (1955)
24. Schmidt, Jeanette P., Siegel, Alan: The spatial complexity of oblivious k-probe hash functions. SIAM J. Comput. **19**(5), 775–786 (1990)

Parameterized Complexity
of the $\mathcal{T}_{h+1}$-Free Edge Deletion Problem

Ajinkya Gaikwad and Soumen Maity[(✉)]

Indian Institute of Science Education and Research, Pune, India
ajinkya.gaikwad@students.iiserpune.ac.in, soumen@iiserpune.ac.in

Abstract. Given an undirected graph $G = (V, E)$ and two integers k and h, we study $\mathcal{T}_{h+1}$-FREE EDGE DELETION, where the goal is to remove at most k edges such that the resulting graph does not contain any tree on $h + 1$ vertices as a (not necessarily induced) subgraph, that is, we delete at most k edges in order to obtain a graph in which every component contains at most h vertices. This is desirable from the point of view of restricting the spread of a disease in transmission networks. Enright and Meeks (Algorithmica, 2018) gave an algorithm to solve $\mathcal{T}_{h+1}$-FREE EDGE DELETION whose running time on an n-vertex graph G of treewidth $\mathsf{tw}(G)$ is bounded by $O((\mathsf{tw}(G)h)^{2\mathsf{tw}(G)}n)$. However, it remains open whether the problem might belong to FPT when parameterized only by the treewidth $\mathsf{tw}(G)$; they conjectured that treewidth alone is not enough, and that the problem is W[1]-hard with respect to this parameterization. We resolve this conjecture by showing that $\mathcal{T}_{h+1}$-FREE EDGE DELETION is indeed W[1]-hard when parameterized by $\mathsf{tw}(G)$ alone. We resolve two additional open questions posed by Enright and Meeks (Algorithmica, 2018) concerning the complexity of $\mathcal{T}_{h+1}$-FREE EDGE DELETION on planar graphs and $\mathcal{T}_{h+1}$-FREE ARC DELETION. We prove that the $\mathcal{T}_{h+1}$-FREE EDGE DELETION problem is NP-complete even when restricted to planar graphs. We also show that the $\mathcal{T}_{h+1}$-FREE ARC DELETION problem is W[2]-hard when parameterized by the solution size on directed acyclic graphs.

1 Introduction

Graph theoretic problems show immense applicability in real-life scenarios such as those involving transportation, flows, relationships and complex systems. The spread of a disease, for instance, can be modeled via such mathematical verbiage. A way to describe this idea is as follows: Livestock are often carriers of various pathogens, and thus their movement, among other reasons, plays a key role in inducing a pandemic amongst humans or livestock [9]. To point at a particular instance, the cause of the 2001 FMD epidemic in the UK can be traced to long-range movements of sheep combined with local transmissions amongst the flocks [9,14]. Thus, an effective way of mitigating the spread of livestock diseases could be modelled by studying the underlying transmission networks based on the

H. Fernau and K. Jansen (Eds.): FCT 2023, LNCS 14292, pp. 221–233, 2023.
https://doi.org/10.1007/978-3-031-43587-4_16

routes of cattle movement. Such models can allow for early detection and better management of disease control strategies [11].

More precisely, we consider a graph with its nodes as livestock farms and edges denoting common routes of livestock movement between farms. Using this graph, we can identify certain edges, or routes, such that connected components of the network obtained by deleting these edges are manageably small in size. Then, these deleted edges will precisely correspond to those trade routes which require more disease surveillance, vaccination stops, movement controls etc., required for disease management. In essence, we have divided our disease control strategies to a few routes and smaller manageable networks. Naturally, for maximum efficiency, one would also like to minimise the number of edges being deleted, and this provides us enough information to chalk out a graph theoretic problem, which we shall describe in the next section. It should be noted that the damaging effect such pandemics have on public health, economies, and businesses, essentially validates the need for such a study.

Many properties that might be desirable from the point of view of restricting the spread of a disease can be expressed in terms of forbidden subgraphs: delete edges so that each connected component in the resulting graph has at most h vertices, is equivalent to edge-deletion to a graph avoiding all trees on $h+1$ vertices. One question of particular relevance to epidemiology would be the complexity of the problem on planar graphs; this would be relevant for considering the spread of a disease based on the geographic location (in situations where a disease is likely to be transmitted between animals in adjacent fields) [5]. Furthermore, in practice animal movement networks can capture more information when considered as directed graphs. The natural generalisation of $\mathcal{T}_{h+1}$-FREE EDGE DELETION to directed graphs in this contexts follows: given a directed graph and a positive integer k, the goal is to verify whether it is possible to delete at most k edges from a given directed graph so that the maximum number of vertices reachable from any given starting vertex is at most h. The $\mathcal{T}_{h+1}$-FREE EDGE DELETION problem on planar graph and directed graphs were introduced by Enright and Meeks [5]. Exploiting information on the direction of movements might allow more efficient algorithms for $\mathcal{T}_{h+1}$-FREE EDGE DELETION; a natural first question would be to consider whether there exists an efficient algorithm to solve this problem on directed acyclic graphs.

1.1 Notations and Definitions

Unless otherwise stated, all graphs are simple, undirected, and loopless. Let $G = (V, E)$ be a graph. We denote by $V(G)$ and $E(G)$ its vertex and edge sets respectively. For a vertex $v \in V(G)$, let $N_G(v) = \{y \in V(G) : (v, y) \in E(G)\}$, and $d_G(v) = |N_G(v)|$ denote its *open neighborhood* and *degree* respectively. Let $E' \subseteq E(G)$ be a set of edges of G. Then the graph obtained by deleting E' from G, denoted by $G \setminus E'$, is the subgraph of G containing the same vertices as G but all the edges of E' removed. We are interested in solving the $\mathcal{T}_{h+1}$-FREE EDGE DELETION problem. This problem is of particular interest because it can be seen as the problem of removing connections so as to obtain a network where each connected component has at most h vertices, an abstract view of numerous real world problems. In this paper, we study $\mathcal{T}_{h+1}$-FREE EDGE DELETION and $\mathcal{T}_{h+1}$-FREE ARC DELETION. We define the problems as follows:

$\mathcal{T}_{h+1}$-FREE EDGE DELETION
Input: An undirected graph $G = (V, E)$, and two positive integers k and h.
Question: Does there exist $E' \subseteq E(G)$ with $|E'| \leq k$ such that each connected component in $G \setminus E'$ has at most h vertices, that is, the graph $G \setminus E'$ does not contain any tree on $h + 1$ vertices as a (not necessarily induced) subgraph?

The natural generalization of this problem to directed graphs in this context would be to consider whether it is possible to delete at most k arcs from a given directed graph so that the maximum number of vertices reachable from any given starting vertex is at most h.

$\mathcal{T}_{h+1}$-FREE ARC DELETION
Input: A directed graph $G = (V, E)$, and two positive integers k and h.
Question: Does there exist $E' \subseteq E(G)$ with $|E'| \leq k$ such that the maximum number of vertices reachable from any given starting vertex is at most h in $G \setminus E'$?

A problem with input size n and parameter k is said to be *fixed-parameter tractable (FPT)* if it has an algorithm that runs in time $\mathcal{O}(f(k)n^c)$, where f is some (usually computable) function, and c is a constant that does not depend on k or n. What makes the theory more interesting is a hierarchy of intractable parameterized problem classes above FPT which helps in distinguishing those problems that are unlikely to be fixed-parameter tractable. We refer to [2] for further details on parameterized complexity.

The graph parameter we discuss in this paper is treewidth. We review the concept of a tree decomposition, introduced by Robertson and Seymour in [16]. Treewidth is a measure of how "tree-like" the graph is.

Definition 1. [4] A *tree decomposition* of a graph $G = (V, E)$ is a tree T together with a collection of subsets X_t (called *bags*) of V labeled by the vertices t of T such that $\bigcup_{t \in V(T)} X_t = V(G)$ and (1) and (2) below hold:

1. For every edge $uv \in E(G)$, there is some $t \in V(T)$ such that $\{u, v\} \subseteq X_t$.
2. (Interpolation Property) If t is a vertex on the unique path in T from t_1 to t_2, then $X_{t_1} \cap X_{t_2} \subseteq X_t$.

Definition 2. [4] The *width* of a tree decomposition is the maximum value of $|X_t| - 1$ taken over all the vertices t of the tree T of the decomposition. The *treewidth* $\mathsf{tw}(G)$ of a graph G is the minimum width among all possible tree decompositions of G.

Definition 3. [4] If the tree T of a tree decomposition is a path, then we say that the tree decomposition is a *path decomposition*. The *pathwidth* pw(G) of a graph G is the minimum width among all possible path decompositions of G.

1.2 Our Results

Enright and Meeks [5] gave an algorithm to solve $\mathcal{T}_{h+1}$-FREE EDGE DELE-
TION whose running time on an n-vertex graph G of treewidth tw(G) is $O((\text{tw}(G)h)^{2\text{tw}(G)}n)$. However, it remains open whether the problem might belong to FPT when parameterized only by the treewidth tw(G); they con-
jectured that treewidth alone is not enough, and that the problem is W[1]-hard with respect to this parameterization. We resolve this conjecture by showing that $\mathcal{T}_{h+1}$-FREE EDGE DELETION is indeed W[1]-hard when parameterized by tw(G) alone. In this paper we also resolve two additional open problems stated by Enright and Meeks (Algorithmica, 2018) concerning the complexity of $\mathcal{T}_{h+1}$-
FREE EDGE DELETION on planar graphs and $\mathcal{T}_{h+1}$-FREE ARC DELETION on directed acyclic graphs. Our results are the following:

- The $\mathcal{T}_{h+1}$-FREE EDGE DELETION problem is W[1]-hard when parameterized by the treewidth of the input graph.
- The $\mathcal{T}_{h+1}$-FREE EDGE DELETION problem is NP-complete even when restricted to planar graphs.
- The $\mathcal{T}_{h+1}$-FREE ARC DELETION problem is W[2]-hard parameterized by the solution size k, even when restricted to directed acyclic graphs.

1.3 Review of Previous Work

Several well-studied graph problems can be formulated as edge-deletion prob-
lems. Yannakakis [19] showed that the edge-deletion problem is NP-complete for the following properties: (1) without cycles of specified length ℓ, for any fixed $\ell \geq 3$, (2) connected and degree-constrained, (3) outer-planar, (4) transitive digraph, (5) line-invertible, (6) bipartite, (7) transitively orientable. Watanabe, Ae, and Nakamra [18] showed that the edge-deletion problem is NP-complete if the required property is finitely characterizable by 3-connected graphs. Natanzon, Shamir and Sharan [15] proved the NP-hardness of edge-deletion prob-
lems with respect to some well-studied classes of graphs. These include per-
fect, chordal, chain, comparability, split and asteroidal triple free graphs. This problem has also been studied in generality under paradigms such as approx-
imation algorithms [6,13] and parameterized complexity [1,10]. Cai [1] showed that edge-deletion to a graph class characterisable by a finite set of forbidden induced subgraphs is fixed-parameter tractable when parameterized by k (the number of edges to delete); he gave an algorithm to solve the problem in time $O(r^{2k} \cdot n^{r+1})$, where n is the number of vertices in the input graph and r is the maximum number of vertices in a forbidden induced subgraph. FPT algorithms have been obtained for the problem of determining whether there are k edges whose deletion results in a split graph [8] and to chain, split, threshold, and

co-trivially perfect graphs [10]. Given a graph $G = (V, E)$ and a set $\mathcal{F}$ of forbidden subgraphs, Enright and Meeks [5] studied the $\mathcal{F}$-FREE EDGE DELETION problem, where the goal is to remove a minimum number of edges such that the resulting graph does not contain any $F \in \mathcal{F}$ as a (not necessarily induced) subgraph. They gave an algorithm for the $\mathcal{F}$-FREE EDGE DELETION problem with running time $2^{O(|\mathcal{F}|\mathsf{tw}(G)^r)}n$ where $\mathsf{tw}(G)$ is the treewidth of the input graph G and r is the maximum number of vertices in any element of $\mathcal{F}$; this is a significant improvement on Cai's algorithm [1] but does not lead to a practical algorithm for addressing real world problems. Gaikwad and Maity [7] complemented this result by showing that $\mathcal{F}$-FREE EDGE DELETION is W[1]-hard when parameterized by $\mathsf{tw}(G) + |\mathcal{F}|$. They also showed that $\mathcal{F}$-FREE EDGE DELETION is W[2]-hard when parameterized by the combined parameters solution size, the feedback vertex set number and pathwidth of the input graph. The special case of this problem in which $\mathcal{F}$ is the set of all trees on $h + 1$ vertices is of particular interest from the point of view of the control of disease in livestock. Enright and Meeks [5] have proved that T_{h+1}-FREE EDGE DELETION is NP-hard for every $h \geq 3$. Gaikwad and Maity [7] showed that the T_{h+1}-FREE EDGE DELETION is fixed-parameter tractable when parameterized by the vertex cover number of the input graph. Enright and Meeks [5] have derived an improved algorithm for this special case, running in time $O((\mathsf{tw}(G)h)^{2\mathsf{tw}(G)}n)$. However, it remained open whether the problem might belong to FPT when parameterized only by the treewidth; they conjectured that treewidth alone is not enough, and that the problem is W[1]-hard with respect to this parameterization. In this paper, we resolve this conjecture. Gaikwad and Maity [7] gave a polynomial kernel for T_{h+1}-FREE EDGE DELETION parameterized by $k + h$ where k is the solution size. As T_{h+1}-FREE EDGE DELETION is NP-hard for every $h \geq 3$ [5], it is not possible to obtain an FPT-algorithm parameterized by h alone (unless P=NP). However, the parameterized complexity of this problem when parameterized by k alone remains open. In this paper we study the complexity of this problem on planar and directed acyclic graphs.

2   W[1]-Hardness of T_{h+1}-FREE EDGE DELETION Parameterized by Treewidth

To show W[1]-hardness of T_{h+1}-FREE EDGE DELETION, we reduce from MINIMUM MAXIMUM OUTDEGREE, which is known to be W[1]-hard parameterized by the treewidth of the graph [17]. An *orientation* of an undirected graph is an assignment of a direction to each of its edges. The MINIMUM MAXIMUM OUTDEGREE Problem (MMO) takes as input an undirected, edge-weighted graph $G = (V, E, w)$, where V, E, and w denote the set of vertices of G, the set of edges of G, and an edge-weight function $w : E \rightarrow Z^+$, respectively, and asks for an orientation of G that minimizes the resulting maximum weighted outdegree taken over all vertices in the oriented graph. More formally

---

MINIMUM MAXIMUM OUTDEGREE
Input: An undirected edge-weighted graph $G = (V, E, w)$, where w denote an edge-weight function $w : E \to Z^+$ where the edge weights w are given in unary, and a positive integer r.
Question: Is there an orientation of the edges of G such that, for each $v \in V(G)$, the sum of the weights of outgoing edges from v is at most r?

Theorem 1. *The T_{h+1}-FREE EDGE DELETION problem is W[1]-hard when parameterized by the treewidth of the input graph.*

Proof. Let $I = (G = (V, E, w), r)$ be an instance of MINIMUM MAXIMUM OUT-DEGREE. We construct an instance $I' = (G', k)$ of T_{h+1}-FREE EDGE DELETION the following way. See Fig. 1. The construction of G' starts with $V(G') := V(G)$ and then add the following new vertices and edges.

1. For each edge $(u, v) \in E(G)$, create a set of $w(u, v)$ vertices $V_{uv} = \{x_1^{uv}, \ldots, x_{w(u,v)}^{uv}\}$. Make u and v adjacent to every vertex of V_{uv}. For every $1 \le i \le w(u, v) - 1$, introduce an edge (x_i^{uv}, x_{i+1}^{uv}) and an edge $(x_{w(u,v)}^{uv}, x_1^{uv})$.
2. Let $\omega = \sum\limits_{e \in E(G)} w(e)$ and $h = 4\omega$. For each $u \in V(G)$, we add a set $V_u = \{x_1^u, \ldots, x_{h-r-1}^u\}$ of $h - r - 1$ new vertices and make them adjacent to u.
3. We set $k = \omega$.

Clearly I' can be computed in time polynomial in the size of I. We now show that the treewidth of G' is bounded by a function of the treewidth of G. We do so by modifying an optimal tree decomposition T of G as follows:

– For every edge (u, v) of G, we take an arbitrary node in T whose bag X contains both u and v; add to this node a chain of nodes $1, 2, \ldots, w(u, v) - 1$ such that the bag of node i is

$$X \cup \{x_1^{uv}, x_i^{uv}, x_{i+1}^{uv}\}.$$

– For every edge $u \in V(G)$, we take an arbitrary node in T whose bag X contains u. Add to this node a chain of nodes $1, 2, \ldots h - (r + 1)$ such that the bag of node i is $X \cup \{x_i^u\}$ where $x_i^u \in V_u$.

It is easy to verify that the result is a valid tree decomposition of G' and its width is at most the treewidth of G plus three. Now we show that our reduction is correct. That is, we prove that $I = (G = (V, E, w), r)$ is a yes instance of MINIMUM MAXIMUM OUTDEGREE if and only if $I' = (G', k)$ is a yes instance of T_{h+1}-FREE EDGE DELETION. Let D be the directed graph obtained by an orientation of the edges of G such that for each vertex the sum of the weights of outgoing edges is at most r. We claim that the set of edges

$$E' = \bigcup_{(u,v) \in E(D)} \{(v, x) \mid x \in V_{uv}\} \subseteq E(G')$$

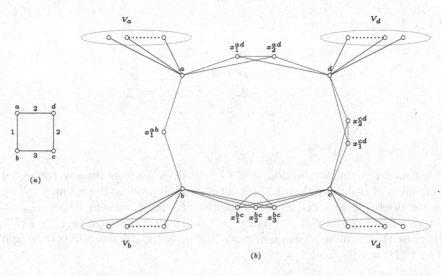

Fig. 1. The reduction from MINIMUM MAXIMUM OUTDEGREE to T_{h+1}-FREE EDGE DELETION in Theorem 1. (a) An instance (G, r) of MINIMUM MAXIMUM OUTDEGREE with $r = 3$. The orientation $(a, d), (d, c), (c, b), (b, a)$ satisfies the property that for each $v \in V(G)$, the sum of the weights of outgoing edges from v is at most 3. (b) The graph G' produced by the reduction algorithm.

is a solution of I'. In Fig. 1, the orientation $(a, d), (d, c), (c, b), (b, a)$ satisfies the property that for each $v \in V(G)$, the sum of the weights of outgoing edges from v is at most 3. Therefore $E(D) = \{(a, d), (d, c), (c, b), (b, a)\}$ and $E' = \{(d, x) \mid x \in V_{ad}\} \cup \{(d, x) \mid x \in V_{dc}\} \cup \{(b, x) \mid x \in V_{bc}\} \cup \{(a, x) \mid x \in V_{ab}\}$. Note that the edges of D are directed. Clearly, we have $|E'| = \omega$. We need to show that $\widetilde{G'} = G' \setminus E'$ does not contain any connected components of size $h + 1$. Observe that every connected component in $\widetilde{G'}$ contains exactly one vertex from $V(G)$. For each $u \in V(\widetilde{G'}) \cap V(G)$, let C_u be the component of $\widetilde{G'}$ that contains u. Then $C_u = \{u\} \cup V_u \cup \bigcup_{(u,v) \in E(D)} V_{uv}$. For each $u \in V(D)$, let w_{out}^u denote the sum of the weights of outgoing edges of vertex u in D. Note that for every $u \in V(G)$, $\left| \bigcup_{(u,v) \in E(D)} V_{uv} \right| = w_{\text{out}}^u \leq r$ and $|V_u| = h - (r + 1)$. Therefore we have $|C_u| \leq 1 + h - (r + 1) + r = h$.

For the reverse direction, let $E' \subseteq E(G')$ be a solution for I', that is, $|E'| = \omega$ and $G' \setminus E'$ does not contain any connected component of size more than h. We first claim that deletion of E' from G' destroys all paths between any pair of vertices $u, v \in V(G) \cap V(\widetilde{G'})$. For the sake of contradiction, let us assume that there is a path between u and v in $\widetilde{G'} = G' \setminus E'$. Note that u (resp. v) is adjacent to $h - (r + 1)$ vertices of V_u (resp. V_v) in G'. If there is a path between u and v in $G' \setminus E'$ then we get a connected component C_{uv} consists of u, v and at least $|V_u| + |V_v| - \omega$ vertices of $V_u \cup V_v$. The reason is this. If E' contains s edges between u and V_u or between v and V_v, then C_{uv} contains $|V_u| + |V_v| - s$ vertices of $V_u \cup V_v$. Thus, we have

$$
\begin{aligned}
|C_{uv}| &\geq 2h - 2(r+1) - \omega + 2 \\
&= 8\omega - 2r - \omega \\
&\geq 8\omega - 2\omega - \omega \qquad \text{as } r < \omega \\
&= 5\omega \\
&\geq 4\omega + 1 \\
&= h + 1
\end{aligned}
$$

This contradict the assumption that $G' \setminus E'$ does not contain any connected component of size more than h. This concludes the proof of the claim.

As we delete at most ω edges, a solution E' must contain either $E_{uv} = \{(u,x) \mid x \in V_{uv}\}$ or $E_{vu} = \{(v,x) \mid x \in V_{uv}\}$ for every edge $(u,v) \in E(G)$; otherwise there will be a path between u and v. We now define a directed graph D by $V(D) = V(G)$ and

$$
E(D) = \Big\{ (u,v) \mid E_{vu} \subseteq E' \Big\} \bigcup \Big\{ (v,u) \mid E_{uv} \subseteq E' \Big\}.
$$

We claim that for each vertex x in D the sum of the weights of outgoing edges is at most r. For the sake of contradiction, suppose there is a vertex x in D for which $w_{out}^x > r$. In this case, we observe that x is adjacent to $h - (r+1) + w_{out}^x \geq h - (r+1) + (r+1) = h$ vertices in graph $\widetilde{G'} = G' \setminus E'$. This is a contradiction as vertex x and its h neighbours form a connected component of size at least $h + 1$, which is a forbidden graph in $\widetilde{G'}$. □

3 $\mathcal{T}_{h+1}$-Free Edge Deletion on Planar Graphs

Enright and Meeks [5] have discussed the importance of studying $\mathcal{T}_{h+1}$-Free Edge Deletion on planar graphs. We show that $\mathcal{T}_{h+1}$-Free Edge Deletion remains NP-complete even when restricted to planar graphs. To prove this, we give a polynomial time reduction from Multiterminal Cut. The Multiterminal Cut problem can be defined as follows: Given a graph $G = (V, E)$, a set $T = \{t_1, t_2, ..., t_p\}$ of p specified vertices or terminals, and a positive weight $w(e)$ for each edge $e \in E$, find a minimum weight set of edges $E' \subseteq E$ such that the removal of E' from E disconnects each terminal from all the others. Dahlhaus et al. [3] proved the following result:

Theorem 2. [3] If p is not fixed, the Multiterminal Cut problem for planar graphs is NP-hard even if all edge weights are equal to 1.

Theorem 3. The $\mathcal{T}_{h+1}$-Free Edge Deletion problem is NP-complete even when restricted to planar graphs.

Proof. It is easy to see that the problem is in NP. In order to obtain the NP-hardness result for the $\mathcal{T}_{h+1}$-Free Edge Deletion problem, we obtain a polynomial reduction from the Mutliterminal Cut problem on planar graphs with all edge weights equal to 1. Let $I = (G, T = \{t_1, t_2, ..., t_p\}, \ell)$ be an instance of MULTITERMINAL CUT. The objective in MULTITERMINAL CUT is to find a set $E' \subseteq E$ of at most ℓ edges such that the removal of E' from E disconnects each terminal from all the others. We produce an equivalent instance $I' = (G', k, h)$ of $\mathcal{T}_{h+1}$-FREE EDGE DELETION in the following way. Start with $G = G'$ and then add the following new vertices and edges. For each $t \in T$, we introduce a set V_t of $\frac{h+1}{2} + \ell$ vertices and make them adjacent to t. We take $h = 100n^3$. This completes the construction of G'. We set $k = \ell$. Let us now show that I and I' are equivalent instances.

Assume first that there exists a set $E' \subseteq E(G)$ of at most ℓ edges such that the removal of E' from E disconnects each terminal from all the others. We claim that the same set $E' \subseteq E(G')$ is a solution of I'. That is, we show that

$$\widetilde{G'} = G' \setminus E'$$

does not contain any connected component of size $h + 1$. For each $t \in T$, let C_t be the component of $\widetilde{G'}$ that contains t. Note that t is adjacent to every vertex in V_t and some vertices in $V(G)$. Therefore the size of C_t is at most $n + \frac{100n^3 + 1}{2} + \ell$. This is true because there is no path between t and any vertex in $V_{t'} \cup \{t'\}$ for all $t' \in T$, $t \neq t'$. Thus, we have

$$|C_t| \leq n + \frac{100n^3 + 1}{2} + \ell$$

$$\leq n + \frac{100n^3 + 1}{2} + |E(G)|$$

$$\leq n + \frac{100n^3 + 1}{2} + \binom{n}{2}$$

$$\leq 100n^3$$

$$= h$$

Hence the size of each component in $\widetilde{G'}$ is at most h.

Conversely, suppose that there exists a set $E' \subseteq E(G')$ of k edges such that $\widetilde{G'} = G' \setminus E'$ does not contain any connected component of size $h + 1$. We claim that there is no path between t_i and t_j in $\widetilde{G'}$ for all $1 \leq i, j \leq k$ and $i \neq j$. For the sake of contradiction, assume that there is a path between terminals t_i and t_j in $\widetilde{G'}$. Note that t_i and t_j are each adjacent to $\frac{h+1}{2} + \ell$ many pendent vertices in G', and we have deleted at most $k = \ell$ many edges. Therefore the connected component containing t_i and t_j contains at least $h + 1$ vertices, which is a contradiction. This concludes the proof of the claim.

We now claim $S = E' \cap E(G)$ is a solution of I. That is, we claim that $G \setminus S$ disconnects each terminal from all the others. Note that $|S| \leq |E'| \leq \ell$. For the sake of contradiction, assume that there is a path between two terminals t_i and t_j in $G \setminus S$. Then there is also a path between t_i and t_j in $G' \setminus E'$. Note that if there exists a path between two terminals t_i and t_j in $G' \setminus E'$ then clearly we get a connected components of size $h + 1$, a contradiction. Therefore $G \setminus S$ disconnects each terminal from all the other terminals, and hence I is a yes-instance. $\qquad\square$

4   W[2]-Hardness of $\mathcal{T}_{h+1}$-FREE ARC DELETION Parameterized by Solution Size

Enright and Meeks [5] explained the importance of studying $\mathcal{T}_{h+1}$-FREE ARC DELETION. A directed acyclic graph (DAG) is a directed graph with no directed cycles. One natural problem mentioned in [5] is to consider whether there exists an efficient algorithm to solve this problem on directed acyclic graphs. In this section, we show that the problem is W[2]-hard parameterized by the solution size k, even when restricted to directed acyclic graphs (DAG). We prove this result via a reduction from HITTING SET. In the HITTING SET problem we are given as input a family $\mathcal{F}$ over a universe U, together with an integer k, and the objective is to determine whether there is a set $B \subseteq U$ of size at most k such that B has nonempty intersection with all sets in $\mathcal{F}$. It is proved in [2] (Theorem 13.28) that HITTING SET problem is W[2]-hard parameterized by the solution size.

Theorem 4. *The $\mathcal{T}_{h+1}$-FREE ARC DELETION problem is W[2]-hard parameterized by the solution size k, even when restricted to directed acyclic graphs.*

Proof. Let $I = (U, \mathcal{F}, k)$ be an instance of HITTING SET where $U = \{x_1, x_2, \ldots, x_n\}$. We create an instance $I' = (G', k', h)$ of $\mathcal{T}_{h+1}$-FREE ARC DELETION the following way. For every $x \in U$, create two vertices v_x and v'_x and add a directed edge (v_x, v'_x). For every $F \in \mathcal{F}$, create one vertex v_F. Next, we add

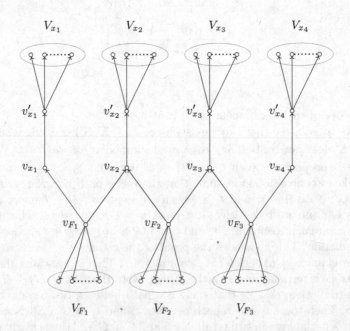

Fig. 2. The graph in the proof of Theorem 4 constructed from HITTING SET instance $U = \{x_1, x_2, x_3, x_4\}$, $F = \{\{x_1, x_2\}, \{x_2, x_3\}, \{x_3, x_4\}\}$ and $k = 2$.

a directed edge (v_F, v_x) if and only if $x \in F$. For each $x \in U$, we add a set V_x of $\frac{h}{n}$ many new vertices and add a directed edge from v'_x to every vertex of V_x. We specify the value of h at the end of the construction. For each vertex $F \in \mathcal{F}$, we add a set V_F of $(h+1) - \sum_{x \in F} |V_x|$ new vertices and add a directed edge from v_F to every vertex of V_F. Finally, we set $k' = k$ and $h = n^c$ for some large constant c. This completes the construction of G'. Next, we show that I and I' are equivalent instances.

Let us assume that there exists a subset $S \subseteq U$ such that $|S| \le k$ and $S \cap F \ne \emptyset$ for all $F \in \mathcal{F}$. We claim that every vertex in $\widetilde{G'} = G' \setminus \bigcup_{x \in S} (v_x, v'_x)$ can reach at most h vertices. Let us assume that there exists a vertex in $\widetilde{G'}$ which can reach more than h vertices. Clearly that vertex must be v_F for some $F \in \mathcal{F}$. Without loss of generality assume that $x_1 \in S \cap F$. As we have removed the edge (v_{x_1}, v'_{x_1}) from G', clearly v_F cannot reach any vertex in V_{x_1}. Note that in such a case v_F cannot reach more than h vertices as $h = n^c$ for some large constant. In particular, v_F can reach at most $h + 1 - (\sum_{x \in F} |V_x|) + (\sum_{x \in F \setminus \{x_1\}} |V_x|) < h$ vertices (Fig. 2).

In the other direction, let us assume that there exists a set $E' \subseteq E(G')$ such that $|E'| \le k$ and every vertex in $\widetilde{G'} = G' \setminus E'$ can reach at most h vertices. First we show that, given a solution E' we can construct another solution E'' such that $E'' \subseteq \bigcup_{x \in U} (v_x, v'_x)$ and $|E''| \le |E'|$. To do this, we observe that the only vertices that can possibly reach more than h vertices are v_F. Note that if E' contains an edge of the form (v_F, u) for some $u \in V_F$ then we can replace it by an arbitrary edge (v_x, v'_x) for some $x \in F$. This will allow us to disconnect at least $\frac{h}{n}$ vertices from v_F rather than just 1. Similar observation can be made for edges of type (v_F, v_x) for some $x \in F$ by replacing it with edge (v_x, v'_x). Therefore, we can assume that $E'' \subseteq \bigcup_{x \in U}^{n} (v_x, v'_x)$. Next, we show that if there exists a vertex v_F such that for every $x \in F$ we have $(v_x, v'_x) \notin E''$ then v_F can reach $h+1$ vertices. Clearly, v_F can reach V_F, $\{v_x \mid x \in F\}$ and also $\{V_x \mid x \in F\}$. Due to construction, this set is of size more than h. This implies that for every $F \in \mathcal{F}$, there exists an edge (v_x, v'_x) for some $x \in F$ which is included in E''. As $|E''| \le k$, we can define $S = \{x \mid (v_x, v'_x) \in E''\}$. Due to earlier observations, S is a hitting set of size at most k.

5   Conclusions and Open Problems

The main contributions in this paper are that the $\mathcal{T}_{h+1}$-FREE EDGE DELETION problem is W[1]-hard when parameterized by the treewidth of the input graph. Thus we resolved a conjecture stated by Enright and Meeks [5] concerning the complexity of $\mathcal{T}_{h+1}$-FREE EDGE DELETION parameterized by the treewidth of the input graph. We also studied the following important open questions stated in [5]: The $\mathcal{T}_{h+1}$-FREE EDGE DELETION problem is NP-complete even when

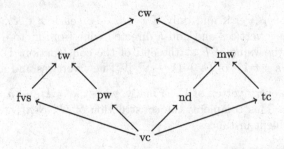

Fig. 3. Relationship between vertex cover (vc), neighbourhood diversity (nd), twin cover (tc), modular width (mw), feedback vertex set (fvs), pathwidth (pw), treewidth (tw) and clique width (cw). Arrow indicate generalizations, for example, treewidth generalizes both feedback vertex set and pathwidth.

restricted to planar graphs; and the T_{h+1}-FREE ARC DELETION problem is W[2]-hard parameterized by the solution size k, even when restricted to directed acyclic graphs. However, it remains open whether T_{h+1}-FREE EDGE DELETION problem is FPT when parameterized by the solution size k. See Fig. 3 for a schematic representation of the relationship between selected graph parameters [12]. Note that $A \rightarrow B$ means that there exists a function f such that for all graphs, $f(A(G)) \geq B(G)$; therefore the existence of an FPT algorithm parameterized by B implies the existence of an FPT algorithm parameterized by A, and conversely, any negative result parameterized by A implies the same negative result parameterized by B. Gaikwad and Maity [7] proved that the T_{h+1}-FREE EDGE DELETION problem is fixed-parameter tractable when parameterized by the vertex cover number of the input graph. Here we have proved that the T_{h+1}-FREE EDGE DELETION problem is W[1]-hard when parameterized by the treewidth of the input graph. The parameterized complexity of the T_{h+1}-FREE EDGE DELETION problem remains open when parameterized by other structural parameters such as feedback vertex set, pathwidth, treedepth, neighbourhood diversity, cluster vertex deletion set, modular width etc.

References

1. Cai, L.: Fixed-parameter tractability of graph modification problems for hereditary properties. Inf. Process. Lett. **58**(4), 171–176 (1996)
2. Cygan, M., et al.: Parameterized Algorithms. Springer, Cham (2015). https://doi.org/10.1007/978-3-319-21275-3
3. Dahlhaus, E., Johnson, D.S., Papadimitriou, C.H., Seymour, P.D., Yannakakis, M.: The complexity of multiterminal cuts. SIAM J. Comput. **23**(4), 864–894 (1994)
4. Downey, R.G., Fellows, M.R.: Parameterized Complexity. Springer, New York (2012). https://doi.org/10.1007/978-1-4612-0515-9
5. Enright, J., Meeks, K.: Deleting edges to restrict the size of an epidemic: a new application for treewidth. Algorithmica **80**(6), 1857–1889 (2018)
6. Fujito, T.: A unified approximation algorithm for node-deletion problems. Discret. Appl. Math. **86**(2), 213–231 (1998)

7. Gaikwad, A., Maity, S.: Further parameterized algorithms for the f-free edge deletion problem. Theor. Comput. Sci. (2022)
8. Ghosh, E., et al.: Faster parameterized algorithms for deletion to split graphs. Algorithmica **71**(4), 989–1006 (2015)
9. Gibbens, J.C., et al.: Descriptive epidemiology of the 2001 foot-and-mouth disease epidemic in great Britain: the first five months. Veterinary Rec. **149**(24), 729–743 (2001)
10. Guo, J.: Problem kernels for NP-complete edge deletion problems: split and related graphs. In: Tokuyama, T. (ed.) ISAAC 2007. LNCS, vol. 4835, pp. 915–926. Springer, Heidelberg (2007). https://doi.org/10.1007/978-3-540-77120-3_79
11. Kerr, B., et al.: Networks and the epidemiology of infectious disease. Interdisc. Perspect. Infect. Dis. **2011**, 284909 (2011)
12. Knop, D., Masařík, T., Toufar, T.: Parameterized complexity of fair vertex evaluation problems. In: MFCS (2019)
13. Lund, C., Yannakakis, M.: On the hardness of approximating minimization problems. J. ACM **41**(5), 960–981 (1994)
14. Mansley, L.M., Dunlop, P.J., Whiteside, S.M., Smith, R.G.H.: Early dissemination of foot-and-mouth disease virus through sheep marketing in February 2001. Veterinary Rec. **153**(2), 43–50 (2003)
15. Natanzon, A., Shamir, R., Sharan, R.: Complexity classification of some edge modification problems. Discret. Appl. Math. **113**(1), 109–128 (2001)
16. Robertson, N., Seymour, P.D.: Graph minors. iii. planar tree-width. J. Comb. Theory Ser. B **36**(1), 49–64 (1984)
17. Szeider, S.: Not so easy problems for tree decomposable graphs. CoRR, abs/1107.1177 (2011). http://arxiv.org/abs/1107.1177, arXiv:1107.1177
18. Watanabe, T., Ae, T., Nakamura, A.: On the np-hardness of edge-deletion and -contraction problems. Discret. Appl. Math. **6**(1), 63–78 (1983)
19. Yannakakis, M.: Node-and edge-deletion np-complete problems. In: Proceedings of the Tenth Annual ACM Symposium on Theory of Computing, STOC '78, pp. 253–264. Association for Computing Machinery, New York, NY, USA (1978)

On the Parallel Complexity of Group Isomorphism via Weisfeiler–Leman

Joshua A. Grochow[1] and Michael Levet[2(✉)]

[1] University of Colorado Boulder, Boulder, CO 80309, USA
joshua.grochow@colorado.edu
[2] College of Charleston, Charleston, SC 29492, USA
levetm@cofc.edu

Abstract. In this paper, we show that the constant-dimensional Weisfeiler–Leman algorithm for groups (Brachter & Schweitzer, LICS 2020) can be fruitfully used to improve parallel complexity upper bounds on isomorphism testing for several families of groups. In particular, we show:

- Groups with an Abelian normal Hall subgroup whose complement is $O(1)$-generated are identified by constant-dimensional Weisfeiler–Leman using only a constant number of rounds. This places isomorphism testing for this family of groups into L; the previous upper bound for isomorphism testing was P (Qiao, Sarma, & Tang, STACS 2011).
- We use the individualize-and-refine paradigm to obtain a quasiSAC^1 isomorphism test for groups without Abelian normal subgroups, previously only known to be in P (Babai, Codenotti, & Qiao, ICALP 2012).
- We extend a result of Brachter & Schweitzer (ESA, 2022) on direct products of groups to the parallel setting. Namely, we also show that Weisfeiler–Leman can identify direct products in parallel, provided it can identify each of the indecomposable direct factors in parallel. They previously showed the analogous result for P.

We finally consider the count-free Weisfeiler–Leman algorithm, where we show that count-free WL is unable to even distinguish Abelian groups in polynomial-time. Nonetheless, we use count-free WL in tandem with bounded non-determinism and limited counting to obtain a new upper bound of $\beta_1 \mathsf{MAC}^0(\mathsf{FOLL})$ for isomorphism testing of Abelian groups. This improves upon the previous $\mathsf{TC}^0(\mathsf{FOLL})$ upper bound due to Chattopadhyay, Torán, & Wagner (*ACM Trans. Comput. Theory*, 2013).

Keywords: Group Isomorphism · Graph Isomorphism · Weisfeiler–Leman · Descriptive Complexity

ML thanks Keith Kearnes for helpful discussions, which led to a better understanding of the Hella-style pebble game. ML also wishes to thank Richard Lipton for helpful discussions regarding previous results. We wish to thank J. Brachter and P. Schweitzer for helpful feedback. JAG was partially supported by NSF award DMS-1750319 and NSF CAREER award CCF-2047756 and during this work. ML was partially supported by J. Grochow startup funds.

H. Fernau and K. Jansen (Eds.): FCT 2023, LNCS 14292, pp. 234–247, 2023.
https://doi.org/10.1007/978-3-031-43587-4_17

1   Introduction

The GROUP ISOMORPHISM problem (GPI) takes as input two finite groups G and H, and asks if there exists an isomorphism $\varphi : G \to H$. When the groups are given by their multiplication (a.k.a. Cayley) tables, it is known that GPI belongs to NP∩coAM. The generator-enumerator algorithm, attributed to Tarjan in 1978 [63], has time complexity $n^{\log_p(n)+O(1)}$, where n is the order of the group and p is the smallest prime dividing n. In more than 40 years, this bound has escaped largely unscathed: Rosenbaum [67] (see [57, Sec. 2.2]) improved this to $n^{(1/4)\log_p(n)+O(1)}$. And even the impressive body of work on practical algorithms for this problem, led by Eick, Holt, Leedham-Green and O'Brien (e. g., [11,12,19,29]) still results in an $n^{\Theta(\log n)}$-time algorithm in the general case (see [77, Page 2]). In the past several years, there have been significant advances on algorithms with worst-case guarantees on the serial runtime for special cases of this problem including Abelian groups [49,68,74], direct product decompositions [50,76], groups with no Abelian normal subgroups [5,6], coprime and tame group extensions [9,34,56,66], low-genus p-groups and their quotients [16,58], Hamiltonian groups [25], and groups of almost all orders [28].

In addition to the intrinsic interest of this natural problem, a key motivation for the GROUP ISOMORPHISM problem is its close relation to the GRAPH ISOMORPHISM problem (GI). In the Cayley (verbose) model, GPI reduces to GI [79], while GI reduces to the succinct GPI problem [41,62] (recently simplified [40]). In light of Babai's breakthrough result that GI is quasipolynomial-time solvable [4], GPI in the Cayley model is a key barrier to improving the complexity of GI. Both verbose GPI and GI are considered to be candidate NP-intermediate problems, that is, problems that belong to NP, but are neither in P nor NP-complete [55]. There is considerable evidence suggesting that GI is not NP-complete [2,4,17,47,53,69]. As verbose GPI reduces to GI, this evidence also suggests that GPI is not NP-complete. It is also known that GI is strictly harder than GPI under AC^0 reductions [20]. Torán showed that GI is DET-hard [72], which provides that PARITY is AC^0-reducible to GI. On the other hand, Chattopadhyay, Torán, and Wagner showed that PARITY is not AC^0-reducible to GPI [20]. To the best of our knowledge, there is no literature on lower bounds for GPI in the Cayley table model. The absence of such lower bounds begs the question of how much existing polynomial-time isomorphism tests can be parallelized, even for special cases for GPI.

Despite GPI in the Cayley table model being strictly easier than GI under AC^0-reductions, there are several key approaches in the GI literature such as parallelization and individualization that have received comparatively little attention in the setting of GPI—see the discussion of Related Work on Page 8. In this paper, using Weisfeiler–Leman for groups [13] as our main tool, we begin to bring both of these techniques to bear on GPI. As a consequence, we also make advances in the descriptive complexity theory of finite groups.

Main Results. In this paper, we show that Weisfeiler–Leman serves as a key subroutine in developing efficient parallel isomorphism tests.

Brachter & Schweitzer [13] actually introduced three different versions of WL for groups. While they are equivalent in terms of pebble complexity up to constant factors, their round complexity may differ by up to an additive $O(\log n)$ (details to appear in the full version), and their parallel complexities differ. Because of these differences we are careful to specify which version of WL for groups each result uses.

As we are interested in both the Weisfeiler–Leman dimension and the number of rounds, we introduce the following notation.

Definition 1. *Let $k \geq 2$ and $r \geq 1$ be integers, and let $J \in \{I, II, III\}$. The (k, r)-WL Version J algorithm for groups is obtained by running k-WL Version J for r rounds. Here, the initial coloring counts as the first round.*

We first examine coprime extensions of the form $H \ltimes N$ where N is Abelian. When either H is elementary Abelian or H is $O(1)$-generated, Qiao, Sarma, & Tang [66] gave a polynomial-time isomorphism test for these families of groups, using some nontrivial representation theory. Here, as a proof of concept that WL can successfully use and parallelize some representation theory (which was not yet considered in [13,14]), we use WL to improve their result's parallel complexity in the case that H is $O(1)$-generated. We remark below about the difficulties in extending WL to handle the case that H is Abelian (without restricting the number of generators).

Theorem 1. *Groups of the form $H \ltimes N$, where N is Abelian, H is $O(1)$-generated, and $|H|$ and $|N|$ are coprime are identified by $(O(1), O(1))$-WL Version II. Consequently, isomorphism between a group of the above form and arbitrary groups can be decided in L.*

Remark 1. Despite Qiao, Sarma, and Tang giving a polynomial-time algorithm for case where H and N are coprime, N is arbitrary Abelian, and H is elementary Abelian (no restriction on number of generators for H or N), we remark here on some of the difficulties we encountered in getting WL to extend beyond the case of H being $O(1)$-generated. When H is $O(1)$-generated, we may start by pebbling the generators of H. After this, by Taunt's Lemma [71], all that is left is to identify the multiset of H-modules appearing in N. In contrast, when H is not $O(1)$-generated, this strategy fails quite badly: if only a small subset of H's generators are pebbled, then it leaves open automorphisms of H that could translate one H-module structure to another. But the latter translation-under-automorphism problem is equivalent to the *entire* problem in this family of groups (see, e.g., [66, Theorem 1.2]).

This same difficulty is encountered even when using the more powerful *second* Ehrenfeucht–Fraïssé pebble game in Hella's [42,43] hierarchy, in which Spoiler may pebble two elements per turn instead of just one. This second game in Hella's hierarchy is already quite powerful: it identifies semisimple groups using only $O(1)$ pebbles and $O(1)$ rounds [33]. It seems plausible to us that with only $O(1)$ pebbles, neither ordinary WL nor this second game in Hella's hierarchy identifies coprime extensions where both H, N are Abelian with no restriction on the number of generators.

We next parallelize a result of Brachter & Schweitzer [14], who showed that Weisfeiler–Leman can identify direct products in polynomial-time provided it can also identify the indecomposable direct factors in polynomial-time. Specifically, we show:

Theorem 2. *For all $G = G_1 \times \cdots \times G_d$ with the G_i directly indecomposable, and all $k \geq 5$, if $(k, O(\log^c n))$-WL Version II identifies each G_i for some $c \geq 1$, then $(k + 1, O(\log^c n))$-WL identifies G.*

More specifically, we show that for $k \geq 5$ and $r(n) \in \Omega(\log n)$, if a direct product G is not distinguished from some group H by (k, r)-WL Version II, then H is a direct product, and there is some direct factor of H that is not distinguished from some direct factor of G by $(k - 1, r)$-WL.

Prior to Theorem 2, the best-known upper bound on computing direct product decompositions was P [50,76]. While Weisfeiler–Leman does not return explicit direct factors, it can implicitly compute a direct product decomposition in $O(\log n)$ rounds, which is sufficient for parallel isomorphism testing. In light of the parallel WL implementation due to Grohe & Verbitsky, our result effectively provides that WL can decompose direct products in TC^1.

We next consider groups without Abelian normal subgroups. Using the individualize and refine paradigm, we obtain a new upper bound of quasiSAC^1 for not only deciding isomorphisms, but also listing isomorphisms. While this does not improve upon the upper bound of P for isomorphism testing [6], this does parallelize the previous bound of $n^{\Theta(\log \log n)}$ runtime for listing isomorphisms [5].

Theorem 3. *Let G be a group without Abelian normal subgroups, and let H be arbitrary. We can test isomorphism between G and H using an SAC circuit of depth $O(\log n)$ and size $n^{\Theta(\log \log n)}$. Furthermore, all such isomorphisms can be listed in this bound.*

Remark 2. The key idea in proving Theorem 3 is to prescribe an isomorphism between $\mathrm{Soc}(G)$ and $\mathrm{Soc}(H)$ (as in [5]), and then use Weisfeiler–Leman to test in L whether the given isomorphism of $\mathrm{Soc}(G) \cong \mathrm{Soc}(H)$ extends to an isomorphism of $G \cong H$. The procedure from [5] for choosing all possible isomorphisms between socles is easily seen to parallelize; our key improvement is in the parallel complexity of testing whether such an isomorphism of socles extends to the whole groups.

Previously, this latter step was shown to be polynomial-time computable [5, Proposition 3.1] via membership checking in the setting of permutation groups. Now, although membership checking in permutation groups is in NC [3], the proof there uses several different group-theoretic techniques, and relies on the Classification of Finite Simple Groups (see the end of the introduction of [3] for a discussion). Furthermore, there is no explicit upper bound on which level of the NC hierarchy these problems are in, just that it is $O(1)$. Thus, it does not appear that membership testing in the setting of permutation groups is known to be even AC^1-computable. So already, our quasiSAC^1 bound is new (the quasi-polynomial size comes only from parallelizing the first step). Furthermore, Weisfeiler–Leman provides a much simpler algorithm; indeed, although we also rely on the fact that all finite simple groups are 2-generated (a result only known via CFSG), this is the only consequence of CFSG that we use, and it is only used in the proof of correctness, not in the algorithm itself. We note, however, that although WL improves the parallel complexity of these particular instances of membership testing, it requires access to the multiplication table for the underlying group, so this technique cannot be leveraged for more general membership testing in permutation groups.

In the case of serial complexity, if the number of simple direct factors of Soc(G) is just slightly less than maximal, even listing isomorphism can be done in FP [5]. Under the same restriction, we get an improvement in the parallel complexity to FL:

Corollary 1 (Cf. [5, Corollary 4.4]). *Let G be a group without Abelian normal subgroups, and let H be arbitrary. Suppose that the number of non-Abelian simple direct factors of Soc(G) is $O(\log n/\log\log n)$. Then we can decide isomorphism between G and H, as well as list all such isomorphisms, in FL.*

It remains open as to whether isomorphism testing of groups without Abelian normal subgroups is even in NC.

Given the lack of lower bounds on GPI, and Grohe & Verbitsky's parallel WL algorithm, it is natural to wonder whether our parallel bounds could be improved. One natural approach to this is via the *count-free* WL algorithm, which compares the set rather than the multiset of colors at each iteration. We show unconditionally that this algorithm fails to serve as a polynomial-time isomorphism test for even Abelian groups.

Theorem 4. *There exists an infinite family $(G_n, H_n)_{n\geq 1}$ where $G_n \not\cong H_n$ are Abelian groups of the same order and count-free WL requires dimension $\geq \Omega(\log |G_n|)$ to distinguish G_n from H_n.*

Remark 3. Even prior to [18], it was well-known that the count-free variant of Weisfeiler–Leman failed to place GI into P [46]. In fact, count-free WL fails to distinguish almost all graphs [31,44], while two iterations of the standard counting 1-WL almost surely assign a unique label to each vertex [7,8]. In light of the equivalence between count-free WL and the logic FO (first-order logic *without* counting quantifiers), this rules out FO as a viable logic to capture P on unordered graphs. Finding such a logic is a central open problem in Descriptive Complexity Theory. On ordered structures such a logic was given by Immerman [45] and Vardi [73].

Theorem 4 establishes the analogous result, ruling out FO as a candidate logic to capture P on unordered groups. This suggests that some counting may indeed be necessary to place GPI into P. As DET is the best known lower bound for GI [72], counting is indeed necessary for GI. There are no such lower bound known for GPI. Furthermore, the work of [20] shows that GPI is not hard (under AC^0-reductions) for any complexity class that can compute PARITY, such as DET.

Determining which families of groups can(not) be identified by count-free WL remains an intriguing open question.

While count-free WL is not sufficiently powerful to compare the multiset of colors, it turns out that $O(\log \log n)$-rounds of count-free $O(1)$-WL Version III will distinguish two elements of different orders. Thus, the multiset of colors computed by the count-free $(O(1), O(\log \log n))$-WL Version III for non-isomorphic Abelian groups G and H will be different. We may use $O(\log n)$ non-deterministic bits to guess the color class where G and H have different multiplicities, and then an MAC^0 circuit to compare said color class. This yields the following.

Theorem 5. *Abelian Group Isomorphism is in* $\beta_1 \mathsf{MAC}^0(\mathsf{FOLL})$.

Remark 4. We note that this and Theorem 3 illustrate uses of WL for groups as a *subroutine* in isomorphism testing, which is how it is so frequently used in the case of graphs. To the best of our knowledge, the only previous uses of WL as a subroutine for GPI were in [15,59]. In particular, Theorem 5 motivated follow-up work by Collins & Levet [22,23], who leveraged count-free WL Version I in a similar manner to obtain novel parallel complexity bounds for isomorphism testing of several families of groups. Most notably, they improved the complexity of isomorphism testing for the *CFI groups* from TC^1 [13] to $\beta_1\mathsf{MAC}^0(\mathsf{FOLL})$. The CFI groups are highly non-trivial, arising via Mekler's construction [40,62] from the CFI graphs [18].

Remark 5. The previous best upper bounds for isomorphism testing of Abelian groups are linear time [49,68,74] and $\mathsf{L} \cap \mathsf{TC}^0(\mathsf{FOLL})$ [20]. As $\beta_1\mathsf{MAC}^0(\mathsf{FOLL}) \subseteq \mathsf{TC}^0(\mathsf{FOLL})$, Theorem 5 improves the upper bound for isomorphism testing of Abelian groups.

Methods. We find the comparison of methods at least as interesting as the comparison of complexity. Here discuss at a high level the methods we use for our main theorems above, and compare them to the methods of their predecessor results.

For Theorem 1, its predecessor in Qiao–Sarma–Tang [66] leveraged a result of Le Gall [56] on testing conjugacy of elements in the automorphism group of an Abelian group. (By further delving into the representation theory of Abelian groups, they were also able to solve the case where H and N are coprime and both are Abelian without any restriction on number of generators; we leave that as an open question in the setting of WL.) Here, we use the pebbling game. Our approach is to first pebble generators for the complement H, which fixes an isomorphism between H and its image. For groups that decompose as a coprime extension of H and N, the isomorphism type is completely determined by the multiplicities of the indecomposable H-module direct summands ([71]). So far, this is the same group-theoretic structure leveraged by Qiao, Sarma, and Tang [66]. However, we then use the representation-theoretic fact that, since $|N|$ and $|H|$ are coprime, each indecomposable H-module is generated by a single element (details to appear in the full version); this is crucial in our setting, as it allows Spoiler to pebble that one element in the WL pebbling game. Then, as the isomorphism of H is fixed, we show that any subsequent bijection that Duplicator selects must restrict to H-module isomorphisms on each indecomposable H-submodule of N that is a direct summand.

For Theorem 3, solving isomorphism of semisimple groups took a series of two papers [5,6]. Our result is really only a parallel improvement on the first of these (we leave the second as an open question). In Babai *et al.* [5], they used CODE EQUIVALENCE techniques to identify semisimple groups where the minimal normal subgroups have a bounded number of non-Abelian simple direct factors, and to identify general semisimple groups in time $n^{O(\log \log n)}$. In contrast, WL—along with individualize-and-refine in the second case—provides a single, combinatorial algorithm that is able to detect the same group-theoretic structures leveraged in previous works to solve isomorphism in these families.

In parallelizing Brachter & Schweitzer's direct product result in Theorem 2, we use two techniques. The first is simply carefully analyzing the number of rounds used in many of the proofs. In several cases, a careful analysis of the rounds used was not sufficient to get a strong parallel result. In those cases, we use the notion of *rank*, which may be of independent interest and have further uses.

Given a subset C of group elements, the C-rank of $g \in G$ is the minimal word-length over C required to generate g. If C is easily identified by Weisfeiler–Leman, then WL can identify $\langle C \rangle$ in $O(\log n)$ rounds. This is made precise (and slightly stronger) with our Rank Lemma:

Lemma 1 (Rank lemma). *If $C \subseteq G$ is distinguished by (k, r)-WL, then any bijection f chosen by Duplicator must respect C-rank, in the sense that $\mathrm{rk}_C(g) = \mathrm{rk}_{f(C)}(f(g))$ for all $g \in G$, or Spoiler can win with $k+1$ pebbles and $\max\{r, \log d + O(1)\}$ rounds, where $d = diam(Cay(\langle C \rangle, C)) \le |\langle C \rangle| \le |G|$.*

One application of our Rank Lemma is that WL identifies verbal subgroups where the words are easily identified. Given a set of words $w_1(x_1, \ldots, x_n), \ldots, w_m(\boldsymbol{x})$, the corresponding *verbal subgroup* is the subgroup generated by $\{w_i(g_1, \ldots, g_n) : i = 1, \ldots, m, g_j \in G\}$. One example that we use in our results is the commutator subgroup. If Duplicator chooses a bijection $f : G \to H$ such that $f([x, y])$ is not a commutator in H, then Spoiler pebbles $[x, y] \mapsto f([x, y])$ and wins in two additional rounds. Thus, by our Rank Lemma, if Spoiler does not map the commutator subgroup $[G, G]$ to the commutator subgroup $[H, H]$, then Duplicator wins with 1 additional pebble and $O(\log n)$ additional rounds.

Brachter & Schweitzer [14] obtained a similar result about verbal subgroups using different techniques. Namely, they showed that if WL assigns a distinct coloring to certain subsets $S_1, \ldots, S_t$, then WL assigns a unique coloring to the set of group elements satisfying systems of equations over $S_1, \ldots, S_t$. They analyzed the WL colorings directly. As a result, it is not clear how to compose their result with the pebble game. For instance, while their result implies that if Duplicator does not map $f([G, G]) = [H, H]$ then Spoiler wins, it is not clear how Spoiler wins nor how quickly Spoiler can win. Our result addresses these latter two points more directly. Recall that the number of rounds is the crucial parameter affecting both the parallel complexity and quantifier depth.

Related Work. There has been considerable work on efficient parallel (NC) isomorphism tests for graphs [1,26,27,30,37,39,48,54,60,75]. In contrast with the work on serial runtime complexity, the literature on the space and parallel complexity for GPI is quite minimal. Around the same time as Tarjan's $n^{\log_p(n)+O(1)}$-time algorithm for GPI [63], Lipton, Snyder, and Zalcstein showed that GPI $\in$ SPACE$(\log^2(n))$ [61]. This bound has been improved to β_2AC^1 (AC^1 circuits that receive $O(\log^2(n))$ non-deterministic bits as input) [78], and subsequently to β_2L $\cap$ β_2FOLL $\cap$ β_2SC^2 [20,70]. In the case of Abelian groups, Chattopadhyay, Torán, and Wagner showed that GPI $\in$ L $\cap$ TC^0(FOLL) [20]. Tang showed that isomorphism testing for groups with a bounded number of generators can also be done in L [70].

Combinatorial techniques, such as individualization with Weisfeiler–Leman refinement, have also been incredibly successful in GI, yielding efficient isomorphism tests for several families [10,21,36–39,52]. Weisfeiler–Leman is also a key subroutine in Babai's quasipolynomial-time isomorphism test [4]. Despite the successes of such combinatorial techniques, they are known to be insufficient to place GI into P [18,64]. In contrast, the use of combinatorial techinques for GPI is relatively new [13–15,59], and it is a central open problem as to whether such techniques are sufficient to improve even the long-standing upper-bound of $n^{\Theta(\log n)}$ runtime.

Examining the distinguishing power of the counting logic $\mathcal{C}_k$ serves as a measure of descriptive complexity for groups. In the setting of graphs, the descriptive complexity has been extensively studied, with [35] serving as a key reference in this area. There has been recent work relating first order logics and groups [65], as well as work examining the descriptive complexity of finite abelian groups [32]. However, the work on the descriptive complexity of groups is scant compared to the algorithmic literature on GPI.

2   Conclusion

We combined the parallel WL implementation of Grohe & Verbitsky [39] with the WL for groups algorithms due to Brachter & Schweitzer [13] to obtain an efficient parallel canonization procedure for several families of groups, including: (i) coprime extensions $H \ltimes N$ where N is Abelian and H is $O(1)$-generated, and (ii) direct products, where WL can efficiently identify the indecomposable direct factors.

We also showed that the individualize-and-refine paradigm allows us to list all isomorphisms of semisimple groups with an SAC circuit of depth $O(\log n)$ and size $n^{O(\log \log n)}$. Prior to our paper, no parallel bound was known. And in light of the fact that multiplying permutations is FL-complete [24], it is not clear that the techniques of Babai, Luks, & Seress [3] can yield circuit depth $o(\log^2 n)$.

Finally, we showed that $\Omega(\log(n))$-dimensional count-free WL is required to identify Abelian groups. It follows that count-free WL fails to serve as a polynomial-time isomorphism test even for Abelian groups. Nonetheless, count-free WL distinguishes group elements of different orders. We leveraged this fact to obtain a new $\beta_1 \mathsf{MAC}^0(\mathsf{FOLL})$ upper bound on isomorphism testing of Abelian groups.

Our work leaves several directions for further research that we believe are approachable and interesting.

Question 1. Show that coprime extensions of the form $H \ltimes N$ with both H, N Abelian have constant WL-dimension (the WL analogue of [66]). More generally, a WL analogue of Babai–Qiao [9] would be to show that when $|H|, |N|$ are coprime and N is Abelian, the WL dimension of $H \ltimes N$ is no more than that of H (or the maximum of that of H and a constant independent of N, H).

Question 2. Is the WL dimension of semisimple groups bounded?

It would be of interest to address this question even in the non-permuting case when $G = \mathrm{PKer}(G)$. Alternatively, establish an upper bound of $O(\log \log n)$ for the WL dimension of semisimple groups. These questions would form the basis of a WL analogue of [5], without needing individualize-and-refine.

For the classes of groups we have studied, when we have been able to give an $O(1)$ bound on their WL-dimension, we also get an $O(\log n)$ bound on the number of rounds needed. The dimension bound alone puts the problem into P, while the bound on rounds puts it into TC^1. A priori, these two should be distinct. For example, in the case of graphs, Kiefer & McKay [51] have shown that there are graphs for which color refinement takes $n - 1$ rounds to stabilize.

Question 3. Is there a family of groups identified by $O(1)$-WL but requiring $\omega(\log n)$ rounds?

We also wish to highlight a question that essentially goes back to [20], who showed that GpI cannot be hard under AC^0 reductions for any class containing PARITY. In Theorem 4, we showed that count-free WL requires dimension $\geq \Omega(\log(n))$ to even identify Abelian groups. This shows that this particular, natural method does not put GpI into $\mathsf{FO}(\mathrm{poly} \log \log n)$, though it does not actually prove GpI $\notin \mathsf{FO}(\mathrm{poly} \log \log n)$, since we cannot rule out clever bit manipulations of the Cayley (multiplication) tables. While we think the latter lower bound would be of significant interest, we think even the following question is interesting:

Question 4 (cf. [20]). Show that GpI does not belong to (uniform) AC^0.

References

1. Arvind, V., Das, B., Köbler, J., Kuhnert, S.: The isomorphism problem for k-trees is complete for logspace. Inf. Comput. **217**, 1–11 (2012). https://doi.org/10.1016/j.ic.2012.04.002
2. Arvind, V., Kurur, P.P.: Graph isomorphism is in SPP. Inf. Comput. **204**(5), 835–852 (2006). https://doi.org/10.1016/j.ic.2006.02.002
3. Babai, L., Luks, E., Seress, A.: Permutation groups in NC. In: STOC 1987. STOC '87, pp. 409–420. Association for Computing Machinery, New York, NY, USA (1987). https://doi.org/10.1145/28395.28439
4. Babai, L.: Graph isomorphism in quasipolynomial time [extended abstract]. In: STOC'16–Proceedings of the 48th Annual ACM SIGACT Symposium on Theory of Computing, pp. 684–697. ACM, New York (2016). https://doi.org/10.1145/2897518.2897542, preprint of full version at arXiv:1512.03547v2 [cs.DS]
5. Babai, L., Codenotti, P., Grochow, J.A., Qiao, Y.: Code equivalence and group isomorphism. In: Proceedings of the Twenty-Second Annual ACM-SIAM Symposium on Discrete Algorithms (SODA11), pp. 1395–1408. SIAM, Philadelphia, PA (2011). https://doi.org/10.1137/1.9781611973082.107
6. Babai, L., Codenotti, P., Qiao, Y.: Polynomial-time isomorphism test for groups with no abelian normal subgroups - (extended abstract). In: International Colloquium on Automata, Languages, and Programming (ICALP), pp. 51–62 (2012). https://doi.org/10.1007/978-3-642-31594-7_5
7. Babai, L., Erdös, P., Selkow, S.M.: Random graph isomorphism. SIAM J. Comput. **9**(3), 628–635 (1980). https://doi.org/10.1137/0209047
8. Babai, L., Kucera, L.: Canonical labelling of graphs in linear average time. In: 20th Annual Symposium on Foundations of Computer Science (SFCS 1979), pp. 39–46 (1979). https://doi.org/10.1109/SFCS.1979.8
9. Babai, L., Qiao, Y.: Polynomial-time isomorphism test for groups with Abelian Sylow towers. In: 29th STACS, pp. 453–464. LNCS, vol. 6651. Springer (2012). https://doi.org/10.4230/LIPIcs.STACS.2012.453
10. Babai, L., Wilmes, J.: Quasipolynomial-time canonical form for Steiner designs. In: STOC 2013, pp. 261–270. Association for Computing Machinery, New York, NY, USA (2013). https://doi.org/10.1145/2488608.2488642
11. Besche, H.U., Eick, B.: Construction of finite groups. J. Symb. Comput. **27**(4), 387–404 (1999). https://doi.org/10.1006/jsco.1998.0258
12. Besche, H.U., Eick, B., O'Brien, E.: A millennium project: constructing small groups. Int. J. Algebra Comput. **12**, 623–644 (2002). https://doi.org/10.1142/S0218196702001115
13. Brachter, J., Schweitzer, P.: On the Weisfeiler-Leman dimension of finite groups. In: Hermanns, H., Zhang, L., Kobayashi, N., Miller, D. (eds.) LICS '20: 35th Annual ACM/IEEE Symposium on Logic in Computer Science, Saarbrücken, Germany, 8–11 July 2020, pp. 287–300. ACM (2020). https://doi.org/10.1145/3373718.3394786
14. Brachter, J., Schweitzer, P.: A systematic study of isomorphism invariants of finite groups via the Weisfeiler-Leman dimension (2022). https://doi.org/10.4230/LIPIcs.ESA.2022.27
15. Brooksbank, P.A., Grochow, J.A., Li, Y., Qiao, Y., Wilson, J.B.: Incorporating Weisfeiler-Leman into algorithms for group isomorphism. arXiv:1905.02518 [cs.CC] (2019)
16. Brooksbank, P.A., Maglione, J., Wilson, J.B.: A fast isomorphism test for groups whose Lie algebra has genus 2. J. Algebra **473**, 545–590 (2017). https://doi.org/10.1016/j.jalgebra.2016.12.007

17. Buhrman, H., Homer, S.: Superpolynomial circuits, almost sparse oracles and the exponential hierarchy. In: Shyamasundar, R. (ed.) FSTTCS 1992. LNCS, vol. 652, pp. 116–127. Springer, Heidelberg (1992). https://doi.org/10.1007/3-540-56287-7_99

18. Cai, J.Y., Fürer, M., Immerman, N.: An optimal lower bound on the number of variables for graph identification. Combinatorica **12**(4), 389–410 (1992). https://doi.org/10.1007/BF01305232, originally appeared in SFCS '89

19. Cannon, J.J., Holt, D.F.: Automorphism group computation and isomorphism testing in finite groups. J. Symb. Comput. **35**, 241–267 (2003). https://doi.org/10.1016/S0747-7171(02)00133-5

20. Chattopadhyay, A., Torán, J., Wagner, F.: Graph isomorphism is not AC^0-reducible to group isomorphism. ACM Trans. Comput. Theory **5**(4), 13 (2013). https://doi.org/10.1145/2540088, preliminary version appeared in FSTTCS '10; ECCC Technical report TR10-117

21. Chen, X., Sun, X., Teng, S.H.: Multi-stage design for quasipolynomial-time isomorphism testing of Steiner 2-systems. In: Proceedings of the Forty-Fifth Annual ACM Symposium on Theory of Computing. STOC '13, pp. 271–280. Association for Computing Machinery, New York, NY, USA (2013). https://doi.org/10.1145/2488608.2488643

22. Collins, N.A.: Weisfeiler-Leman and group isomorphism (2023). Undergraduate Thesis; In-Preparation. University of Coloardo Boulder

23. Collins, N.A., Levet, M.: Count-free Weisfeiler-Leman and group isomorphism (2022). https://doi.org/10.48550/ARXIV.2212.11247

24. Cook, S.A., McKenzie, P.: Problems complete for deterministic logarithmic space. J. Algorithms **8**(3), 385–394 (1987). https://doi.org/10.1016/0196-6774(87)90018-6

25. Das, B., Sharma, S.: Nearly linear time isomorphism algorithms for some non-abelian group classes. In: van Bevern, R., Kucherov, G. (eds.) CSR 2019. LNCS, vol. 11532, pp. 80–92. Springer, Cham (2019). https://doi.org/10.1007/978-3-030-19955-5_8

26. Datta, S., Limaye, N., Nimbhorkar, P., Thierauf, T., Wagner, F.: Planar graph isomorphism is in log-space. In: 2009 24th Annual IEEE Conference on Computational Complexity, pp. 203–214 (2009). https://doi.org/10.1109/CCC.2009.16

27. Datta, S., Nimbhorkar, P., Thierauf, T., Wagner, F.: Graph isomorphism for $K_{3,3}$-free and K_5-free graphs is in Log-space. In: Kannan, R., Kumar, K.N. (eds.) IARCS Annual Conference on Foundations of Software Technology and Theoretical Computer Science. Leibniz International Proceedings in Informatics (LIPIcs), vol. 4, pp. 145–156. Schloss Dagstuhl-Leibniz-Zentrum fuer Informatik, Dagstuhl, Germany (2009). https://doi.org/10.4230/LIPIcs.FSTTCS.2009.2314

28. Dietrich, H., Wilson, J.B.: Group isomorphism is nearly-linear time for most orders. In: 2021 IEEE 62nd Annual Symposium on Foundations of Computer Science (FOCS), pp. 457–467 (2022). https://doi.org/10.1109/FOCS52979.2021.00053

29. Eick, B., Leedham-Green, C.R., O'Brien, E.A.: Constructing automorphism groups of p-groups. Comm. Algebra **30**(5), 2271–2295 (2002). https://doi.org/10.1081/AGB-120003468

30. Elberfeld, M., Schweitzer, P.: Canonizing graphs of bounded tree width in logspace. ACM Trans. Comput. Theory **9**(3) (2017). https://doi.org/10.1145/3132720

31. Fagin, R.: Probabilities on finite models. J. Symb. Logic **41**(1), 50–58 (1976). https://doi.org/10.2307/2272945

32. Gomaa, W.: Descriptive complexity of finite abelian groups. IJAC **20**, 1087–1116 (2010). https://doi.org/10.1142/S0218196710006047

33. Grochow, J.A., Levet, M.: On the descriptive complexity of groups without Abelian normal subgroups (2022). https://doi.org/10.48550/ARXIV.2209.13725

34. Grochow, J.A., Qiao, Y.: Polynomial-time isomorphism test of groups that are tame extensions - (extended abstract). In: Algorithms and Computation - 26th International Symposium, ISAAC 2015, Nagoya, Japan, 9–11 December 2015, Proceedings, pp. 578–589 (2015). https://doi.org/10.1007/978-3-662-48971-0_49

35. Grohe, M.: Descriptive Complexity, Canonisation, and Definable Graph Structure Theory, Lecture Notes in Logic, vol. 47. Association for Symbolic Logic, Ithaca, NY; Cambridge University Press, Cambridge (2017). https://doi.org/10.1017/9781139028868

36. Grohe, M., Kiefer, S.: A linear upper bound on the Weisfeiler-Leman dimension of graphs of bounded genus. In: Baier, C., Chatzigiannakis, I., Flocchini, P., Leonardi, S. (eds.) 46th International Colloquium on Automata, Languages, and Programming (ICALP 2019). Leibniz International Proceedings in Informatics (LIPIcs), vol. 132, pp. 117:1–117:15. Schloss Dagstuhl-Leibniz-Zentrum fuer Informatik, Dagstuhl, Germany (2019). https://doi.org/10.4230/LIPIcs.ICALP.2019.117

37. Grohe, M., Kiefer, S.: Logarithmic Weisfeiler-Leman Identifies All Planar Graphs. In: Bansal, N., Merelli, E., Worrell, J. (eds.) 48th International Colloquium on Automata, Languages, and Programming (ICALP 2021). Leibniz International Proceedings in Informatics (LIPIcs), vol. 198, pp. 134:1–134:20. Schloss Dagstuhl - Leibniz-Zentrum für Informatik, Dagstuhl, Germany (2021). https://doi.org/10.4230/LIPIcs.ICALP.2021.134

38. Grohe, M., Neuen, D.: Isomorphism, canonization, and definability for graphs of bounded rank width. Commun. ACM **64**(5), 98–105 (2021). https://doi.org/10.1145/3453943

39. Grohe, M., Verbitsky, O.: Testing graph isomorphism in parallel by playing a game. In: Bugliesi, M., Preneel, B., Sassone, V., Wegener, I. (eds.) ICALP 2006. LNCS, vol. 4051, pp. 3–14. Springer, Heidelberg (2006). https://doi.org/10.1007/11786986_2

40. He, X., Qiao, Y.: On the Baer-Lovász-Tutte construction of groups from graphs: isomorphism types and homomorphism notions. Eur. J. Combin. **98**, 103404 (2021). https://doi.org/10.1016/j.ejc.2021.103404

41. Heineken, H., Liebeck, H.: The occurrence of finite groups in the automorphism group of nilpotent groups of class 2. Arch. Math. (Basel) **25**, 8–16 (1974). https://doi.org/10.1007/BF01238631

42. Hella, L.: Definability hierarchies of generalized quantifiers. Ann. Pure Appl. Logic **43**(3), 235–271 (1989). https://doi.org/10.1016/0168-0072(89)90070-5

43. Hella, L.: Logical hierarchies in PTIME. Inf. Comput. **129**(1), 1–19 (1996). https://doi.org/10.1006/inco.1996.0070

44. Immerman, N.: Upper and lower bounds for first order expressibility. J. Comput. Syst. Sci. **25**(1), 76–98 (1982). https://doi.org/10.1016/0022-0000(82)90011-3

45. Immerman, N.: Relational queries computable in polynomial time. Inf. Control **68**(1–3), 86–104 (1986). https://doi.org/10.1016/S0019-9958(86)80029-8

46. Immerman, N., Lander, E.: Describing graphs: a first-order approach to graph canonization. In: Selman, A.L. (ed.) Complexity Theory Retrospective, pp. 59–81. Springer, New York (1990). https://doi.org/10.1007/978-1-4612-4478-3_5

47. Impagliazzo, R., Paturi, R., Zane, F.: Which problems have strongly exponential complexity? J. Comput. Syst. Sci. **63**(4), 512–530 (2001). https://doi.org/10.1006/jcss.2001.1774

48. Jenner, B., Köbler, J., McKenzie, P., Torán, J.: Completeness results for graph isomorphism. J. Comput. Syst. Sci. **66**(3), 549–566 (2003). https://doi.org/10.1016/S0022-0000(03)00042-4

49. Kavitha, T.: Linear time algorithms for abelian group isomorphism and related problems. J. Comput. Syst. Sci. **73**(6), 986–996 (2007). https://doi.org/10.1016/j.jcss.2007.03.013

50. Kayal, N., Nezhmetdinov, T.: Factoring groups efficiently. In: Albers, S., Marchetti-Spaccamela, A., Matias, Y., Nikoletseas, S., Thomas, W. (eds.) ICALP 2009. LNCS, vol. 5555, pp. 585–596. Springer, Heidelberg (2009). https://doi.org/10.1007/978-3-642-02927-1_49

51. Kiefer, S., McKay, B.D.: The iteration number of colour refinement. In: Czumaj, A., Dawar, A., Merelli, E. (eds.) 47th International Colloquium on Automata, Languages, and Programming, ICALP 2020, 8–11 July 2020, Saarbrücken, Germany (Virtual Conference). LIPIcs, vol. 168, pp. 73:1–73:19. Schloss Dagstuhl - Leibniz-Zentrum für Informatik (2020). https://doi.org/10.4230/LIPIcs.ICALP.2020.73

52. Kiefer, S., Ponomarenko, I., Schweitzer, P.: The Weisfeiler-Leman dimension of planar graphs is at most 3. J. ACM **66**(6) (2019). https://doi.org/10.1145/3333003

53. Köbler, J., Schöning, U., Torán, J.: Graph isomorphism is low for pp. Comput. Complex. **2**, 301–330 (1992). https://doi.org/10.1007/BF01200427

54. Köbler, J., Verbitsky, O.: From invariants to canonization in parallel. In: Hirsch, E.A., Razborov, A.A., Semenov, A., Slissenko, A. (eds.) CSR 2008. LNCS, vol. 5010, pp. 216–227. Springer, Heidelberg (2008). https://doi.org/10.1007/978-3-540-79709-8_23

55. Ladner, R.E.: On the structure of polynomial time reducibility. J. ACM **22**(1), 155–171 (1975). https://doi.org/10.1145/321864.321877

56. Le Gall, F.: Efficient isomorphism testing for a class of group extensions. In: Proceedings of 26th STACS, pp. 625–636 (2009). https://doi.org/10.4230/LIPIcs.STACS.2009.1830

57. Le Gall, F., Rosenbaum, D.J.: On the group and color isomorphism problems. arXiv:1609.08253 [cs.CC]

58. Lewis, M.L., Wilson, J.B.: Isomorphism in expanding families of indistinguishable groups. Groups - Complexity - Cryptology **4**(1), 73–110 (2012). https://doi.org/10.1515/gcc-2012-0008

59. Li, Y., Qiao, Y.: Linear algebraic analogues of the graph isomorphism problem and the Erdös-Rényi model. In: 2017 IEEE 58th Annual Symposium on Foundations of Computer Science (FOCS), pp. 463–474 (2017). https://doi.org/10.1109/FOCS.2017.49

60. Lindell, S.: A logspace algorithm for tree canonization (extended abstract). In: Proceedings of the Twenty-Fourth Annual ACM Symposium on Theory of Computing. STOC '92, pp. 400–404. Association for Computing Machinery, New York, NY, USA (1992). https://doi.org/10.1145/129712.129750

61. Lipton, R.J., Snyder, L., Zalcstein, Y.: The complexity of word and isomorphism problems for finite groups. Yale University, Department of Computer Science Research Report # 91 (1977). https://apps.dtic.mil/dtic/tr/fulltext/u2/a053246.pdf

62. Mekler, A.H.: Stability of nilpotent groups of class 2 and prime exponent. J. Symb. Logic **46**(4), 781–788 (1981). https://doi.org/10.2307/2273227

63. Miller, G.L.: On the $n^{\log n}$ isomorphism technique (a preliminary report). In: Proceedings of the Tenth Annual ACM Symposium on Theory of Computing. STOC '78, pp. 51–58. Association for Computing Machinery, New York, NY, USA (1978). https://doi.org/10.1145/800133.804331

64. Neuen, D., Schweitzer, P.: An exponential lower bound for individualization-refinement algorithms for graph isomorphism. In: Diakonikolas, I., Kempe, D., Henzinger, M. (eds.) Proceedings of the 50th Annual ACM SIGACT Symposium on Theory of Computing, STOC 2018, Los Angeles, CA, USA, 25–29 June 2018, pp. 138–150. ACM (2018). https://doi.org/10.1145/3188745.3188900

65. Nies, A., Tent, K.: Describing finite groups by short first-order sentences. Israel J. Math. **221**(1), 85–115 (2017). https://doi.org/10.1007/s11856-017-1563-2

66. Qiao, Y., Sarma, J.M.N., Tang, B.: On isomorphism testing of groups with normal Hall subgroups. In: Proceedings of 28th STACS, pp. 567–578 (2011). https://doi.org/10.4230/LIPIcs.STACS.2011.567

67. Rosenbaum, D.J.: Bidirectional collision detection and faster deterministic isomorphism testing. arXiv:1304.3935 [cs.DS] (2013)

68. Savage, C.: An $O(n^2)$ algorithm for abelian group isomorphism. Technical report. North Carolina State University (1980)

69. Schöning, U.: Graph isomorphism is in the low hierarchy. J. Comput. Syst. Sci. **37**(3), 312–323 (1988). https://doi.org/10.1016/0022-0000(88)90010-4

70. Tang, B.: Towards Understanding Satisfiability, Group Isomorphism and Their Connections. Ph.D. thesis, Tsinghua University (2013). http://papakonstantinou.org/periklis/pdfs/bangsheng_thesis.pdf

71. Taunt, D.R.: Remarks on the isomorphism problem in theories of construction of finite groups. Math. Proc. Cambridge Philos. Soc. **51**(1), 16–24 (1955). https://doi.org/10.1017/S030500410002987X

72. Torán, J.: On the hardness of graph isomorphism. SIAM J. Comput. **33**(5), 1093–1108 (2004). https://doi.org/10.1137/S009753970241096X

73. Vardi, M.Y.: The complexity of relational query languages (extended abstract). In: Lewis, H.R., Simons, B.B., Burkhard, W.A., Landweber, L.H. (eds.) Proceedings of the 14th Annual ACM Symposium on Theory of Computing, 5–7 May 1982, San Francisco, California, USA, pp. 137–146. ACM (1982). https://doi.org/10.1145/800070.802186

74. Vikas, N.: An $O(n)$ algorithm for abelian p-group isomorphism and an $O(n \log n)$ algorithm for abelian group isomorphism. J. Comput. Syst. Sci. **53**(1), 1–9 (1996). https://doi.org/10.1006/jcss.1996.0045

75. Wagner, F.: Graphs of bounded treewidth can be canonized in AC^1. In: Kulikov, A., Vereshchagin, N. (eds.) CSR 2011. LNCS, vol. 6651, pp. 209–222. Springer, Heidelberg (2011). https://doi.org/10.1007/978-3-642-20712-9_16

76. Wilson, J.B.: Existence, algorithms, and asymptotics of direct product decompositions, I. Groups Complex. Cryptol. **4**(1) (2012). https://doi.org/10.1515/gcc-2012-0007

77. Wilson, J.B.: The threshold for subgroup profiles to agree is logarithmic. Theory Comput. **15**(19), 1–25 (2019). https://doi.org/10.4086/toc.2019.v015a019

78. Wolf, M.J.: Nondeterministic circuits, space complexity and quasigroups. Theor. Comput. Sci. **125**(2), 295–313 (1994). https://doi.org/10.1016/0304-3975(92)00014-I

79. Zemlyachenko, V.N., Korneenko, N.M., Tyshkevich, R.I.: Graph isomorphism problem. J. Soviet Math. **29**(4), 1426–1481 (1985). https://doi.org/10.1007/BF02104746

The Complexity of (P_k, P_ℓ)-Arrowing

Zohair Raza Hassan[✉], Edith Hemaspaandra, and Stanisław Radziszowski

Rochester Institute of Technology, Rochester, NY 14623, USA
zh5337@rit.edu, {eh,spr}@cs.rit.edu

Abstract. For fixed nonnegative integers k and ℓ, the (P_k, P_ℓ)-Arrowing problem asks whether a given graph, G, has a red/blue coloring of $E(G)$ such that there are no red copies of P_k and no blue copies of P_ℓ. The problem is trivial when $\max(k, \ell) \leq 3$, but has been shown to be coNP-complete when $k = \ell = 4$. In this work, we show that the problem remains coNP-complete for all pairs of k and ℓ, except $(3, 4)$, and when $\max(k, \ell) \leq 3$.

Our result is only the second hardness result for (F, H)-Arrowing for an infinite family of graphs and the first for 1-connected graphs. Previous hardness results for (F, H)-Arrowing depended on constructing graphs that avoided the creation of too many copies of F and H, allowing easier analysis of the reduction. This is clearly unavoidable with paths and thus requires a more careful approach. We define and prove the existence of special graphs that we refer to as "transmitters." Using transmitters, we construct gadgets for three distinct cases: 1) $k = 3$ and $\ell \geq 5$, 2) $\ell > k \geq 4$, and 3) $\ell = k \geq 4$. For (P_3, P_4)-Arrowing we show a polynomial-time algorithm by reducing the problem to 2SAT, thus successfully categorizing the complexity of all (P_k, P_ℓ)-Arrowing problems.

Keywords: Graph arrowing · Ramsey theory · Complexity

1 Introduction and Related Work

Often regarded as the study of how order emerges from randomness, Ramsey theory has played an important role in mathematics and computer science; it has applications in several diverse fields, including, but not limited to, game theory, information theory, and approximation algorithms [17]. A key operator within the field is the arrowing operator: given graphs F, G, and H, we say that $G \rightarrow (F, H)$ (read, G *arrows* F, H) if every red/blue coloring of G's edges contains a red F or a blue H. In this work, we categorize the computational complexity of evaluating this operator when F and H are fixed path graphs. The problem is defined formally as follows.

Problem 1 ((F, H)-Arrowing). Let F and H be fixed graphs. Given a graph G, does $G \rightarrow (F, H)$?

H. Fernau and K. Jansen (Eds.): FCT 2023, LNCS 14292, pp. 248–261, 2023.
https://doi.org/10.1007/978-3-031-43587-4_18

The problem is clearly in coNP; a red/blue coloring of G's edges with no red F's and no blue H's forms a certificate that can be verified in polynomial time since F and H are fixed graphs. We refer to such a coloring as an (F, H)-good coloring. The computational complexity of (F, H)-Arrowing has been categorized for a number of pairs (F, H), with a significant amount of work done in the 80 s and 90 s. Most relevant to our work is a result by Rutenburg, who showed that (P_4, P_4)-Arrowing is coNP-complete [18], where P_n is the path graph on n vertices. Burr showed that (F, H)-Arrowing is in P when F and H are star graphs or when F is a matching [5]. Using "senders"—graphs with restricted (F, H)-good colorings introduced a few years earlier by Burr et al. [6,7], Burr showed that (F, H)-Arrowing is coNP-complete when F and H are members of Γ_3, the family of all 3-connected graphs and K_3. The generalized (F, H)-Arrowing problem, where F and H are also part of the input, was shown to be Π_2^p-complete by Schaefer [19].[1] Aside from categorizing complexity, the primary research avenue concerned with the arrowing operator is focused on finding minimal—with different possible definitions of minimal—graphs for which arrowing holds. The smallest orders of such graphs are referred to as Ramsey numbers. Folkman numbers are defined similarly for graphs with some extra structural constraints. We refer the interested reader to surveys by Radziszowski [16] and Bikov [4] for more information on Ramsey numbers and Folkman numbers, respectively.

Our work provides the first complexity result for (F, H)-Arrowing for an infinite family of graphs since Burr's Γ_3 result from 1990. It is important to note that Burr's construction relies on that fact that contracting less than three vertices between pairs of 3-connected graphs does not create new copies of said graph. Let F be 3-connected and $u, v \in V(F)$. Construct G by taking two copies of F and contracting u across both copies, then contracting v across both copies. Observe that no new copies of F are constructed in this process; if a new F is created then it must be disconnected by the removal of the two contracted vertices, contradicting F's 3-connectivity. This process does not work for paths since contracting two path graphs will always make several new paths across the vertices of both paths. Thus, we require a more careful approach when constructing the gadgets necessary for our reductions. We focus on the problem defined below and prove a dichotomy theorem categorizing the problem to be in P or be coNP-complete. We note that such theorems for other graph problems exist in the literature, e.g., [1,8,11,14].

Problem 2 ((P_k, P_ℓ)-Arrowing). Let k and ℓ be fixed integers such that $2 \leq k \leq \ell$. Given a graph G, does $G \rightarrow (P_k, P_\ell)$?

Theorem 1. (P_k, P_ℓ)-Arrowing is coNP-complete for all k and ℓ unless $k = 2$, $(k, \ell) = (3, 3)$, or $(k, \ell) = (3, 4)$. For these exceptions, the problem is in P.

Before this, the only known coNP-complete case for paths was when $k = \ell = 4$ [18]. Despite being intuitively likely, generalizing the hardness result to larger

---

[1] $\Pi_2^p = \text{coNP}^{\text{NP}}$, the class of all problems whose complements are solvable by a nondeterministic polynomial-time Turing machine having access to an NP oracle [15].

paths proved to be an arduous task. Our proof relies on proving the existence of graphs with special colorings—we rely heavily on work by Hook [12], who categorized the (P_k, P_ℓ)-good colorings of the largest complete graphs which do not arrow (P_k, P_ℓ). After showing the existence of these graphs, the reduction is straightforward. The polynomial-time cases are straightforward (Theorem 2) apart from the case where $(k, \ell) = (3, 4)$, wherein we reduce the problem to 2SAT (Theorem 3).

The rest of this paper is organized as follows. We present the necessary preliminaries in Sect. 2. The proof for Theorem 1 is split into Sects. 3 (the polynomial-time cases) and 4 (the coNP-complete cases). We conclude in Sect. 5.

2   Preliminaries

All graphs discussed in this work are simple and undirected. $V(G)$ and $E(G)$ denote the vertex and edge set of a graph G, respectively. We denote an edge in $E(G)$ between $u, v \in V(G)$ as (u, v). For two disjoint subsets $A, B \subset V(G)$, $E(A, B)$ refers to the edges with one vertex in A and one vertex in B. The neighborhood of a vertex $v \in V(G)$ is denoted as $N(v)$ and its degree as $d(v) := |N(v)|$. The path, cycle, and complete graphs on n vertices are denoted as P_n, C_n, and K_n, respectively. The complete graph on n vertices missing an edge is denoted as $K_n - e$. Vertex contraction is the process of replacing two vertices u and v with a new vertex w such that w is adjacent to all remaining neighbors $N(u) \cup N(v)$.

An (F, H)-good coloring of a graph G is a red/blue coloring of $E(G)$ where the red subgraph is F-free, and the blue subgraph is H-free. We say that G is (F, H)-good if it has at least one (F, H)-good coloring. When the context is clear, we will omit (F, H) and refer to the coloring as a good coloring.

Formally, a coloring for G is defined as function $c : E(G) \rightarrow \{\text{red}, \text{blue}\}$ that maps edges to the colors red and blue. For an edge (u, v) and coloring c, we denote its color as $c(u, v)$.

3   Polynomial-Time Cases

In this section, we prove the P cases from Theorem 1. Particularly, we describe polynomial-time algorithms for (P_2, P_ℓ)-Arrowing and (P_3, P_3)-Arrowing (Theorem 2) and provide a polynomial-time reduction from (P_3, P_4)-Arrowing to 2SAT (Theorem 3).

Theorem 2. (P_k, P_ℓ)-Arrowing is in P when $k = 2$ and when $k = \ell = 3$.

Proof. Let G be the input graph. Without loss of generality, assume that G is connected (for disconnected graphs, we run the algorithm on each connected component).

Case 1 ($k = 2$). Coloring any edge in G red will form a red P_2. Thereby, the entire graph must be colored blue. Thus, a blue P_ℓ is avoided if and only if G is P_ℓ-free, which can be checked by brute force, since ℓ is constant.

Case 2 $(k = \ell = 3)$. Note that in any (P_3, P_3)-good coloring of G, edges of the same color cannot be adjacent; otherwise, a red or blue P_3 is formed. Thus, we can check if G is (P_3, P_3)-good similarly to how we check if a graph is 2-colorable: arbitrarily color an edge red and color all of its adjacent edges blue. For each blue edge, color its neighbors red and for each red edge, color its neighbors blue. Repeat this process until all edges are colored or a red or blue P_3 is formed. This algorithm is clearly polynomial-time.   □

The proof that (P_3, P_4)-Arrowing is in P consists of two parts. A preprocessing step to simplify the graph (using Lemmas 1 and 2), followed by a reduction to 2SAT, which was proven to be in P by Krom in 1967 [13].

Problem 3 (2SAT). Let ϕ be a CNF formula where each clause has at most two literals. Does there exist a satisfying assignment of ϕ?

Lemma 1. *Suppose G is a graph and $v \in V(G)$ is a vertex such that $d(v) = 1$ and v's only neighbor has degree at most two. Then, G is (P_3, P_4)-good if and only if $G - v$ is (P_3, P_4)-good.*

Proof. Let u be the neighbor of v. If $d(u) = 1$, the connected component of v is a K_2 and the statement is trivially true. If $d(u) = 2$, let w be the other neighbor of u, i.e., the neighbor that is not v. Clearly, if G is (P_3, P_4)-good, then $G - v$ is (P_3, P_4)-good. We now prove the other direction. Suppose we have good coloring of $G - v$. It is immediate that we can extend this to a good coloring of G by coloring (v, u) (the only edge adjacent to v) red if (u, w) is colored blue, and blue if (u, w) is colored red.   □

Lemma 2. *Suppose G is a graph and there is a P_4 in G with edges $(v_1, v_2), (v_2, v_3)$, and (v_3, v_4) such that $d(v_1) = d(v_2) = d(v_3) = d(v_4) = 2$. Then, G is (P_3, P_4)-good if and only if $G - v_2$ is (P_3, P_4)-good.*

Proof. If (v_1, v_4) is an edge, then the connected component of v_2 is a C_4 and the statement is trivially true. If not, let $v_0, v_5 \notin \{v_1, v_2, v_3, v_4\}$ be such that (v_0, v_1) and (v_4, v_5) are edges. Note that it is possible that $v_0 = v_5$. Clearly, if G is (P_3, P_4)-good then $G - v_2$ is (P_3, P_4)-good. For the other direction, suppose c is a (P_3, P_4)-good coloring of $G - v_2$. We now construct a coloring c' of G. We color all edges other than $(v_1, v_2), (v_2, v_3)$, and (v_3, v_4) the same as c. The colors of the remaining three edges are determined by the coloring of (v_0, v_1) and (v_4, v_5) as follows.

- If $c(v_0, v_1) = c(v_4, v_5) = $ red, then $c'(v_1, v_2), c'(v_2, v_3), c'(v_3, v_4) = $ blue, red, blue.
- If $c(v_0, v_1) = c(v_4, v_5) = $ blue, then $c'(v_1, v_2), c'(v_2, v_3), c'(v_3, v_4) = $ red, blue, red.
- If $c(v_0, v_1) = $ red and $c(v_4, v_5) = $ blue, then $c'(v_1, v_2), c'(v_2, v_3), c'(v_3, v_4) = $ blue, blue, red.
- If $c(v_0, v_1) = $ blue and $c(v_4, v_5) = $ red, then $c'(v_1, v_2), c'(v_2, v_3), c'(v_3, v_4) = $ red, blue, blue.

Since the cases above are mutually exhaustive, this completes the proof. □

Theorem 3. (P_3, P_4)-*Arrowing is in* P.

Proof. Let G be the input graph. Let G' be the graph obtained by repeatedly removing vertices v described in Lemma 1 and vertices v_2 described in Lemma 2 until no more such vertices exist. As implied by said lemmas, $G' \rightarrow (P_3, P_4)$ if and only if $G \rightarrow (P_3, P_4)$. Thus, it suffices to construct a 2SAT formula ϕ such that ϕ is satisfiable if and only if G' is (P_3, P_4)-good.

Let r_e be a variable corresponding to the edge $e \in E(G')$, denoting that e is colored red. We construct a formula ϕ, where a solution to ϕ corresponds to a coloring of G'. For each P_3 in G', with edges (v_1, v_2) and (v_2, v_3), add the clause $\left(\overline{r_{(v_1,v_2)}} \vee \overline{r_{(v_2,v_3)}}\right)$. Note that this expresses "no red P_3's." For each P_4 in G', with edges $(v_1, v_2), (v_2, v_3)$, and (v_3, v_4):

1. If $(v_2, v_4) \in E(G')$, add the clause $\left(r_{(v_1,v_2)} \vee r_{(v_3,v_4)}\right)$.
2. If $(v_2, v_4) \notin E(G')$ and $d(v_2) > 2$, then add the clause $\left(r_{(v_2,v_3)} \vee r_{(v_3,v_4)}\right)$.

It is easy to see that the conditions specified above must be satisfied by each good coloring of G', and thus G' being (P_3, P_4)-good implies that ϕ is satisfiable. We now prove the other direction by contradiction. Suppose ϕ is satisfied, but the corresponding coloring c is not (P_3, P_4)-good. It is immediate that red P_3's cannot occur in c, so we assume that there exists a blue P_4, with edges $e = (v_1, v_2), f = (v_2, v_3)$, and $g = (v_3, v_4)$ such that $r_e = r_f = r_g =$ false in the satisfying assignment of ϕ. Without loss of generality, assume that $d(v_2) \geq d(v_3)$.

- If $d(v_2) > 2$, ϕ would contain clause $r_e \vee r_g$ or $r_f \vee r_g$. It follows that $d(v_2) = d(v_3) = 2$.
- If $d(v_1) = 1$, v_1 would have been deleted by applying Lemma 1. It follows that $d(v_1) > 1$. Similarly, $d(v_4) > 1$.
- If $d(v_1) > 2$, then there exists a vertex v_0 such that $(v_0, v_1), (v_1, v_2), (v_2, v_3)$ are a P_4 in G', $d(v_1) > 2$ and $(v_1, v_3) \notin E(G')$ (since $d(v_3) = 2$). This implies that ϕ contains clause $r_e \vee r_f$, which is a contradiction. It follows that $d(v_1) = 2$. Similarly, $d(v_4) = 2$.
- So, we are in the situation that $d(v_1) = d(v_2) = d(v_3) = d(v_4) = 2$. But then v_2 would have been deleted by Lemma 2.

Since the cases above are mutually exhaustive, this completes the proof.   □

4   coNP-Complete Cases

In this section, we discuss the coNP-complete cases in Theorem 1. In Sect. 4.1, we describe how NP-complete SAT variants can be reduced to (P_k, P_ℓ)-Nonarrowing (the complement of (P_k, P_ℓ)-Arrowing: does there exist a (P_k, P_ℓ)-good coloring of G?). The NP-complete SAT variants are defined below.

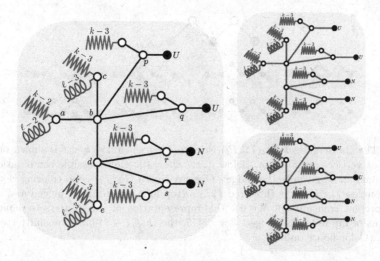

Fig. 1. The variable gadget for (P_k, P_ℓ)-Nonarrowing when $4 \leq k < \ell$ is shown on the left. The output vertices are filled in. Red jagged lines and blue spring lines represent (k, ℓ, x)-red- and (k, ℓ, x)-blue-transmitters, respectively, where the value of x is shown on the top, and the vertex the lines are connected to are the strict endpoints of the monochromatic paths. Observe that when (a, b) is red, other edges adjacent to b must be blue to avoid a red P_k. This, in turn, causes neighbors p and q to have incoming blue $P_{\ell-1}$'s, and vertices marked **U** are now strict endpoints of red P_{k-1}'s. Moreover, edges adjacent to d (except (b, d)) must be red to avoid blue P_ℓ's. Thus, r and s are strict endpoints of red P_{k-1}'s, causing the vertices marked **N** to be strict endpoints of blue P_3's. A similar pattern is observed when (a, b) is blue. Note that for $k \leq 4$, the $(k, \ell, k-3)$-red-transmitter can be ignored. On the right, the two kinds of (P_k, P_ℓ)-good colorings of the gadget are shown. (Color figure online)

Problem 4 ((2,2)-3SAT [3]). Let ϕ be a CNF formula where each clause contains exactly three distinct variables, and each variable appears only four times: twice unnegated and twice negated. Does there exist a satisfying assignment of ϕ?

Problem 5 (Positive NAE E3SAT-4 [2]). Let ϕ be a CNF formula where each clause is an NAE-clause (a clause that is satisfied when its literals are not all true or all false) containing exactly three (not necessarily distinct) variables, and each variable appears at most four times, only unnegated. Does there exist a satisfying assignment for ϕ?

Our proofs depend on the existence of graphs we refer to as "transmitters," defined below. These graphs enforce behavior on special vertices which are *strict endpoints* of red or blue paths. For a graph G and coloring c, we say that v is a strict endpoint of a red (resp., blue) P_k in c if k is the length of the longest red (resp., blue) path that v is the endpoint of. We prove the existence of these graphs in Sect. 4.2.

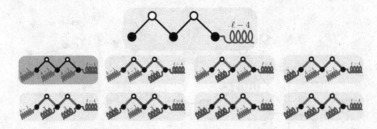

Fig. 2. The clause gadget for (P_k, P_ℓ)-Nonarrowing when $4 \leq k < \ell$ is shown on top. The input vertices are filled in. Below it, we show the eight possible combinations of inputs that can be given to the gadget. Observe that a (P_k, P_ℓ)-good coloring is always possible unless the input is three red P_{k-1}'s (top left). As in Fig. 1, jagged and spring lines represent transmitters. We use this representation of transmitters to depict the two forms of input to the gadget. For $\ell \leq 5$, the $(k, \ell, \ell - 4)$-blue-transmitter can be ignored. (Color figure online)

Definition 1. *Let $3 \leq k < \ell$. For an integer $x \in \{2, 3, \ldots, k - 1\}$ (resp., $x \in \{2, 3, \ldots, \ell - 1\}$) a (k, ℓ, x)-red-transmitter (resp., (k, ℓ, x)-blue-transmitter) is a (P_k, P_ℓ)-good graph G with a vertex $v \in V(G)$ such that in every (P_k, P_ℓ)-good coloring of G, v is the strict endpoint of a red (resp., blue) P_x, and is not adjacent to any blue (resp., red) edge.*

Definition 2. *Let $k \geq 3$ and $x \in \{2, 3, \ldots, k - 1\}$. A (k, x)-transmitter is a (P_k, P_k)-good graph G with a vertex $v \in V(G)$ such that in every (P_k, P_k)-good coloring of G, v is either (1) the strict endpoint of a red P_x and not adjacent to any blue edge, or (2) the strict endpoint of a blue P_x and not adjacent to any red edge.*

4.1 Reductions

We present three theorems that describe gadgets to reduce NP-complete variants of SAT to (P_k, P_ℓ)-Nonarrowing.

Theorem 4. *(P_k, P_ℓ)-Arrowing is coNP-complete for all $4 \leq k < \ell$.*

Proof. We reduce $(2, 2)$-3SAT to (P_k, P_ℓ)-Nonarrowing. Let ϕ be the input to $(2, 2)$-3SAT. We construct G_ϕ such that G_ϕ is (P_k, P_ℓ)-good if and only if ϕ is satisfiable. Let VG and CG be the variable and clause gadgets shown in Figs. 1 and 2. VG has four output vertices that emulate the role of sending a truth signal from a variable to a clause. We first look at Fig. 1. The vertices labeled **U** (resp., **N**) correspond to unnegated (resp., negated) signals. Being the strict endpoint of a blue P_3 corresponds to a true signal while being the strict endpoint of a red P_{k-1} corresponds to a false signal. We now look at Fig. 2. When three red P_{k-1} signals are sent to the clause gadget, it forces the entire graph to be blue, forming a blue P_ℓ. When at least one blue P_3 is present, a good coloring of CG is possible.

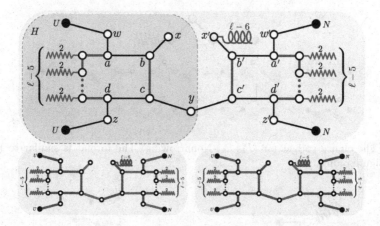

Fig. 3. The variable gadget for (P_3, P_ℓ)-Nonarrowing where $\ell \geq 6$ (top) and its two good colorings (bottom). The variable gadget is a combination of two H's, whose properties we discuss in the proof of Theorem 5. Note that when (a, b) is red in H, then (a', b') is blue in H's copy, and vice versa; if both copies have the same coloring of (a, b), then a red P_3 is formed at y, or a blue P_ℓ is formed from the path from x to x' and the $(3, \ell, \ell - 6)$-blue-transmitter that x' is connected to. When $\ell = 5$, the edge (a, d) is added in H, in lieu of the $\ell - 5$ vertices connected to $(3, \ell, 2)$-red-transmitters. Note that for $\ell \leq 8$, the $(3, \ell, \ell - 6)$-blue-transmitter can be ignored. (Color figure online)

We construct G_ϕ like so. For each variable (resp., clause) in ϕ, we add a copy of VG (resp., CG) to G_ϕ. If a variable appears unnegated (resp., negated) in a clause, a **U**-vertex (resp., **N**-vertex) from the corresponding VG is contracted with a previously uncontracted input vertex of the CG corresponding to said clause. The correspondence between satisfying assignments of ϕ and good colorings of G_ϕ is easy to see.   □

Theorem 5. (P_3, P_ℓ)-*Arrowing is* coNP-*complete for all* $\ell \geq 5$.

Proof. We proceed as in the proof of Theorem 4. The variable gadget is shown in Fig. 3. Blue (resp., red) P_2's incident to vertices marked **U** and **N** correspond to true (resp., false) signals. The clause gadget is the same as Theorem 4's, but the good colorings are different since the inputs are red/blue P_2's instead. These colorings are illustrated in the full version of this paper [10].

Suppose $\ell \geq 6$. Let H be the graph circled with a dotted line in Fig. 3. We first discuss the properties of H. Note that any edge adjacent to a red P_2 must be colored blue to avoid a red P_3. Let $v_1, v_2, \ldots, v_{\ell-5}$ be the vertices connected to $(3, \ell, 2)$-red-transmitters such that v_1 is adjacent to a. Observe that (a, b) and (c, d) must always be the same color; if, without loss of generality, (a, b) is red and (c, d) is blue, a blue P_ℓ is formed via the sequence $a, v_1, \ldots, v_{\ell-5}, d, c, b, x$. In the coloring where (a, b) and (c, d) are blue, the vertices $a, v_1, \ldots, v_{\ell-5}, d, c, b$ form a blue $C_{\ell-1}$, and all edges going out from the cycle must be colored red to

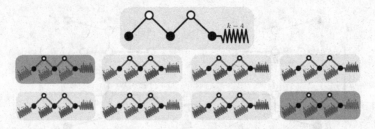

Fig. 4. The clause gadget for (P_k, P_k)-Nonarrowing. The format is similar to Fig. 2. (Color figure online)

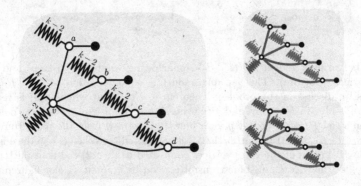

Fig. 5. The variable gadget for (P_k, P_k)-Nonarrowing. Observe that the transmitters connected to v must have different colors; otherwise, a red or blue $P_{k-1+k-2-1}$ is formed, which is forbidden when $k \geq 4$. When the $(k, k-1)$-transmitter is red, v's other neighboring edges must be blue. Thus, vertices $a, b, c,$ and d are strict endpoints of blue P_{k-1}'s, causing the output vertices (filled) to be strict endpoints of red P_{k-1}'s. A similar situation occurs when the $(k, k-1)$-transmitter is blue. Both (P_k, P_k)-good colorings are shown on the right. (Color figure online)

avoid blue P_ℓ's. This forces the vertices marked **U** to be strict endpoints of blue P_2's. If (a, b) and (c, d) are red, $w, a, v_1, \ldots, v_{\ell-5}, d, z$ forms a blue $P_{\ell-1}$, forcing the vertices marked **U** to be strict endpoints of red P_2's. Moreover, (x, b) and (y, c) must also be blue.

With these properties of H in mind, the functionality of the variable gadget described in Fig. 3's caption is easy to follow. The $\ell = 5$ case uses a slightly different H, also described in the caption.   □

Theorem 6. (P_k, P_k)-*Arrowing is* coNP-*complete for all* $k \geq 4$.

Proof. For $k = 4$, Rutenburg showed that the problem is coNP-complete by providing gadgets that reduce from an NAE SAT variant [18]. For $k \geq 5$, we take a similar approach and reduce Positive NAE E3SAT-4 to (P_k, P_k)-Nonarrowing using the clause and variable gadgets described in Figs. 4 and 5. The variable gadget has four output vertices, all of which are unnegated. Without loss of

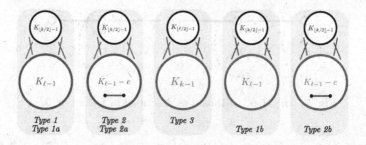

Fig. 6. Illustrations of (P_k, P_ℓ)-good colorings of $K_{R(P_k,P_\ell)-1}$. (Color figure online)

generality, we assume that blue P_{k-1}'s correspond to true signals. The graph G_ϕ is constructed as in the proofs of Theorems 4 and 5. Our variable gadget is still valid when $k = 4$, but the clause gadget does not admit a (P_4, P_4)-good coloring for all the required inputs. In the full version of this paper, we show a different clause gadget that can be used to show the hardness of (P_4, P_4)-Arrowing using our reduction. □

4.2 Existence of Transmitters

Our proofs for the existence of transmitters are corollaries of the following.

Lemma 3. *For integers k, ℓ, where $3 \leq k < \ell$, $(k, \ell, k-1)$-red-transmitters exist.*

Lemma 4. *For all $k \geq 3$, $(k, k-1)$-transmitters exist.*

Below, we will prove Lemma 3 for the case where k is even. The odd case and the proof for Lemma 4 are discussed in the full version of this paper [10]. We construct these transmitters by carefully combining copies of complete graphs. The Ramsey number $R(P_k, P_\ell)$ is defined as the smallest number n such that $K_n \to (P_k, P_\ell)$. We know that $R(P_k, P_\ell) = \ell + \lfloor k/2 \rfloor - 1$, where $2 \leq k \leq \ell$ [9]. In 2015, Hook characterized the (P_k, P_ℓ)-good colorings of all "critical" complete graphs: $K_{R(P_k,P_\ell)-1}$. We summarize Hook's results below.[2]

Theorem 7 (Hook [12]). *Let $4 \leq k < \ell$ and $r = R(P_k, P_\ell) - 1$. The possible (P_k, P_ℓ)-good colorings of K_r can be categorized into three types. In each case, $V(G)$ is partitioned into sets A and B. The types are defined as follows:*

- *Type 1. Let $|A| = \lfloor k/2 \rfloor - 1$ and $|B| = \ell - 1$. Each edge in $E(B)$ must be blue, and each edge in $E(A, B)$ must be red. Any coloring of $E(A)$ is allowed.*
- *Type 2. Let $|A| = \lfloor k/2 \rfloor - 1$ and $|B| = \ell - 1$, and let $b \in E(B)$. Each edge in $E(B) \setminus \{b\}$ must be blue, and each edge in $E(A, B) \cup \{b\}$ must be red. Any coloring of $E(A)$ is allowed.*

---

[2] We note that Hook's ordering convention differs from ours, i.e., they look at (P_ℓ, P_k)-good colorings. Moreover, they use m and n in lieu of k and ℓ.

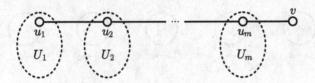

Fig. 7. An (H, u, m)-thread as described in Definition 3.

– *Type 3. Let $|A| = \lfloor \ell/2 \rfloor - 1$ and $|B| = k - 1$. Each edge in $E(B)$ must be blue, and each edge in $E(A, B)$ must be red. Any coloring of $E(A)$ is allowed.*

Moreover, the types of colorings allowed vary according to the parity of k. If k is even, then K_r can only have Type 1 colorings. If k is odd and $\ell > k + 1$, then K_r can only have Type 1 and 2 colorings. If k is odd and $\ell = k + 1$, then K_r can have all types of colorings.

For the case where $k = \ell$, K_r can have Type 1 and 2 colorings as described in the theorem above. Due to symmetry, the colors in these can be swapped and are referred to as Type 1a, 1b, 2a, and 2b colorings. The colorings described have been illustrated in Fig. 6. We note the following useful observation.

Observation 1 *Suppose $\ell > k \geq 4$ and $r = R(P_k, P_\ell) - 1$.*

– *In Type 1 (P_k, P_ℓ)-good colorings of K_r: (1) each vertex in B is a strict endpoint of a blue $P_{\ell-1}$, (2) when k is even (resp., odd), each vertex in B is a strict endpoint of a red P_{k-1} (resp., P_{k-2}), and (3) when k is even (resp., odd), each vertex in A is a strict endpoint of a red P_{k-2} (resp., P_{k-3}).*
– *In Type 2 (P_k, P_ℓ)-good colorings of K_r: (1) each vertex in B is a strict endpoint of a blue $P_{\ell-1}$, (2) each vertex in B is a strict endpoint of a red P_{k-1}, and (3) each vertex in A is a strict endpoint of a red P_{k-2}.*
– *In Type 3 (P_k, P_ℓ)-good colorings of K_r: (1) each vertex in B is a strict endpoint of a red P_{k-1}, (2) each vertex in B is a strict endpoint of a blue $P_{\ell-1}$, and (3) each vertex in A is a strict endpoint of a blue $P_{\ell-2}$.*

We justify these claims in the full version of this paper [10], wherein we also formally define the colorings K_r when $k = \ell$ and justify a similar observation. Finally, we define a special graph that we will use throughout our proofs.

Definition 3 ((H, u, m)-thread). *Let H be a graph, $u \in V(H)$, and $m \geq 1$ be an integer. An (H, u, m)-thread G, is a graph on $m|V(H)|+1$ vertices constructed as follows. Add m copies of H to G. Let $U_i \subset V(G)$ be the vertex set of the ith copy of H, and u_i be the vertex u in H's i^{th} copy. Connect each u_i to u_{i+1} for each $i \in \{1, 2, \ldots, m - 1\}$. Finally, add a vertex v to G and connect it to u_m. We refer to v as the thread-end of G. This graph is illustrated in Fig. 7.*

Using Theorem 7, Observation 1, and Definition 3 we are ready to prove the existence of $(k, \ell, k-1)$-red-transmitters and $(k, k-1)$-transmitters via construction. Transmitters for various cases are shown in Figs. 8 and 9.

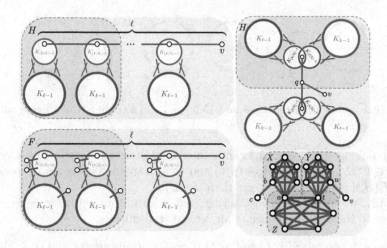

Fig. 8. $(k, \ell, k-1)$-red-transmitters for even k with $\ell > k$, odd k with $\ell > k+1$, and odd k with $\ell = k+1$ are shown on the top-left, bottom-left, and top-right, respectively. The latter construction does not work for the case where $k = 5$, so an alternative construction for a $(5, 6, 4)$-red-transmitter is shown on the bottom-right. The graphs (H and F) described in each case are circled so that the proofs are easier to follow. A good coloring is shown for each transmitter. (Color figure online)

Proof of Lemma 3 when k is even. Let $k \geq 4$ be an even integer and $r = R(P_k, P_\ell) - 1$. In this case, by Theorem 7, only Type 1 colorings are allowed for K_r. The term $A1$-vertex (resp., $B1$-vertex) is used to refer to vertices belonging to set A (resp., B), in a K_r with a Type 1 coloring, as defined in Theorem 7. We first make an observation about the graph H, constructed by adding an edge (u, v) between two disjoint K_r's. Note that u must be an $A1$-vertex, otherwise the edge (u, v) would form a red P_{k-1} or blue $P_{\ell-1}$ when colored red or blue, respectively (Observation 1). Similarly, v must also be an $A1$-vertex. Note that (u, v) must be blue; otherwise, by Observation 1, a red $P_{k-2+k-2}$ is formed, which cannot exist in a good coloring when $k \geq 4$.

We define the $(k, \ell, k-1)$-red-transmitter, G, as the $(K_r, u, \ell-1)$-thread graph, where u is an arbitrary vertex in $V(K_r)$. The thread-end v of G is a strict endpoint of a red P_{k-1}. Let U_i and u_i be the sets and vertices of G as described in Definition 3. From our observation about H, we know that each edge (u_i, u_{i+1}) must be blue. Thus, $u_{\ell-1}$ must be the strict endpoint of a blue $P_{\ell-1}$, implying that $(u_{\ell-1}, v)$ must be red. Since $u_{\ell-1}$ is also a strict endpoint of a red P_{k-2} (Observation 1), v must be the strict endpoint of a red P_{k-1}.

For completeness, we must also show that G is (P_k, P_ℓ)-good. Let A_i and B_i be the sets A and B as defined in Theorem 7 for each U_i. Note that the only edges whose coloring we have not discussed are the edges in each $E(A_i)$. It is easy to see that if each edge in each $E(A_i)$ is colored red, the resulting coloring is (P_k, P_ℓ)-good. This is because introducing a red edge in $E(A_i)$ cannot form a longer red path than is already present in the graph, i.e., any path going

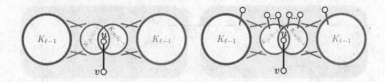

Fig. 9. $(k, k-1)$-transmitters for even k (left) and odd k (right). (Color figure online)

through an edge $(p, q) \in E(A_i)$ can be increased in length by selecting a vertex from $r \in E(B_i)$ using the edges (p, r) and (r, q) instead. This is always possible since $|E(B_i)|$ is sufficiently larger than $|E(A_i)|$.   □

Finally, we show how constructing (red-)transmitters where $x = k - 1$ is sufficient to show the existence of all defined transmitters.

Corollary 1. *For valid k, ℓ, and x, (k, ℓ, x)-blue-transmitters and (k, ℓ, x)-red-transmitters exist.*

Proof. Let H be a $(k, \ell, k - 1)$-red-transmitter where $u \in V(H)$ is the strict endpoint of a red P_{k-1} in all of H's good colorings. For valid x, the $(H, u, x - 1)$-thread graph G is a (k, ℓ, x)-blue-transmitter, where the thread-end v is the strict endpoint of a blue P_x in all good colorings of G; to avoid constructing red P_k's each edge along the path of u_i's is forced to be blue by the red P_{k-1} from H, where u_i is the vertex u in the ith copy of H as defined in Definition 3.

To construct a (k, ℓ, x)-red-transmitter, we use a similar construction. Let H be a $(k, \ell, \ell - 1)$-blue-transmitter where $u \in V(H)$ is the strict endpoint of a blue $P_{\ell-1}$ in all good colorings of H. For valid x, the (H, u, x)-thread graph G is a $(k, \ell, x - 1)$-red-transmitter, where the thread-end v is the strict endpoint of a red P_x in all good colorings of G.   □

Corollary 2. *For valid k and x, (k, x)-transmitters exist.*

Proof. Let H be a $(k, k-1)$-transmitter where $u \in V(H)$ is the strict endpoint of a red/blue P_{k-1} in all of H's good colorings. For valid x, the $(k, u, x - 1)$-thread graph G is a (k, x)-transmitter, where the thread-end v is the strict endpoint of a red or blue P_x in all good colorings of G. Let u_i be the vertex as defined in Definition 3. Each u_i is the strict endpoint of P_{k-1} of the same color; otherwise, the edge between two u's cannot be colored without forming a red or blue P_k. Thus, each such edge must be colored red (resp., blue) by the blue (resp., red) P_{k-1} coming from H.   □

5   Conclusion and Future Work

A major and difficult goal is to classify the complexity for (F, H)-Arrowing for all fixed F and H. We conjecture that in this much more general case a dichotomy theorem still holds, with these problems being either in P or coNP-complete. This seems exceptionally difficult to prove. To our knowledge, all known dichotomy theorems for graphs classify the problem according to one fixed graph, and the polynomial-time characterizations are much simpler than in our case. We see this paper as an important first step in accomplishing this goal.

Acknowledgments. This work was supported in part by grant NSF-DUE-1819546. We would like to thank the anonymous reviewers for their valuable comments.

References

1. Achlioptas, D.: The complexity of G-free colorability. Discrete Math. **165–166**, 21–30 (1997)
2. Antunes Filho, I.T.F.: Characterizing Boolean Satisfiability Variants. Ph.D. thesis, Massachusetts Institute of Technology (2019)
3. Berman, P., Karpinski, M., Scott, A.: Approximation Hardness of Short Symmetric Instances of MAX-3SAT. ECCC (2003)
4. Bikov, A.: Computation and Bounding of Folkman Numbers. Ph.D. thesis, Sofia University "St. Kliment Ohridski" (2018)
5. Burr, S.A.: On the computational complexity of Ramsey-type problems. Math. Ramsey Theory, Algorithms Comb. **5**, 46–52 (1990)
6. Burr, S.A., Erdős, P., Lovász, L.: On graphs of Ramsey type. Ars Combin. **1**(1), 167–190 (1976)
7. Burr, S.A., Nešetřil, J., Rödl, V.: On the use of senders in generalized Ramsey theory for graphs. Discrete Math. **54**(1), 1–13 (1985)
8. Fortune, S., Hopcroft, J., Wyllie, J.: The directed subgraph homeomorphism problem. Theor. Comput. Sci. **10**, 111–121 (1980)
9. Gerencsér, L., Gyárfás, A.: On Ramsey-type problems. Ann. Univ. Sci. Budapest. Eötvös Sect. Math **10**, 167–170 (1967)
10. Hassan, Z., Hemaspaandra, E., Radziszowski, S.: The complexity of (P_k, P_ℓ)-arrowing. CoRR abs/2307.10510 (2023). https://arxiv.org/abs/2307.10510
11. Hell, P., Nešetřil, J.: On the complexity of H-coloring. J. Comb. Theory, Ser. B **48**, 92–110 (1990)
12. Hook, J.: Critical graphs for $R(P_n, P_m)$ and the star-critical Ramsey number for paths. Discuss. Math. Graph Theory **35**(4), 689–701 (2015)
13. Krom, M.R.: The decision problem for a class of first-order formulas in which all disjunctions are binary. Math. Logic Q. **13**(1–2), 15–20 (1967)
14. Le, H.O., Le, V.B.: Complexity of the cluster vertex deletion problem on H-free graphs. In: MFCS 2022, vol. 241, pp. 68:1–68:10 (2022)
15. Meyer, A.R., Stockmeyer, L.J.: The equivalence problem for regular expressions with squaring requires exponential space. In: IEEE SWAT, pp. 125–129 (1972)
16. Radziszowski, S.: Small Ramsey Numbers. Electron. J. Comb. DS1, 1–116 (2021). https://www.combinatorics.org/
17. Rosta, V.: Ramsey theory applications. Electron. J. Comb. DS13, 1–43 (2004). https://www.combinatorics.org/
18. Rutenburg, V.: Complexity of generalized graph coloring. In: Gruska, J., Rovan, B., Wiedermann, J. (eds.) MFCS 1986. LNCS, vol. 233, pp. 573–581. Springer, Heidelberg (1986). https://doi.org/10.1007/BFb0016284
19. Schaefer, M.: Graph Ramsey theory and the polynomial hierarchy. J. Comput. Syst. Sci. **62**, 290–322 (2001)

On Computing a Center Persistence Diagram

Yuya Higashikawa[1], Naoki Katoh[1], Guohui Lin[2], Eiji Miyano[3], Suguru Tamaki[1], Junichi Teruyama[1], and Binhai Zhu[4]($\boxtimes$)

[1] School of Social Information Science, University of Hyogo, Kobe, Japan
{higashikawa,tamak,junichi.teyuyama}@sis.u-hyogo.ac.jp
[2] Department of Computing Science, University of Alberta, Edmonton, AB, Canada
guohui@ualberta.ca
[3] Department of Artificial Intelligence, Kyushu Institute of Technology, Iizuka, Japan
miyano@ces.kyutech.ac.jp
[4] Gianforte School of Computing, Montana State University, Bozeman, MT 59717, USA
bhz@montana.edu

Abstract. Given a set of persistence diagrams $\mathcal{P}_1, ..., \mathcal{P}_m$, for the data reduction purpose, one way to summarize their topological features is to compute the *center* $\mathcal{C}$ of them first under the bottleneck distance. Here we mainly focus on the two discrete versions when points in $\mathcal{C}$ could be selected with or without replacement from all P_i's. (We will briefly discuss the continuous case, i.e., points in $\mathcal{C}$ are arbitrary, which turns out to be closely related to the 3-dimensional geometric assignment problem). For technical reasons, we first focus on the case when $|P_i|$'s are all the same (i.e., all have the same size n), and the problem is to compute a center point set C under the bottleneck matching distance. We show, by a non-trivial reduction from the Planar 3D-Matching problem, that this problem is NP-hard even when $m = 3$ diagrams are given. This implies that the general center problem for persistence diagrams under the bottleneck distance, when all P_i's possibly have different sizes, is also NP-hard when $m \geq 3$. On the positive side, we show that this problem is polynomially solvable when $m = 2$ and admits a factor-2 approximation for $m \geq 3$. These positive results hold for any L_p metric when all P_i's are point sets of the same size, and also hold for the case when all P_i's have different sizes in the L_∞ metric (i.e., for the Center Persistence Diagram problem). This is the best possible in polynomial time for Center Persistence Diagram under the bottleneck distance unless P = NP. All these results hold for both of the discrete versions and the continuous version; in fact, the NP-hardness and approximation results also hold under the Wasserstein distance for the continuous version.

Keywords: Persistence diagrams · Bottleneck distance · Center persistence diagram · NP-hardness · Approximation algorithms

1 Introduction

Computational topology has found a lot of applications in recent years [7]. Among them, persistence diagrams, each being a set of (topological feature) points above and inclusive of the line $Y = X$ in the X-Y plane, have also found various applications, for

H. Fernau and K. Jansen (Eds.): FCT 2023, LNCS 14292, pp. 262–275, 2023.
https://doi.org/10.1007/978-3-031-43587-4_19

instance in GIS [1], in neural science [12], in wireless networks [21], and in prostate cancer research [20]. (Such a topological feature point (b, d) in a persistence diagram, which we will simply call a point henceforth, indicates a topological feature which appears at time b and disappears at time d. Hence $b \leq d$. In the next section, we will present some technical details.) A consequence is that practitioners gradually have a database of persistence diagrams when processing the input data over a certain period of time. It is not uncommon these days that such a database has tens of thousands of persistence diagrams, each with up to several thousands of points. How to process and search these diagrams becomes a new challenge for algorithm designers, especially because the bottleneck distance is typically used to measure the similarity between two persistence diagrams.

In [10] the following problem was studied: given a set of persistence diagrams $\mathcal{P}_1, ..., \mathcal{P}_m$, each with size at most n, how to preprocess them so that each has a key k_i for $i = 1, ..., m$ and for a query persistence diagram Q with key k, an approximate nearest persistence diagram $\mathcal{P}_j$ can be returned by searching the key k in the data structure for all k_i's. A hierarchical data structure was built and the keys are basically constructed using snap roundings on a grid with different resolutions. There is a trade-off between the space complexity (i.e., number of keys stored) and the query time. For instance, if one wants an efficient (polylogarithmic) query time, then one has to use an exponential space; and with a linear or polynomial space, then one needs to spend an exponential query time. Different from traditional problems of searching similar point sets [15], one of the main technical difficulties is to handle points near the line $Y = X$.

In prostate cancer research, one important part is to determine how the cancer progresses over a certain period of time. In [20], a method is to use a persistence diagram for each of the histopathology images (taken over a certain period of time). Naturally, for the data collected over some time interval, one could consider packing a corresponding set of persistence diagrams with a center, which could be considered as a *median* persistence diagram summarizing these persistence diagrams. A sequence of such centers over a longer time period would give a rough estimate on how the prostate cancer progresses. This motivates our research. On the other hand, while the traditional center concept (and the corresponding algorithms) has been used for planar point sets (under the Euclidean distance) [24] and on binary strings (under the Hamming distance) [22]; recently we have also seen its applications in more complex objects, like polygonal chains (under the discrete Frechet distance) [2,18]. In this sense, this paper is also along this line.

Formally, in this paper we consider a way to pack persistence diagrams. Namely, given a set of persistence diagrams $\mathcal{P}_1, ..., \mathcal{P}_m$, how to compute a *center* persistence diagram? Here the metric used is the traditional bottleneck distance and Wasserstein distance (where we first focus on the former). We first describe the case when all $\mathcal{P}_i$'s have the same size n, and later we show how to withdraw this constraint (by slightly increasing the running time of the algorithms). It turns out that the *continuous* case, i.e., when the points in the center can be arbitrary, is very similar to the geometric 3-dimensional assignment problem: Given three points sets P_i of the same size n and each colored with *color-i* for $i = 1..3$, divide points in all P_i's into n 3-clusters (or triangles) such that points in each cluster or triangle have different colors, and some geometric quantity (like the maximum area or perimeter of these triangles) is minimized [13,26]. For our application, we need to investigate discrete versions where points in the center persistence diagram must come from the input diagrams (might be from more than one diagrams). We show that the problem is NP-hard even when $m = 3$ diagrams are given. On the other hand, we show that the problem is polynomially solvable when $m = 2$ and the problem admits a 2-approximation for $m \geq 3$. At the end, we briefly discuss how to adapt the results to Wasserstein distance for the continuous case. The following table summarizes the main results in this paper (Table 1).

Table 1. Results for the Center Persistence Diagram problems under the bottleneck (d_B) and Wasserstein (W_p) distances when $m \geq 3$ diagrams are given.

	Hardness	Inapproximability bound	Approximation factor
d_B, with no replacement	NP-complete	$2 - \varepsilon$	2
d_B, with replacement	NP-complete	$2 - \varepsilon$	2
d_B, continuous	NP-hard	$2 - \varepsilon$	2
W_p, with no replacement	?	?	2
W_p, with replacement	?	?	2
W_p, continuous	NP-hard	?	2

This paper is organized as follows. In Sect. 2, we give some necessary definitions and we also show, as a warm-up, that the case when $m = 2$ is polynomially solvable. In Sect. 3, we prove that the Center Persistence Diagram problem under the bottleneck distance is NP-hard when $m = 3$ via a non-trivial reduction from the Planar three-dimensional Matching (Planar 3DM) problem. In Sect. 4, we present the factor-2 approximation algorithm for the problem (when $m \geq 3$). In Sect. 5, we conclude the paper with some open questions. Due to space constraint, we leave out quite some details, which can be found in [16].

2 Preliminaries

We assume that the readers are familiar with the standard terms in algorithms, like approximation algorithms [4], and NP-completeness [11].

2.1 Persistence Diagram

Homology is a machinery from algebraic topology which gives the ability to count the number of holes in a k-dimensional simplicial complex. For instance, let X be a simplicial complex, and let the corresponding k-dimensional homology be $H_k(X)$, then the dimension of $H_0(X)$ is the number of path connected components of X and $H_1(X)$ consists of loops in X, each is a 'hole' in X. It is clear that these numbers are invariant under rigid motions (and almost invariant under small numerical perturbations) on the original data, which is important in many applications. For further details on the classical homology theory, readers are referred to [14], and to [7,25] for additional information on computational homology. It is well known that the k-dimensional homology of X can be computed in polynomial time [7,8,25].

Ignoring the details for topology, the central part of persistence homology is to track the birth and death of the topological features when computing $H_k(X)$. These features give a *persistence diagram* (containing the birth and death times of features as pairs (b, d) in the extended plane). See Fig. 1 for an example. Note that as a convention, the line $Y = X$ is included in each persistence diagram, where points on the line $Y = X$ provide infinite multiplicity, i.e., a point (t, t) on it could be considered as a dummy feature which is born at time t then immediately dies. Formally, a persistence diagram $\mathcal{P}$ is composed of a set P of planar points (each corresponding to a topological feature) above the line $Y = X$, as well as the line $Y = X$ itself, i.e., $\mathcal{P} = P \cup \{(x, y) | y = x\}$. Due to the infinite multiplicity on $Y = X$, there is always a bijection between $\mathcal{P}_i$ and $\mathcal{P}_j$, even if P_i and P_j have different sizes.

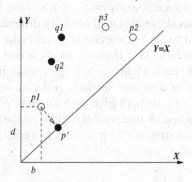

Fig. 1. Two persistence diagrams $\mathcal{P}$ and $\mathcal{Q}$, with feature point sets $P = \{p_1, p_2, p_3\}$ and $Q = \{q_1, q_2\}$ respectively. A point $p_1 = (b, d)$ means that it is born at time b and it dies at time d. The projection of p_1 on $Y = X$ gives p'.

Given two persistence diagrams $\mathcal{P}_i$ and $\mathcal{P}_j$, each with $O(n)$ points, the *bottleneck distance* between them is defined as follows:

$$d_B(\mathcal{P}_i, \mathcal{P}_j) = \inf_{\phi}\{\sup_{x \in \mathcal{P}_i} \|x - \phi(x)\|_\infty, \phi : \mathcal{P}_i \to \mathcal{P}_j \text{ is a bijection}\}.$$

Similarly, the *p-Wasserstein* distance is defined as

$$W_p(\mathcal{P}_i, \mathcal{P}_j) = \left(\inf_{\phi} \sum_{x \in \mathcal{P}_i} \|x - \phi(x)\|_\infty^p \right)^{1/p}, \phi : \mathcal{P}_i \to \mathcal{P}_j \text{ is a bijection.}$$

We refer the readers to [7] for further information regarding persistence diagrams. Our extended results regarding Wasserstein distance can be found in [16], hence we focus only on the bottleneck distance here.

For point sets P_1, P_2 of the same size, we will also use $d_B^p(P_1, P_2)$ to represent their bottleneck matching distance, i.e., let β be a bijection between P_1 and P_2,

$$d_B^p(P_1, P_2) = \min_{\beta} \max_{a \in P_1} d_p(a, \beta(a)).$$

Here, $d_p(-)$ is the distance under the L_p metric. As we mainly cover the case $p = 2$, we will use $d_B(P_i, P_j)$ instead of $d_B^2(P_i, P_j)$ henceforth. Note that in comparing persistence diagrams, the L_∞ metric is always used. For our hardness constructions, all the valid clusters form either horizontal or vertical segments, hence the distances under L_2 and L_∞ metrics are all equal in our constructions.

While the bottleneck distance between two persistence diagrams is continuous in its original form, it was shown that it can be computed using a discrete method [7], i.e., the traditional geometric bottleneck matching [9], in $O(n^{1.5} \log n)$ time. In fact, it was shown that the multiplicity property of the line $Y = X$ can be used to compute the bottleneck matching between two diagrams $\mathcal{P}_1$ and $\mathcal{P}_2$ more conveniently — regardless of their sizes [7]. This can be done as follows. Let P_i be the set of feature points in $\mathcal{P}_i$. Then project points in P_i perpendicularly on $Y = X$ to have P_i' respectively, for $i = 1, 2$. (See also Fig. 1.) It was shown that the bottleneck distance between two diagrams $\mathcal{P}_1$ and $\mathcal{P}_2$ is exactly equal to the bottleneck (bipartite) matching distance, in the L_∞ metric, between $P_1 \cup P_2'$ and $P_2 \cup P_1'$. Here the weight or cost of an edge $c(u, v)$, with $u \in P_2'$ and $v \in P_1'$, is set to zero; while $c(u, v) = \|u - v\|_\infty$, if $u \in P_1$ or $v \in P_2$. The p-Wasserstein distance can be computed similarly, using a min-sum bipartite matching between $P_1 \cup P_2'$ and $P_2 \cup P_1'$, with all edge costs raised to c^p. (Kerber, et al. showed that several steps of the bottleneck matching algorithm can be further simplified [19].) Later, we will extend this construction for more than two diagrams.

2.2 Problem Definition

Throughout this paper, for two points $p_1 = (x_1, y_1)$ and $p_2 = (x_2, y_2)$, we use $d_p(p_1, p_2)$ to represent the L_p distance between p_1 and p_2, which is $d_p(p_1, p_2) = (|x_1 - x_2|^p + |y_1 - y_2|^p)^{1/p}$, for $p < \infty$. When $p = \infty$, $d_\infty(p_1, p_2) = \|p_1 - p_2\|_\infty = \max\{|x_1 - x_2|, |y_1 - y_2|\}$. We will mainly focus on L_2 and L_∞ metrics, for the former, we simplify it as $d(p_1, p_2)$.

Definition 1. The Center Persistence Diagram Problem under the Bottleneck Distance (CPD-B)

    Instance: A set of m persistence diagrams $\mathcal{P}_1, ..., \mathcal{P}_m$ with the corresponding feature point sets $P_1, ..., P_m$ respectively, and a real value r.

    Question: Is there a persistence diagram $\mathcal{Q}$ such that $\max_i d_B(\mathcal{Q}, \mathcal{P}_i) \leq r$?

Note that we could have three versions, depending on $\mathcal{Q}$. We mainly focus on the discrete version when the points in $\mathcal{Q}$ are selected with no replacement from the multiset $\cup_{i=1..m} P_i$. It turns out that the other discrete version, i.e., the points in $\mathcal{Q}$ are selected with replacement from the set $\cup_{i=1..m} P_i$, is different from the first version but all the results can be carried over with some simple twist. We will briefly cover the *continuous* case, i.e., when points $\mathcal{Q}$ are arbitrary; as we covered earlier in the introduction, when

$m = 3$, this version is very similar to the geometric three-dimensional assignment problem [13,26].

We will firstly consider two simplified versions of the corresponding problem.

Definition 2. The m-Bottleneck Matching Without Replacement Problem

Instance: *A set of m planar point sets $P_1, ..., P_m$ such that $|P_1| = \cdots = |P_m| = n$, and a real value r.*

Question: *Is there a point set Q, with $|Q| = n$, such that any $q \in Q$ is selected from the multiset $\cup_i P_i$ with no replacement and $\max_i d_B(Q, P_i) \leq r$?*

Definition 3. The m-Bottleneck Matching With Replacement Problem

Instance: *A set of m planar point sets $P_1, ..., P_m$ such that $|P_1| = \cdots = |P_m| = n$, and a real value r.*

Question: *Is there a point set Q, with $|Q| = n$, such that any $q \in Q$ is selected from the set $\cup_i P_i$ with replacement and $\max_i d_B(Q, P_i) \leq r$?*

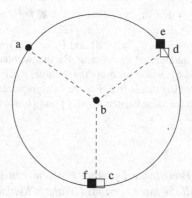

Fig. 2. An example with $P_1 = \{a, b\}$, $P_2 = \{c, d\}$ and $P_3 = \{e, f\}$, with all the points (except b) on a unit circle and b being the center of the circle. For the 'without replacement' version, the optimal solution is $Q_1 = \{b, c\}$, where b covers the 3-cluster $\{a, d, e\}$, c covers the 3-cluster $\{b, c, f\}$ and the optimal covering radius is 1. For the 'with replacement' version, the optimal solution could be the same, but could also be $\{b, b\}$.

It turns out that these two problems are really to find center points in Q to cover m-clusters with an optimal covering radius r, with each cluster being composed of m points, one each from P_i. For $m = 3$, this is similar to the geometric three-dimensional assignment problem which aims at finding m-clusters with certain criteria [13,26]. However, the two versions of the problem are slightly different from the geometric three-dimensional assignment problem. The main difference is that in these discrete versions a cluster could be covered by a center point which does not belong to the cluster. See Fig. 2 for an example. Also, note that the two versions themselves are slightly different; in fact, their solution values could differ by a factor of 2 (see Fig. 3).

Note that we could define a continuous version in which the condition on q is withdrawn and this will be briefly covered at the end of each section. In fact, we focus more on the optimization versions of these problems. We will show that 3-Bottleneck Matching, for both the discrete versions, is NP-hard, immediately implying CPD-B is NP-hard for $m \geq 3$. We then present a 2-approximation for the m-Bottleneck Matching Problem and later we will show how to make some simple generalization so the 'equal size' condition can be withdrawn for persistence diagrams — this implies that CPD-B also admits a 2-approximation for $m \geq 3$. We will focus on the 'without replacement' version in our writing, and later we will show how to generalize it to the 'with replacement' version at the end of each section. Henceforth, we will refer to the 'without replacement' version simply as m-Bottleneck Matching unless otherwise specified.

At first, we briefly go over polynomial time solutions for the cases when $m = 2$, which can be handled using maximum flow [23], bipartite matching [17] and geometric bottleneck matching [9]. (Details can be found in [16].) We summarize as follows.

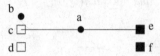

Fig. 3. An example with $P_1 = \{a, b\}$, $P_2 = \{c, d\}$ and $P_3 = \{e, f\}$, with all the points on a unit line segment and a being the midpoint of the segment. For the 'without replacement' version, the optimal solution is $Q_1 = \{a, b\}$, where a covers the 3-cluster $\{a, c, f\}$, b covers the 3-cluster $\{b, d, e\}$ and the optimal covering radius is 1. For the 'with replacement' version, the optimal solution is $Q_2 = \{a, a\}$, with the same clusters $\{a, c, f\}$ and $\{b, d, e\}$, and the optimal covering radius being 1/2.

Theorem 1. *The Center Persistence Diagram Problem can be solved in polynomial time, for $m = 2$ and for all the three versions ('Without Replacement', 'With Replacement' and continuous versions).*

In the next section, we will consider the case for $m = 3$.

3   3-Bottleneck Matching Is NP-Complete

We will first focus on the L_2 metric in this section and at the end of the proof it should be seen that the proof also works for the L_∞ metric. For $m = 3$, we can color points in P_1, P_2 and P_3 in color-1, color-2 and color-3. Then, in this case, the problem is really to find n disks centered at n points from $P_1 \cup P_2 \cup P_3$, with smallest radii r_i^* ($i = 1..n$) respectively, such that each disk contains exactly 3 points of different colors (possibly including the center of the disk); moreover, $\max_{i=1..n} r_i^*$ is bounded from above by a given value r. We also say that these 3 points form a *cluster*.

It is easily seen that (the decision version of) 3-Bottleneck Matching is in NP. Once the n guessed disks are given, the problem is then a max-flow problem, which can be verified in polynomial time.

We next show that Planar 3-D Matching (Planar 3DM) can be reduced to 3-Bottleneck Matching in polynomial time. The former is a known NP-complete problem [6]. In 3DM, we are given three sets of elements E_1, E_2, E_3 (with $|E_1| = |E_2| = |E_3| = \gamma$) and a set T of n triples, where $T \in \mathcal{T}$ implies that $T = (a_1, a_2, a_3)$ with $a_i \in E_i$. The problem is to decide whether there is a set S of γ triples such that each element in E_i appears exactly once in (the triples of) S. The Planar 3DM incurs an additional constraint: if we embed elements and triples as points on the plane such that there is an edge between an element a and a triple T iff a appears in T, then the resulting graph is planar.

An example for Planar 3DM is as follows: $E_1 = \{1, 2\}, E_2 = \{a, b\}, E_3 = \{x, y\}$, and $\mathcal{T} = \{(1, a, x), (2, b, x), (2, b, y), (1, b, y)\}$. The solution is $S = \{(1, a, x), (2, b, y)\}$.

Given an instance for Planar 3DM and a corresponding planar graph G with $O(n)$ vertices, we first convert it to a planar graph with degree at most 3. This can be done by replacing a degree-d element node x in G with a path of d nodes $x_1, ..., x_d$, each with degree at most 3 and the connection between x and a triple node T is replaced by a connection from x_i to T for some i (see also Fig. 4). We have a resulting planar graph $G' = (V(G'), E(G'))$ with degree at most 3 and with $O(n)$ vertices. Then we construct a rectilinear embedding of G' on a regular rectilinear grid with a unit grid length, where each vertex $u \in V(G')$ is embedded at a grid point and an edge $(u, v) \in E(G')$ is embedded as an intersection-free path between u and v on the grid. It is well-known that such an embedding can be computed in $O(n^2)$ time [27].

Let x be in E_1 or a black node ($\bullet$ in Fig. 4) with degree d in G. In the rectilinear embedding of G', the paths from x_i to x_{i+1} ($i = 1, ..., d - 1$) will be the basis of the element gadget for x. (Henceforth, unless otherwise specified, everything we talk about in this reduction is referred to the rectilinear embedding of G'.) We put a copy of $\bullet$ at each (grid point) x_i as in Fig. 4. (If the path from x_i to x_{i+1} is of length greater than one, then we put $\bullet$ at each grid point on the path from x_i to x_{i+1}.)

We now put color-2 and color-3 points ($\square$ and $\blacksquare$) at 1/3 and 2/3 positions at each grid edge which is contained in some path in an element gadget (in the embedding of G'). These points are put in a way such that it is impossible to use a discrete disk centered at a $\bullet$ point with radius 1/3 to cover three points with different colors. These patterns are repeated to reach a *triple gadget*, which will be given later. Note that this construction is done similarly for elements y and z, except that the grid points in the element gadgets for y and z are permuted, i.e., are of color-2 ($\square$) and color-3 ($\blacksquare$) respectively. To be more precise, we proceed similarly for vertices in E_2 and E_3 by permuting the colors, that is, for vertices in E_2, color-2 points are placed at the grid points, and color-1 and color-3 points are placed on the grid edges; and for vertices in E_3, color-3 points are placed on the grid points, and color-1 and color-2 points are placed on the grid edges.

Lemma 1. *In an element gadget for x, exactly one x_i is covered by a discrete disk (i.e., a disk centered at a colored grid point) of radius 1/3 out of the gadget.*

Proof. Throughout the proof, we refer to Fig. 4. Let x be in E_1 and colored by color-1 (e.g., $\bullet$). In the rectilinear embedding, let the number of grid edges between x_1 and x_d

be D. Then, the total number of points on the path from x_1 to x_d, of colors 1, 2 and 3, is $3D + 1$. By the placement of color-2 and color-3 points in the gadget for x, exactly $3D$ points of them can be covered by D discrete disks of radii 1/3 (centered either at color-2 or color-3 points in the gadget). Therefore, exactly one of x_i must be covered by a discrete disk centered at a point out of the gadget.   □

When x_i is covered by a discrete disk of radius 1/3 centered at a point out of the gadget x, we also say that x_i is *pulled out* of x.

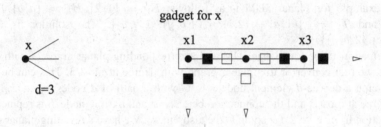

Fig. 4. The gadget for element x.

We now illustrate how to construct a triple gadget $T = (x, y, z)$. It is basically a grid point on which we put three points with different colors. (In Fig. 5, we simply use a ▲ representing such a triple gadget.) The interpretation of T being selected in a solution for Planar 3DM is that the three colored points at ▲ is covered by a disk of radius zero, centered at one of these three points. When one of these three points at ▲ is covered by a disk of radius 1/3 centered at some other points (on the path from one of the elements x, y or z to T), we say that such a point is *pulled out* of the triple gadget T by the corresponding element gadget.

Lemma 2. *In a triple gadget for $T = (x, y, z)$, to cover the three points representing T using discrete disks of radii at most 1/3, either all the three points are pulled out of the triple gadget T by the three element gadgets respectively, or none is pulled out. In the latter case, these three points can be covered by a discrete disk of radius zero.*

Proof. Throughout the proof, we refer to Fig. 5. At the triple gadget T, if only one point (say ●) is pulled out or two points (say, ● and □) are pulled out, then the remaining points in the triple, □ and ■ or ■ respectively, could not be properly covered by a discrete disk of radius 1/3 — such a disk would not be able to cover a cluster of exactly three points of distinct colors. Therefore, either all the three points associated with T are pulled out by the three corresponding element gadgets, hence covered by three different discrete disks of radii 1/3; or none of these three points is pulled out. Clearly, in the latter case, these three points associated with T can be covered by a discrete disk of radius zero, as a cluster.   □

In Fig. 5, we show the case when x would not pull any point out of the gadget for T. By Lemma 1, y and z would do the same, leading $T = \langle x, y, z \rangle$ to be selected in a solution S for Planar 3DM. Similarly, in Fig. 6, x would pull a • point out of T. Again, by Lemma 1, y and z would pull □ and ■ points (one each) out of T, which implies that T would not be selected in a solution S for Planar 3DM.

We hence have the following theorem, whose proof can be found in [16].

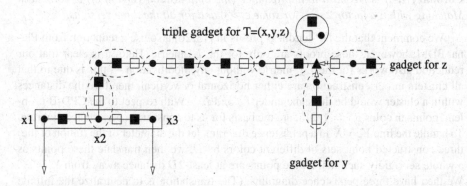

Fig. 5. The triple gadget for $T = \langle x, y, x \rangle$ (represented as ▲, which is really putting three element points on a grid point). In this case the triple $\langle x, y, z \rangle$ is selected in the final solution (assuming operations are similarly performed on y, z). Exactly one of x_i (in this case x_2) is pulled out of the gadget for the element x.

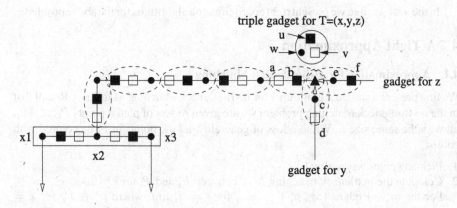

Fig. 6. The triple gadget for $T = \langle x, y, x \rangle$ (represented as ▲, which is really putting three element points on a grid point). In this case the triple $\langle x, y, z \rangle$ would not be selected in the final solution. Note that the black round point in the triple gadget is pulled out by the element x, and the other two points are pulled out similarly by the element y and z.

Theorem 2. *The decision versions of 3-Bottleneck Matching for both the 'Without Replacement' and 'With Replacement' cases are NP-complete, and the decision version of the continuous 3-Bottleneck Matching is NP-hard.*

Note that in the above proof, if Planar 3DM does not have a solution, then we need to use discrete disks of radii at least 2/3 to have a valid solution for 3-Bottleneck Matching. This implies that finding a factor-$(2-\varepsilon)$ approximation for (the optimization version of) 3-Bottleneck Matching remains NP-hard.

Corollary 1. *It is NP-hard to approximate (the optimization version of) 3-Bottleneck Matching within a factor $2 - \varepsilon$, for some $\varepsilon > 0$ and for all the three versions.*

We comment that the NP-hardness proofs in [13,26] also use a reduction from Planar 3DM; however, those proofs are only for the L_2 metric. Here, it is clear that our reduction also works for the L_∞ metric without any modification — this is due to that all clusters in our construction are either horizontal or vertical, therefore the distances within a cluster would be the same under L_2 and L_∞. With respect to the CPD-B problem, points in color-i, $i = 1, 2, 3$, are the basis for us to construct a persistence diagram. To handle the line $Y = X$ in a persistence diagram, let the diameter of the (union of the) three constructed point sets of different colors be $\hat{D}$, we then translate these points as a whole set rigidly such that all the points are at least $2\hat{D}$ distance away from $Y = X$. We then have three persistence diagrams. (The translation is to neutralize the infinite multiplicity of $Y = X$, i.e., to enforce that all points on $Y = X$ can be ignored when computing the bottleneck distance between the corresponding persistence diagrams.) Hence, we have the following corollary.

Corollary 2. *It is NP-hard to approximate (the optimization version of) Center Persistence Diagram problem under the bottleneck distance for $m \geq 3$ within a factor $2 - \varepsilon$, for some $\varepsilon > 0$ and for all the three versions.*

In the next section, we present tight approximation algorithms for the above problems.

4   A Tight Approximation

4.1   Approximation for m-Bottleneck Matching

We first present a simple Algorithm 1 for m-Bottleneck Matching as follows. Recall that in the m-Bottleneck Matching problem we are given m sets of planar points $P_1, ..., P_m$, all with the same size n. Without loss of generality, let the points in P_i be colored with color-i.

1. Pick any color, say, color-1.
2. Compute the bottleneck matching $M_{1,i}$ between P_1 and P_i for $i = 2, ..., m$.
3. For the $m - 1$ edges $(p_{j_1}^1, p_{j_i}^i) \in M_{1,i}$ for $i = 2, ...m$, where $p_y^x \in P_x$ for $x = 1, ..., m$, form a cluster $\{p_{j_1}^1, p_{j_2}^2, ..., p_{j_m}^m\}$ with $p_{j_1}^1$ as its center.

We comment that the algorithm itself is similar to the one given for $m = 3$ in [13], which has a different objective function (i.e., minimizing the maximum perimeter of clusters). We show next that Algorithm 1 is a factor-2 approximation for m-Bottleneck Matching. Surprisingly, the main tool here is the triangle inequality of a metric. Note that we cannot only handle for any given $m \geq 3$, we also need some twist in the proof a bit later for the three versions of the Center Persistence Diagram problem, where the diagrams could have different sizes. We have the following theorem, whose proof can again be found in [16].

Theorem 3. Algorithm 1 *is a polynomial time factor-2 approximation for m-Bottleneck Matching for all the three versions (i.e., 'Without Replacement', 'With Replacement' and continuous versions).*

4.2 Generalization to the Center Persistence Diagram Problem Under the Bottleneck Distance

First of all, note that the above approximation algorithm works for m-Bottleneck Matching when the metric is L_∞. Hence, obviously it works for the case when the input is a set of m persistence diagrams (all having the same size), whose (feature) points are all far away from $Y = X$, and the metric is the bottleneck distance. (Recall that, when computing the bottleneck distance between two persistence diagrams using a projection method, we always use the L_∞ metric to measure the distance between two points.)

We next show how to generalize the factor-2 approximation algorithm for m-Bottleneck Matching to the Center Persistence Diagram problem, first for $m = 3$. Note that we are given m persistence diagrams $\mathcal{P}_1, \mathcal{P}_2, ...,$ and $\mathcal{P}_m$, with the corresponding non-diagonal point sets being $P_1, P_2, ...,$ and P_m respectively. Here the sizes of P_i's could be different and we assume that the points in P_i are of color-i for $i = 1, ..., m$.

Given a point $p \in P_i$, let $\tau(p)$ be the (perpendicular) projection of p on the line $Y = X$. Consequently, let $\tau(P_i)$ be the projected points of P_i on $Y = X$, i.e.,

$$\tau(P_i) = \{\tau(p)|p \in P_i\}.$$

When $m = 2$, i.e., when we are only given $\mathcal{P}_1$ and $\mathcal{P}_2$, not necessarily of the same size, it was shown by Edelsbrunner and Harer that $d_B(\mathcal{P}_1, \mathcal{P}_2) = d_B^\infty(P_1 \cup \tau(P_2), P_2 \cup \tau(P_1))$ [7]. (Note that $|P_1 \cup \tau(P_2)| = |P_2 \cup \tau(P_1)|$.) We next generalize this result. For $i \in M = \{1, 2, ..., m\}$, let $M(-i) = \{1, 2, ..., i - 1, i + 1, ..., m\}$. We have the following lemma, whose proof can be found in [16].

Lemma 3. *Let $\{i, j\} \subseteq M = \{1, 2, ..., m\}$. Let $\tau_i(P_k)$ be the projected points of P_k on $Y = X$ such that these projected points all have color-i, with $k \in M, i \neq k$. Then,*

$$d_B(\mathcal{P}_i, \mathcal{P}_j) = d_B^\infty(P_i \bigcup_{k \in M(-i)} \tau_i(P_k), P_j \bigcup_{k \in M(-j)} \tau_j(P_k)).$$

The implication of the above lemma is that the approximation algorithm in the previous subsection can be used to compute the approximate center of m persistence diagrams. The algorithm can be generalized by simply projecting each point of color-i, say $p \in P_i$, on $Y = X$ to have $m - 1$ projection points with every color k, where $k \in M(-i)$. Then we have m augmented sets P_i'', $i = 1, ..., m$, of distinct colors, but with the same size $\sum_{l=1..m} |P_l|$. Finally, we simply run Algorithm 1 over $\{P_1'', P_2'', ..., P_m''\}$, with the distance in the L_∞ metric, to have a factor-2 approximation. We leave out the details for the analysis as at this point all we need is the triangle inequality of the L_∞ metric.

Theorem 4. *There is a polynomial time factor-2 approximation for the Center Persistence Diagram problem under the bottleneck distance with m input diagrams for all the three versions (i.e., 'Without Replacement', 'With Replacement' and continuous versions).*

Proof. The analysis of the approximation factor is identical with Theorem 3. However, when m is part of the input, each of the augmented point set P_i'', $i = 1..., m$, has a size $\sum_{l=1..m} |P_l| = O(mn)$. Therefore, the running time of the algorithm increases to $O((mn)^{1.5} \log(mn))$, which is, nonetheless, still polynomial.   □

It is interesting to raise the question whether these results still hold if the p-Wasserstein distance is used, which we depict in the appendix due to space constraint. In a nutshell, the NP-hardness proof for the continuous case can be carried over with minor modifications to the corresponding case under the Wasserstein distance; moreover, the 2-approximation can also be adapted to all the three cases under the Wasserstein distance. However, different from under the bottleneck distance, under the Wasserstein distance a lot of questions still remain open.

5 Concluding Remarks

In this paper, we study systematically the Center Persistence Diagram problem under both the bottleneck and p-Wasserstein distances. Under the bottleneck distance, the results are tight as we have a $2 - \varepsilon$ inapproximability lower bound and a 2-approximation algorithm (in fact, for all the three versions). Under the p-Wasserstein distance, unfortunately, we only have the NP-hardness for the continuous version and a 2-approximation, how to reduce the gap poses an interesting open problem. In fact, a similar question of obtaining some APX-hardness result was posed in [5] already, although the (min-sum) objective function there is slightly different. For the discrete cases under the p-Wasserstein distance, it is not even known whether the problems are NP-hard.

References

1. Ahmed, M., Fasy, B., Wenk, C.: Local persistent homology based distance between maps. In: Proceedings of 22nd ACM SIGSPATIAL International Conference on Advances in GIS (SIGSPATIAL 2014), pp. 43–52 (2014)
2. Buchin, K., et al.: Approximating (k, l)-center clustering for curves. In: Proceedings of 30th ACM-SIAM Symposium on Discrete Algorithms (SODA 2019), pp. 2922–2938 (2019)
3. Burkard, R., Dell'Amico, M., Martello, S.: Assignment Problems. SIAM (2009)
4. Cormen, T., Leiserson, C., Rivest, R., Stein, C.: Introduction to Algorithms, second edition, MIT Press, Cambridge (2001)
5. Custic, A., Klinz, B., Woeginger, G.: Geometric versions of the three-dimensional assignment problem under general norms. Discrete Optim. **18**, 38–55 (2015)
6. Dyer, M., Frieze, A.: Planar 3DM is NP-complete. J. Algorithms **7**, 174–184 (1986)
7. Edelsbrunner, H., Harer, J.: Computational Topology: An Introduction. American Mathematical Soc. (2010)
8. Edelsbrunner, H., Letscher, D., Zomorodian, A.: Topological persistence and simplification. Disc. Comp. Geom. **28**, 511–513 (2002)

9. Efrat, A., Itai, A., Katz, M.: Geometry helps in bottleneck matching and related problems. Algorithmica **31**(1), 1–28 (2001)

10. Fasy, B., He, X., Liu, Z., Micka, S., Millman, D., Zhu,B.: Approximate nearest neighbors in the space of persistence diagrams. CoRR abs/1812.11257 (2018)

11. Garey, M.R., Johnson, D.S.: Computers and Intractability: A Guide to the Theory of NP-Completeness. Freeman, W. H (1979)

12. Giusti, C., Pastalkova, E., Curto, C., Itskov, V.: Clique topology reveals intrinsic geometric structure in neural correlations. Proc. Nat. Acad. Sci. **112**(44), 13455–13460 (2015)

13. Goossens, D., Polyakovskiy, S., Spieksma, F., Woeginger, G.: The approximability of three-dimensional assignment problems with bottleneck objective. Optim. Lett. **4**, 7–16 (2010)

14. Hatcher, A.: Algebraic Topology. Cambridge University Press, Cambridge (2001)

15. Heffernan, P., Schirra, S.: Approximate decision algorithms for point set congruence. Comput. Geom. Theor. Appl. **4**(3), 137–156 (1994)

16. Higashikawa, Y., et al.: On computing a center persistence diagram. CoRR abs/1910.01753 (2019)

17. Hopcroft, J., Karp, R.: An $n^{5/2}$ algorithm for maximum matchings in bipartite graphs. SIAM J. Comput. **2**(4), 225–231 (1973)

18. Indyk, P.: Approximate nearest neighbor algorithms for Frechet distance via product metrics. In: Proceedings of the 18th Annual ACM Symposium on Computational Geometry (SoCG 2002), pp. 102–106 (2002)

19. Kerber, M., Morozov, D., Nigmetov, A.: Geometry helps to compare persistence diagrams. In: Proceedings of the 18th Workshop on Algorithm Engineering and Experiments (ALENEX 2016), pp. 103–112, SIAM (2016)

20. Lawson, P., Schupbach, J., Fasy, B., Sheppard, J.: Persistent homology for the automatic classification of prostate cancer aggressiveness in histopathology images. In: Proceedings of Medical Imaging: Digital Pathology 2019, pp. 109560G (2019)

21. Le, N-K., Martins, P., Decreusefond, L., Vergne, A.: Simplicial homology based energy saving algorithms for wireless networks. In: 2015 IEEE International Conference on Communication Workshop (ICCW), pp. 166–172 (2015)

22. Li, M., Ma, B., Wang, L.: On the closest string and substring problems. J. ACM **49**(2), 157–171 (2002)

23. Malhotra, V.M., Kumar, M.P., Maheshwari, S.N.: An $O(|V|^3)$ algorithm for finding maximum flows in networks. Info. Process. Lett. **7**(6), 277–278 (1978)

24. Masuyama, S., Ibaraki, T., Hasegawa, T.: The computational complexity of the m-center problems. Transac. of IEICE. **E64**(2), 57–64 (1981)

25. Kaczynski, T., Mischaikow, K., Mrozek, M.: Computational Homology. AMS, vol. 157. Springer, New York (2004). https://doi.org/10.1007/b97315

26. Spieksma, F., Woeginger, G.: Geometric three-dimensional assignment problems. Eur. J. Oper. Res. **91**, 611–618 (1996)

27. Valiant, L.G.: Universality considerations in VLSI circuits. IEEE Trans. Comput. **30**(2), 135–140 (1981)

Robust Identification in the Limit
from Incomplete Positive Data

Philip Kaelbling[1(✉)], Dakotah Lambert[2], and Jeffrey Heinz[3]

[1] Department of Computer Science, Wesleyan University, Middletown, USA
pkaelbling@wesleyan.edu
[2] Université Jean Monnet Saint-Étienne, CNRS, Institut d Optique Graduate School,
Laboratoire Hubert Curien UMR 5516, Saint-Étienne, France
dakotahlambert@acm.org
[3] Department of Linguistics and Institute for Advanced Computational Science,
Stony Brook University, Stony Brook, USA
jeffrey.heinz@stonybrook.edu

Abstract. Intuitively, a learning algorithm is robust if it can succeed despite adverse conditions. We examine conditions under which learning algorithms for classes of formal languages are able to succeed when the data presentations are systematically incomplete; that is, when certain kinds of examples are systematically absent. One motivation comes from linguistics, where the phonotactic pattern of a language may be understood as the intersection of formal languages, each of which formalizes a distinct linguistic generalization. We examine under what conditions these generalizations can be learned when the only data available to a learner belongs to their intersection. In particular, we provide three formal definitions of robustness in the identification in the limit from positive data paradigm, and several theorems which describe the kinds of classes of formal languages which are, and are not, robustly learnable in the relevant sense. We relate these results to classes relevant to natural language phonology.

Keywords: identification in the limit · grammatical inference ·
regular languages · model theory · locally testable · piecewise testable

1 Introduction

This paper presents an analysis of Gold-style learning [8] of formal languages from systematically deficient data and the conclusions one can draw from three different definitions of correctness. For our purposes, the omissions in the data arise from other constraints. We specifically consider data presentations which are the intersection of two languages, one of which is the target of learning.

The analysis is illustrated with, and motivated by, classes of formal languages that are both computationally natural and of particular interest to natural language phonology [11]. These classes are well-studied subregular classes which often have multiple characterizations, including language-theoretic, automata-theoretic, logical, and algebraic. The classes used to exemplify this work include the Strictly Local languages [20], the Strictly Piecewise languages [23], and the Tier-Based Strictly Local languages [13,18].

© The Author(s), under exclusive license to Springer Nature Switzerland AG 2023
H. Fernau and K. Jansen (Eds.): FCT 2023, LNCS 14292, pp. 276–290, 2023.
https://doi.org/10.1007/978-3-031-43587-4_20

As an example, suppose we are interested in learning the formal language L containing all strings which do not contain bb as a substring. As explained in more detail in Sect. 2, a positive data presentation for this language would eventually include strings like $babaaca$ (because it does not contain the bb substring). Now suppose the observable sequences are also subject to a constraint that words must not contain a b preceding c at any distance. In this case, the word $babaaca$ would *not* be part of the data presentation. Is it still possible to learn L if such words are never presented?

We provide three formal definitions of robustness in the identification in the limit from positive data learning paradigm, and several theorems which describe the kinds of classes of formal languages which are, and are not, robustly learnable in the relevant sense.[1]

We opt to explore a modification of Gold-style instead of the Probably Approximately Correct learning framework (PAC 26) in order to avoid the issue of defining a distance between formal languages. In the PAC framework, data is drawn from a stationary distribution, and a learner is required to be reasonably correct based on the data presented to it. This data could be considered "deficient" if the distribution poorly represents the target concept. As discussed by Eyraud et al. [5], PAC is not necessarily well-suited for the problem of learning formal languages. The approximate nature of correctness in PAC requires a notion of distance between formal languages to judge the quality of a proposed solution. There are many feasible metrics [4,22,25], and the PAC results are expected to be sensitive to the chosen metric. We choose to study robust learning in a model where this is not a concern.

Generally, research on identification in the limit from positive data in the presence of data inaccuracies have identified the following three types [16, chap. 8].

1. Noisy data. A data presentation for a formal language L includes intrusions from the complement of L.
2. Incomplete data. A data presentation for a formal language L omits examples from L. That is, if E represents the set of omitted examples, the presentation is actually a text for $L - E$ rather than for L itself.
3. Imperfect data. Data presentations for a formal language L which both includes intrusions from the complement of L and omits examples from L.

In this work we only study the identification in the limit from incomplete positive data. Fulk and Jain [7] study the problem of learning from incomplete data when there are finitely many data points omitted, which is unlike the case we consider where there can be infinitely many omitted examples. On the other hand, Jain [14] considers cases where there are infinitely many omissions. This work, like

---

[1] The notion of robustness studied here is different from the one studied by Case et al. [3]. There, a class is "robustly learnable" if and only if its effective transformations are learnable too. As such, their primary interest is classes "outside the world of the recursively enumerable classes." This paper uses the term "robustly learnable" to mean learnable despite the absence of some positive evidence.

that of Fulk and Jain [7], establishes hierarchies of classes that are exactly learnable or not in the presence of inaccurate data. While our first theorem regards exact identification, our other theorems relax that requirement. Additionally, Freivalds et al. [6] and Jain et al. [15] consider learning from a finite set of examples which contains at least all the *good examples*, which intuitively are well-chosen illustrations of the language. As learners must succeed with finitely many examples, this scenario potentially omits infinitely many. The scenario we consider, however, does not make provision for good examples.

As just mentioned, our strongest definition of correctness requires a learner to recover exactly the target language on a text drawn from the intersection of two languages. A key result under this definition is that of its strength, namely that very few classes of languages are independent of interference under it.

Our main result comes under our second notion of correctness, strong robustness, which requires a learner only to recover a language compatible with the target grammar when restricted to the intersection. Under this definition, we show that classes of languages identifiable with string extension learners [10,12] are strongly robust in the presence of interference from all other classes of languages.

Finally, we present our weakest correctness definition, weak robustness, removing the prior requirement of learning a language in the correct concept class. Under this definition the class of Tier-Based Strictly Local languages is robustly learnable, specifically by the algorithm presented by Lambert [17].

More generally, the results here are related to the question of whether two learnable classes of languages C and D imply a successful learning algorithm for the class of languages formed by their pointwise intersection $\{L_C \cap L_D : L_C \in C, L_D \in D\}$. In the case of identification in the limit from positive data, the answer in the general case is negative.[2] However, the results above – in particular strong robustness – help us understand the conditions sufficient for this situation to occur. In this way, this work helps take a step towards a compositional theory of language learning.

2   Background

2.1   Identification in the Limit

Gold [8] introduced a number of different definitions of what it means to learn a formal language. In this work, we concern ourselves only with the notion of learnability in the limit from positive data (ILPD), which is also called explanatory learning from text [16].

Let Σ denote a fixed finite set of symbols and Σ^* the set of all strings of finite length greater than or equal to zero. A formal language L (a constraint) is a subset of Σ^*.

A **text** t can be thought of as a function from the natural numbers to Σ^*. Let $\vec{t}_n$ represent the sequence $\langle t_0, t_1, \ldots, t_{n-1} \rangle$, the length-$n$ initial segment of a

---

[2] Alexander Clark (personal communication) provides a counterexample. Let $C = \{L_\infty, L_1, \ldots\}$ where $L_n = \{a^m : 0 < m < n\} \cup \{b^{n+1}\}$ and $L_\infty = a^+ \cup \{b\}$. Let $D = \{a^*\}$. Both classes are ILPD-learnable but $\{L_C \cap L_D : L_C \in C, L_D \in D\}$ is not.

text t. Note $\vec{t}_n$ is always of finite size. Let $\mathbb{T}$ denote all texts and $\mathbb{\vec{T}}$ represent the collection of finite initial segments of these texts. Using the notation for sequences instead of functions, we write t_i instead of $t(i)$. For a text t, let $\mathrm{CT}(t) = \{w : \exists n \in \mathbb{N}, t_n = w\}$, and similarly $\mathrm{CT}(\vec{t}_n) = \{w : 0 \leq i < n, t_i = w\}$. For a given language L, we say t is a text for L if and only if $\mathrm{CT}(t) = L$. The set of all texts for a language L is denoted $\mathbb{T}_L$.

Osherson et al. [21] discuss a modification that allows a text to exist for the empty language: t_n may be either an element of L or a distinct symbol $\odot$ representing a lack of data. Consequently, the empty language has exactly one text; for each $n \in \mathbb{N}, t_n = \odot$. We denote this text with $t^\odot$.

We denote with $\mathbb{G}$ a collection of grammars, by which we mean a set of finitely-sized representations, each of which is associated with a formal language in a well-defined way. If a grammar $G \in \mathbb{G}$ is associated with a formal language L, we write $\mathcal{L}(G) = L$, and say that G recognizes, accepts, or generates L.

An ILPD-learner is a function $\varphi \colon \mathbb{\vec{T}} \to \mathbb{G}$. The learner **converges on** t iff there is some grammar $G \in \mathbb{G}$ and some $i \in \mathbb{N}$ such that for all $j > i$, $\mathcal{L}(\varphi(\vec{t}_j)) = \mathcal{L}(G)$. If, for all texts t for a language L, it holds that φ converges on t to a grammar G such that $\mathcal{L}(G) = L$, then φ **identifies** L **in the limit**. If this holds for all languages of a class C, then φ identifies C in the limit and can be called a C-ILPD-learner.

Angluin [1] proved the following theorem.

Theorem 1. *Let C be a collection of languages which is indexed by some computable function. C is ILPD-learnable iff there exists a computably enumerable family of finite sets S such that for each $L_i \in C$, there exists a finite $S_i \subseteq L_i$ such that for any $L_j \in C$ which contains S_i, $L_j \not\subseteq L_i$.*

The finite set S_i is called a **telltale set** for L_i with respect to C.

2.2   String Extension Learning

Heinz [10] introduced string extension learning, a general type of set-based ILPD-learning algorithm based solely on the information contained in strings. A generalization of this technique is discussed here; for another generalization, see Heinz et al. [12]. The learner is defined as follows:

$$\varphi(\vec{t}_i) = \begin{cases} \varnothing & \text{if } i = 0,\, t_i = \odot \\ \varnothing \oplus f(t_i) & \text{if } i = 0,\, t_i \neq \odot \\ \varphi(\vec{t}_{i-1}) & \text{if } i \neq 0,\, t_i = \odot \\ \varphi(\vec{t}_{i-1}) \oplus f(t_i) & \text{otherwise,} \end{cases}$$

where f is a function that extracts information from a string and $\oplus$ is an operation for inserting information into a grammar.

Finally there is an interpretation relation $\models$, describing the language represented by the grammar. The statement $w \models G$ means that w satisfies the interpretation of G given by this relation. The language of the grammar G then

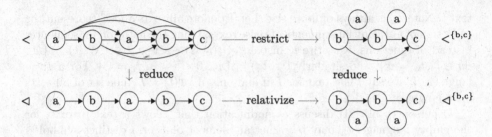

Fig. 1. Piecewise, local, and tier-based factors [18].

is $\mathcal{L}(G) = \{w : w \models G\}$. If it is the case that $w \models (G \oplus f(w))$ for all G and w, then φ is **consistent**. Often, a learner is defined with the following constraints: f extracts a set of factors of some sort, grammars are sets of the same type, $\oplus$ is set union, and $w \models G$ iff $f(w) \subseteq G$. Such a learner is consistent.

This may appear to simply refer to any incremental learner, but we add one further restriction: the class of string extension learners is defined to be the subfamily of these learners that are guaranteed to converge on any text. Every learner defined by Heinz et al. [12] satisfies this property, as does the incremental learner defined by Lambert [17] that will be explored further in Sect. 3.3.

2.3   Model-Theoretic Factors and Related Formal Language Classes

Lambert et al. [19] discuss a model-theoretic notion of factors. A simplification, sufficient for the present discussion, involves only a collection of symbol-labeled domain elements along with a binary relation between them. The relation induces a graph. A k-factor of a model is a collection of k nodes connected by the transitive closure of the relevant binary relation. Grammars and formal languages can be defined in terms of such k-factors.

Different binary relations give rise to different k-factors. Figure 1 shows some examples for the word $ababc$. The precedence relation $(<)$ (upper left) yields several 3-factors: $aab, aac, aba, abb, abc, bab, bac$. The sucessor relation $(\lhd)$ (lower left) yields fewer 3-factors: aba, bab, abc. The precedence relation can be restricted to a tier $T \subseteq \Sigma$ of salient symbols. In Fig. 1, $T = \{b, c\}$. It follows that the tier-precedence relation $(<^{\{b,c\}})$ (upper right) yields only the 2-factors: bb, bc. The tier-successor relation $(\lhd^{\{b,c\}})$ (lower right), is a binary relation relating positions on the tier to the positions that are "next" on the tier. Hence, it yields the 2-factors bb, bc.

Consider grammars G which are sets of k-factors and say $w \models G$ only if the k-factors in w are a subset of G. Such a definition for the relations $\lhd, <, \lhd^T$ yields the classes of formal languages that are testable in the strict sense: strictly k-local (SL) [20], strictly k-piecewise (SP) [9,23], or tier-based strictly k-local (TSL) [13,18], respectively.[3] Such classes are string extension learnable where f maps w to its k-factors, and $\oplus$ is set union [10].

---

[3] Technically, local classes need to be augmented with symbols marking word edges.

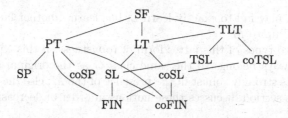

Fig. 2. A hierarchy by subclass of subregular classes.

Next consider grammars G which are sets of sets of k-factors and say $w \models G$ only if the k-factors in w are an element of G. Such a definition for the $\lhd, <,$ and $<^T$ relations yields the testable classes: locally k-testable (LT) [20], piecewise k-testable (PT) [24], or tier-based locally k, T-testable (TLT) [18], respectively. These classes are string extension learnable where f maps w to its k-factors, and $\oplus$ is set insertion [10].

The model-theoretic perspective combined provides a uniform way to characterize these well-studied classes. It also fits well into the generalized string extension scheme above because both the functions f and the operation $\oplus$ are understood simply in terms of k-factors.

The aforementioned results hold for classes where the parameters k, T are fixed. It is of special interest in linguistics to learn the family of k, T-TSL languages when k is fixed but T is not. Lambert [17] provides an incremental learning algorithm for this class of languages, and in Sect. 3.3, it is shown that this class is robustly learnable in a weak sense.

The aforementioned language classes and others are shown in Fig. 2 including some complement classes indicated with the prefix 'co'. Many other subregular classes exist, but only these few will be discussed in this work. For more details, readers are referred to Lambert et al. [19]. SF is star-free [20].

3 Robustness

When two or more constraints interact, the intersection of their licensed sets may no longer provide enough data to learn the constraints. Formally, we are interested in whether languages in C can be ILPD-learned from texts that are systematically deficient in some way.

A rare sort of robustness is when the individual constraints are retrievable exactly from the intersection, entirely unaffected by the increased sparsity of data. Consider any $L \in C$ and any other language M. This sort of robustness would mean that L is ILPD-learnable on all texts for $L \cap M$. In this case, there is a learner φ which can still exactly learn L despite interference from M.

If L might be unrecoverable, there could still be a guarantee that one can recover a grammar whose language produces the same intersection. This sort of robustness would mean that a learner φ, on any text for $L \cap M$, need only converge to a grammar G such that $\mathcal{L}(G) \cap M = L \cap M$. In this case, there is a learner φ which may fail to exactly learn L, but learns another language which is 'good enough' up to M.

We study two types of this latter form of robustness. If this equivalent constraint $\mathcal{L}(G)$ always belongs to the same class C as the original constraint L, then that class is strongly robust in the presence of the M; else the robustness is only weak. This section discusses these notions in order of decreasing strength.

3.1 Unaffectedness

The strongest form of robustness is that in which constraints are guaranteed to be extractable without loss of information from the interacting pattern.

Definition 1. *A class C is **unaffected by** another class D iff there is a learning function φ such that for all languages $L \in C$ and all languages $M \in D$, φ converges to a grammar for L on all texts for $L \cap M$.*

C is **affected by** D iff it is not unaffected by D. The strictness of this criterion is suggested by the following theorem.

Theorem 2. *Every ILPD-learnable class which includes two languages is affected by any class D where $\varnothing \in D$.*

Proof. Let C be a class which contains distinct languages L_1, L_2 and let D be a class containing $\varnothing$. There is only one text for $L_1 \cap \varnothing = L_2 \cap \varnothing = \varnothing$, which is $t^\circledcirc$. If a learner exists which correctly converges to L_1 on $t^\circledcirc$, it would not correctly converge to L_2 on this same text and vice versa.   □

Nearly every class in Fig. 2 contains the empty set. Over a non-empty alphabet, the sole exception is the class of cofinite languages coFIN, which may exclude only finitely many strings. However, even the cofinite languages can be shown to affect classes with general properties.

Theorem 3. *Every ILPD-learnable class which includes two distinct finite languages is affected by coFIN.*

Proof. Let C be a class which contains two distinct finite language L_1 and L_2. Because they are finite, their complements $\complement L_1$ and $\complement L_2$ belong to coFIN. The intersection $L_1 \cap \complement L_1 = L_2 \cap \complement L_2 = \varnothing$ has but a single text: $t^\circledcirc$. If a learner exists which correctly converges to L_1 on $t^\circledcirc$, it would not correctly converge to L_2 on this same text and vice versa.   □

Because both FIN and SP contain both the empty language and at least one nonempty finite language (Σ^k for nonzero k), neither they nor any of their superclasses can be unaffected by either the FIN or coFIN classes.

The SL and SP classes are not saved from being affected by coFIN even if restricted to their subclasses containing only infinite languages. Let L be the language of all and only those words over Σ which, if longer than n symbols, do not contain a specific symbol $a \in \Sigma$. In other words, a appears only in words shorter than n symbols. Further let M be the cofinite language containing all and only those words over Σ of length at least n. The intersection of L and M is $(\Sigma - \{a\})^{\geq n}$, an $(n+2)$-SL proper subset of L and which contains all and only those piecewise factors (subsequences) over $\Sigma - \{a\}$. For $k < n+2$, all and only those local factors (substrings) over this same alphabet are present in the language. In any case, because a does not appear in the data, it will be forbidden. Therefore for any parameters, the SL and SP learners will converge on some superset of this intersection which contains no instances of a, rather than on L itself.

In general, ILPD-learnable classes are affected by certain overlapping classes.

Theorem 4. *If L is a language in an* ILPD-*learnable class C, and $M \subset L$ belongs to $C \cap D$ for some class D, then C is affected by D.*

Proof. Let C and D be language classes such that C is ILPD-learnable and contains a language L and D overlaps with C such that $C \cap D$ contains a language $M \subset L$. Then $L \cap M = M$ and, since C is ILPD-learnable and M is in C, the C-learner must converge to M on a text for $L \cap M$. □

Corollary 1. *An* ILPD-*learnable class that contains a language L is affected by all subclasses of itself that contain any smaller language $M \subset L$.*

Our main result on unaffectedness is a characterization of which classes C are not affected by which classes D. We prove this result by adapting Angluin's [1980] characterization of the ILPD-learnable classes. Following Osherson et al. [21], we obtain this result via an adaptation of Blum and Blum's [1975].

Theorem 5. *Let $L, M \subseteq \Sigma^*$ and suppose φ is a learning function which identifies L on all texts for $L \cap M$. Letting $\mathbb{T}_{L \cap M}$ denote all texts for $L \cap M$, then there is some $\sigma \in \vec{\mathbb{T}}_{L \cap M}$ such that*

1. $\mathrm{CT}(\sigma) \subseteq L \cap M$
2. $\varphi(\sigma) = G$ where $\mathcal{L}(G) = L$.
3. $\forall \tau \in \vec{\mathbb{T}}_{L \cap M}[\mathrm{CT}(\tau) \subseteq L \cap M \to \varphi(\sigma\tau) = \varphi(\sigma)]$.

In other words, if a learner identifies a language L in the limit on texts from $L \cap M$, then there is some point in each text from which the learner is 'locked' into a particular grammatical hypothesis.

Proof. The proof is by contradiction. If the theorem is not true, it must be the case that for every $\sigma \in \vec{\mathbb{T}}_{L \cap M}$ such that (1) and (2) above are true, there is a $\tau \in \vec{\mathbb{T}}_{L \cap M}$ such that $\mathrm{CT}(\tau) \subseteq L \cap M$, but $\varphi(\sigma\tau) \neq \varphi(\sigma)$.

If this is true, then it is possible to construct a positive text for $L \cap M$ with which φ fails to converge, thus contradicting the initial assumption that φ identifies L in the limit on all texts for $L \cap M$. It will be helpful to consider some text t for $L \cap M$. Construct the new text q recursively as follows. Let $q^{(0)} = t_0$. Note that $\mathrm{CT}(q^{(0)})$ is a subset of $L \cap M$. $q^{(n)}$ is determined by the following cases:

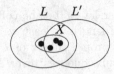

Fig. 3. Dots represent a telltale set for L, distinguishing it from L', despite interference from some third language M whose intersection with L is X.

Case 1. $\varphi(q^{(n-1)}) = G$ where $\mathcal{L}(G) = L$. Then by the reduction assumption we know that there exists some τ_n such that $\mathrm{CT}(\tau_n) \subseteq L \cap M$ and $\varphi(q^{(n-1)}\tau_n) \neq G$. Let $q^{(n)} = q^{(n-1)}\tau_n t_n$, and note that $\mathrm{CT}(q^{(n)})$ is a subset of $L \cap M$.

Case 2. $\varphi(q^{(n-1)}) = G$ where $\mathcal{L}(G) \neq L$. Then let $q^{(n)} = q^{(n-1)}t_n$. As in the other case, $\mathrm{CT}(q^{(n)})$ is a subset of $L \cap M$.

Observe that $\mathrm{CT}(q) = L \cap M$ and thus q is a text for $L \cap M$. This is because t is a text for $L \cap M$ and an element of t is added to q at every step in its construction. However, φ fails to converge on q because for every $i \in \mathbb{N}$ such that $\varphi(q^{(i)}) = G$ where $L = \mathcal{L}(G)$, there is a later point $q^{(i+1)}$, where $\varphi(q^{(i+1)})$ does not equal G by the construction above (Case 1). Therefore, we contradict the original assumption that φ identifies L on all texts for $L \cap M$ and the reductio assumption is false, proving the theorem.   □

Now one can state a property of all classes C which are unaffected by another class D. A crucial concept is the **telltale set despite interference** of a language in some class, defined below and demonstrated in Fig. 3.

Definition 2. *Any finite $S \subset \Sigma^*$ is a **telltale set** of a language $L \in C$ **despite interference** from $M \in D$ iff $S \subseteq L \cap M$ and for any $L' \in C$ such that $L' \cap M$ contains S, it holds that $L' \not\subseteq L$.*

If a learner guesses language L upon observing a telltale set for L despite interference from M, then it is guaranteed that the learner has guessed the smallest language in C which contains the sample. Thus the learner has not overgeneralized as no other language in the class of languages which includes the sample is strictly contained within L.

Theorem 6. *Let C, D be collections of languages which are both indexed by some computable functions. C is unaffected by D iff there exists a computably enumerable family of finite sets S such that for each $L_i \in C$ and $M_j \in D$, there exists a finite $S_{i,j} \subseteq L_i \cap M_j$ such that $S_{i,j}$ is a telltale set for L_i despite interference from M_j.*

Proof. ($\Rightarrow$) Suppose C is unaffected by D. Then there exists φ which for all $L \in C$ and $M \in D$ identifies L despite interference from M. By Theorem 5, there is a locking sequence σ for L where $\mathrm{CT}(\sigma) \subseteq L \cap M$. We show that the $\mathrm{CT}(\sigma)$ is a telltale set for L despite interference from M. First, as locking sequences are finite, $\mathrm{CT}(\sigma)$ is finite too. Now for contradiction assume that there is some $L' \in C$

such that $\mathrm{CT}(\sigma) \subseteq L'$, and $L' \subset L$. Then, per Theorem 5, φ fails to identify L' on a text t for L' where t begins with $\sigma\tau$, as $\varphi(\sigma) = G$ where $\mathcal{L}(G) = L$.

($\Leftarrow$) Assume that for every $L \in C$ and $M \in D$, L has a telltale set S despite interference from M, and further assume some enumeration of grammars and of these sets. Let X be the first (only) telltale set such that $X \subseteq \mathrm{CT}(\vec{t_i})$ and let $G = \varphi(\vec{t_i})$ be the first grammar in the enumeration such that $X \subseteq \mathrm{CT}(\vec{t_i}) \subseteq \mathcal{L}(G)$ if such objects exist, otherwise let X and G be the first set and grammar in their respective enumerations.

Now consider any $L \in C, M \in D$, any text t for $L \cap M$ and let G be the n-th grammar in the enumeration, but the first such that $\mathcal{L}(G) = L$. As S is finite, there is an i_1 such that $S \subseteq \mathrm{CT}(\vec{t_{i_1}}) \subseteq \mathcal{L}(G)$. Thus for all $j \geq i_1$, $\varphi(\vec{t_j})$ returns G unless there is some G' earlier in the enumeration such that $\mathcal{L}(G') \in C$, and S' is a telltale set for $\mathcal{L}(G')$ despite interference from M and $S' \subseteq \mathrm{CT}(\vec{t_{i_1}}) \subseteq \mathcal{L}(G')$.

However, we can find $i_2 \geq i_1$ which ensures that no such G' exists. Suppose there is some G' earlier in the enumeration such that $S' \subseteq \mathrm{CT}(\vec{t_{i_1}}) \subseteq \mathcal{L}(G')$. Then $\mathcal{L}(G')$ cannot properly include L because S' is a telltale set for $\mathcal{L}(G')$ and both $L, \mathcal{L}(G') \in C$. Thus there must be some sentence s in $L \cap M$ that is not in $\mathcal{L}(G') \cap M$. As t is a text for $L \cap M$, there is a k such that $s \in \mathrm{CT}(\vec{t_k})$.

Thus for any $j \geq k$, $\varphi(\vec{t_j}) \neq G'$ since $s \notin \mathcal{L}(G')$ and thus $\mathrm{CT}(\vec{t_j}) \not\subseteq \mathcal{L}(G')$. It follows that for each G_m (such that $\mathcal{L}(G_m) \in C$) which occurs earlier in the enumeration than G (i.e. $m < n$), there is some k_m such that $\mathrm{CT}(\vec{t_{k_m}}) \not\subseteq \mathcal{L}(G_m)$. There are only finitely many grammars before G in the enumeration and so by letting i_2 be the largest element of $\{i_1\} \cup \{k_m : 0 \leq m < n\}$, we guarantee that for any $j \geq i_2$, $\varphi(\vec{t_j}) = G$. $\qquad\square$

In short, for each $L \in C$ and for each $M \in D$, there must be a telltale set for L contained with $L \cap M$. This highlights the difficulty of this paradigm. The only classes unaffected by others to our knowledge are the singleton language classes $\{L\}$, which are unaffected by every class D. Future work involves identification of non-trivial C, D of linguistic interest such that C that is unaffected by D.

3.2   Strong Robustness

There are few cases of classes being unaffected by another. Yet this raises a question: should we care if the learned constraint is incorrect only on data that it cannot encounter? Learning a language consistent with the data should suffice.

Definition 3. *A class C is **strongly robust in the presence of** another class D iff there exists a learning function φ such that for all languages $L \in C$ and $M \in D$, there exists a grammar G such that $\mathcal{L}(G) \in C$, $\mathcal{L}(G) \cap M = L \cap M$, and φ converges to G on all texts for $L \cap M$.*

Theorem 7. *If a class C is intersection-closed (i.e. closed under finitary intersection) and string extension learnable by a learner φ which for any initial segment of a text $\vec{t_i}$ guarantees as output a unique minimum grammar $G = \varphi(\vec{t_i})$*

where $\mathcal{L}(G) \in C$ such that $\mathrm{CT}(\vec{t}_i) \subseteq \mathcal{L}(\varphi(\vec{t}_i))^4$, then C is strongly robust in the presence of any class D.

Proof. Let C be an intersection-closed, string extension learnable class whose associated learner φ guarantees a unique minimum grammar whose language is in C and compatible with the received text. That is, given any text t it holds that for any initial segment $\vec{t}_i$ of t we have $\mathrm{CT}(\vec{t}_i) \subseteq \mathcal{L}(\varphi(\vec{t}_i))$ and there is no grammar $X \neq \varphi(\vec{t}_i)$ such that $\mathcal{L}(X) \in C$ and $\mathrm{CT}(\vec{t}_i) \subseteq \mathcal{L}(X) \subseteq \mathcal{L}(\varphi(\vec{t}_i))$ Further, let $L \in C$, let M be any language, and let G be the grammar obtained by applying φ to a text drawn from $L \cap M$.

If $L \subset \mathcal{L}(G)$ then $\mathcal{L}(G)$ is not the minimal language in C compatible with the data, contradicting the assumption.

Suppose by way of contradiction that $\mathcal{L}(G) \cap M \neq L \cap M$. If $\mathcal{L}(G) \cap M \subset L \cap M$ then there exists some $w \in L \cap M$ such that $w \not\models G$. But w is in the text, violating the assumption that φ is compatible with the data it receives. Then it must be that there is some $v \models G$ such that $v \in M - L$, and notably v cannot appear in the text. The language $\mathcal{L}(G) \cap L$ is in C by intersection-closure, is a subset of $\mathcal{L}(G)$ by definition, and does not contain v; this violates the assumption that φ returns a grammar for the smallest language compatible with the text.

The only remaining option is that $\mathcal{L}(G) \cap M = L \cap M$. As M was unrestricted, it follows that C is strongly robust in the presence of any class D. $\square$

Each of the FIN, SL, LT, SP, and PT classes are intersection-closed and, when appropriately parameterized, string extension learnable in a way that guarantees a unique minimum language consistent with the text [10]. Therefore each of these classes is strongly robust in the presence of any class.

Corollary 2. *A intersection-closed class of ILPD-learnable languages C is strongly robust in the presence of any of its subclasses $C' \subseteq C$.*

Proof. Let C and D be classes of languages such that $D \subseteq C$, where C is ILPD-learnable and intersection-closed. Let $L \in C$ and $M \in D$. Then the intersection $L \cap M$ is in C. As C is ILPD-learnable, the intersection is learned exactly. $\square$

If two classes, A and B, are string-extension learnable by φ_A and φ_B, respectively, then one can define a string-extension learner for their pointwise intersection, $A \cap\!\!\!\!\!\!\!\cap B = \{a \cap b : a \in A, b \in B\}$, as follows:

$$f(w) = \langle f_A(w), f_B(w) \rangle$$
$$\langle G_A, G_B \rangle \oplus \langle x, y \rangle = \langle G_A \oplus_A x, G_B \oplus_B y \rangle$$
$$w \models \langle G_A, G_B \rangle \iff w \models G_A \wedge w \models G_B.$$

The learner thus defined is a **pointwise string extension learner** for $A \cap\!\!\!\!\!\!\!\cap B$.

For example, the intersection closure of the TSL class, MTSL, is pointwise string extension learnable. Given the alphabet Σ over which the text is drawn,

---

^{4} Note that this is a stronger guarantee than consistency.

construct $2^{|\Sigma|}$ k-SL learners in parallel, one for each subset of Σ. Each of these learners will be responsible for learning the constraints over its associated tier, by first projecting to that subset of Σ the words it encounters, then extracting the local factors of the result. Such a learner is not particularly efficient; for an alphabet of ten unique symbols, this results in 1,024 parallel SL learners.

Theorem 8. *If A and B are intersection-closed and string extension learnable, and both A and B are strongly robust in the presence of the other pointwise intersected with some third class C, then the class $A \cap\!\!\!\!\cap B$ is strongly robust in the presence of C.*

Proof. Let A and B be string extension learnable classes such that A is strongly robust in the presence of $B \cap\!\!\!\!\cap C$ and B is strongly robust in the presence of $A \cap\!\!\!\!\cap C$. Let φ_A and φ_B be the learners for A and B, respectively. Finally, let $L = L_A \cap L_B$ be some language in $A \cap\!\!\!\!\cap B$ and let $L' \in C$. Given some text t drawn from $L \cap L'$, $\mathcal{L}(\varphi_A(t)) \cap L_B \cap L' = L_A \cap L_B \cap L' = L \cap L'$, and $\mathcal{L}(\varphi_B(t)) \cap L_A \cap L' = L_A \cap L_B \cap L' = L \cap L'$ by strong robustness. The pointwise string extension learner φ for $A \cap\!\!\!\!\cap B$ exists such that $\mathcal{L}(\varphi(t)) = \mathcal{L}(\varphi_A(t)) \cap \mathcal{L}(\varphi_B(t)) = L \cap L'$. Therefore $A \cap\!\!\!\!\cap B$ is strongly robust in the presence of C. $\qquad\square$

Suppose that A and B are classes that satisfy the conditions of Theorem 7. That is, they are string extension learnable in a way that guarantees as output a unique minimum language in the respective class, compatible with the text they were given. Then they are strongly robust in the face of any interactions, by that theorem. It then follows immediately from Theorem 8 that their pointwise intersection $A \cap\!\!\!\!\cap B$ is similarly strongly robust in the presence of any interactions. However, we cannot turn this around and make strong claims about A or B based on the learnability of $A \cap\!\!\!\!\cap B$. Consider the case where A contains the empty language and B is the singleton class containing only the empty language. Then $A \cap\!\!\!\!\cap B = B$, which as a singleton class is unaffected by any other class C, no matter what A is. Furthermore, suppose A and C are identical and intersection-closed, but not ILPD-learnable. Concretely, suppose $A = C = \text{Reg}$, the class of all regular languages. Then A cannot be strongly robustly learnable in the presence of C, because it is not learnable in the first place.

3.3 Weak Robustness

An ILPD-learner only guarantees convergence to a language in its target class when presented with a text for such a language. When given a text from a language not in the target class, the result can be anything, even a lack of convergence. A weaker form of robustness might then be a guarantee that the learner will necessarily converge to some language consistent with the data, even if that language is not in the target class C.

Definition 4. *A class C is **weakly robust in the presence of** another class D iff there exists a learning function φ such that for all languages $L \in C$ and $M \in D$, there exists a grammar G such that $\mathcal{L}(G) \cap M = L \cap M$, and φ converges to G on all texts for $L \cap M$.*

This suggests the existence of a third class X, a superclass of C, where X is strongly robust in the presence of D.

As a concrete example of weak robustness, we shall consider $C = D = $ TSL. Membership in TSL is closure under suffix substitution on some tier T, and under insertion and deletion of elements not on that tier. That is, for $x \in T^k$, if $u_1 x u_2 \in L$, $v_1 x v_2 \in L$, then $u_1 x v_2 \in L$, and if $u_1 a u_2 \in L$ for $a \notin T$ then $u_1 u_2 \in L$ and vice versa. Let $\Sigma = \{a, b, c\}$, L be the language forbidding ab on the $\{a, b\}$ tier, M be that forbidding bc on the $\{b, c\}$ tier. The intersection $L \cap M$ is not TSL for any tier T. No letter is freely insertable or deletable in $L \cap M$, so $T = \{a, b, c\}$. Notice that $b(a^k)a \in L \cap M$ and $a(a^k)c \in L \cap M$. If it were TSL, then we would expect by suffix-substitution that $b(a^k)c \in L \cap M$, but it is not, as $b(a^k)c \notin M$. It follows that $L \cap M \notin$ TSL.

Recall the earlier discussion on pointwise intersections. Suppose that A and B are classes such that $A \pitchfork B$ is ILPD-learnable. We have already noted that this provides no guarantees regarding the robustness or even learnability of A or B in the presence of some third class C. However, we can state that A and B are weakly robust in the presence of one another, as the learner for $A \pitchfork B$ is by definition guaranteed to converge exactly on texts from the intersection. In fact, this example is just such a case. For $A = B = $ TSL, their intersection is (a subclass of) MTSL and therefore learnable by the algorithm which uses $2^{|\Sigma|}$ k-SL learners operating in parallel mentioned earlier. We have thus shown that TSL is weakly robust in the presence of itself.

4 Conclusions

We motivated and discussed the notion of learning from data systematically lacking in completeness. The result is four categories of learnability. The strongest, unaffectedness, provides a guarantee that a telltale set for the target language remains present despite interference. This requires that the correct generalizations be made even for data that can never appear. Strong robustness, while weaker than unaffectedness, makes a more reasonable guarantee: the learned language is only necessarily consistent with the target on data that can naturally occur in the face of the other constraints. Weak robustness keeps this more reasonable guarantee, but allows the learner to use grammars outside the target class. Finally, if none of these hold, the class is not robust.

We showed that each of the FIN, SL, LT, SP, and PT classes are strongly robust in the presence of any other class. On the other hand, we showed that the tier-based strictly local class of constraints, while quite natural for descriptive phonology, fails to be even strongly robust in the case where the relevant tier is unknown. Yet it has superclasses that are strongly robust in the presence of some types of interference. Such a quality makes this class weakly robust: one might fail to learn the target grammar, but learn instead a compatible grammar from the superclass. In the case of TSL, the relevant superclass was MTSL.

Each of these robustness categories is parameterized by the class from which interfering constraints are drawn. A class may be strongly robust in the presence of one class, yet not robust at all when faced with another. Open questions include characterizing the strongly and weakly robust learning paradigms. It would also be interesting to consider the effects of interference from other constraints when learning from good examples [6,15], as well as the problem of learning from data presentations which misrepresent the target language by including examples that do not belong to it.

Finally, the strongly robust learning paradigm provides a sufficient condition for when the pointwise intersection of two learnable classes of languages C and D is itself also learnable. The fact that each of the FIN, SL, LT, SP, and PT classes are strongly robust in the presence of any other class implies that classes of languages which must satisfy constraints from more than one of these classes are also learnable. To put it another way, some learnable classes of languages can be factored into simpler learnable classes. We hope this work helps lead to a more fully developed *compositional* theory of language learning.

Acknowledgements. We acknowledge support from the Data + Computing = Discovery summer REU program at the Institute for Advanced Computational Science at Stony Brook University, supported by the NSF under award 1950052.

References

1. Angluin, D.: Inductive inference of formal languages from positive data. Inf. Control **45**(2), 117–135 (1980)
2. Blum, L., Blum, M.: Toward a mathematical theory of inductive inference. Inf. Control **28**(2), 125–155 (1975)
3. Case, J., Jain, S., Stephan, F., Wiehagen, R.: Robust learning-rich and poor. J. Comput. Syst. Sci. **69**(2), 123–165 (2004)
4. Clark, A., Lappin, S.: Linguistic Nativism and the Poverty of the Stimulus. Wiley-Blackwell (2011)
5. Eyraud, R., Heinz, J., Yoshinaka, R.: Efficiency in the identification in the limit learning paradigm. In: Heinz, J., Sempere, J.M. (eds.) Topics in Grammatical Inference, pp. 25–46. Springer, Heidelberg (2016). https://doi.org/10.1007/978-3-662-48395-4_2
6. Freivalds, R., Kinber, E., Wiehagen, R.: On the power of inductive inference from good examples. Theoret. Comput. Sci. **110**(1), 131–144 (1993)
7. Fulk, M., Jain, S.: Learning in the presence of inaccurate information. Theoret. Comput. Sci. **161**, 235–261 (1996)
8. Gold, E.M.: Language identification in the limit. Inf. Control **10**(5), 447–474 (1967)
9. Haines, L.H.: On free monoids partially ordered by embedding. J. Combinatorial Theory **6**(1), 94–98 (1969)
10. Heinz, J.: String extension learning. In: Proceedings of the 48th Annual Meeting of the Association for Computational Linguistics, pp. 897–906. Association for Computational Linguistics, Uppsala, Sweden (July 2010)
11. Heinz, J.: The computational nature of phonological generalizations. In: Hyman, L., Plank, F. (eds.) Phonological Typology, Phonetics and Phonology, vol. 23, chap. 5, pp. 126–195. Mouton de Gruyter (2018)
12. Heinz, J., Kasprzik, A., Kötzing, T.: Learning in the limit with lattice-structured hypothesis spaces. Theoret. Comput. Sci. **457**, 111–127 (2012)

13. Heinz, J., Rawal, C., Tanner, H.G.: Tier-based strictly local constraints for phonology. In: Proceedings of the 49th Annual Meeting of the Association for Computational Linguistics: Short Papers, vol. 2, pp. 58–64. Association for Computational Linguistics, Portland (2011)
14. Jain, S.: Program synthesis in the presence of infinite number of inaccuracies. J. Comput. Syst. Sci. **53**, 583–591 (1996)
15. Jain, S., Lange, S., Nessel, J.: On the learnability of recursively enumerable languages from good examples. Theoret. Comput. Sci. **261**, 3–29 (2001)
16. Jain, S., Osherson, D., Royer, J.S., Sharma, A.: Systems That Learn: An Introduction to Learning Theory, 2nd edn. The MIT Press (1999)
17. Lambert, D.: Grammar interpretations and learning TSL online. In: Proceedings of the Fifteenth International Conference on Grammatical Inference. Proceedings of Machine Learning Research, vol. 153, pp. 81–91, August 2021
18. Lambert, D.: Relativized adjacency. Journal of Logic, Language and Information, May 2023
19. Lambert, D., Rawski, J., Heinz, J.: Typology emerges from simplicity in representations and learning. J. Lang. Modelling **9**(1), 151–194 (2021)
20. McNaughton, R., Papert, S.A.: Counter-Free Automata. MIT Press (1971)
21. Osherson, D.N., Stob, M., Weinstein, S.: Systems That Learn. MIT Press, Cambridge (1986)
22. Pin, J.E.: Profinite methods in automata theory. In: 26th International Symposium on Theoretical Aspects of Computer Science STACS 2009, February 2009
23. Rogers, J., et al.: On languages piecewise testable in the strict sense. In: Ebert, C., Jäger, G., Michaelis, J. (eds.) MOL 2007/2009. LNCS (LNAI), vol. 6149, pp. 255–265. Springer, Heidelberg (2010). https://doi.org/10.1007/978-3-642-14322-9_19
24. Simon, I.: Piecewise testable events. In: Brakhage, H. (ed.) GI-Fachtagung 1975. LNCS, vol. 33, pp. 214–222. Springer, Heidelberg (1975). https://doi.org/10.1007/3-540-07407-4_23
25. Smetsers, R., Volpato, M., Vaandrager, F., Verwer, S.: Bigger is not always better: on the quality of hypotheses in active automata learning. In: Clark, A., Kanazawa, M., Yoshinaka, R. (eds.) The 12th International Conference on Grammatical Inference. Proceedings of Machine Learning Research, vol. 34, pp. 167–181. PMLR, Kyoto, Japan, 17–19 Sep 2014
26. Valiant, L.G.: A theory of the learnable. Commun. ACM **27**(11), 1134–1142 (1984)

Cordial Forests

Feston Kastrati[1], Wendy Myrvold[2], Lucas D. Panjer[3], and Aaron Williams[4(✉)]

[1] Bard College at Simon's Rock, Great Barrington, USA
fkastrati13@simons-rock.edu
[2] University of Victoria, Victoria, Canada
wendym@uvic.ca
[3] Amazon Web Services, Seattle, USA
lucas@panjer.org
[4] Williams College, Williamstown, USA
aaron.williams@williams.edu

Abstract. We prove that a forest is cordial if and only if it does not have 4k+2 components and every vertex has odd-degree.

Keywords: Cordial labeling · graceful labeling · graceful tree conjecture

1 Introduction

Consider a graph whose vertices are labeled with 0s and 1s. Let its edges be labeled with the difference of their incident vertex labels taken modulo 2. In other words, an edge is given an *induced label* of 0 if its two incident vertices have the same label, or 1 if its two incident vertices have different labels. If the number of vertices labeled 0 and 1 are equal or off-by-one, and the number of edges labeled 0 and 1 are equal or off-by-one, then the labeling is said to be *cordial*, and the graph is said to be *cordial*. See Fig. 1 for an example.

More formally, a *binary-labeled graph* is a graph in which the vertices are labeled with 0 and 1, and each edge is labeled with the absolute difference of its endpoints. Let $n_x(G)$ and $m_x(G)$ be the number of vertices and edges labeled x in a labeled graph G, respectively. The number of extra vertex and edge labels that equal 1 are respectively denoted with Δs as follows,

$$\Delta_{\mathsf{v}}(G) = n_1(G) - n_0(G) \text{ and } \Delta_{\mathsf{e}}(G) = m_1(G) - m_0(G),$$

and we omit G from this notation when context allows.

Definition 1. *A binary-labeled graph is* cordial *if*

$$\Delta_{\mathsf{v}} \in \{-1, 0, 1\} \text{ and } \Delta_{\mathsf{e}} \in \{-1, 0, 1\}.$$

Cahit introduced cordiality in 1987 and it is a weaker version of both graceful and harmonious labelings [1]. The famous *Graceful Tree Conjecture* (or Ringel-Kotzig Conjecture) asserts that all trees are graceful [5], and it has attracted

H. Fernau and K. Jansen (Eds.): FCT 2023, LNCS 14292, pp. 291–303, 2023.
https://doi.org/10.1007/978-3-031-43587-4_21

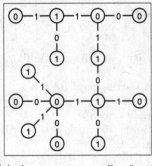
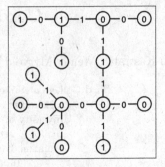

(a) $\Delta_e = m_1 - m_0 = 7 - 6 = 1.$ (b) $\Delta_e = m_1 - m_0 = 6 - 7 = -1.$

Fig. 1. Every tree is cordial. Here a tree T is given two different cordial labelings. In both cases, the vertex labels are balanced: $\Delta_v = n_1 - n_0 = 7 - 7 = 0$; (a) has a surplus of one 1-labeled edge, and (b) has a surplus of one 0-labeled edge.

considerable attention over the past 50 years (e.g., see the survey by Edwards and Howard [2]). Trees have been proven to have other types of labels (e.g., 3-equitable by Speyer and Szanislo [7]). Variations of cordial labelings have also been considered, including Hovey's introduction of A-cordial graphs. For a broad overview of graph labels see Gallian's dynamic survey [3].

By 1999 it was known that every tree is cordial [1] and that some forests are not cordial [6]. However, there has never been a characterization of which forests are cordial. In this paper, we provide the missing characterization. We define an *oddity* to be a graph in which every vertex has odd-degree. An *m-oddity* is a forest with m connected components in which every vertex has odd-degree. In other words, an *m*-oddity is a forest oddity with m components.

Theorem 1. *A forest is cordial if and only if it is not a* $(4k + 2)$*-oddity.*

Theorem 1 is illustrated in Fig. 2. The negative direction of Theorem 1 (i.e., the forests that are not cordial) follows from a general theorem of Seoud and Maqusoud [6]. For the positive direction (i.e., the forests that are cordial), we construct a suitable labeling. Since every tree has one component, Theorem 1 generalizes the fact that every tree is cordial [1].

1.1 Parity Conditions for the Graceful Tree Conjecture

While the focus of this paper is on cordiality, our underlying motivation is to understand parity requirements for proving (or disproving) the Graceful Tree Conjecture. To illustrate this point, consider the tree with $n = 14$ vertices and $m = 13$ edges in Fig. 3. The tree has been given two partial labelings that we hope to extend to a *graceful labeling*. In other words, we want to continue labeling the graph so that the vertices have distinct labels from $\{0, 1, \ldots, n-1\}$ and their absolute differences result in unique induced edge labels from $\{1, 2, \ldots, m\}$.

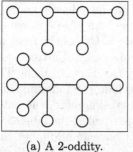   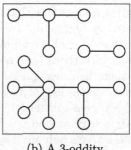

(a) A 2-oddity.   (b) A 3-oddity.

Fig. 2. Characterizing cordial forests. The forest in (a) is not cordial since it is a $(4k+2)$-oddity for $k = 0$ (i.e., every vertex has odd-degree and its number of components is 2 mod 4). The forest in (b) is a $(4k + 3)$-oddity, so it is cordial.

Theorem 1 implies that only one of the two starting points in Fig. 3 can hope to succeed, while the other is doomed to fail. To understand why this is the case, first note that deleting the labeled edge e in the tree gives the 2-oddity from Fig. 2a. Now suppose that this edge e is given the induced label $8 - 3 = 5$, as in Fig. 3a. Since this induced label is odd, and the full set of induced edge labels in any graceful labeling of the tree is $\{1, 2, \ldots, 13\}$, it must be that the remaining $13 - 1 = 12$ edges are equally split between even and odd induced labels. Therefore, if we were to take the vertex labels and the remaining edge labels modulo two, then we would have a cordial labeling for the forest in Fig. 2a. But this is not possible by the negative direction of Theorem 1.

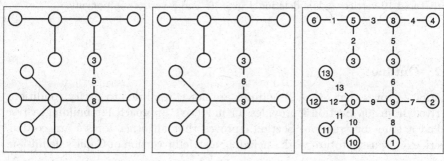

(a) A partial labeling where edge e has an odd induced label of $8 - 3 = 5$.

(b) A partial labeling where edge e has an even induced label of $9 - 3 = 6$.

(c) A completed graceful labeling starting from (b).

Fig. 3. Application to graceful labelings. The partial labeling in (a) cannot be extended to a graceful labeling. This is because the deletion of the labeled edge e gives the 2-oddity in Fig. 2a, and the remaining labels are cordial when taken modulo 2. Alternatively, e is the tree's only even-splitter (see Fig. 4a), so it cannot have an odd label. In contrast, the labeling in (b) is completed in (c).

To better understand this type of argument, we introduce a concept that is illustrated in Fig. 4. In a tree T, an *even-splitter* is an edge e in which both components of $T - e$ have an even number of vertices. Obviously, trees with an odd number of vertices have no such edges. Less obviously, if a tree has i even splitters, then their deletion gives an $(i + 1)$-oddity (see Sect. 3). Using Theorem 1 and the previous line of reasoning, we can conclude the following: If a tree T has $4k + 1$ even-splitters, then in any graceful labeling of T, the number of even-splitters with an odd label is not $2k + 1$. In particular, the labeled edge in Fig. 2a is the tree's only even-splitter, so it cannot have an odd label. Stronger parity restrictions for the Graceful Tree Conjecture are discussed in Sect. 5.

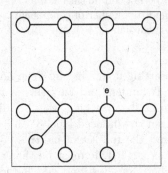
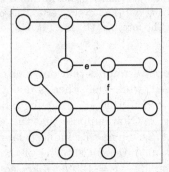

(a) A tree with one even-splitter e. Its deletion gives the 2-oddity in Figure 2a.

(b) A tree with two even-splitters e, f. Their deletion gives the 3-oddity in Figure 2b.

Fig. 4. Even-splitters. In (a) the edge e is an even-splitter since its deletion gives components with 6 and 8 vertices. Similarly, in (b) the deletion of e gives components with 4 and 10 vertices, while deleting f gives 6- and 8-vertex components.

1.2   Outline

Section 2 proves that $(4k + 2)$-oddities are not cordial due to parity conditions derived from [6]. Section 3 provides an inductive approach for building a tree called a twin-construction. Section 4 proves that all other forests are cordial. Section 5 discusses future work, including a stricter version of Cahit's definition.

2   Parity Conditions

In this section we consider the forests that are not cordial. We begin by restating a result by Seoud and Maqusoud [6].

Theorem 2. *If a graph G has n vertices and m edges with $n + m \equiv 2 \bmod 4$, then G is not cordial.*

Now we can obtain our negative result for cordial forests.

Lemma 1. *A $(4k+2)$-oddity is not cordial for all $k \geq 0$.*

Proof. Suppose that F that is a $(4k+2)$-oddity for some $k \geq 0$ with n vertices and m edges. Since F has $4k+2$ components, we have $n-m = 4k+2$. Consider the following series of equalities.

$$n - m = 4k + 2$$
$$n - m + 2m = 4k + 2 + 2m$$
$$n + m = 2(2k) + 2 + 2m$$
$$n + m = 2(2k + m) + 2$$
$$n + m = 2(2k') + 2$$
$$n + m = 4k' + 2$$

Since k' is an integer, we can conclude that $n+m$ is congruent to 2 (mod 4). Therefore, F is not cordial by Theorem 2.   □

3 Twin-Constructions

This section provides an unusual inductive definition of a tree, which will become useful when we want to label vertices and edges as equitably as possible. The method was developed independently, but is similar to one used by Cahit [1].

Our inductive approach relies on two *twin-addition* operations. In both cases we suppose that T is a tree with vertex set V such that $x \in V$ and $y, z \notin V$.

1. A *twin-leaf* addition $\mathsf{leaf}(x, y, z)$ adds y and z with edges xy and xz to T.
2. A *twin-path* addition $\mathsf{path}(x, y, z)$ adds y and z with edges xy and yz to T.

These two addition operations are illustrated in Fig. 5.

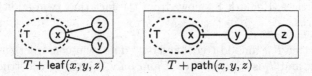

$$T + \mathsf{leaf}(x, y, z) \qquad\qquad T + \mathsf{path}(x, y, z)$$

Fig. 5. The two twin-addition operations, where the initial tree T is shown within the dotted perimeter.

A *twin-construction* of a non-empty tree T is a sequence of twin-additions that creates T; the construction starts from a single vertex or a pair of vertices connected by an edge that we call a *base vertex* or *base edge*, respectively. Theorem 3 proves that every tree has a twin-construction, and Fig. 6 shows that these constructions are not unique for a given tree.

Theorem 3. *Every non-empty tree T can be built by starting from a base vertex or a base edge followed by a sequence of twin-additions.*

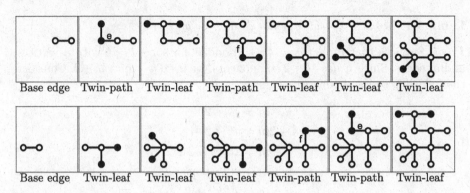

Fig. 6. Two different twin-constructions of the tree from Fig. 4b. Each step adds a twin-leaf or twin-path to an existing vertex. The tree's even-splitters e and f are labeled when they are added as the first edge on a twin-path addition.

Proof. If T has one or two vertices, then it is a base vertex or a base edge, respectively. Otherwise, consider a longest path in T whose third-last, second-last, and last vertices on one end are u, v, and w, respectively. There are two cases based on the degree of v. In the first case, vertex v has degree two in T and then $T = T' + \text{path}(u, v, w)$ for some tree T'. In the second case, vertex v has degree at least three in T. Because w is an endpoint of a longest path in T, there has to be at least one more leaf $x \neq u$ that is adjacent to vertex v. Hence, $T = T' + \text{leaf}(v, w, x)$ for some tree T'. In both cases, T' has two fewer vertices, so the result follows by induction on the number of vertices in a tree.   □

We frequently refer to the parity of the number of vertices of a tree in the remainder of this article. For this reason, we define an *even-tree* to be a tree that has an even number of vertices and an *odd-tree* to be a tree that has an odd number of vertices. Remark 1 follows from the fact that twin-additions always add two vertices.

Remark 1. A tree T is an odd-tree if and only if its twin-constructions start with a base vertex, and T is an even-tree if and only if its twin-constructions start with a base edge.

An *odd-degree tree* is a tree T in which every vertex of T has odd degree. In other words, it is a 1-oddity. Theorem 4 proves that these trees have special twin-constructions.

Theorem 4. *A tree is an odd-degree tree (1-oddity) if and only if it is an even-tree and its twin-constructions use only twin-leaf additions.*

Proof. A base edge is an odd-degree tree, and twin-leaf additions add two vertices of degree one without changing the parities of the degrees of the vertices already in the tree. Therefore, twin-constructions that start from a base edge and use only twin-leaf additions will produce an odd-degree tree. On the other hand, a base vertex has even degree, and a twin-path addition $\mathsf{path}(u, v, w)$ adds the degree two vertex v. Further twin-path additions can change the degree of v so that it is odd, but at the same time they add one new vertex of even degree. Therefore, a twin-construction that starts from a base vertex or includes a twin-path addition will create a tree with an even-degree vertex. □

The next lemma shows that the even-splitters of a tree can be characterized in terms of twin-constructions.

Lemma 2. *For any twin-construction of an even-tree, the even-splitters of the tree are the edges (u, v) that are added with a twin-path addition $\mathsf{path}(u, v, w)$ for some vertex w.*

Proof. Suppose T' is an even-tree, and consider $T = T' + \mathsf{path}(u, v, w)$. Notice that (u, v) is an even-splitter in T, whereas (v, w) is not an even-splitter in T. Similarly, in $T = T' + \mathsf{leaf}(x, y, z)$ neither (x, y) nor (y, z) are even-splitters. In other words, the only edges of twin-additions that are even-splitters when they are added to an even-tree are the (u, v) edges in $\mathsf{path}(u, v, w)$ additions.

To complete the proof consider an arbitrary edge e in an arbitrary tree T. Notice that twin-additions to T do not change the parities of the vertices in the two components in $T - e$, since both new vertices are added to the same component of $T - e$. Hence an edge is an even-splitter if and only if it is an even-splitter when initially added in the twin-construction. □

A simple corollary of these last two results is that the odd-degree trees are precisely the even-trees without even-splitters.

Corollary 1. *An even-tree T is an odd-degree tree if and only if T has no even-splitters.*

Proof. By Theorem 4 the twin-constructions of an even-tree T use only twin-leaves. Therefore, T has no even-splitters by Lemma 2. □

An *odd-degree forest* is a forest that contains only odd-degree vertices. In other words, an odd-degree forest is a collection of odd-degree trees; if the forest contains c such trees, then it is a c-oddity. Now we ask the following question:

> When can a set of edges S be deleted from a tree T to create an odd-degree forest $T - S$? In other words, when is $T - S$ an $(|S| + 1)$-oddity?

For example, deleting the edge sets $\{e\}$ and $\{e, f\}$ in Figs. 4a–4b result in the 2-oddity and 3-oddity in Figs. 2a–2b, respectively. The next theorem proves that the above question is answered precisely when the deleted edges are the even-splitters of an even-tree T. This generalizes Corollary 1, which covers the special case when T has no even-splitters.

Theorem 5. *Suppose that T is a tree and S is a subset of its edges. The forest $T - S$ is an odd-degree forest (or more precisely, an $(|S| + 1)$-oddity) if and only if T is an even-tree and S is its set of even-splitters.*

Proof. By the handshaking lemma, graphs have an even number of odd-degree vertices. In addition, T has the same number of vertices as $T - S$. Therefore, $T - S$ can only be an odd-degree forest if T is an even-tree. We assume that T is an even-tree and we prove the theorem in three cases.

Suppose that S is the set of even-splitters of T. Consider a twin-construction of T and the corresponding edges that belong to $T - S$. By Lemma 2, the construction begins with a base edge that is in $T - S$. Furthermore, each step of the twin-construction extends an existing odd-tree by adding both edges from some leaf(u, v, w) addition, or starts a new odd-tree containing the isolated edge (v, w) from some path(u, v, w) addition. Therefore, by Remark 1 and Theorem 4, each tree in $T - S$ is an odd-degree tree.

Suppose that e is not an even-splitter of T and $e \in S$. Therefore, $T - e$ contains two odd-trees. Deleting further edges will always result in an odd-tree, and odd-trees have at least one even-degree vertex.

Suppose that S is a strict subset of the even-splitters of T. Consider the even-splitters of T that are not in S, with respect to a twin-construction of T. Let (u, v) be the last even-splitter that is added during the twin-construction such that $(u, v) \notin S$. We complete the proof by arguing that v has even-degree in $T - S$. By Lemma 2, (u, v) was added by a twin-path path(u, v, w) and $(v, w) \notin S$. Therefore, v is incident with both (u, v) and (v, w) in $T - S$. Each subsequent twin-leaf addition leaf(v, x, y) has $(v, x), (v, y) \notin S$ by our choice of S. Thus, each addition increases the degree of v in $T - S$ by two. On the other hand, each subsequent twin-path addition path(v, x, y) has $(v, x) \in S$ and hence does not contribute to the degree of v in $T - S$. □

4   Characterization of Cordial Forests

This section completes our characterization of the forests that are cordial by constructing suitable labelings for the forests that are not $(4k + 2)$-oddities. We describe six different types of labelings for trees in Sect. 4.1. Then we use these labelings as building blocks for labeling forests in Sects. 4.2 and 4.3.

Our constructions are summarized by Tables 1, 2 and 3. These tables use images to facilitate the calculation of Δ_v and Δ_e values, and our convention for these images is explained at the end of Sect. 4.1.

4.1   Labeling Trees

A twin-addition to a binary-labeled tree is *equitable* if its added vertices and edges are labeled such that the following two points hold.

1. One vertex has label 0 and one vertex has label 1.
2. One edge has induced label 0 and one edge has induced label 1.

The definition implies that a twin-leaf addition leaf(u, v, w) is equitable when v and w are labeled oppositely. Similarly, a twin-path addition path(x, y, z) is equitable when x and y are labeled equally, and z is labeled oppositely. The four types of equitable additions are illustrated in Fig. 7.

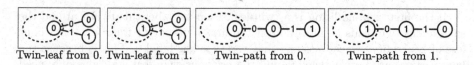

Twin-leaf from 0.   Twin-leaf from 1.   Twin-path from 0.   Twin-path from 1.

Fig. 7. Equitable twin-additions. The new vertices and edges are drawn outside of the dotted perimeter. In each case, the new vertex labels and edge labels are balanced (i.e., a single 0 label and a single 1 label).

When an equitable twin-addition is added to a tree its Δ_v and Δ_e values do not change, since the new 0 and 1 labels offset each other.

Remark 2. If T is a binary-labeled tree, and T' is obtained by applying an equitable twin-addition to T, then $\Delta_v(T) = \Delta_v(T')$ and $\Delta_e(T) = \Delta_e(T')$.

Our approach to labeling trees is to label the base vertex or base edge in a particular way, and then to extend the labeling equitably using Remark 2. The following lemma describes the five distinct types of labelings that result from this approach.

Lemma 3. *All odd-trees have binary labelings such that*

$$\text{(i) } (\Delta_v, \Delta_e) = (1, 0) \text{ and (ii) } (\Delta_v, \Delta_e) = (-1, 0).$$

All non-empty even-trees have binary labelings such that

$$\text{(iii) } (\Delta_v, \Delta_e) = (2, -1), \text{ (iv) } (\Delta_v, \Delta_e) = (0, 1), \text{ and (v) } (\Delta_v, \Delta_e) = (-2, -1).$$

Proof. This follows from Remarks 1 and 2 by labeling the bases as (i) ①, (ii) ⓪, (iii) ①—o—①, (iv) ⓪—1—① , and (v) ⓪—o—⓪ .   □

To prove our main result, we need a sixth type of binary-labeled tree which cannot be constructed by Lemma 3. A tree has *mixed-degree* if it has at least one vertex of odd-degree and at least one vertex of even-degree. (Non-empty trees contain a leaf of degree one, so mixed-degree trees are non-empty trees containing an even-degree vertex.)

Lemma 4. *If T is a mixed-degree even-tree, then T has a binary labeling such that*

$$\text{(vi) } (\Delta_v, \Delta_e) = (0, -1).$$

Proof. Suppose T is a mixed-degree even-tree and consider an arbitrary twin-construction for T. By Remark 1, the twin-construction begins with a base edge. We label the base as ⓪—o—⓪ so the base has $(\Delta_v, \Delta_e) = (-2, -1)$. Since T is a mixed-degree even-tree, Theorem 4 implies that one of the twin-additions in the twin-construction is a twin-path. The proof now considers two cases.

If the first twin-addition is a twin-path, then we label it so that the resulting tree has labels ⓪—o—⓪—1—①—o—① , and so $(\Delta_v, \Delta_e) = (0, -1)$ after this addition. From this point, we complete the labeling using equitable twin-additions.

If the first twin-addition is a twin-leaf, then we label this twin-leaf and the subsequent twin-leafs equitably until the first twin-path $\mathsf{path}(x, y, z)$. Without loss of generality, we can assume that vertex x has label 0. This is because the base vertices are both labeled 0, and the previous twin-leaf additions only need to label its two new vertices oppositely to be equitable. Now we label y and z with 1 as below.

Therefore, $(\Delta_v, \Delta_e) = (-2 + 2, -1 + 0) = (0, -1)$ after this addition. Again, we complete the labeling using equitable twin-additions.   □

In the remainder of this section we combine our six types of binary-labeled trees in various ways and compute the Δ_v and Δ_e values of the resulting forest. To facilitate these calculations we represent types (i)–(vi) so that the Δ values are immediately apparent. To explain our representations, recall that Lemma 3 grows each tree equitably from a labeled base. Therefore, the Δ values of the final tree equal the Δ values of the labeled base. Thus, the labelings of type (i)–(v) in Lemma 3 can be represented simply by the labeled base. For example, we use ①–∘–① to refer to an arbitrary even-tree that is labeled so that $(\Delta_v, \Delta_e) = (2, -1)$. Notice that the Δ values follow from the image since there are two vertices of label 1 and zero vertices of label 0, and so $\Delta_v = 2 - 0 = 0$. Similarly there are zero edges labeled 1 and one edge labeled 0, so $\Delta_e = 0 - 1 = -1$.

In Lemma 4, we use ⓪–∘–⓪–1–①–∘–① to represent mixed-degree even-trees labeled so that $(\Delta_v, \Delta_e) = (0, -1)$. This image represents the labeling of the base edge and first twin-path in the proof of Lemma 4. Again the Δ values can be determined from the image since every other twin-addition is equitable.

Table 1 summarizes our six types of binary-labeled trees.

Table 1. Six types of binary-labeled trees. Each row gives a labeled base (i.e., a base vertex or a base edge), the type of non-empty tree, and the Δ values obtained by building the tree using equitable twin-additions. The representation in (vi) shows labels for the tree's base edge and its first twin-path (see Lemma 4).

	Base Label	Tree Type	Δ_v	Δ_e
	Binary-Labeled Trees			
(i)	①	odd-tree	1	0
(ii)	⓪	odd-tree	-1	0
(iii)	①–∘–①	even-tree	2	-1
(iv)	⓪–1–①	even-tree	0	1
(v)	⓪–∘–⓪	even-tree	-2	-1
(vi)	⓪–∘–⓪–1–①–∘–①	mixed-degree even-tree	0	-1

4.2 Δ-Neutral Forests

If a binary-labeled forest F consists of binary-labeled trees $T_1, T_2, \ldots, T_k$, then

$$\Delta_v(F) = \sum_{i=1}^{k} \Delta_v(T_i) \text{ and } \Delta_e(F) = \sum_{i=1}^{k} \Delta_e(T_i). \tag{1}$$

A forest has a Δ-*neutral* binary-labeling if $(\Delta_v, \Delta_e) = (0,0)$. Section 4.2 will partition forests into various subsets of trees and label each subset; the Δ-neutral labeled subsets are helpful in this context since they do not contribute to the overall Δ values by (1).

We describe four types of Δ-neutral forests in Lemma 5 and their binary-labelings are summarized by Table 2. To verify that each type of forest is Δ-neutral, (1) allows us to simply count the vertex and edges labels in the graphical representations of the constituent trees. For example, Table 2 (c) has a forest of two odd-trees and two-even trees which are binary-labeled using ⓪ , ⓪ , ①–o–① , and ⓪–o–① according to Table 1. Note that there are three vertices labeled 1 and three vertices labeled 0 among these four images, so $\Delta_v = 3 - 3 = 0$. Similarly, there is one edge labeled 1 and one edge labeled 0, so $\Delta_e = 1 - 1 = 0$. Thus, every forest labeled in this way is Δ-neutral.

Lemma 5. *A forest has a Δ-neutral labeling if it consists of (a) two odd-trees, (b) four even-trees, (c) two even-trees and two odd-trees, or (d) two even-trees including at least one mixed-degree even-tree.*

Proof. The forest types in Table 2 match the forest types given in the statement of the lemma. In particular, row (d) includes the labeling of one mixed-degree even-tree from Lemma 4. The (Δ_v, Δ_e) values of the rows can computed as follows: (a) $(1-1, 0) = (0,0)$, (b) $(4-4, 2-2) = (0,0)$, (c) $(3-3, 1-1) = (0,0)$, and (d) $(3-3, 2-2) = (0,0)$. $\qquad\square$

Table 2. Four types of Δ-neutral forests. Each row gives the number of odd-trees d and even-trees e in the forest, and Δ-neutral labels according to Table 1. For example, row (b) shows that a forest of four even-trees has a Δ-neutral labeling by labeling one tree using (iii), two trees using (iv), and one tree using (v), as per Table 1. [†]Assumes that one of the even-trees has mixed-degree.

Δ-Neutral Binary-Labeled Forests

	d	e	Binary-Labeled Trees			
(a)	2	0	(i): ①	(ii): ⓪		
(b)	0	4	(iii): ①–o–①	(iv): ⓪–1–①	(iv): ⓪–1–①	(v): ⓪–o–⓪
(c)	2	2	(ii): ⓪	(ii): ⓪	(iii): ①–o–①	(iv): ⓪–1–①
(d)	0	$2^\dagger$	(iv): ⓪–1–①	(vi): ⓪–o–⓪–1–①–o–①		

4.3   Labeling Forests

Now we prove that all forests that are not $(4k + 2)$-oddies are cordial. Our proof reduces a given forest by giving Δ-neutral binary-labels to various subsets of trees from Table 2. For example, repeated use of subsets of type (a) and (b) can reduce the forest down to d odd-trees and e even-trees with $0 \leq d \leq 1$ and $0 \leq e \leq 3$. However, we must avoid Δ-neutral reductions that terminate with $d = 0$ odd-trees and $e = 2$ odd-degree even-trees. This is because the remaining unlabeled forest would be a $(4k + 2)$-oddity, and thus by Lemma 1 and (1) we could not complete the labeling to one that is cordial. For this reason we will also make strategic use the Δ-neutral forests of type (c) and (d) in Table 2. More specifically, we will use (c) and (d) at most once in total. After our Δ-neutral reductions, we will be left with a small forest that can be labeled according to one of the cases summarized by Table 3.

Theorem 6. *If F is a forest that is not a $(4k + 2)$-oddity, then F is cordial.*

Proof. Assume that F is a forest that is not a $(4k + 2)$-oddity with d odd-trees and e even-trees. We binary-label the trees of F in three steps. The first two steps use Δ-neutral labels, so the overall Δ-values for F are determined solely from the third step by (1). After each step, we update d and e to equal the number of unlabeled odd-trees and even-trees that remain in F, respectively.

The first step only involves forests with $e \bmod 4 = 2$. Since F is not a $(4k+2)$-oddity, at least one of the following is true:

- F has at least one mixed-degree even-tree, or
- F has at least one odd-tree $(d \geq 1)$.

We complete the first step using one of three cases. If F has at least one mixed-degree even-tree, then we label it and another even-tree using (d) in Table 2. Otherwise, if F has at least two odd-trees $(d \geq 2)$, then we label two even-trees and two odd-trees using (c) in Table 2. In both of these two cases the number of unlabeled even-trees is reduced by two, and so $e \bmod 4 = 0$ at the end of the first step. In the third case, there is a single odd-tree $(d = 1)$ and we do nothing. Therefore, the following implication holds at the end of the first step

$$e \bmod 4 = 2 \implies d = 1. \tag{2}$$

In other words, if the first step wasn't able to change the number of even-trees e from being equal to 2 modulo 4, then the forest contains exactly $d = 1$ odd-tree.

In the second step we apply row (a) in Table 2 as many times as possible, and then we apply row (b) in Table 2 as many times as possible. Since (a) consists of two odd-trees and (b) consists of four even-trees, we have $d \leq 1$ and $e \leq 3$ at the end of the second step. Furthermore, the implication in (2) still holds at the end of the second step.

In the third step we label the remaining trees. Since $d \leq 1$ and $e \leq 3$, there are up to eight cases to consider. However, if $d = 0$ and $e = 0$, then the forest has already been given a suitable labeling. In addition, we can ignore the case of $d = 0$ and $e = 2$ since the implication in (2) ensures that we avoid it. The labelings for the remaining six cases are given by the rows of Table 3.   □

Table 3. Six types of cordial forests. The number of odd-trees in the forest is d and its number of even-trees is e. Each row gives the binary-labels for the trees according to Table 1 and the resulting Δ values. Note that there is no row for $d = 0$ and $e = 0$ (as it is an empty forest) and $d = 1$ and $e = 2$ (by avoidance).

Cordial Forests

	d	e	Binary-Labeled Trees				Δ_v	Δ_e
1.	1	0	(ii): ⓪				-1	0
2.	0	1	(iv): ⓪-1-①				0	1
3.	1	1	(i): ①	(v): ⓪-0-⓪			-1	-1
4.	1	2	(ii): ①	(iv): ⓪-1-①	(v): ⓪-0-⓪		-1	0
5.	0	3	(iv): ⓪-1-①	(v): ⓪-0-⓪	(vi): ①-0-①		0	-1
6.	1	3	(ii): ⓪	(iii): ①-0-①	(iv): ⓪-1-①	(v): ⓪-0-⓪	-1	-1

5   Final Remarks

Cahit's definition of cordiality allows the number of 0 and 1 labels to be equal, or off by one in either direction. However, if one were to take the vertex and edge labels of a graceful labeling modulo two, then any surplus would need to involve an extra 1 for edge labels, or an extra 0 for vertex labels. In future work (also see [4]), we'll introduce a stricter version of cordiality which requires this higher level of precision. We'll also derive a strengthening of the graceful labeling application discussed in Sect. 1.1: In any graceful labeling of a tree, an even number of its even-splitters have an odd label. (In the case of Fig. 3, there is one even-splitter, so it cannot have an odd label.)

References

1. Cahit, I.: Cordial graphs: a weaker version of graceful and harmonious graphs. Ars Combin. **23**, 201–207 (1987)
2. Cairnie, N., Edwards, K.: The computational complexity of cordial and equitable labelling. Discret. Math. **216**, 29–34 (2000)
3. Gallian, J.A.: A dynamic survey of graph labeling. Electron. J. Combin. 5: Dynamic Survey **6**, 260 (2013). (electronic), 1998
4. Kastrati, F.: Graceful and cordial forests: A computational investigation of graph labelings. Bachelor's thesis, Bard College at Simon's Rock (1991)
5. Rosa, A.: On certain valuations of the vertices of a graph. In: Theory of Graphs (Internat. Sympos., Rome, 1966), pp. 349–355. Gordon and Breach, New York (1967)
6. Seoud, M.A., Abdel Maqsoud, A.E.I.: On cordial and balanced labelings of graphs. J. Egyptian Math. Soc. **7**, 127–135 (1999)
7. Speyer, D.E., Szanislo, Z.: Every tree is 3-equitable. Discret. Math. **220**, 283–289 (2000)

Vertex Ordering with Precedence Constraints

Jeff Kinne[1], Akbar Rafiey[2], Arash Rafiey[1(✉)], and Mohammad Sorkhpar[1]

[1] Math and Computer Science, Indiana State University, Terre Haute, IN, USA
jkinne@cs.indstate.edu, arash.rafiey@indstate.edu,
msorkhpar@sycamores.indstate.edu
[2] Computer Science, University of California San Diego, San Diego, CA, USA
arafiey@ucsd.edu

Abstract. We study *bipartite graph ordering problem*, which arises in various domains such as production management, bioinformatics, and job scheduling with precedence constraints. In the bipartite vertex ordering problem, we are given a bipartite graph $H = (B, S, E)$ where each vertex in B has a cost and each vertex in S has a profit. The goal is to find a minimum K together with an ordering $<$ of the vertices of H, so that $i < j$ whenever $i \in B$ is adjacent to $j \in S$. Moreover, at each sub-order the difference between the costs and profits of the vertices in the sub-order does not exceed K.

The bipartite ordering problem is NP-complete when the weights are one, and the bipartite graph H is a bipartite circle graph. This restricted version was used in the study of the secondary structure of RNA in [11].

Thus, we seek exact algorithms for solving the bipartite ordering problem in classes with simpler structures than bipartite circle graphs. We give non-trivial polynomial time algorithms for finding the optimal solutions for bipartite permutation graphs, bipartite trivially perfect graphs, bipartite cographs, and trees. There are still several classes of bipartite graphs for which the ordering problem could be polynomial, such as bipartite interval graphs, bipartite convex graphs, bipartite chordal graphs, etc.

In addition, we formulate the problem as a linear programming (LP) model and conduct experiments on random instances. We did not find any example with an integrality gap of two or more when limited to bipartite circle graphs with unit weights. No example with an integrality gap of more than 5/2 was found for arbitrary bipartite graphs with random weights. It would be interesting to investigate the possibility of designing a constant approximation algorithm for this problem.

Keywords: Vertex ordering · Bipartite graph classes · Precedence constraints · Energy barrier

1 Introduction and Problem Definition

In this paper, we introduce the bipartite graph ordering problem, motivated by a studying energy barrier problem for transitioning from one DNA secondary

H. Fernau and K. Jansen (Eds.): FCT 2023, LNCS 14292, pp. 304–317, 2023.
https://doi.org/10.1007/978-3-031-43587-4_22

structure (one folding) to another secondary DNA structure (with the same sequence and different folding) [14].

The authors of [11] looked at the energy barrier problem as a combinatorial problem on bipartite graphs, and they proved that the problem is NP-complete even on circle bipartite graphs[1] where the input weights are one.

Although the energy barrier problem is NP-complete, several algorithms have been developed to solve it. In [6,7] heuristic methods were given. In [17,18], the authors have focused on exact algorithms that take exponential time to solve the problem. The running time of the algorithm in [18] is $n^{O(K)}$, where K is the minimum energy required for this transformation. The worst-case time complexity of the algorithm in [17] is $O(|H|2^{|H|})$, where $|H|$ is the Hamming distance between the two input structures.

The bipartite ordering problem can be viewed as a variation of job scheduling problems with precedence constraints. The goal of our problem is to find the minimum initial budget required so that the vertices of the given bipartite graph are ordered, respecting the precedence and non-negative budget constraints. Job scheduling problems with precedence constraints have received much attention in theoretical computer science and applied mathematics due to their real-world applications in supply chain and production management. The aim of scheduling problems with precedence constraints is to order the jobs while respecting the precedence constraint. The objective function, however, can be different for different scenarios. Most of the work on job scheduling with precedence constraints has focused on minimizing the weighted completion time of the jobs in the single-processor or multi-processor setting [1,2,13,19]. The general problem of finding an ordering of the jobs to schedule that respects the precedence constraints and minimizes the weighted completion time, or cost is NP-complete. Therefore, some approximation algorithms and the hardness of approximation results have been studied to solve the scheduling problem with precedence constraints [1,2,19]. There are also some special classes of scheduling problems with precedence constraints that one can find an exact solution in polynomial time [3,10].

Our Results: In this work, we develop algorithms for some special graph classes; trivially perfect bipartite graphs, bipartite cographs, and bipartite permutation graphs; that admit polynomial-time exact solutions for the bipartite graph ordering problem.

We briefly mention that these classes of bipartite graphs have been considered in other optimization problems. Trivially perfect bipartite graphs play an important role in studying the list homomorphism problem. The authors of [5] showed that the list homomorphism problem could be solved in logarithmic space for these bipartite graphs. They were also considered in the fixed parametrized version of the list homomorphism problem in [4]. The subclass of trivially perfect

---

[1] A circle bipartite graph can be represented as two sets A, B where the vertices in A are a set of non-crossing arcs on a real line and the vertices in B are a set of non-crossing arcs from a real line; there is an edge between a vertex in A and a vertex in B if their arcs cross.

bipartite graphs called *laminar family bipartite graphs* was considered in [15] to obtain a polynomial time approximation scheme (PTAS) for special instances of a job scheduling problem. Each problem instance in [15] is a bipartite graph $H = (J, M, E)$ where J is a set of jobs, and M is a set of machines. For every pair of jobs $u, v \in J$, the set of machines that can process u, v are either disjoint or one is a subset of the other. Bipartite permutation graphs, also known as proper interval bipartite graphs are of interest in graph homomorphism problems [9], and in energy production applications where resources (in our case B vertices) can be assigned (bought) and used (sold) within some successive time steps [12]. There are recognition algorithms for bipartite permutation graphs [9,16].

1.1   Problem Definition

We are given a bipartite graph $H = (B, S, E)$, where $B \cup S$ is the set of vertices, and E is the set of edges, a subset of $B \times S$. Each vertex $u \in B$ has a negative cost p_u, and each vertex $v \in S$ has a positive cost p_v. The goal of the bipartite graph ordering problem for H is to find a minimum value $bg(H)$ and an ordering $v_1 < v_2 < \cdots < v_n$ of the vertices of H that satisfies:

- *Precedence constraints*: if $(v_i, v_j) \in E$, with $v_i \in B$ and $v_j \in S$ then $v_i < v_j$.
- *Budget constraints*: for every sub-order of the vertices $v_1 < v_2 < \cdots < v_r$, $r < |B \cup S|$, we have $bg(H) + \sum_{k=1}^{k=r} p_{v_k} \geq 0$.

We often use the term *process first* (*process next*) for a subset of vertices of H, and we mean order them before (after) some other vertices of H in the final total ordering. Throughout the paper we denote the input instance by $H = (B, S, E)$ and we assume the cost of vertices in B are negative and the costs of vertices in S are positive. Figure 1 describes an example of the problem when the costs $p_v = 1$, $v \in S$ and $p_u = -1$, $u \in B$.

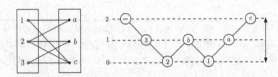

Fig. 1. The left graph is a bipartite circle graph with $B = \{1, 2, 3\}$, and $S = \{a, b, c\}$. Ordering $3, 2, b, 1, a, c$ and $bg(H) = 2$ give an optimal solution when the weights are 1 and -1.

The bipartite graph ordering problem is a natural variation of scheduling problems with precedence constraints. It can be used to model the purchase of supplies and production of goods when purchasing in bulk. Another way to view the problem is that the items in B are training sessions that employees must complete before employees (vertices in S) can begin to work.

No Bound on the Value of $bg(G)$ When the Weights are One. In what follows, we introduce a class of bipartite graphs G with maximum degree at most $\sqrt{|G|}$ ($|G|$ number of vertices) while $bg(G)$ is greater than $|G|/2$. Let $\mathcal{P}$ be a projective plane of order p^2+p+1 with p prime. The projective plane of order $n = p^2+p+1$ consists of n lines, each consisting of precisely $p+1$ points, and n points which each are intersected by precisely $p+1$ lines. We construct a bipartite graph with each vertex in B corresponding to a line from the projective plane, each vertex in S corresponding to a point from the projective plane and a connection from $b \in B$ to $s \in S$ if the point corresponding to s is contained in the line corresponding to B. Vertices in B are given weight -1, and vertices in S are given weight 1. Note that the degree of each vertex in B is $p+1$. One can observe that the neighborhood of every set of $p+1$ vertices in S is at least $p^2 - \binom{p}{2}$. Therefore, to process the first $p+1$ vertices in S we need to process their neighborhood which decreases the budget by at least $p^2 - \binom{p}{2} + p > n/2$; implying that $bg(G) > n/2$.

1.2   Warm-Up (Simple Cases)

In this subsection, we consider simple instances of the problem. This gives a better understanding of the problem and its difficulty. We provide this section to assist the reader in developing an intuition for the problem.

Let $H = (B, S, E)$ be a bipartite graph, and let X be a subset of vertices in H. $\|X\|$ refers to the mass of set X defined by $\sum_{x \in X} |p_x|$, where p_x is the cost of vertex x. We often consider X to be entirely in B or entirely in S.

Proposition 1. *Let $H = (B, S, E)$ be an instance of the bipartite ordering problem where H is a disjoint union of bicliques (bipartite cliques), and random weights. Then computing $bg(H)$ is a polynomial-time task.*

Proof. First, we note that if H is a biclique, then $bg(H) = \|B\|$. Now we consider the case where our graph is a disjoint union of bicliques $H_1, H_2, ..., H_m$ where each $H_i = (B_i, S_i, E_i)$ is a biclique. We start with value $K = \sum_{j=1}^{j=m} \|B_j\|$ as initial budget. Intuition suggests that we should first process those H_i with $\|S_i\| \geq \|B_i\|$, which we call positive sets. If multiple positive sets exist, we process the H_i with minimum $\|B_i\|$ and increase K by $\|S_i\| - \|B_i\|$. Then we are left with bicliques $H_i = (B_i, S_i)$ where $\|B_i\| > \|S_i\|$, which we call negative set.

In processing the remaining negative sets, the budget steadily goes down. As we shall see momentarily, we should process the H_i with the largest $\|S_i\|$ first and decrease K by $\|S_i\| - \|B_i\|$. Suppose on the contrary that $\|S_i\| > \|S_j\|$ but an optimal strategy *opt* processes H_j right before H_i. If K is the budget before this step we first have that $K - \|B_j\| + \|S_j\| \geq \|B_i\|$ because otherwise, there would not be sufficient budget after processing H_j to process H_i. Since we assumed that $\|S_i\| > \|S_j\|$ we have $K - \|B_i\| + \|S_i\| \geq \|B_j\|$. Thus, we could first process H_i and then H_j. We have thus given a method to compute an optimal strategy for a disjoint union of bicliques: first process positive sets in decreasing order of $\|B_i\|$, and then process negative bicliques in decreasing order of $\|S_i\|$.

Suppose during this process K' is the minimum value of the current budget. Thus, $bg(H) = \sum_{j=1}^{j=m} \|B_j\| - K'$.   □

Notice that when bipartite graph H consists of disjoint paths each of length 4 (path P_5) together with random weights, the approach in Proposition 1 does not work, giving some indication of the difficulty of the problem.

Next, we assume the input graph is a tree, and the costs are $p_v = 1$, $v \in S$ and $p_u = -1$, $u \in B$.

Proposition 2. *Let $T = (B, S, E)$ be a tree with weights one. Then $bg(T) = \|B\| - \|S\| + 1$ and finding an optimal ordering is a polynomial time task.*

Proof. Note that any leaf has a single neighbor (or none if it is an isolated vertex). We can thus immediately process any leaf $j \in S$ by processing its parent in the tree and then processing j. This requires an initial budget of only 1. After processing all leaves in S, we are left with a forest where all leaves are in B. We can first remove from consideration any disconnected vertices in B (these can, without loss of generality, be processed last). We are left with a forest H'. We next take a vertex $j_1 \in S$ (which is not a leaf because all leaves in S have already been processed) and process all of its neighbors. After processing j_1 we return 1 unit to the budget. Since H' is a forest, the neighborhood of j_1 has an intersection at most 1 with the neighborhood of any other sold vertex in S. Because we have already processed all leaves in S, we know that only j_1 can be processed after processing its neighbors.

After processing j_1, we may be left with some leaves in S. If so, we deal with these as above. We note that if removing the neighborhood of j_1 does create any leaves in S, then each of these has at least one vertex in B that is its neighbor and is not the neighbor of any of the other leaves in S. When no leaves remain, we pick a vertex $j_2 \in S$ and deal with it as we did j_1.

This process is repeated until all of H' is processed. We note that after initially dealing with all leaves in S, we gain at most a single leaf in S at a time. That is, the budget initially increases as we process vertices in S and process their parents in the tree, and then the budget goes down progressively, only ever temporarily going up by a single unit each time a vertex in S is processed. Note that the budget initially increases, and then once it is decreasing only a single vertex in S is processed at a time. This implies that the budget required for our strategy is $\|B\| - \|S\| + 1$, the best possible budget for T with weights 1 and -1.

2 Definitions and Concepts

In this section, we define key terms and concepts that are relevant to algorithms solving the bipartite graph ordering problem.

Let $H = (B, S, E)$ be a bipartite graph. For a subset, $I \subseteq B$, let $N^*(I)$ be the set of all vertices in S whose entire neighborhood lies in I. For example, in Fig. 2, $N^*(J) = F$.

Definition 1 (Prime set). *We say a set $I \subseteq B$ is* prime *if there exists set $X \subset S$, where $N(X) = I$ and there is no X' with $N(X') \subset I$. Equivalently, I is prime if $N^*(I)$ is non-empty and for every $I' \subset I$, $N^*(I')$ is empty.*

In Fig. 2, J and I are primes, but $I \cup I_1$ is not prime since $I_1 \subset I \cup I_1$, and $N^*(I_1) = O \neq \emptyset$. Other examples for prime sets in Fig. 2 are $J_1 \cup J_2$, J, I, I_1 with $N^*(J_1 \cup J_2) = D$, $N^*(J) = F$, $N^*(I) = L$, and $N^*(I_1) = Q$. Note that the bipartite graph induced by a prime set I and $N^*(I)$ is a bipartite clique. For $X \subset B$, let $H[X \cup N^*(X)]$ be the induced subgraph of H by $X \cup N^*(X)$.

Definition 2 (Positive/Negative set). *A set $I \subseteq B$ is called* positive *if $\|I\| \leq \|N^*(I)\|$ and it is* negative *if $\|I\| > \|N^*(I)\|$.*

Definition 3 (Positive minimal set). *A set $I \subseteq B$ is called* positive minimal *if I is positive, and for every other positive subset I' of I we have $bg(H[I' \cup N^*(I')]) \geq bg(H[I \cup N^*(I)])$.*

For the given graph in Fig. 2, I_1 is the only positive minimal set where $N^*(I_1) = O$ contains 7 vertices. Note that, in general, there can be more than one positive minimal set. Positive minimal sets are key in algorithms solving the general case of bipartite graph ordering because these are precisely the sets that we can process first, as can be seen from Lemma 2. In the graph of Fig. 2, the positive set I_1 is the first to be processed.

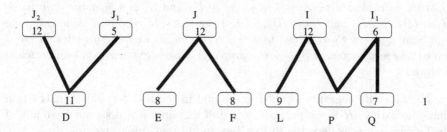

Fig. 2. Each bold line shows a complete connection, i.e. the induced sub-graph by $I \cup L$ is a biclique. The numbers in the boxes are the number of vertices. The sets J_1, J_2, J, I, I_1 are the vertices B, with each vertex having weight -1. The sets D, E, F, L, P, O are the vertices in S, with each vertex having weight 1.

Fixing an Order for B in Instance $H = (B, S, E)$: Let $\prec$ be an arbitrary order of the vertices in B. We order the positive minimal subsets of B, in the lexicographic order $\prec_L$. This means for two sets $A_1 \subset B$ and $A_2 \subseteq B$, $A_1 \prec_L A_2$ (A_1 is before A_2) if the smallest element of A_1 (according to $\prec$) say a_1 is before the smallest element A_2, say a_2. If $a_1 = a_2$ then $A_1 \prec_L A_2$ if $A_1 \setminus \{a_1\} \prec_L A_2 \setminus \{a_2\}$.

Definition 4 (Closure). *For $I \subseteq B$ of instance $H = (B, S, E)$, let $c\ell(I) = \cup_{i=1}^r I_i \cup I$ where each $I_i \subseteq B$, $1 \leq i \leq r$ is the lexicographically first positive* minimal subset in $H_i = H \setminus (\cup_{j=0}^{i-1} I_j \cup N^*(\cup_{j=0}^{i-1} I_j))$ $(I_0 = I)$ such that in H_i we have $bg(I_i) \leq bg(H) - \cup_{j=0}^{i-1} \|I_j\| + \cup_{j=0}^{i-1} \|N^*(I_j)\|$. Here r is the number of times the process of ordering a positive minimal set can be repeated after I.

Note that $cl(I)$ could be only I, in this case $r = 0$. For instance, consider Fig. 2. In the graph induced by $\{J, J_1, J_2, I, D, E, F, L\}$ we have $cl(J) = J \cup J_1$.

In what follows, we define trivially perfect bipartite graphs, bipartite cographs, and bipartite permutation graphs. We discuss the key properties that are used in our algorithm for solving the bipartite graph ordering problem on these graph classes.

Definition 5 (Trivially perfect bipartite graph). *A bipartite graph $H = (B, S, E)$ is called trivially perfect if it can be constructed using the following operations. If H_1 and H_2 are trivially perfect bipartite graphs, then the disjoint union of H_1 and H_2 is trivially perfect. If $H_1 = (B_1, S_1, E_1)$ and $H_2 = (B_2, S_2, E_2)$ are trivially perfect bipartite graphs then by joining every vertex in S_1 to every vertex in B_2, the resulting bipartite graph is trivially perfect. Notice that a bipartite graph with one vertex is trivially perfect.*

Bipartite graph H is trivially perfect if and only if it does not contain any of the following as an induced sub-graph: C_6, P_6 [5].

Definition 6 (Bipartite cograph). *A bipartite graph $H = (B, S, E)$ is called cograph if it can be constructed using the following operations. If H_1 and H_2 are bipartite cographs then the disjoint union of H_1 and H_2 is a bipartite cograph. If $H_1 = (B_1, S_1, E_1)$ and $H_2 = (B_2, S_2, E_2)$ are bipartite cographs, their complete join—where every S_1 is joined to every vertex in B_2 and every vertex in B_1 is joined to every vertex in S_2—is a cograph. A bipartite graph with one vertex is a cograph.*

The bipartite cographs studied in [8], and in terms of forbidden obstruction characterization, H is a bipartite cograph if and only if it does not have any of the following graphs depicted in Fig. 3 as an induced sub-graph.

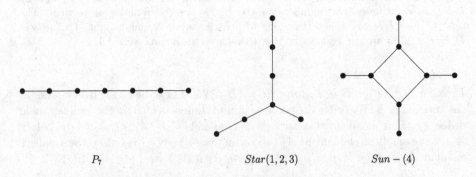

P_7 $Star(1, 2, 3)$ $Sun - (4)$

Fig. 3. Forbidden constructions for bipartite cographs.

An example of each type of graph is given in Fig. 4. In the left figure (trivially perfect) $I = \{I_1, I_2\}$ and $J = \{I_2, I_3\}$ are prime sets. On the right figure (bipartite cograph) prime sets are $R_1 = \{J_1, J_2, J_3\}, R_2 = \{J_1, J_2, J_4\}, R_3 = \{J_3, J_4, J_1\}, R_4 = \{J_3, J_4, J_2\}$ are prime sets.

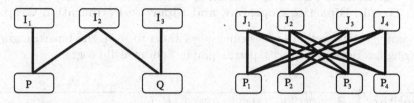

Fig. 4. Each bold line shows a complete connection, i.e. the induced sub-graph by $I_1 \cup P$ is a biclique (complete bipartite graph)

Definition 7 (Bipartite permutation graph). *A bipartite graph $H = (B, S, E)$ is called permutation graph (proper interval bipartite graph) if there exists an ordering $b_1, b_2, \ldots, b_p$ of the vertices in B, and an ordering $s_1, s_2, \ldots, s_q$ of the vertices in S such that if $s_i b_j$ and $s_{i'} b_{j'}$ are edges of H and $j' < j$ and $i < i'$ then $s_i b_{j'}, s_{i'} b_j \in E(H)$. This ordering is called* min-max *ordering [9].*

3 Polynomial Time Cases

We mention the following lemma which is correct for solving the general case of bipartite graphs. Therefore, identifying the prime subsets of B, would lead us to an optimal solution according to the following lemma.

Lemma 1. *Let $H = (B, S, E)$ be a bipartite graph without isolated vertices. Then, there is an optimal strategy to compute $bg(H)$ starting by a prime set.*

Proof. Let $u_1, u_2, \ldots, u_n$ be an optimal ordering that does not start with a prime set. Suppose $u_i, 2 \leq i \leq n$, is the first vertex in S. Let $M = \{u_1, u_2, \ldots, u_{i-1}\}$. Let $I \subseteq M$ be the smallest set with $N^*(I) \neq \emptyset$. Note that such I exists since all the adjacent vertices to u_i are among vertices in M. Observe that changing the processing order on vertices in M does not harm optimality. Therefore, we can modify the order, by placing the vertices in I first, without changing the budget. In addition, we can order $N^*(I)$ immediately after the vertices in I.

We continue this section by two lemmas about the positive minimal subsets which are used in designing our polynomial time algorithms, and they can also be used in designing a heuristic.

Lemma 2. *Let $H = (B, S, E)$ be a bipartite graph that can be processed with $bg(H) = K$. If H contains a positive minimal set I with $bg(I) \leq K$ then there is a strategy for H with budget K that begins by processing a positive minimal subset of H.*

Lemma 3. *Suppose that I^+ is a positive subset with $bg(H[I^+ \cup N^*(I^+)]) > K$ and I^- is a negative subset where $bg(H[I^- \cup N^*(I^-)]) \leq K$ and $I^+ \cap I^- \neq \emptyset$. If $bg(H[I^+ \cup I^- \cup N^*(I^+ \cup I^-)]) \leq K$ then $I^+ \cup I^-$ forms a positive subset.*

3.1   Polynomial Time Algorithm for Trivially Perfect Bipartite Graph, Bipartite Cographs, and Bipartite Permutation Graphs

We continue this section by designing algorithms to solve the bipartite graph ordering for Trivially perfect bipartite graphs and bipartite cograph.

Algorithm 1. BUDGETTRIVIALLYPERFECT (H, K)

1: **Input:** Trivially perfect bipartite graph $H = (B, S, E)$, integer K, and decomposition tree T for H

2: **Output:** "True" if we can process H with budget at most K, otherwise "False".

3: **if** $S = \emptyset$ and $K \geq 0$ **then return** True

4: **if** H is a bipartite clique and $\|B\| \leq K$ **then** process H by ordering vertices in B first and then ordering vertices in S after and **return** True

5: **if** H is constructed by join operation between $H_1 = (B_1, S_1)$ and $H_2 = (B_2, S_2)$ **then**

        ▷ $bg(H_1), bg(H_2)$ already computed and B_1 and S_2 induce a bipartite clique.

6:     **if** $bg(H_1) > K$ **then return** False;

7:     **else if** $bg(H_2) > K - \|B_1\| + \|S_1\|$ **then return** False;

8:     **else** first process H_1 then process H_2 and **return** True,

9: **if** H is constructed by union of H_1 and H_2 **then**

10:     **return** COMBINE(H_1, H_2, K)

Our algorithm to solve $bg(H)$ for trivially perfect bipartite graphs and bipartite cographs centers around constructing H as in Definition 5. We view this construction as a tree of operations that are performed to build up the final bipartite graph, and where the leaves of the tree of operations are *bicliques*. If H is not connected, then the root operation in the tree is a disjoint union, and each of its connected components is a trivially perfect bipartite graph (respectively, bipartite cograph). If H is connected, then the root operation is a join. It is easy to construct a decomposition tree for a given trivially perfect bipartite graph. We traverse the decomposition tree in a bottom-up manner. Algorithm 1, takes the input bipartite graph H with weights and a decomposition tree for $H = (B, S, E)$, together with integer K, and it returns yes together with an ordering if $bg(H) \leq K$. To guess the right value for $bg(H)$, we do a binary search between 1 and $\|B\|$.

At each node of the decomposition tree, we assume the optimal budgets for its children have been computed and stored for the graph associated with a particular tree node. If H is constructed by union operation, it requires a merging procedure, which is given in Algorithm 2 called **Combine**. **Combine** takes optimal solutions of two trivially perfect (respectively co-bipartite) graphs and return an optimal strategy for their union. We give the description of our algorithm and prove its correctness. Recall that we assume every vertex in B has at least one neighbor.

Algorithm 2. COMBINE (H_1, H_2, K)

1: **Input:** K and optimal strategies for $H_1 = (B_1, S_1, E_1), H_2 = (B_2, S_2, E_2)$
2: **Output:** "True" if we can process $H = H_1 \cup H_2$ with budget at most K,
        otherwise "False"
3: **if** H_1 is empty **then** process H_2 if $bg(H_2) \leq K$ and return True, else return False.
4: **if** H_2 is empty **then** process H_1 if $bg(H_1) \leq K$ and return True, else return False.
5: **while** $\exists$ positive minimal set I in $H_1 \cup H_2$ with $bg(I) \leq K$ **do**
6:     Process $c\ell(I)$ and $N^*(I)$.
7:     **if** $I \subset H_1$ **then** Set $H_1 \setminus (c\ell(I) \cup N^*(c\ell(N^*(I)))$
8:     **else** Set $H_2 \setminus (c\ell(I) \cup N^*(c\ell(N^*(I))))$
9:     Set $K \leftarrow K - \|c\ell(I)\| + \|N^*(c\ell(I))\|$.
10: Let J_1 be the first prime set in an optimal solution for H_1 and J_2 be the first prime set in optimal solution for H_2.
11: **if** $\|J_1\| > K$ OR $bg(H_2) > K - \|c\ell(J_1)\| + \|N^*(c\ell(J_1))\|$ **then**
12:     Process $c\ell(J_2)$ and $N^*(c\ell(J_2))$
13:     Call COMBINE$(H_1, H_2 \setminus (c\ell(J_2) \cup N^*(c\ell(J_2))), K - \|c\ell(J_2)\| + \|N^*(c\ell(J_2))\|)$.
14: **else**
15:     Process $c\ell(J_1)$ and $N^*(c\ell(J_1))$
16:     Call COMBINE $(H_1 \setminus (c\ell(J_1) \cup N^*(c\ell(J_1))), H_2, K - \|c\ell(J_1)\| + \|N^*(c\ell(J_1))\|)$.

Theorem 1. *For trivially perfect bipartite graph H with n vertices the* BUD-GETTRIVIALLYPERFECT *algorithm runs in $O(n^2)$ and correctly decides if H can be processed with budget K (Algorithm 1 and Algorithm 2).*

Our algorithm for computing $bg(H)$ when H is bipartite cograph is similar to Algorithm 1. The main difference is in the way we deal with bipartite cograph $H = (B, S, E)$ when it is constructed from two bipartite cographs $H_1 = (B_1, S_1, E_1)$ and $H_2 = (B_2, S_2, E_2)$ by join operation. Recall that in the join operation for bipartite cographs, $H[B_1 \cup S_2]$ and $H[B_2 \cup S_1]$ are bipartite cliques. Observe that, in this case, there are two possibilities for processing H:

- first process entire B_2 then solve the problem for H_1 with budget $K - \|B_2\|$, and at the end process S_2, or
- first process entire B_1 then solve the problem for H_2 with budget $K - \|B_1\|$, and at the end process S_1.

For the case when H is constructed from H_1 and H_2 by union operation, we call COMBINE Algorithm 2. The proof of correctness is almost identical to the proof of Theorem 1.

Theorem 2. *$bg(H)$ can be found in polynomial time for bipartite cograph H.*

Let $H = (B, S, , E)$ be a bipartite permutation graph. Notice that by definition 7, the neighborhood of each vertex in S and B form an interval. Note that the class of circle bipartite graphs $G = (X, Y)$, for which obtaining the optimal budget is NP-complete, contains the class of bipartite permutation graphs. Let $B[i, j]$ denote the interval of vertices $b_i, b_{i+1}, \ldots, b_j$ in B. We compute the optimal budget for every $B[i, j]$. In order to compute $bg(H[B[i, j] \cup N^*(B[i, j])])$ we assume that the optimal strategy starts with some sub-interval J of $B[i, j]$ and it processes $c\ell(J)$, which is indeed an interval. This is because of the property of the min-max ordering. We are left with two disjoint instances, B_1 and B_2 possibility with some vertices in S with neighbors in $B_1 \cup c\ell(J)$ or in $B_2 \cup c\ell(J)$.

We then argue how to combine the optimal solutions of B_1 and B_2 and obtain an optimal strategy for $B[i, j] \setminus c\ell(J)$. We must consider every possible prime interval J in the range $B[i, j]$ and take the minimum budget. For details, see Algorithm 3.

Algorithm 3. BUDGETPERMUTATION (H, K)

1: **Input:** Bipartite permutation graph $G = (B, S, E)$ with ordering $<$ on vertices in
    B, S i.e. $b_1 < b_2 < \cdots < b_n, s_1 < s_2 < \cdots < s_m$ which is a min-max ordering
2: **Output:** Computing the budget for G and optimal strategy
3: **for** $i = 1$ to $i = n - 1$ **do**
4:     **for** $j = 1$ to $j \leq n - i$ **do**
5:         Let $H' = (B[j, j + i], N^*(B[j, j + i]))$
6:         Let K' be the minimum number s.t. Optimal-Budget(H', K') is True
7:         Set $bg(H') = K'$ and process H' be according to Optimal-Budget(H', K')
8:         Let S_r be the set of vertices with neighbors in both $B[i, i+j], B[i+j+1, n])$
9:         Set $H_r = H' \cup S_r$
10:       Let K' be the minimum number s.t. Optimal-Budget(H_r, K') is True
11:       Set $bg(H_r) = K'$ and process H_r be according to Optimal-Budget(H_r, K')
12:       Let S_l be the set of vertices with neighbors in both $B[1, i - 1], B[i, j + i]$
13:       Set $H_l = H' \cup S_l$
14:       Let K' be the minimum number s.t. Optimal-Budget(H_l, K') is True
15:       Set $bg(H_l) = K'$ and process H_l be according to Optimal-Budget(H_l, K')

Theorem 3. BIPARTITE ORDERING PROBLEM *on a bipartite permutation graph with n vertices is solved in time* $O(n^6 \log \|B\|)$.

We heavily use the min-max ordering property to find $bg(H)$ when $H = (B, S, E)$ is a bipartite permutation graph. The next natural superclass of bipartite permutation graphs is the class of convex bipartite graphs. A bipartite graph H is convex if the vertices are ordered in B so that the neighborhood of each vertex in S is an interval.

Problem 1. Let H be a convex bipartite graph. Is it polynomial to decide the optimal value of $bg(H)$?

1: **function** OPTIMAL-BUDGET($H = (B, S), K$)
2:     **Input:** Bipartite permutation graph $H = (B, S, E)$ with ordering $<$ on vertices in B, S
3:     **Output:** Process H with budget at most K, otherwise "False"
4:     **if** $S = \emptyset$ and $K \geq 0$ OR H is a bipartite clique and $\|B\| \leq K$ **then** process H **return** True
5:     **if** for every prime $I \subseteq B$, $\|I\| > K$ **then return** False
6:     **while** $\exists$ positive prime set I with $bg(I) \leq K$ **do**
7:         Process $cl(I) \cup N^*(cl(I))$ and set $H \leftarrow H \setminus cl(I) \cup N^*(cl(I))$.
8:         Set $K \leftarrow K - \|cl(I)\| + \|N^*(cl(I))\|$
9:     **for** every prime interval I of H **do**
10:         Set $B_1 = \{b_1, b_2, \ldots, b_i\}$ and $B_2 = \{b_j, \ldots, b_n\}$ where b_{i+1} is the first vertex of $cl(I)$ and b_{j-1} is the last vertex of $cl(I)$ in the ordering $<$
11:         Let S_i, $i = 1, 2$ be the set of vertices in S that have neighbors in B_i
                                                                    ▷ $S_1 \cap S_2 = \emptyset$
12:         Let $H_1 = H[B_1 \cup S_1]$ and $H_2 = H[B_2 \cup S_2]$.
13:         Set Flag=Combine($H_1, H_2, K - \|cl(I)\| + \|N^*(cl(I))\|$)
14:         **if** Flag=True **then**
15:             **return** Process of $cl(I)$ together with process of $H \setminus (cl(I) \cup N^*(cl(I)))$ by Combine Algorithm.

When the instance $T = (B, S, E)$ is a tree with arbitrary weights. It is easy to see that every prime and positive set form a sub-tree in T; hence, we can find all the positive sets in polynomial time. Moreover, once a prime set I and $N^*(I)$ is removed from T, the remaining becomes a forest. Suppose we have $bg(T_1), bg(T_2), \ldots, bg(T_r)$, where $T_1, T_2, \ldots, T_r$ are disjoint sub-trees in $T \setminus (I \cup N^*(I))$. Now we can use the COMBINE Algorithm 2 to combine the optimal strategy of T_1 and T_2 and obtain an optimal strategy for $T_1 \cup T_2$, and then the optimal strategy for $T_1 \cup T_2 \cup T_3$, and eventually an optimal strategy for T and $bg(T)$. Therefore, we have the following proposition.

Proposition 3. *Let $T = (B, S, E)$ together with arbitrary weights be an instance of the bipartite graph ordering problem. If T is a tree, then $bg(T)$ can be computed in polynomial time.*

4   Linear Program Formulation of the Problem

Let $H = (B, S, E)$ together with the weight be an instance of the bipartite ordering problem. Suppose there is an ordering $<$ on the vertices of a given bipartite graph $H = (B, S, E)$ in which for every edge uv of H ($u \in B$ and $v \in S$) u is before v in $<$. Then we obtain a strategy for solving the budget minimization on H and decide whether the budget would be K or smaller. With a given value K for $bg(H)$, we translate this ordering process into a linear program as follows. For every pair of vertices $u, v \in B \cup S$, we define variable $0 \leq X_{u,v} \leq 1$. We interpret $X_{u,v} = 1$ (in integral solution) as placing u before

v in a total ordering. The linear program defines as follows. Minimize K such that:

$$\forall u \in B, v \in S, \ \text{with } uv \in E, \quad X_{u,v} = 1$$
$$\forall u, v \in B \cup S, u \neq v, \quad X_{u,v} + X_{v,u} = 1$$
$$\forall u, v, w \in B \cup S, \ u \neq v, \quad X_{u,v} + X_{v,w} + X_{w,u} \geq 1$$
$$\forall y \in B \cup S, \quad K + \sum_{u \in B} p_u X_{u,y} + \sum_{v \in S} p_v X_{v,y} \geq 0$$
$$K \geq \min_{v \in S}\{ \sum_{u \in N(v)} |p_u| \}$$
$$\forall u, v \in S \ \ if \ \ N(u) \subset N(v) \ \ then \quad X_{u,v} = 1$$
$$\forall u, v \in B, \ \ if \ \ N(u) \subset N(v) \ \ then \quad X_{v,u} = 1$$
$$\forall u, v, w \in B \cup S \ \ with \ \ w \notin \{u, v\} \ \ if \ \ N(u) = N(v) \ \ then \ \ X_{u,w} = X_{v,w}$$

There is a one-to-one correspondence between the optimal solutions of the $bg(H)$ and integral solutions of the above LP. The following table shows the result of our experiment. We have run the LP on random graphs and integer LP on those samples (each having 50 vertices) and taken the maximum ratio of the integral LP by optimal fractional LP.

Graph type	Circle bipartite graph	Circle bipartite graph with weight	General bipartite graph	General bipartite graph with weight
Max Ratio	1.818181818	2.813949433	2.179503945	4.311947725
Number of samples	1943	677	1687	932

We pose the following problem.

Problem 2. Does the bipartite ordering problem admit a constant factor approximation?

Conclusion and Future Works. The bipartite ordering problem has several applications, there are several open problems which leaves the door open for future research.

References

1. Ambühl, C., Mastrolilli, M., Mutsanas, N., Svensson, O.: On the approximability of single-machine scheduling with precedence constraints. Math. Oper. Res., 653–669 (2011)

2. Ambuhl, C., Mastrolilli, M., Svensson, O.: Inapproximability results for sparsest cut, optimal linear arrangement, and precedence constrained scheduling. In: 48th Annual IEEE Symposium on Foundations of Computer Science (FOCS 2007), pp. 329–337. IEEE (2007)

3. Berger, A., Grigoriev, A., Heggernes, P., van der Zwaan, R.: Scheduling unit-length jobs with precedence constraints of small height. Oper. Res. Lett. **42**(2), 166–172 (2014)

4. Chitnis, R., Egri, L., Marx, D.: List H-coloring a graph by removing few vertices. In: Bodlaender, H.L., Italiano, G.F. (eds.) ESA 2013. LNCS, vol. 8125, pp. 313–324. Springer, Heidelberg (2013). https://doi.org/10.1007/978-3-642-40450-4_27

5. Egri, L., Krokhin, A., Larose, B., Tesson, P.: The complexity of the list homomorphism problem for graphs. In: Theory of Computing Systems, pp. 143–178 (2012)

6. Flamm, C., Hofacker, I.L., Maurer-Stroh, S., Stadler, P.F., Zehl, M.: Design of multistable rna molecules. Rna, pp. 254–265 (2001)

7. Geis, M., et al.: Folding kinetics of large rnas. J. Mol. Biol., 160–173 (2008)

8. Giakoumakis, V., Vanherpe, J.-M.: Bi-complement reducible graphs. Advances in Applied Mathematics, pp. 389–402 (1997)

9. Gutin, G., Hell, P., Rafiey, A., Yeo, A.: A dichotomy for minimum cost graph homomorphisms. Eur. J. Combinatorics, pp. 900–911 (2008)

10. Johannes, B.: On the complexity of scheduling unit-time jobs with or-precedence constraints. Oper. Res. Lett. **33**(6), 587–596 (2005)

11. Maňuch, J., Thachuk, C., Stacho, L., Condon, A.: NP-completeness of the direct energy barrier problem without pseudoknots. In: Deaton, R., Suyama, A. (eds.) DNA 2009. LNCS, vol. 5877, pp. 106–115. Springer, Heidelberg (2009). https://doi.org/10.1007/978-3-642-10604-0_11

12. Mastrolilli, M., Stamoulis, G.: Restricted max-min fair allocations with inclusion-free intervals. In: Gudmundsson, J., Mestre, J., Viglas, T. (eds.) COCOON 2012. LNCS, vol. 7434, pp. 98–108. Springer, Heidelberg (2012). https://doi.org/10.1007/978-3-642-32241-9_9

13. Möhring, R.H., Skutella, M., Stork, F.: Scheduling with and/or precedence constraints. SIAM J. Comput., 393–415 (2004)

14. Morgan, S.R., Higgs, P.G.: Barrier heights between ground states in a model of rna secondary structure. J. Phys. A Math. General, 3153 (1998)

15. Muratore, G., Schwarz, U.M., Woeginger, G.J.: Parallel machine scheduling with nested job assignment restrictions. Oper. Res. Lett., 47–50 (2010)

16. Spinrad, J., Brandstädt, A., Stewart, L.: Bipartite permutation graphs. Discrete Appl. Math., 279–292 (1987)

17. Takizawa, H., Iwakiri, J., Terai, G., Asai, K.: Finding the direct optimal rna barrier energy and improving pathways with an arbitrary energy model. Bioinformatics **36**, 227–235 (2020)

18. Thachuk, C., Maňuch, J., Rafiey, A., Mathieson, L.-A., Stacho, L., Condon, A.: An algorithm for the energy barrier problem without pseudoknots and temporary arcs. In: Biocomputing 2010, pages 108–119. World Scientific (2010)

19. Woeginger, G.J.: On the approximability of average completion time scheduling under precedence constraints. Discrete Appl. Math., 237–252 (2003)

Forwards- and Backwards-Reachability for Cooperating Multi-pushdown Systems

Chris Köcher[1]([✉])[iD] and Dietrich Kuske[2]

[1] Max Planck Institute for Software Systems, Kaiserslautern, Germany
ckoecher@mpi-sws.org
[2] Technische Universität Ilmenau, Ilmenau, Germany
dietrich.kuske@tu-ilmenau.de

Abstract. A cooperating multi-pushdown system consists of a tuple of pushdown systems that can delegate the execution of recursive procedures to sub-tuples; control returns to the calling tuple once all sub-tuples finished their task. This allows the concurrent execution since disjoint sub-tuples can perform their task independently. Because of the concrete form of recursive descent into sub-tuples, the content of the multi-pushdown does not form an arbitrary tuple of words, but can be understood as a Mazurkiewicz trace.

For such systems, we prove that the backwards reachability relation efficiently preserves recognizability, generalizing a result and proof technique by Bouajjani et al. for single-pushdown systems. While this preservation does not hold for the forwards reachability relation, we can show that it efficiently preserves the rationality of a set of configurations; the proof of this latter result is inspired by the work by Finkel et al. It follows that the reachability relation is decidable for cooperating multi-pushdown systems in polynomial time and the same holds, e.g., for safety and liveness properties given by recognizable sets of configurations.

Keywords: Reachability · Formal Verification · Pushdown Automaton · Distributed System

1 Introduction

In this paper, we introduce the model of cooperating multi-pushdown systems[1] and study the reachability relation for such systems. To explain the idea of a cooperating multi-pushdown system, we first look at well-studied pushdown systems. They model the behavior of a sequential recursive program and possess a control state as well as a pushdown. The top symbol of the pushdown stores the execution context, e.g., parameters and local variables, the state can be used

---

[1] A more descriptive name would be "cooperating systems of pushdown systems", but we refrain from using this term.

This work was done while Chris Köcher was affiliated with the Technische Universität Ilmenau.

© The Author(s), under exclusive license to Springer Nature Switzerland AG 2023
H. Fernau and K. Jansen (Eds.): FCT 2023, LNCS 14292, pp. 318–332, 2023.
https://doi.org/10.1007/978-3-031-43587-4_23

to return values from a subroutine to the calling routine. Such a system can, depending on the state and the top symbol, do three types of moves: it can call a subroutine (i.e., change state and top symbol and add a new symbol on top of the pushdown), it can do an internal action (i.e., change state and top symbol), and it can return from a subroutine (i.e., delete the top symbol and store the necessary information into the state). This leads to the unifying definition of a transition that, depending on state and top symbol, changes state and replaces the top symbol by a (possibly empty) word.

A cooperating multi-pushdown system consists of a finite family of pushdown systems (indexed by a set P). Cooperation is realized by the formation of temporary coalitions that perform a possibly recursive subroutine in a joint manner. Suppose the system is in a configuration where $C \subseteq P$ forms one of the coalitions. The execution context of the joint task is distributed between the top symbols of the pushdowns from the coalition and can only be changed in all these components at once. As above, there are three types of moves depending on the top symbols and the states of the systems from the coalition. First, a (further) subroutine can be called on a sub-coalition $C_0 \subseteq C$. Even more, several subroutines can be called in parallel on disjoint sub-coalitions of C. This is modeled as a change of states and top symbols of C and addition of some further symbols on the pushdowns from subsets of C. Internal actions of the coalition C can change the (common) top symbol as well as the states of the systems that form the coalition C. Similarly, a return move deletes the common top symbol and changes the states of the systems from C, in this moment, the coalition C is dissolved and the systems from C are free to be assigned to new coalitions and tasks by the calling routine. Since several, mutually disjoint coalitions can exist and operate at any particular moment, the cooperating multi-pushdown system is a non-sequential model.

Since a cooperating multi-pushdown system consists of several pushdown systems, a configuration consists of a tuple of local states and a tuple of pushdown contents; the current division into coalitions is modeled by the top symbols of the pushdowns: any component forms a coalition with all components that have the same top symbol a on their stack. Since all these occurrences of the letter a can only change at once, there is some dependency in the tuple of pushdown contents of a configuration. It turns out to be convenient and fruitful to understand such a "consistent" tuple of pushdown contents as a Mazurkiewicz trace. Since the set of all Mazurkiewicz traces forms a monoid, we can define recognizable and rational sets of traces and therefore of configurations: Both these classes of sets of traces enjoy finite representations (by asynchronous automata [19] and NFAs, resp.) that allow to decide membership, any recognizable set is rational but not *vice versa*, any singleton is both, recognizable and rational, and inclusion of a rational set (and therefore in particular of a recognizable set) in a recognizable set is efficiently decidable (but not *vice versa*).

As our main results, we obtain that (1) backwards reachability efficiently preserves the *recognizability* of sets of configurations while (2) forwards reachability efficiently preserves the *rationality*.[1] We also show that asynchronous multi-pushdown systems (a slight generalization of our model) can model 2-pushdown systems and therefore have an undecidable reachability relation.

From our positive results, we infer that the reachability relation as well as certain safety and liveness properties are decidable in polynomial time. Furthermore, the first result implies that EF-model checking is decidable, although one only obtains a non-elementary complexity bound.

Related Work. Corresponding results for pushdown systems can be found in [6,11] where rationality and recognizability coincide [14]; Finkel et al. gave a simple algorithm proving that the forwards reachability relation preserves the recognizability while this preservation under the backwards reachability relation was shown by Bouajjani et al. Our proof of (1) generalizes the one by Bouajjani et al. while the work by Finkel et al. inspired our proof of (2).

Other forms of multi-pushdown systems have been considered by different groups of authors, e.g., [1–5,7,8,12,15,16,18]. These alternative models may contain a central control or, similarly to our cooperating systems, local control states. The models can have a fixed number of processes and pushdowns or they are allowed to spawn or terminate other processes. Local processes can differ in their communication mechanism, e.g., by rendevouz or FIFO-channels. The decidability results concern logical formulas of some form or bounded model checking problems.

Mazurkiewicz traces as a form of storage mechanism have been considered by Hutagalung et al. in [13], where multi-buffer systems were studied.

2   Preliminaries

For $R \subseteq S^2$ and $s, t \in S$, let $sR := \{t \in S \mid s\,R\,t\}$ and $Rt := \{s \in S \mid s\,R\,t\}$.

For $n \in \mathbb{N}$, $[n] = \{1, \ldots, n\}$. Let $(S_i)_{i \in [n]}$ be a tuple of sets, $I, J \subseteq [n]$ be two disjoint sets, and $\overline{s} = (s_i)_{i \in [n]}$ and $\overline{t}$ be tuples from $\prod_{i=1}^{n} S_i$. We write $\overline{s}\!\restriction_I = (s_i)_{i \in I} \in \prod_{i \in I} S_i$ for the restriction of $\overline{s}$ to the components in I and $(\overline{s}\!\restriction_I, \overline{t}\!\restriction_J)$ for the joint tuple $\overline{r} \in \prod_{i \in I \cup J} S_i$ with $\overline{r}\!\restriction_I = \overline{s}\!\restriction_I$ and $\overline{r}\!\restriction_J = \overline{t}\!\restriction_J$.

For a word $w \in A^*$, we write $\mathrm{Alph}(w)$ for the set of letters occurring in w.

A *non-deterministic finite automaton* or *NFA* is a tuple $\mathfrak{A} = (Q, A, I, \delta, F)$ where Q is a finite set of *states*, A is an alphabet, $I, F \subseteq Q$ are sets of *initial* and *accepting* states, respectively, and $\delta \subseteq Q \times A \times Q$ is a set of *transitions*; its size $\|\mathfrak{A}\|$ is $|Q| + |A|$. We write $Q_1 \xrightarrow{w}_{\mathfrak{A}} Q_2$ if there is a run from some state $p \in Q_1$ to some state $q \in Q_2$ labeled with w in $\mathfrak{A}$; $\{p\} \xrightarrow{w}_{\mathfrak{A}} \{q\}$ is abbreviated $p \xrightarrow{w}_{\mathfrak{A}} q$. The *language accepted by* $\mathfrak{A}$ is $L(\mathfrak{A}) := \{w \in A^* \mid I \xrightarrow{w}_{\mathfrak{A}} F\}$.

We will model the contents of our multi-pushdown systems with the help of Mazurkiewicz traces; for a comprehensive survey of this topic we refer to [10]. Traces were first studied in [9] as "heaps of pieces" and later introduced

---

[1] The full version of this paper also shows that backwards (forwards) reachability does not preserve rationality (recognizability, resp.).

into computer science by Mazurkiewicz to model the behavior of a distributed system [17]. The fundamental idea is that any letter $a \in A$ is assigned a set of *locations* or *processes* $a\mathcal{L} \subseteq P$ it operates on (where P is some set):

A *distributed alphabet* is a triple $\mathcal{D} = (A, P, \mathcal{L})$ where A and P are two alphabets of *letters* and *processes*, respectively, and $\mathcal{L} \subseteq A \times P$ associates letters to processes such that $a\mathcal{L} \neq \emptyset$ for each $a \in A$. In this paper, $\mathcal{D}$ will always denote a distributed alphabet $(A, P, \mathcal{L})$.

For a word $w \in A^*$ we denote the set of processes associated with w by $w\mathcal{L} := \bigcup_{a \in \mathrm{Alph}(w)} a\mathcal{L} \subseteq P$. In particular, we set $\varepsilon\mathcal{L} := \emptyset$. By $\pi_i : A^* \to A_i^*$ we denote the *projection* onto $A_i := \mathcal{L}i$ (the alphabet of all letters associated to process i), i.e., the monoid morphism with $\pi_i(a) = a$ for $a \in A_i$ and $\pi_i(b) = \varepsilon$ for $b \in A \setminus A_i$.

Since $\pi_i : A^* \to A_i^*$ is a monoid morphism for all $i \in [n]$, also the mapping

$$\overline{\pi} : A^* \to \prod_{i \in P} A_i^* : w \mapsto (\pi_i(w))_{i \in P}$$

is a monoid morphism. For $w \in A^*$, we call $\overline{\pi}(w)$ the *(Mazurkiewicz) trace* induced by w. The *trace monoid* is the submonoid of $\prod_{i \in P} A_i^*$ with universe $\mathbb{M}(\mathcal{D}) = \{\overline{\pi}(w) \mid w \in A^*\}$; its elements are *traces* and its subsets are *trace languages*.

We call two words $v, w \in A^*$ with $v\mathcal{L} \cap w\mathcal{L} = \emptyset$ *independent* and denote this fact by $v \parallel w$. We can see that $v \parallel w$ implies $\overline{\pi}(vw) = \overline{\pi}(wv)$.

Let $\mathfrak{A} = (Q, A, I, \delta, F)$ be an NFA. The *accepted trace language* of $\mathfrak{A}$ is $T(\mathfrak{A}) := \{\overline{\pi}(w) \mid I \xrightarrow{w}_{\mathfrak{A}} F\}$. In other words, $T(\mathfrak{A})$ is the image of $L(\mathfrak{A})$ under the morphism $\overline{\pi}$. A trace language $L \subseteq \mathbb{M}(\mathcal{D})$ is called *rational* if there is an NFA $\mathfrak{A}$ with $T(\mathfrak{A}) = L$, i.e., iff L is the image of some regular language in A^* under the morphism $\overline{\pi}$. A trace language L is *recognizable* iff its preimage under the morphism $\overline{\pi}$, i.e. $\{w \in A^* \mid \overline{\pi}(w) \in L\}$, is regular. Clearly, any recognizable trace language is rational. The converse implication holds only in case any two letters are dependent.

A finite automaton that reads letters of a distributed alphabet should consist of components for all $i \in P$ such that any letter $a \in A$ acts only on the components from $a\mathcal{L}$. This idea leads to the following definition of an asynchronous automaton. But first, we fix a particular notation: For a tuple $(Q_i)_{i \in P}$ of finite sets Q_i, we write $\mathbf{Q}$ for the direct product $\prod_{i \in P} Q_i$.

Definition 2.1. *An* asynchronous automaton *or* AA *is an NFA* $\mathfrak{A} = (Q, A, I, \delta, F)$ *where* $Q = \mathbf{Q}$ *is the product of finite sets* Q_i *of local states and where, for every* $(\overline{p}, a, \overline{q}) \in \delta$ *and* $\overline{r} \in \prod_{i \in P \setminus a\mathcal{L}} Q_i$, *we have*

$$(i)\ \overline{p}\!\restriction_{P \setminus a\mathcal{L}} = \overline{q}\!\restriction_{P \setminus a\mathcal{L}}\ and\ (ii)\ ((\overline{p}\!\restriction_{a\mathcal{L}}, \overline{r}), a, (\overline{q}\!\restriction_{a\mathcal{L}}, \overline{r})) \in \delta.$$

Here, (i) ensures that any a-transition of $\mathfrak{A}$ only modifies components from $a\mathcal{L}$ while the other components are left untouched, and (ii) guarantees that a-transitions are insensitive to the local states of the components in $P \setminus a\mathcal{L}$.

Every asynchronous automaton accepts a recognizable trace language. Conversely, every recognizable trace language $L \subseteq \mathbb{M}(\mathcal{D})$ is accepted by some deterministic asynchronous automaton [19].

3   Introducing Cooperating Multi-pushdown Systems

An AA consists of several NFAs that synchronize by joint actions. In a similar manner, we will now consider several pushdown systems.

Recall that a pushdown system (or PDS) consists of a control unit (that can be in any of finitely many control states) and a pushdown (that can hold words over the pushdown alphabet A). Its transitions read the top letter a from the pushdown, write a word w onto it, and change the control state. In our model, we have a pushdown system $\mathfrak{P}_i$ for every $i \in P$ whose pushdown alphabet is A_i. These systems synchronize by the letters read and written onto their pushdown.

Definition 3.1. *An* asynchronous multi-pushdown system *or* aPDS *is a tuple* $\mathfrak{P} = (Q, \Delta)$ *where* $Q = \mathbf{Q}$ *holds for some finite sets* Q_i *of local states and* $\Delta \subseteq Q \times A \times A^* \times Q$ *is a finite set of transitions such that, for each transition* $(\overline{p}, a, w, \overline{q}) \in \Delta$ *and* $\overline{r} \in \prod_{i \in P \setminus aw \, \mathcal{L}} Q_i$, *we have*

$$(i) \; \overline{p}\upharpoonright_{P \setminus aw \, \mathcal{L}} = \overline{q}\upharpoonright_{P \setminus aw \, \mathcal{L}} \; and \; (ii) \; ((\overline{p}\upharpoonright_{aw \, \mathcal{L}}, \overline{r}), a, w, (q\upharpoonright_{aw \, \mathcal{L}}, \overline{r})) \in \Delta \, .$$

Its size $\|\mathfrak{P}\|$ *is* $|Q| + k \cdot |\Delta|$ *where* $k - 1$ *is the maximal length of a word written by any of the transitions (i.e.,* $\Delta \subseteq Q \times A \times A^{<k} \times Q$).

The set of configurations $\mathrm{Conf}_{\mathfrak{P}}$ *of* $\mathfrak{P}$ *equals* $Q \times \mathbb{M}(\mathcal{D})$. *For two configurations* $(\overline{p}, \pi(u)), (\overline{q}, \pi(v)) \in \mathrm{Conf}_{\mathfrak{P}}$ *we set* $(\overline{p}, \pi(u)) \vdash (\overline{q}, \pi(v))$ *if there is a transition* $(\overline{p}, a, w, \overline{q}) \in \Delta$ *and a word* $x \in A^*$ *with* $\pi(u) = \pi(ax)$ *and* $\pi(v) = \pi(wx)$. *The reflexive and transitive closure of* $\vdash$ *is the reachability relation* $\vdash^*$.

Let C *and* D *be sets of configurations.*

- *We write* $C \vdash^* D$ *if there are* $c \in C$ *and* $d \in D$ *with* $c \vdash^* d$.
- *The set* C *is* rational *(recognizable, resp.) if, for all* $\overline{q} \in Q$, *the trace language* $C_{\overline{q}} := \{\pi(u) \mid (\overline{q}, \pi(u)) \in C\}$ *is rational (recognizable, resp.).*
- $\mathrm{pre}_{\mathfrak{P}}(C) := \{c \in \mathrm{Conf}_{\mathfrak{P}} \mid c \vdash C\}$ *and* $\mathrm{post}_{\mathfrak{P}}(C) := \{d \in \mathrm{Conf}_{\mathfrak{P}} \mid C \vdash d\}$ *are the sets of predecessors/successors of configurations from* C, *and*

$$\mathrm{pre}_{\mathfrak{P}}^*(C) := \bigcup_{k \in \mathbb{N}} \mathrm{pre}_{\mathfrak{P}}^k(C) \; and \; \mathrm{post}_{\mathfrak{P}}^*(C) := \bigcup_{k \in \mathbb{N}} \mathrm{post}_{\mathfrak{P}}^k(C)$$

are the sets of configurations backwards *(forwards, resp.) reachable from some configuration in* C.

The reachability relation for configurations of asynchronous multi-pushdown systems is, in general, undecidable:

Theorem 3.2. *There exists an aPDS with undecidable reachability relation* $\vdash^*$.

Proof. We start with a classical 2-pushdown system $\mathfrak{P}$ with an undecidable reachability relation (its set of states is Q and the two pushdowns use disjoint alphabets A_1 and A_2). Let $A = A_1 \cup A_2 \cup \{\top\}$ and $P = [2]$. We consider the distributed alphabet $\mathcal{D}$ with $a \mathcal{L} = \{i\}$ for $a \in A_i$ and $\top \mathcal{L} = \{1, 2\}$.

We simulate $\mathfrak{P}$ by an aPDS $\mathfrak{P}'$ over $\mathcal{D}$ as follows. The first process of $\mathfrak{P}'$ stores the state of the simulated system $\mathfrak{P}$ together with a letter from A_1 or

ε, i.e., $Q_1 = Q(A_1 \cup \{\varepsilon\})$, the second process can store a letter from A_2 or the empty word, i.e., $Q_2 = A_2 \cup \{\varepsilon\}$.

A transition $(p, (a, b), (u, v), q)$ of $\mathfrak{P}$ (that replaces a and b by u and v on the two pushdowns) is simulated by three transitions of the aPDS: $((p\varepsilon, .), a, \varepsilon, (pa, .))$ reads a from the first pushdown and stores it in the first local state; $((., \varepsilon), b, \top, (., b))$ reads b from the second pushdown, stores it in the second local state, and puts $\top$ onto both pushdowns; finally, $((pa, b), \top, uv, (q\varepsilon, \varepsilon))$ replaces $\top$ by uv (i.e., $\pi_1(uv) = u$ is written onto the first pushdown and $\pi_2(uv) = v$ onto the second). $\qquad\square$

To obtain a model with a decidable reachability relation, we therefore have to restrict aPDS.[2] To this aim, we require that any transition can only write onto pushdowns it reads from.

Definition 3.3. *A* cooperating multi-pushdown system *or* cPDS *is an aPDS* $\mathfrak{P} = (\mathbf{Q}, \Delta)$ *with* $w\,\mathcal{L} \subseteq a\,\mathcal{L}$ *for each transition* $(\overline{p}, a, w, \overline{q}) \in \Delta$.

Example 3.4. Suppose $\mathcal{D} = (\{a, b, c\}, \{1, 2\}, \{(a, 1), (a, 2), (b, 1), (c, 2)\})$. We consider the cPDS $\mathfrak{P}$ from Fig. 1 where edges from global state $\overline{p}$ to global state $\overline{q}$ labeled $a \mid w$ visualize global transitions $(\overline{p}, a, w, \overline{q})$. The set of global states of $\mathfrak{P}$ is the product $\{p_1, q_1\} \times \{p_2, q_2\}$. Additionally, the transitions reading b and c only depend on process 1 and 2, resp. Since $b\,\mathcal{L}, c\,\mathcal{L} \subseteq a\,\mathcal{L}$, any global transition $(\overline{p}, x, w, \overline{q})$ satisfies $w\,\mathcal{L} \subseteq x\,\mathcal{L}$, i.e., $\mathfrak{P}$ is, indeed, a cPDS.

The following sequence is a run of $\mathfrak{P}$ from $((p_1, p_2), \overline{\pi}(ac))$ to $((q_1, q_2), \overline{\pi}(bb))$:

$$((p_1, p_2), \overline{\pi}(ac)) \vdash ((q_1, p_2), \overline{\pi}(abc)) \vdash ((q_1, p_2), \overline{\pi}(abbc))$$
$$\vdash ((q_1, q_2), \overline{\pi}(bbc)) \vdash ((q_1, q_2), \overline{\pi}(bb)).$$

In order to decide the reachability relation, we will compute, from a set of configurations C, the set $\mathrm{pre}^*_{\mathfrak{P}}(C)$. To represent possibly infinite sets of configurations, we use $\mathfrak{P}$-asynchronous automata (defined next).

Definition 3.5. *Let* $\mathfrak{P} = (\mathbf{Q}, \Delta)$ *be a cPDS. A* $\mathfrak{P}$-asynchronous automaton *or* $\mathfrak{P}$-AA *is an AA* $\mathfrak{A} = (\mathbf{S}, A, \emptyset, \delta, F)$ *such that* $Q_i \subseteq S_i$ *for all* $i \in P$.

The $\mathfrak{P}$-AA $\mathfrak{A}$ *accepts the following set* $C(\mathfrak{A})$ *of configurations of* $\mathfrak{P}$:

$$\{(\overline{q}, \overline{\pi}(w)) \in \mathrm{Conf}_{\mathfrak{P}} \mid \overline{q} \in \mathbf{Q}, \overline{q} \xrightarrow{w}_{\mathfrak{A}} F\}.$$

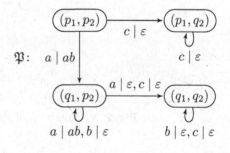

Fig. 1. The cPDS $\mathfrak{P}$ from Example 3.4.

---

[2] The proof of Theorem 3.2 shows that requiring aw to be connected for any transition $(\overline{p}, a, w, \overline{q})$ does not yield decidability.

By the very definition, any set $C(\mathfrak{A})$ is a recognizable set of configurations. Conversely, suppose $C \subseteq \mathrm{Conf}_{\mathfrak{P}}$ is recognizable such that, by definition, all the languages $C_{\overline{q}}$ are recognizable. Then we can represent each of the languages $C_{\overline{q}}$ by an AA $\mathfrak{A}_{\overline{q}}$. Since $\overline{q}$ is a P-tuple, we can assume, without loss of generality, that $\overline{q}$ is the only initial state of the AA $\mathfrak{A}_{\overline{q}}$. Following Bouajjani et al. [6], we can further assume that all these AAs differ in their initial state, only. Thus, we obtain the following result.

Observation 3.6. *Let* $\mathfrak{P} = (\mathbf{Q}, \Delta)$ *be a cPDS. A set of configurations* $C \subseteq \mathrm{Conf}_{\mathfrak{P}}$ *is recognizable if, and only if, there is a* $\mathfrak{P}$-*AA* $\mathfrak{A}$ *with* $C(\mathfrak{A}) = C$.

4 Computing the Backwards Reachable Configurations

In this section we want to compute the backwards reachable configurations in a cPDS $\mathfrak{P}$. The main result of this section states that the mapping $\mathrm{pre}^*_{\mathfrak{P}}$ effectively preserves the recognizability of sets of configurations.

Theorem 4.1. *Let* $\mathfrak{P} = (\mathbf{Q}, \Delta)$ *be a cPDS and* $C \subseteq \mathrm{Conf}_{\mathfrak{P}}$ *be a recognizable set of configurations. Then the set* $\mathrm{pre}^*_{\mathfrak{P}}(C)$ *is recognizable.*

Even more, from $\mathcal{D}$, $\mathfrak{P}$, *and a* $\mathfrak{P}$-*AA* $\mathfrak{A}^{(0)}$, *one can construct in polynomial time a* $\mathfrak{P}$-*AA* $\mathfrak{A}$ *that accepts the set* $\mathrm{pre}^*_{\mathfrak{P}}(C(\mathfrak{A}^{(0)}))$.

The rest of this section is devoted to the proof of this result.

Adapting ideas by Bouajjani et al. [6] from NFAs to AA, we construct a $\mathfrak{P}$-AA $\mathfrak{A}$ that accepts the set $\mathrm{pre}^*_{\mathfrak{P}}(C(\mathfrak{A}^{(0)}))$ of configurations backwards reachable from $C(\mathfrak{A}^{(0)})$. To this aim, we will inductively add new transitions to the $\mathfrak{P}$-AA $\mathfrak{A}^{(0)} = (\mathbf{S}, A, \emptyset, \delta^{(0)}, F)$, but leave the sets of states, initial states, and accepting states unchanged. We can assume (and this assumption is crucial for the correctness of the construction) that the automaton cannot enter a local state from the cPDS $\mathfrak{P}$, i.e., we have $q_i \in S_i \setminus Q_i$ for any $(\overline{p}, a, \overline{q}) \in \Delta$ and any $i \in a\mathcal{L}$.

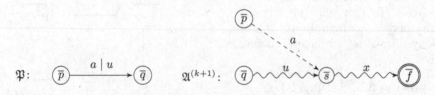

Fig. 2. Visualization of the construction of $\mathfrak{A}^{(k+1)}$.

Suppose that we already constructed the $\mathfrak{P}$-AA $\mathfrak{A}^{(k)}$. To obtain $\mathfrak{A}^{(k+1)}$ from $\mathfrak{A}^{(k)}$, we just add all transitions $(\overline{p}, a, \overline{s})$ with $(\overline{p}, a, u, \overline{q}) \in \Delta$ and $\overline{q} \xrightarrow{u}_{\mathfrak{A}^{(k)}} \overline{s}$ for some $\overline{q} \in \mathbf{Q}$ and $u \in A^*$ (see Fig. 2). Note that $(\overline{p}, a, u, \overline{q}) \in \Delta$ as well as the run $\overline{q} \xrightarrow{u}_{\mathfrak{A}^{(k)}} \overline{s}$ operate on components from $a\mathcal{L} \supseteq u\mathcal{L}$, only. Hence, the same applies to the new transition $(\overline{p}, a, \overline{s})$ ensuring that $\mathfrak{A}^{(k+1)}$ is asynchronous (this argument requires $a\mathcal{L} \supseteq u\mathcal{L}$ and would therefore not work for aPDS).

The "limit" of this construction is the $\mathfrak{P}$-AA $\mathfrak{A}^{(\infty)} = (\mathbf{S}, A, \emptyset, \delta^{(\infty)}, F)$ with $\delta^{(\infty)} = \bigcup_{k \in \mathbb{N}} \delta^{(k)}$.

Example 4.2. Recall the cPDS $\mathfrak{P}$ from Example 3.4. In Fig. 3 we depict our algorithm on input $\mathfrak{P}$ and the set of configurations $C = \{((q_1, q_2), \varepsilon)\}$. A $\mathfrak{P}$-AA $\mathfrak{A}^{(0)} = (S_1 \times S_2, A, \emptyset, \delta, F)$ accepting this set is depicted in the left.

In $\mathfrak{A}^{(1)}$, we have $(q_1, p_2) \xrightarrow{ab}_{\mathfrak{A}^{(1)}} (q_1, q_2)$ (depicted in bold and red) and, in $\mathfrak{P}$, we have the transition $((p_1, p_2), a, ab, (q_1, p_2)) \in \Delta$. The definition of $\delta^{(2)}$ implies that $((p_1, p_2), a, (q_1, q_2))$ is a new local transition.

The construction terminates with $\mathfrak{A}^{(2)}$ which is a $\mathfrak{P}$-AA that accepts the set of configurations backwards reachable from $C = \{((q_1, q_2), \varepsilon)\}$.

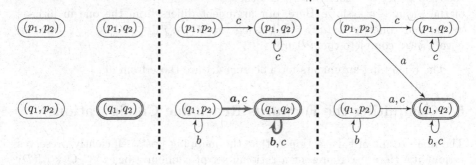

Fig. 3. The $\mathfrak{P}$-AA $\mathfrak{A}^{(0)}$, $\mathfrak{A}^{(1)}$, and $\mathfrak{A}^{(2)}$ (from left to right) from Example 4.2. (Color figure online)

Now, we show $C(\mathfrak{A}^{(\infty)}) = \mathrm{pre}^*_{\mathfrak{P}}(C(\mathfrak{A}^{(0)}))$. First, by induction on $k \in \mathbb{N}$, one can easily prove $\mathrm{pre}^k_{\mathfrak{P}}(C(\mathfrak{A}^{(0)})) \subseteq C(\mathfrak{A}^{(k)})$ (which ensures the inclusion "$\supseteq$"). On the other hand, Example 4.2 shows that the converse inclusion $C(\mathfrak{A}^{(k)}) \subseteq \mathrm{pre}^k_{\mathfrak{P}}(C(\mathfrak{A}^{(0)}))$ does not necessarily hold. The following lemma is the central argument in the proof of the inclusion "$\subseteq$" as it allows to infer $C(\mathfrak{A}^{(k)}) \subseteq \mathrm{pre}^*_{\mathfrak{P}}(C(\mathfrak{A}^{(0)}))$ for all $k \in \mathbb{N}$.

Lemma 4.3. *Let $k \in \mathbb{N}$, $v \in A^*$, $\overline{p} \in \mathbf{Q}$, and $\overline{s} \in \mathbf{S}$ with $\overline{p} \xrightarrow{v}_{\mathfrak{A}^{(k)}} \overline{s}$. Then there are a·global state $\overline{r} \in \mathbf{Q}$ and a word $w \in A^*$ with the following properties:*

(a) $(\overline{p}, \pi(v)) \vdash^* (\overline{r}, \pi(w))$ and (b) $\overline{r} \xrightarrow{w}_{\mathfrak{A}^{(0)}} \overline{s}$.

Proof Idea. We only indicate where our proof differs significantly from a similar one from [6] for pushdown systems. In general, the lemma is shown by induction on k, and the significant difference occurs in the induction step. So assume the lemma holds for k. To prove it for $k+1$, one proceeds by induction on the length of the word v. Again, the significant difference occurs in the induction step. Hence, we assume that the lemma holds for $k + 1$ and any word of length at most n and we will prove it for a word $v = v'a \in A^{n+1}$ with $a \in A$ and $v' \in A^n$.

So let $\overline{p} \in \mathbf{Q}$ and $\overline{s} \in \mathbf{S}$ such that $\overline{p} \xrightarrow{v}_{\mathfrak{A}^{(k+1)}} \overline{s}$. Since $\overline{p} \xrightarrow{v'a}_{\mathfrak{A}^{(k+1)}} \overline{s}$, there is some global state $\overline{s'} \in \mathbf{S}$ with $\overline{p} \xrightarrow{v'}_{\mathfrak{A}^{(k+1)}} \overline{s'} \xrightarrow{a}_{\mathfrak{A}^{(k+1)}} \overline{s}$.

Since $|v'| = n$, the inductive hypothesis provides a global state $\overline{q'} \in \mathbf{Q}$ and a word $w' \in A^*$ with $(\overline{p}, \pi(v')) \vdash^* (\overline{q'}, \pi(w'))$ and $\overline{q'} \xrightarrow{w'} \mathfrak{A}^{(0)}$ $\overline{s'}$. Note that the former implies in particular $(\overline{p}, \pi(v)) = (\overline{p}, \pi(v'a)) \vdash^* (\overline{q'}, \pi(w'a))$. The difficult case is if $\overline{s'} \xrightarrow{a} \mathfrak{A}^{(k)} \overline{s}$ does not hold (see Fig. 4 where edges $\xrightarrow{w} \mathfrak{A}^{(\ell)}$ are denoted by $\xrightarrow{w}_\ell$). Then $(\overline{s'}, a, \overline{s})$ is a new transition in $\delta^{(k+1)}$ which implies $s_i' \in Q_i$ for all $i \in a\mathcal{L}$. Recall that the $\mathfrak{P}$-AA $\mathfrak{A}^{(0)}$ cannot enter a local state

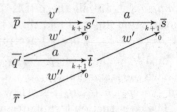

Fig. 4.

of the pushdown system. Here, our argument differs from the one in [6]), as we can only infer $w'\mathcal{L} \cap a\mathcal{L} = \emptyset$. Setting $\overline{t} = (\overline{s}\restriction_{a\mathcal{L}}, \overline{q'}\restriction_{w'\mathcal{L}}, \overline{s'}\restriction_{P \setminus w'a\mathcal{L}})$, one can nevertheless complete the picture. □

The remaining arguments for Theorem 4.1 are those from [6].

5   Computing the Forwards Reachable Configurations

The main result of this section is that the mapping $\text{post}^*_{\mathfrak{P}}$ efficiently preserves rationality. Here, we represent a rational set of configurations $C \subseteq \mathbf{Q} \times \mathbb{M}(\mathcal{D})$ by a tuple $\overline{\mathfrak{A}}$ of NFAs $\mathfrak{A}_{\overline{q}}$ for $\overline{q} \in \mathbf{Q}$ that, for all global states $\overline{q}$, accept the trace language $T(\mathfrak{A}_{\overline{q}}) = C_{\overline{q}} = \{\pi(w) \mid (\overline{q}, \pi(w)) \in C\}$. If this is the case, we say "the tuple $\overline{\mathfrak{A}}$ accepts C".

Theorem 5.1. *Let $\mathfrak{P}$ be a cPDS and $C \subseteq \text{Conf}_{\mathfrak{P}}$ be rational. Then $\text{post}^*_{\mathfrak{P}}(C)$ is rational. In particular, we can compute a tuple of NFAs accepting $\text{post}^*_{\mathfrak{P}}(C)$ from $\mathfrak{P}$ and a tuple of NFAs accepting C. If $\mathcal{D}$ is fixed, this construction is possible in polynomial time.*

The proof of this theorem is inspired by the work by Finkel et al. [11]. To explain its idea and particularities, we first start with a classical pushdown system $\mathfrak{P} = (Q_1, \Delta)$. Suppose there are transitions (p, a, bv, q) and (q, b, ε, r) implying $(p, ax) \vdash (q, bvx) \vdash (r, vx)$ for any word x. If we add the transition (p, a, v, r) to Δ that allows to go from (p, ax) to (r, vx) in one step, the reachability relation does not change. We keep adding such "shortcuts" and call the resulting pushdown system $\mathfrak{P}^{(\infty)}$. Then, any run of the original system $\mathfrak{P}$ can be simulated by a run of the system with shortcuts $\mathfrak{P}^{(\infty)}$ that first shortens the pushdown and then writes onto the pushdown. It follows that, for pushdown systems $\mathfrak{P}$, the mapping $\text{post}^*_{\mathfrak{P}}$ preserves rationality.

The crucial point of the above construction is that any run of the system $\mathfrak{P}^{(\infty)}$ can be brought into some "simple form" by using shortcuts. Here, "simple form" means that it consists of two phases: the pushdown decreases properly in every step of the first phase and does not decrease in any step of the second phase.

Our strategy in the proof of Theorem 5.1 will extend the above idea:

1. First, one demonstrates that Theorem 5.1 holds for "homogeneous" systems that formalize and strengthen the two types of phases from above:

A cPDS $\mathfrak{P} = (\mathbf{Q}, \Delta)$ is *homogeneous* if one of the following holds.
(1) All transitions $(\bar{p}, a, w, \bar{q}) \in \Delta$ satisfy $w = \varepsilon$.
(2) There is $X \subseteq P$ such that all transitions $(\bar{p}, a, w, \bar{q}) \in \Delta$ satisfy $a\mathcal{L} = X$ and $w \neq \varepsilon$.

This means, $\mathfrak{P}$ is homogeneous if either no transition writes anything or if all transitions read exactly from the same subset $X \subseteq P$ of processes and write at least one letter. In particular, in the second case we have $a\mathcal{L} = b\mathcal{L}$ for each pair of transitions $(\bar{p}, a, v, \bar{q}), (\bar{r}, b, w, \bar{s}) \in \Delta$ (but not necessarily $a = b$).
2. Using the result on homogeneous systems, one demonstrates Theorem 5.1 for "saturated" systems, i.e., systems where no new "shortcuts" can be added:

A cPDS $\mathfrak{P} = (\mathbf{Q}, \Delta)$ is *saturated* if $(\bar{p}, a, ubv, \bar{q}), (\bar{q}, b, \varepsilon, \bar{r}) \in \Delta$ with $u \parallel b$ implies $(\bar{p}, a, uv, \bar{r}) \in \Delta$.

This step differs significantly from the above arguments from [11] as the main difficulty is to show that the number of "phases" can be bounded (the bound is linear in the number of sets $a\mathcal{L} \subseteq P$ which is bounded by $|A|$).
3. Finally, Proposition 5.6 proves Theorem 5.1 in full generality by showing that any system can be saturated by adding shortcuts.

5.1   Forwards Reachability in Homogeneous Systems

Let $\mathfrak{P} = (\mathbf{Q}, \Delta)$ be a cPDS and let D be a rational set of configurations.

If $\Delta \subseteq \mathbf{Q} \times A \times \{\varepsilon\} \times \mathbf{Q}$, then any transition shortens the pushdowns. Hence, the effect of $\mathrm{post}^*_{\mathfrak{P}}$ is a left quotient of $D_{\bar{q}}$ wrt. a *recognizable* trace language.

Now suppose $X \subseteq P$ and $a\mathcal{L} = X$ as well as $u \neq \varepsilon$ for all transitions $(\bar{p}, a, u, \bar{q}) \in \Delta$. Then, dually to the above case, the effect of $\mathrm{post}^*_{\mathfrak{P}}$ is the concatenation of $D_{\bar{q}}$ and a rational (not necessarily recognizable) trace language.

It follows (in both cases), that $\mathrm{post}^*_{\mathfrak{P}}(D)$ is rational. A closer analysis reveals that also the remaining claims of Theorem 5.1 hold for homogeneous systems.

5.2   Forwards Reachability in Saturated Systems

Recall the constructed pushdown system with just one pushdown $\mathfrak{P}^{(\infty)}$ from the beginning of this section. This pushdown system is saturated. We learned that

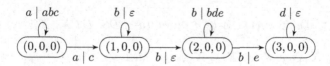

Fig. 5. The cPDS from Example 5.2.

any run in $\mathfrak{P}^{(\infty)}$ can be simulated by a run consisting of two phases: first, the pushdown shortens and then, it increases. The following example shows that this is not possible in systems with more than one pushdown.

Example 5.2. We consider the distributed alphabet $\mathcal{D} = (A, P, \mathcal{L})$ with $A = \{a, b, c, d, e\}$ and $P = \{1, 2, 3\}$ where $a\,\mathcal{L} = P$, $b\,\mathcal{L} = \{1, 2\}$, $c\,\mathcal{L} = \{3\}$, $d\,\mathcal{L} = \{1\}$, and $e\,\mathcal{L} = \{2\}$. Further, let $\mathfrak{P} = (\mathbf{Q}, \Delta)$ be the saturated cPDS from Fig. 5. The following is the only run from the configuration $(\overline{0}, \overline{\pi}(a))$ to the configuration $(\overline{3}, \overline{\pi}(e^4 c^4))$ where we write $\overline{n}$ for the state $(n, 0, 0)$:

$$
\begin{aligned}
(\overline{0}, \overline{\pi}(a)) &\vdash^3 (\overline{0}, \overline{\pi}(abcbcbc)) \vdash (\overline{1}, \overline{\pi}(cbcbcbc)) = (\overline{1}, \overline{\pi}(b^3 c^4)) \\
&\vdash^2 (\overline{2}, \overline{\pi}(bc^4)) \\
&\vdash^3 (\overline{2}, \overline{\pi}(bdededec^4)) \vdash (\overline{3}, \overline{\pi}(edededec^4)) = (\overline{3}, \overline{\pi}(d^3 e^4 c^4)) \\
&\vdash^3 (\overline{3}, \overline{\pi}(e^4 c^4))
\end{aligned}
$$

Note that this run splits into four phases (that correspond to the four lines above); it increases its pushdowns in the first and third and decreases them in the second and fourth.

So far, we used the term "phase" without defining it formally. To be a bit more precise, a "phase" is a run of some maximal homogeneous subsystem of $\mathfrak{P}$. These subsystems are defined next.

Definition 5.3. *Let $\mathfrak{P} = (\mathbf{Q}, \Delta)$ be a cPDS.*

1. *Let $\Delta_\varepsilon = \{(\overline{p}, a, \varepsilon, \overline{q}) \in \Delta\}$ and $\mathfrak{P}_\varepsilon = (\mathbf{Q}, \Delta_\varepsilon)$.*
2. *For $X \subseteq P$, let $\Delta_X = \{(\overline{p}, a, u, \overline{q}) \in \Delta \mid a\,\mathcal{L} = X \text{ and } u \neq \varepsilon\}$ and $\mathfrak{P}_X = (\mathbf{Q}, \Delta_X)$.*

To simplify notation, we write $\vdash_\varepsilon$ for $\vdash_{\mathfrak{P}_\varepsilon}$ and $\vdash_X$ for $\vdash_{\mathfrak{P}_X}$ for any $X \subseteq P$.

Since Δ is the disjoint union of the subsets Δ_ε and Δ_X for $X \subseteq P$, any run of $\mathfrak{P}$ splits uniquely into maximal subruns of these subsystems and these subruns are precisely what we called "phase".

For $X \subseteq P$, set $\Vdash_X = \vdash_\varepsilon^* \circ \vdash_X^+ \subseteq \mathrm{Conf}_{\mathfrak{P}} \times \mathrm{Conf}_{\mathfrak{P}}$. In other words, $c_1 \Vdash_X c_2$ means that the system $\mathfrak{P}$ has a run from c_1 to c_2 that first shortens the pushdowns and then, in the second phase, uses transitions from Δ_X, only. Note that the first (deleting) phase is allowed to be empty while the second (writing) phase is required to be non-empty.

For $\overline{X} = (X_i)_{i \in [n]}$ with $X_i \subseteq P$, set $\Vdash_{\overline{X}} = \Vdash_{X_1} \circ \Vdash_{X_2} \circ \cdots \circ \Vdash_{X_n}$.

The binary relation $\vdash^*$ is the union of all relations $\Vdash_{\overline{X}} \circ \vdash_\varepsilon^*$ for $\overline{X}$ a sequence of subsets of P of arbitrary length. Our next aim is to show that we only need to consider sequences $\overline{X}$ of bounded length. The central lemma proves that, under certain conditions, $\Vdash_{X_0} \circ \Vdash_{\overline{X}}$ is contained in $\Vdash_{\overline{X}}$, i.e., that we can shorten the sequence $X_0 \overline{X}$.

Lemma 5.4. *Let $\mathfrak{P} = (\mathbf{Q}, \Delta)$ be a saturated cPDS. Let $X_0, X_1, \ldots, X_{n+1} \subseteq P$ such that*

  (i) $X_0 \subseteq X_{n+1}$ and (ii) $X_i \not\subseteq X_{n+1}$ for all $1 \leq i \leq n$.
  Then $\Vdash_{(X_0, X_1, \ldots, X_{n+1})} \subseteq \Vdash_{(X_1, \ldots, X_{n+1})}$.

Proof Idea. The central argument of the proof goes as follows: Suppose we have

$$c_0 \vdash c_1 \Vdash_{(X_1, X_2, \ldots, X_n)} \circ \vdash_\varepsilon^* \circ \vdash_{X_{n+1}} d$$

and let $(\overline{p}, a, u, \overline{q}) \in \Delta$ with $a\mathcal{L} = X_0$ denote the transition used in the first step. The proof then proceeds by induction on the length of the word u. If $u = \varepsilon$, we get $c_0 \vdash_\varepsilon c_1$ implying $c_0 \Vdash_{(X_1, \ldots, X_{n+1})} d$. Now let $u \neq \varepsilon$. By (i), the run from c_1 to d reads from its pushdowns, at least once, a letter b with $b\mathcal{L} \cap X_0 \neq \emptyset$; we consider the first such transition t. If $t \in \Delta_\varepsilon$, the choice of t allows to prove that it can be executed at the very beginning (i.e., in the configuration c_1). Using that $\mathfrak{P}$ is saturated, the first two transitions can be combined into one of the form $(\overline{p}, a, u', \overline{q}')$ with $|u'| < |u|$ such that the induction hypothesis is applicable. If $t \notin \Delta_\varepsilon$, property (ii) implies that it is the very last transition (that leads to the configuration d) and that $a\mathcal{L} = b\mathcal{L}$. Using (ii), it follows that the very first transition can be postponed to the last-but-one position implying $c_0 \Vdash_{(X_1, \ldots, X_n)} \circ \vdash_\varepsilon^* \circ \vdash_{X_{n+1}}^2 d$. $\qquad\square$

It follows from the above lemma that $\vdash^*$ is the union of all relations $\Vdash_{\overline{X}}$ where the sets in the sequence $\overline{X}$ are mutually distinct implying that the length of $\overline{X}$ is bounded. Since the subsystems $\mathfrak{P}_\varepsilon$ and $\mathfrak{P}_X$ are homogenous, the arguments from Sect. 5.1 ensure that Theorem 5.1 holds for saturated systems.

5.3 Saturating a System

It remains to transform an arbitrary system into an equivalent saturated one. For a classical pushdown system (with just one pushdown), the idea is very simple: if there are transitions $(\overline{p}, a, bw, \overline{q})$ and $(\overline{q}, b, \varepsilon, \overline{r})$, then adding the transition $(\overline{p}, a, w, \overline{r})$ does not change the behavior and transforms the system closer to a saturated one. In the multi-pushdown setting, the technicalities are a bit more involved: suppose we have the transitions $(\overline{p}, a, cbw, \overline{q})$ and $(\overline{q}, b, \varepsilon, \overline{r})$ with $c\mathcal{L} \cap b\mathcal{L} = \emptyset$, i.e., $b \parallel c$. Then $\overline{\pi}(cbw) = \overline{\pi}(bcw)$, i.e., after doing the first transition (that writes the trace $\overline{\pi}(cbw) = \overline{\pi}(bcw)$), the second transition (eliminating b) can be executed immediately. Therefore, also in this situation, we add the transition $(\overline{p}, a, cw, \overline{r})$ to get closer to a saturated system.

Now, we construct cPDS $\mathfrak{P}^{(k)} = (\mathbf{Q}, \Delta^{(k)})$ for any $k \in \mathbb{N}$ as follows:

- we set $\Delta^{(0)} := \{(\overline{p}, a, \mathrm{lnf}(w), \overline{q}) \mid (\overline{p}, a, w, \overline{q}) \in \Delta\}$.[3]
- To obtain $\Delta^{(k+1)}$, we add to the set $\Delta^{(k)}$ all transitions $(\overline{p}, a, \mathrm{lnf}(uv), \overline{r})$ for which there are a letter $b \in A$ and a global state $\overline{q} \in \mathbf{Q}$ such that $(\overline{p}, a, ubv, \overline{q}), (\overline{q}, b, \varepsilon, \overline{r}) \in \Delta^{(k)}$ and $u \parallel b$.

Let $\Delta^{(\infty)} = \bigcup_{k \geq 0} \Delta^{(k)}$ be the "limit" of the increasing sequence of sets $\Delta^{(k)}$. Note that the length of words written by transitions in $\Delta^{(\infty)}$ is bounded by the length of words written by transitions in $\Delta^{(0)}$; hence $\Delta^{(\infty)}$ is finite.

---

[3] $\mathrm{lnf}(w)$ denotes the lexicographic normal form of the trace $\overline{\pi}(w)$. The use of $\mathrm{lnf}(w)$ instead of w allows to easily prove a polynomial upper bound for the number of transitions.

Example 5.5. Recall the cPDS $\mathfrak{P}$ from Example 3.4. In Fig. 6 we depict our construction of the multi-pushdown systems $\mathfrak{P}^{(k)}$ for $k \in \{0, 1, 2\}$. It can be verified that $\mathfrak{P}^{(2)} = \mathfrak{P}^{(3)}$.

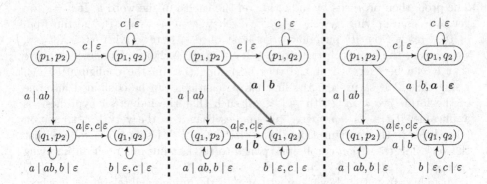

Fig. 6. The cooperating multi-pushdown system $\mathfrak{P} = \mathfrak{P}^{(0)}$, $\mathfrak{P}^{(1)}$, and $\mathfrak{P}^{(2)} = \mathfrak{P}^{(\infty)}$ (from left to right). New transitions are marked in bold and red. (Color figure online)

One can then show that the pair $\mathfrak{P}^{(\infty)} = (\mathbf{Q}, \Delta^{(\infty)})$ is a cPDS and that its reachability relation coincides with that of the original system $\mathfrak{P}$.

Proposition 5.6. *Let $\mathfrak{P}$ be a cPDS. Then, in time polynomial in $\|\mathfrak{P}\|^{|P|}$, one can construct an equivalent saturated system $\mathfrak{P}^{(\infty)}$, i.e., a saturated cPDS with $\vdash^*_{\mathfrak{P}} = \vdash^*_{\mathfrak{P}^{(\infty)}}$.*

6 Summary, Consequences, and Open Questions

We proved that the backwards reachability relation of communicating multi-pushdown systems efficiently preserves the recognizability of a set of configurations and that the forwards reachability relation efficiently preserves rationality. Conversely, one can demonstrate that the backwards reachability relation does not preserve rationality (i.e., there is a cPDS and a rational set C of configurations such that $\text{pre}^*(C)$ is not rational anymore). Similarly, one can demonstrate that the forwards reachability relation does not preserve recognizability.

It is decidable whether a given rational set of traces is contained in a given recognizable set of traces. Hence our positive results allow to decide, for C_1 rational and C_2 recognizable, the following questions.

- $\text{post}^*(C_1) \subseteq C_2$, or, since the class of recognizable trace languages is closed under complementation, $\text{post}^*(C_1) \cap C_2 = \emptyset$. This amounts to a safety property.

  Since singleton sets are both recognizable and rational, this also implies that the reachability relation is decidable.

- $\text{post}^*(C_1) \cap C_2 \neq \emptyset$. Since the set C_2 of configurations with a given global state is recognizable, this implies that the control state reachability problem is decidable for C_1 rational.
- $C_1 \subseteq \text{pre}^*(C_2)$.
- $\text{post}^*(C_1) \subseteq \text{pre}^*(C_2)$ which amounts to a liveness property: From every configuration reachable from C_1, we can reach a configuration from C_2. This property can also by expressed by the EF-formula $C_1 \wedge \neg\text{EF}(\neg\text{EF}\,C_2)$. More generally, EF-model checking is decidable for cPDS, although our results allow to bound the running time only non-elementary.

The next and obvious open question regarding the verification of cPDS, one would have to consider the recurrent reachability, i.e., the question whether, starting from some configuration, there is an infinite run that visits some global state infinitely often. This could then form the basis for algorithms deciding properties that are given by formulas from linear time temporal logics.

Since we can see cPDS as a natural extension of pushdown systems from word semantics to trace semantics, another open problem is to find some generalized context-free grammars accepting the class of languages of cPDS. Additionally, one could compare this new model with other known models for multi-pushdown systems.

References

1. Aiswarya, C., Gastin, P., Narayan Kumar, K.: Controllers for the verification of communicating multi-pushdown systems. In: Baldan, P., Gorla, D. (eds.) CONCUR 2014. LNCS, vol. 8704, pp. 297–311. Springer, Heidelberg (2014). https://doi.org/10.1007/978-3-662-44584-6_21
2. Aiswarya, C., Gastin, P., Narayan Kumar, K.: Verifying communicating multi-pushdown systems via split-width. In: Cassez, F., Raskin, J.-F. (eds.) ATVA 2014. LNCS, vol. 8837, pp. 1–17. Springer, Cham (2014). https://doi.org/10.1007/978-3-319-11936-6_1
3. Atig, M.F., Bollig, B., Habermehl, P.: Emptiness of ordered multi-pushdown automata is 2ETIME-complete. Int. J. Found. Comput. Sci. **28**(8), 945–976 (2017)
4. Babić, D., Rakamarić, Z.: Asynchronously communicating visibly pushdown systems. In: Beyer, D., Boreale, M. (eds.) FMOODS/FORTE 2013. LNCS, vol. 7892, pp. 225–241. Springer, Heidelberg (2013). https://doi.org/10.1007/978-3-642-38592-6_16
5. Bollig, B., Kuske, D., Mennicke, R.: The complexity of model checking multi-stack systems. Theory Comput. Syst. **60**(4), 695–736 (2017)
6. Bouajjani, A., Esparza, J., Maler, O.: Reachability analysis of pushdown automata: application to model-checking. In: Mazurkiewicz, A., Winkowski, J. (eds.) CONCUR 1997. LNCS, vol. 1243, pp. 135–150. Springer, Heidelberg (1997). https://doi.org/10.1007/3-540-63141-0_10
7. Bouajjani, A., Esparza, J., Touili, T.: A generic approach to the static analysis of concurrent programs with procedures. Int. J. Found. Comput. Sci. **14**(4), 551 (2003)

8. Bouajjani, A., Müller-Olm, M., Touili, T.: Regular symbolic analysis of dynamic networks of pushdown systems. In: Abadi, M., de Alfaro, L. (eds.) CONCUR 2005. LNCS, vol. 3653, pp. 473–487. Springer, Heidelberg (2005). https://doi.org/10.1007/11539452_36

9. Cartier, P., Foata, D.: Problèmes combinatoires de commutation et réarrangements. Lecture Notes in Mathematics, vol. 85. Springer, Heidelberg (1969). https://doi.org/10.1007/BFb0079468

10. Diekert, V., Rozenberg, G.: The Book of Traces. World Scientific (1995)

11. Finkel, A., Willems, B., Wolper, P.: A direct symbolic approach to model checking pushdown systems. Electron. Notes Theoretical Comput. Sci. **9**, 27–37 (1997)

12. Heußner, A., Leroux, J., Muscholl, A., Sutre, G.: Reachability analysis of communicating pushdown systems. Log. Methods Comput. Sci. **8**(3) (2012)

13. Hutagalung, M., Hundeshagen, N., Kuske, D., Lange, M., Lozes, É.: Multi-buffer simulations: decidability and complexity. Inf. Comput. **262**(2), 280–310 (2018)

14. Kleene, S.: Representation of events in nerve nets and finite automata. In: Shannon, C., McCarthy, J. (eds.) Automata Studies, pp. 3–40. Annals of Mathematics Studies, vol. 34. Princeton University Press (1956)

15. La Torre, S., Madhusudan, P., Parlato, G.: A robust class of context-sensitive languages. In: LICS 2007, pp. 161–170. IEEE Computer Society (2007)

16. La Torre, S., Napoli, M., Parlato, G.: Reachability of scope-bounded multistack pushdown systems. Inf. Comput. **275**, 104588 (2020)

17. Mazurkiewicz, A.: Concurrent program schemes and their interpretations. DAIMI Rep. Ser. **6**(78) (1977)

18. Qadeer, S., Rehof, J.: Context-bounded model checking of concurrent software. In: Halbwachs, N., Zuck, L.D. (eds.) TACAS 2005. LNCS, vol. 3440, pp. 93–107. Springer, Heidelberg (2005). https://doi.org/10.1007/978-3-540-31980-1_7

19. Zielonka, W.: Notes on finite asynchronous automata. RAIRO - Theor. Inf. Appl. **21**(2), 99–135 (1987)

Shortest Dominating Set Reconfiguration Under Token Sliding

Jan Matyáš Křišťan[1]([✉])(iD) and Jakub Svoboda[2](iD)

[1] Faculty of Information Technology, Czech Technical University in Prague,
Thákurova 9, 160 00, Prague, Czech Republic
kristja6@fit.cvut.cz
[2] Institute of Science and Technology, Klosterneuburg, Austria
jakub.svoboda@ist.ac.at

Abstract. In this paper, we present novel algorithms that efficiently
compute a shortest reconfiguration sequence between two given dominating sets in trees and interval graphs under the TOKEN SLIDING model.
In this problem, a graph is provided along with its two dominating sets,
which can be imagined as tokens placed on vertices. The objective is
to find a shortest sequence of dominating sets that transforms one set
into the other, with each set in the sequence resulting from sliding a
single token in the previous set. While identifying any sequence has been
well studied, our work presents the first polynomial algorithms for this
optimization variant in the context of dominating sets.

Keywords: Reconfiguration · Dominating set · Trees · Interval
graphs · Algorithms

1 Introduction

Reconfiguration problems arise when the goal is to transform one feasible solution into another through a series of small steps, while ensuring that all intermediate solutions remain feasible. These problems have been widely studied in the context of graph problems, such as INDEPENDENT SET [1, 2, 7, 14], DOMINATING SET [3–5, 10, 15, 18], and SHORTEST PATHS [8, 13]. Reconfiguration problems have also been studied in the context of SATISFIABILITY [9, 16]. See [17] for a general survey.

In the case of the DOMINATING SET and other graph problems, the most commonly studied reconfiguration rules are TOKEN JUMPING and TOKEN SLIDING. The feasible solution can be represented by tokens placed on the vertices of a graph. Under TOKEN JUMPING, tokens can be moved one at a time to any other vertex, while under TOKEN SLIDING, tokens can only be moved one at a time to a neighboring vertex.

We focus on the TOKEN SLIDING variant, particularly on finding a shortest reconfiguration sequence. This optimization variant has been extensively studied

---

The original version of this chapter was revised: the funding information was missing. This was corrected. The correction to this chapter is available at
https://doi.org/10.1007/978-3-031-43587-4_31

H. Fernau and K. Jansen (Eds.): FCT 2023, LNCS 14292, pp. 333–347, 2023.
https://doi.org/10.1007/978-3-031-43587-4_24

Table 1. Complexities of problems of reconfiguring dominating sets under TOKEN SLIDING on various graph classes. The decision problem results are due to [4].

Graph class	Decision problem	Optimization variant
Trees	P	$\mathcal{O}(n)$ **(Corollary 1)**
Interval graphs	P	$\mathcal{O}(n^3)$ **(Theorem 2)**
Dually chordal graphs	P	open
Split	PSPACE-complete	$n^{\omega(1)}$
Bipartite	PSPACE-complete	$n^{\omega(1)}$
Planar	PSPACE-complete	$n^{\omega(1)}$

in the context of reconfiguring solutions for SHORTEST PATHS [13], INDEPENDENT SET [11,19], and SATISFIABILITY [16].

Our main contribution is the presentation of two polynomial algorithms for finding a shortest reconfiguration sequence between dominating sets on trees and on interval graphs. This is achieved through a novel approach to finding a reconfiguration sequence, combined with a natural lower bound on their lengths.

We extend the results of Bonamy et al. [4], who have shown that a reconfiguration sequence between dominating sets under TOKEN SLIDING can be found in polynomial time when the input graph is a dually chordal graph. This is a class of graphs encompassing both trees and interval graphs.

We provide a brief overview of the known results in Table 1, along with our contributions. The lower bound for cases where the reachability problem is PSPACE-hard follows from the fact that the reconfiguration sequence must have superpolynomial length in some instances (unless PSPACE = NP), as otherwise, a reconfiguration sequence would serve as a polynomial-sized proof of reachability.

2   Preliminaries

Graphs and Trees. Given a graph G and vertex v, we denote the set of neighbors of v by $N(v)$; moreover, $N[v] = N(v) \cup \{v\}$. Given two vertices v and u, we denote $d_G(v, u)$ as the distance between v and u, that is the number of edges on a shortest path between v and u.

Given a rooted tree T rooted at r and vertex v, we denote: the subtree below v as $T[v]$; the depth of vertex v as $d(v) = d(v, r)$; the parent of v as $p(v)$. Let $\sigma(u, v)$ be the set of vertices that follow u on a shortest path from u to v. We assume that $\sigma(u, u) = \emptyset$.

Multisets. Formally, a multiset H of elements from a base set S is defined as a *multiplicity function* $H : S \rightarrow \mathbb{N} \cup \{0\}$. We define the *support* of H as $\mathrm{Supp}(H) = \{v \mid H(v) \geq 1\}$. Let H and I be multisets, then $H \cap I = \min(H, I)$, $H \cup I = H + I$, $H \setminus I = \max(H - I, 0)$, $H \triangle I = (H \setminus I) \cup (I \setminus H)$. The cardinality is defined as $|H| = \sum_{v \in S} H(v)$ and $v \in H$ if $v \in \mathrm{Supp}(H)$. The Cartesian product $H \times I$ is a multiset of base set $S \times S$ such that $(H \times I)((u, v)) = H(u) \cdot I(v)$ for all $u, v \in S$.

Note that if one of the operands is a set, we can assume that it is a multiset with multiplicities of 1 for all elements in the set.

Graph Problems. Given a graph $G = (V, E)$, a set D of vertices is dominating if every vertex is either in D or a neighbor of a vertex in D. A multiset H is dominating if $\text{Supp}(H)$ is dominating. We say that given a set S of vertices, the vertices with a neighbor in S are *dominated* from S.

For trees, we solve a more general problem called reconfiguration of hitting sets. A *hitting set* of a set system $\mathcal{S}$ is a set H such that for each $S \in \mathcal{S}$ it holds $H \cap S \neq \emptyset$. A multiset H is a hitting set if $\text{Supp}(H)$ is a hitting set.

Reconfiguration Sequence. Given a graph G a multiset D of its vertices representing the placement of tokens, we denote $D(u \to v) = (D \setminus \{u\}) \cup \{v\}$ the multiset resulting from jumping a token on u to v (or sliding a token on u to v if $\{u, v\} \in E(G)$). Given a graph G and a set Π of feasible solutions, we say that a sequence of multisets $D_1, D_2, \ldots, D_\ell$ (of length ℓ) is a *reconfiguration sequence* under TOKEN SLIDING between $D_1, D_\ell \in \Pi$ if

- $D_i \in \Pi$ for all $1 \leq i \leq \ell$,
- $D_{i+1} = D_i(u \to v)$ such that $v \in V(G)$, $u \in D_i$ and $\{u, v\} \in E(G)$ for all $1 \leq i < \ell$.

The sequence can be concisely represented by a sequence of *moves*. Given a multiset D_s, moves $(u_1, v_1), \ldots, (u_{k-1}, v_{k-1})$ induce sequence $D_1, \ldots, D_k$ such that $D_1 = D_s$ and $D_{i+1} = D_i(u_i \to v_i)$ for all $1 \leq i < k$. This allows us to formally give the main problem of this paper.

SHORTEST RECONFIGURATION OF DOMINATING SETS UNDER TOKEN SLIDING
Input: Graph $G = (V, E)$ and two dominating sets D_s and D_t.
Output: Shortest sequence of moves $(u_1, v_1), \ldots, (u_{k-1}, v_{k-1})$ inducing a reconfiguration sequence under TOKEN SLIDING between D_s and D_t.

In the case of trees, we design an algorithm that finds a reconfiguration sequence whenever the feasible solutions can be expressed as hitting sets of a set system $\mathcal{S}$ such that every $S \in \mathcal{S}$ induces a subtree of the input tree T. Several problems can be formulated in terms of such hitting sets.

If $\mathcal{S}$ is the set of all closed neighborhoods of T, then the hitting sets of $\mathcal{S}$ are exactly the dominating sets of T. If $\mathcal{S}$ is the set of all edges, then the hitting sets are exactly all vertex covers of T. An instance of an (unrestricted) vertex multicut is equivalent to a hitting set problem with $\mathcal{S}$ being the set of all paths which must be cut.

The general problem of reconfiguring hitting sets is as follows.

SHORTEST RECONFIGURATION OF HITTING SETS UNDER TOKEN SLIDING
Input: Graph $G = (V, E)$ and two hitting sets H_s and H_t of a set system $\mathcal{S} \subseteq 2^{V(T)}$.
Output: Shortest sequence of moves $(u_1, v_1), \ldots, (u_{k-1}, v_{k-1})$ inducing a reconfiguration sequence under TOKEN SLIDING between H_s and H_t.

Reconfiguration Graph. Given a graph G and an integer k, the *reconfiguration graph* $\mathcal{R}(G, k)$ has as vertices all feasible solutions, in our case dominating multisets, of size k. Two vertices are adjacent whenever one can be reached from the other in a single move, i.e. sliding a token. Note that the shortest reconfiguration of dominating sets under TOKEN SLIDING between D_s and D_t is equivalent to finding a shortest path in $\mathcal{R}(G, |D_s|)$ between D_s and D_t. Furthermore, as each move under TOKEN SLIDING is reversible, the edges of $\mathcal{R}(G, k)$ are undirected. Thus, finding a shortest path from D_s to D_t is equivalent to finding a shortest path from D_t to D_s.

It follows that for D_s and D_t, if $D_s \neq D_t$ and both are in the same connected component of $\mathcal{R}(G, |D_s|)$, then $d_{\mathcal{R}(G, |D_s|)}(D_s, D_t)$ is the minimum number of moves inducing a reconfiguration sequence between D_s and D_t. If G and $|D_s|$ is clear from the context, we consider $\mathcal{R} = \mathcal{R}(G, |D_s|)$.

Interval Graphs. A graph G is an *interval graph* if each vertex v can be mapped to a different closed interval $I(v)$ on the real line so that $v, u \in E$ if and only if $I(v) \cap I(u) \neq \emptyset$. Such a mapping to intervals is called *interval representation*.

We denote the endpoints of an interval $I(v)$ as $\ell(v)$ and $r(v)$ so that $I(v) = [\ell(v), r(v)]$. Every interval graph has an interval representation with unique integer endpoints which can be computed in linear time [6], we assume such representation is available.

We say that interval I is *to the left* of J (or that J is *to the right* of I) if $r(I) < \ell(J)$. Similarly, we say that I is *nested* in J (or that J *contains* I) if $\ell(J) < \ell(I)$, $r(I) < r(J)$. Furthermore, we say that I *left-intersects* J (or that J *right-intersects* I) if $\ell(I) < \ell(J) < r(I) < r(J)$. Note that every pair of intervals is in exactly one of those relations. We say that two vertices u and v of an interval graph are in a given relationship if their intervals $I(u)$ and $I(v)$ in a fixed interval representation are in the given relationship.

3 Lower Bounds on Lengths of Reconfiguration Sequences

We can obtain a lower bound on the length of a reconfiguration sequence by dropping the requirement that the tokens induce a feasible solution (such as a dominating set) at each step. The problem of finding such shortest reconfiguration sequence is polynomial-time solvable by reducing to the minimum-cost matching in bipartite graphs.

Let G be a graph, $D_s, D_t \subseteq V(G)$ be the multisets representing tokens. Then $M \subseteq D_s \times D_t$ is a *matching* between D_s and D_t if for every $v \in D_s$, there is exactly $D_s(v)$ pairs $(v, \cdot) \in M$ and similarly for every $v \in D_t$, there is exactly $D_t(v)$ pairs $(\cdot, v) \in M$. Note that M is a multiset and the same pair may be contained in M multiple times.

We say that $u \in D_s$ and $v \in D_t$ are matched in M if $(u, v) \in M$. We also use $M(u)$ to denote the set of matches of u, that is the vertices v such that $(u, v) \in M$. The *cost* $c(M)$ of the matching M is defined as

$$c(M) = \sum_{(u,v) \in M} d_G(u, v) \cdot M(u, v).$$

We say that a matching has *minimum cost* if its cost is the minimum over all possible matchings between D_s and D_t and denote this cost as $c^*(D_s, D_t)$. We use $\sigma_M(u)$ to denote the vertices which follow u on some shortest path to some match $M(u) \neq u$. Formally

$$\sigma_M(u) = \bigcup_{v \neq u : (u,v) \in M} \sigma(u, v).$$

We define M^{-1} so that $M^{-1}(v, u) = M(u, v)$ for all $u \in D_s, v \in D_t$.

Lemma 1. *Every sequence of moves inducing a reconfiguration sequence between D_s and D_t under* TOKEN SLIDING *has length at least $c^*(D_s, D_t)$.*

Proof. Suppose a reconfiguration sequence using fewer than $c^*(D_s, D_t)$ moves exists. Let M be a matching between D_s and D_t of minimum cost. Then we can track the moves of each token and construct a matching M' between D_s and D_t given by the starting and ending position of each token. Note that the cost of each matched pair is at most the length of the path travelled by the given token. Thus in total the cost of M' is at most the total number of moves used. Hence, we have $c(M', D_s, D_t) < c^*(D_s, D_t)$, a contradiction. $\square$

The following observation shows that in a minimum-cost matching, if a token can be matched with zero cost, we can assume that is the case for all such tokens.

Lemma 2. *For graph G, let D_s, D_t be multisets of the same size and let $I = D_s \cap D_t$. Then there exists minimum-cost matching M in G between D_s and D_t such that for every $v \in I$ we have $M(v, v) = I(v)$.*

Proof. Given a minimum matching M between D_s and D_t and v such that $M(v, v) < I(v)$, we show that we can produce M' of the same cost such that $\sum_{u \in I} M'(u, u) > \sum_{u \in I} M(u, u)$. Note that there exist $(x, v), (v, y) \in M$ with $x \neq v, y \neq v$ as otherwise $M((v, v)) = I(v)$. Then we define

$$M' = (M \setminus \{(x, v), (v, y)\}) \cup \{(v, v), (x, y)\}.$$

We have $c(M') - c(M) = -d_G(x, v) - d_G(v, y) + d_G(v, v) + d_G(x, y) = -d_G(x, v) - d_G(v, y) + d_G(x, y)$. From the triangle inequality $d_G(x, y) \leq d_G(x, v) + d_G(v, y)$, thus we have that the cost of M' is at most the cost of M and thus is minimum. By repeated application, we arrive at minimum-cost matching M^* with $M^*(v, v) = I(v)$ for all v. $\square$

The following observation shows that, given D_s and D_t, if we pick a token in D_s and slide it along an edge to decrease its distance to its match in a minimum-cost matching, the resulting D'_s and D_t have minimum cost of matching of exactly one less than D_s and D_t. Thus if each move in the reconfiguration sequence is of such a kind, the length of the resulting sequence will match the lower bound of Lemma 1.

Lemma 3. *Let M^* be a minimum-cost matching between D_s and D_t, $(u, g) \in M^*$ and $v \in \sigma(u, g)$ a vertex that follows u on a shortest path from u to g. Furthermore, let*

$$M = (M^* \setminus \{(u, g)\}) \cup \{(v, g)\}.$$

Then M is a minimum-cost matching between $D_s(u \to v)$ and D_t. Furthermore, $c^*(D_s, D_t) = c^*(D_s(u \to v), D_t) + 1$.

Proof. From definition $c(M^*) - c(M) = d_G(u, g) - d_G(v, g)$, but v is the vertex on the path from u to g, so $d_G(u, g) - d_G(v, g) = 1$ and $c^*(D_s(u \to v), D_t) \leq c^*(D_s, D_t) - 1$.

Suppose that $c^*(D_s(u \to v), D_t) < c^*(D_s, D_t) - 1$, i.e., there exists matching M' between $D_s(u \to v)$ and D_t such that $c(M') < c(M)$. From M', we construct a matching M'' between D_s and D_t such that $c(M'') < c(M^*)$, which is a contradiction.

Let $x \in M'(v)$ and set $M'' = (M' \setminus \{(v, x)\}) \cup \{(u, x)\}$. The cost $c(M'') \leq c(M') + 1$, since the distance between v and u is 1. That means if $c(M') < c(M)$, then $c^*(D_s, D_t) < c(M^*)$, but M^* is minimum-cost matching.

Therefore, M is a minimum-cost matching between $D_s(u \to v)$ and D_t and $c^*(D_s, D_t) = c^*(D_s(u \to v), D_t) + 1$. □

4 Algorithms for Finding a Shortest Reconfiguration Sequence

4.1 Trees

We present an algorithm that, given a tree T and two hitting sets H_s, H_t of a set system S such that every $S \in S$ induces a subtree of T, finds a shortest reconfiguration sequence between H_s and H_t under TOKEN SLIDING. As dominating sets are exactly the hitting sets of closed neighborhoods, the algorithm finds a shortest reconfiguration sequence between two dominating sets. Note that S need not be provided on the input.

Consider the reconfiguration graph $\mathcal{R}(G, |H_s|)$, whose vertices are all the hitting multisets of S of size $|H_s|$. The high-level idea is to extend two paths in $\mathcal{R}(G, |H_s|)$, one from H_s and another from H_t, until they reach a common configuration. We repeatedly identify a subtree $T[v]$ of the rooted T for which the configurations H_s and H_t are identical, except for v itself. Then, we modify either H_s or H_t by sliding the token (or tokens) on v to its parent, ensuring that H_s and H_t become equal when restricted to $T[v]$.

Algorithm 1 describes the algorithm. We assume that the input tree is rooted in some vertex r. The proof of correctness uses techniques of Sect. 3. While the correctness of the algorithm can be proved without them, we believe this presentation is helpful for understanding the proofs in subsequent sections.

Theorem 1. *Let T be a tree on n vertices and H_s and H_t hitting sets of a set system S in which every $S \in S$ induces a subtree of T. Then RECONFTREE (Algorithm 1) correctly computes a solution to SHORTEST RECONFIGURATION OF HITTING SETS UNDER TOKEN SLIDING. Furthermore, it runs in time $\mathcal{O}(n)$.*

Algorithm 1. Reconfiguration of hitting sets in trees

1: **procedure** RECONFTREE(T, H_s, H_t)
2:     **if** $H_s = H_t$ **then return** $\emptyset$
3:     $v \leftarrow$ vertex v such that $H_s(v) \neq H_t(v)$ and $H_s(u) = H_t(u)$ for all $u \in T(v)$.
4:     **if** $H_s(v) > H_t(v)$ **then**
5:         **return** $(v, p(v)) +$ RECONFTREE$(T, H_s(v \rightarrow p(v)), H_t)$
6:     **else**
7:         **return** RECONFTREE$(T, H_s, H_t(v \rightarrow p(v))) + (p(v), v)$

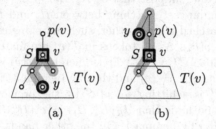

(a)   (b)

Fig. 1. Bold squares represent a token of H_s, bold circles represent a token of H_t, the grey areas represent examples of S.

Proof. We will show that RECONFTREE outputs a sequence of $d_\mathcal{R}(H_s, H_t)$ moves which induces a reconfiguration sequence between the two hitting sets H_s, H_t of $\mathcal{S}$. If $H_s = H_t$, then $d_\mathcal{R}(H_s, H_t) = 0$ and the procedure correctly outputs an empty sequence. Thus assume that $H_s \neq H_t$.

Suppose that T is rooted in r and let v be a vertex such that $H_s(v) \neq H_t(v)$ and $H_s(u) = H_t(u)$ for all $u \in T(v)$. Without loss of generality, assume that $H_s(v) > H_t(v)$ as otherwise, we can swap H_s and H_t.

Claim. $H_s(v \rightarrow p(v))$ is a hitting set of $\mathcal{S}$.

Proof. Suppose that $H'_s = H_s(v \rightarrow p(v))$ is not a hitting set of $\mathcal{S}$. It follows that $\mathrm{Supp}(H_s) \not\subseteq \mathrm{Supp}(H'_s)$ and therefore $H_s(v) = 1$ and $H'_s(v) = 0$ and H'_s is not intersecting only sets $S \in \mathcal{S}$ such that $v \in S$ and $p(v) \notin S$. Furthermore, $H_t(v) = 0$ as $H_t(v) < H_s(v)$.

Let $S \in \mathcal{S}$ be a set not intersecting H'_s and let $y \in S \cap H_t$. Such y distinct from v must exist as H_t is a hitting set of $\mathcal{S}$ and $v \notin H_t$. If $y \in T[v]$, then $y \in H_s$ as $H_s(y) = H_t(y)$ by the choice of v, which contradicts H'_s not intersecting S. This case is shown in Fig. 1a.

Therefore $y \in T \setminus T[v]$. Note that the path connecting v with y must visit $p(v)$, thus as S induces a subtree and contains u and y, it contains $p(v)$ as well and therefore S intersects H'_s. This case is shown in Fig. 1b. $\square$

Claim. The number of moves outputted by RECONFTREE (T, H_s, H_t) is equal to $d_\mathcal{R}(H_s, H_t)$.

Proof. We first claim that if H_s, H_t are two hitting sets of S with the same size, then $d_R(H_s, H_t) = c^*(H_s, H_t)$. Furthermore, we show that a move from v to $p(v)$ decreases the cost of a minimum-cost matching between H_s and H_t by one, where v is a vertex such that $H_s(v) \neq H_t(v)$ and $H_s(u) = H_t(u)$ for all $u \in T(v)$. This together implies that each outputted move decreases the distance in the reconfiguration graph by one.

We prove the claim by induction on $c^*(H_s, H_t)$ that $d_R(H_s, H_t) = c^*(H_s, H_t)$ for any hitting sets H_s, H_t of the same size. First note that $c^*(H_s, H_t) = 0$ if and only if $H_s = H_t$. Now, suppose that $c^*(H_s, H_t) \geq 1$.

Let M^* be a minimum-cost matching between H_s and H_t such that tokens with distance 0 are matched to each other, such matching exists by Lemma 2.

Let $H'_s = H_s(v \to p(v))$. As all tokens in $T(v)$ are matched by M^* only to the same vertex, it holds $M^*(v) \subseteq V \setminus T[v]$. Therefore $p(v)$ is the next vertex on the path from v to some $g \in M^*(v)$ and thus by Lemma 3 it holds $c^*(H'_s, H_t) = c^*(H_s, H_t) - 1$. As H'_s is a hitting set of S by the previous claim, it follows from the induction hypothesis that $d_R(H'_s, H_t) = c^*(H'_s, H_t)$. Now, note that $d_R(H_s, H_t) \geq c^*(H_s, H_t)$ by Lemma 1. On the other hand,

$$d_R(H_s, H_t) \leq d_R(H'_s, H_t) + 1 = c^*(H'_s, H_t) + 1 = c^*(H_s, H_t)$$

as H_s can be reached from H'_s by a single token slide. This concludes the proof of the inductive step.

As $d_R(H'_s, H_t) = d_R(H_s, H_t) - 1$, each call of the algorithm decreases the distance between the hitting sets by one and also outputs one move. Thus the resulting reconfiguration sequence is shortest possible. □

We now describe how to implement Algorithm 1 so that it achieves the linear running time. Note that we assume that the input H_s and H_t of the initial call of RECONFTREE are subsets of $V(T)$ and therefore $|H_s|, |H_t| \leq n$. Next, we show how to compress the output to $\mathcal{O}(n)$ size. Whenever $|H_s(v) - H_t(v)| > 1$, we can perform all $|H_s(v) - H_t(v)|$ moves from v to $p(v)$ at once and output them as a triple $(v, p(v), |H_s(v) - H_t(v)|)$ if $H_s(v) > H_t(v)$ or $(p(v), v, |H_s(v) - H_t(v)|)$ in case $H_s(v) < H_t(v)$.

Note that with this optimization, the vertex v on line 3 is distinct for each call of RECONFTREE. Furthermore, we can fix in advance the order in which we pick candidates of v on line 3 by ordering the vertices of T by their distance from r in decreasing order. This is correct as the depth of the lowest vertex satisfying the condition of line 3 cannot increase in the subsequent calls. Then, the process of finding v on line 3 has total runtime of $\mathcal{O}(n)$ over the course of the whole algorithm. □

Corollary 1. *Let T be a tree on n vertices and D_s, D_t dominating sets of T such that $|D_s| = |D_t|$. Algorithm 1 finds a shortest reconfiguration sequence between D_s and D_t under* TOKEN SLIDING *in $\mathcal{O}(n)$ time.*

In general, the length of the reconfiguration sequence can be up to $\Omega(n^2)$, for instance when $\Omega(n)$ tokens are required to move from one end of a path to

---

Algorithm 2. Reconfiguration of dominating sets in interval graphs

---

1: **procedure** RECONFIG(G, D_s, D_t, M)
2:    **if** $D_s = D_t$ **then return** $\emptyset$
3:    **if** $\exists (u,v) \in M, u' \in \sigma(u,v)$ such that $D_s(u \to u')$ is dominating **then**
4:       **return** $(u \to u') + $ RECONFIG$(G, D_s(u \to u'), D_t)$
5:    **if** $\exists (u,v) \in M, v' \in \sigma(v,u)$ such that $D_t(v \to v')$ is dominating **then**
6:       **return** RECONFIG$(G, D_s, D_t(v \to v')) + (v' \to v)$
7:    $M' \leftarrow$ FIXMATCHING(G, D_s, D_t, M)
8:    **return** RECONFIG(G, D_s, D_t, M')
9: **procedure** FIXMATCHING(G, D_s, D_t, M)
10:    $v \in D_s \triangle D_t$ with minimum possible $r(v)$.
11:    **if** $v \in D_t$ **then**
12:       **return** FIXMATCHING$(G, D_t, D_s, M^{-1})^{-1}$ ▷ Symmetric, swap D_s, D_t
13:    Find $y \in D_t \setminus D_s, y' \in M(y), v' \in M(v)$ such that $D_s' = D_s(v \to y)$ is dominating and $M' = M \setminus \{(v,v'),(y',y)\} \cup \{(v,y),(y',v')\}$ is a minimum-cost matching
14:    between D_s and D_t.
15:    **return** M'

---

the other end, as each must move to a distance of at least $\Omega(n)$. However, when this happens, a lot of tokens move by one edge and we can move them at the same time, so the running time of the algorithm can be smaller than the number of moves.

4.2 Interval Graphs

In this section, we describe a polynomial-time algorithm for finding a shortest reconfiguration sequence between two dominating sets under the TOKEN SLIDING model in interval graphs. As with trees, we demonstrate that the distance between two dominating sets in interval graphs is equal to the lower bound established in Lemma 1. Our approach involves a minimum-cost matching between the dominating sets D_s and D_t to identify a valid move. The key insight of this algorithm is that we can always recalculate the minimum-cost matching to enable sliding at least one token along a shortest path towards its corresponding match.

Algorithm 2 outlines the algorithm. A minimum-cost matching M between D_s and D_t is assumed to be provided on the input.

The bulk of the proof consists of showing that the procedure FIXMATCHING is correct, in particular that the call on line 13 succeeds. First, we present a technical lemma related to shortest paths in interval graphs.

Lemma 4. *Let $P = (v_1, v_2, \ldots, v_k)$ be a shortest path between v_1 and v_k in an interval graph with $r(v_1) < r(v_k)$ and $k \geq 3$. Then v_{i+1} right-intersects v_i and v_{i+2} does not intersect v_i for all $i \in \{1, \ldots, k-2\}$.*

Proof. If for some $i \in \{1, \ldots, k-2\}$ v_{i+2} intersects v_i, then we can create a shorter path from v_1 to v_k by removing v_{i+1} from P, contradicting P being a shortest path.

Suppose that for some $i \in \{1, \ldots, k-2\}$ it holds $r(v_{i+1}) < r(v_i)$. Note that a shortest path contains no nested intervals with a possible exception of v_1 and v_k, as every other nested interval can be removed to make the path shorter. Thus v_{i+1} left-intersects v_i. Let v_j be the first next vertex after v_i such that $r(v_i) < r(v_j)$. If none such exists, then v_i must intersect v_k and thus the path can be made shorter. Otherwise we show that v_j intersects v_i. If it does not, then $\ell(v_j) > r(v_i)$. But for the path to be connected, another interval v_a must cover $[r(v_i), \ell(v_j)]$. Such interval either has $r(v_j) < r(v_a)$, thus v_j is nested in v_a or $r(v_i) < r(v_a) < r(v_j)$, contradicting the choice of v_j. $\qquad\square$

The following lemma shows that we can efficiently recompute the minimum-cost matching to ensure that for some token a valid move across a shortest path to its match will be available.

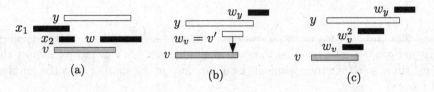

Fig. 2. Grey represents tokens D_s, white represents tokens D_t.

Lemma 5. *The call of* FixMatching *on line 7 returns a minimum-cost matching M' between D_s and D_t such that at least one the following holds.*

- *There is $(u, v) \in M', u' \in \sigma(u, v)$ such that $D_s(u \to u')$ is dominating,*
- *there is $(u, v) \in M', v' \in \sigma(v, u)$ such that $D_t(v \to v')$ is dominating.*

Proof. The idea of the proof is in showing that if no token can move along a shortest path to its match, then there is always a way to modify the matching which does not increase cost and makes moving along a shortest path possible for at least one token. In particular, we need to show that the operation of finding y on line 13 always succeeds and the constructed M' is a minimum-cost matching between D_s and D_t.

As the algorithm has not finished on line 2, it holds $D_s \neq D_t$. Let M be a minimum-cost matching between D_s and D_t. If for some $(u, v) \in M$, $w \in \sigma(u, v)$, $w' \in \sigma(v, u)$ $D_s(u \to w)$ or $D_t(v \to w')$ is dominating, then we would not have reached line 7. Therefore, assume that for every $(u, v) \in M$, $w \in \sigma(u, v)$, $w' \in \sigma(v, u)$ neither $D_s(u \to w)$ nor $D_t(v \to w')$ is dominating.

Let $(v, v') \in M$ such that $v \neq v'$ and $\min(r(v), r(v'))$ is minimum possible. Without loss of generality, assume that $r(v) < r(v')$ as otherwise, we can swap D_s and D_t.

Claim. For every $w \in \sigma_M(v)$, $I(w)$ right-intersects $I(v)$.

Proof. Suppose that $I(w)$ contains $I(v)$. Then $D_s(v \rightarrow w)$ is dominating, a contradiction. Now suppose that $I(v)$ contains $I(w)$. Then by Lemma 4 it holds $(v, w) \in M$, which implies that $v \in \sigma(w, v)$ and $D_t(w \rightarrow v)$ is dominating as $N[v] \subseteq N[w]$, a contradiction.

The remaining case is that $I(w)$ left-intersects $I(v)$. Then again, by Lemma 4 it holds $(v, w) \in M$ and this contradicts the choice of v as $r(w) < r(v)$.   □

Now, let $w \in \sigma_M(v)$ be a fixed vertex and consider why $D_s(v \rightarrow w) = D'_s$ is not dominating. Let $x_1, \ldots, x_k \subset N(v) \setminus N(w)$ be the vertices that are not dominated by D'_s.

Claim. There exists $y \in N(v) \cap (D_t \setminus D_s)$ such that all x_i are adjacent to y.

Proof. First, we will show that $I(x_i)$ is to the left of $I(w)$ for all x_i. Note that as each no x_i is adjacent to w, $I(x_i)$ is either to the left or to the right of $I(w)$.

Suppose there is some $I(x_i)$ to the left of $I(w)$ and some $I(x_j)$ to the right of $I(w)$, then $I(v)$ contains $I(w)$, which as previously argued may not be the case. The remaining case is that all $I(x_i)$ are to the right of $I(w)$, which would imply that $I(w)$ left-intersects $I(v)$, which again was shown not to hold. Therefore, each $I(x_i)$ is to the left of $I(w)$. This further implies that $\ell(v) < r(x_i) < \ell(w)$, thus each $I(x_i)$ is either nested in $I(v)$ or left-intersects $I(v)$.

Observe that each x_i is adjacent to some $y_i \in D_t \setminus D_s$ and $r(v) < r(y_i)$ by the choice of v. Therefore, there exists $y \in D_t \setminus D_s$ such that $I(y)$ contains $\min(r(x_1), \ldots, r(x_k))$. Together, we get

$$\ell(y) < \min_{1 \leq i \leq k} (r(x_i)) \leq \max_{1 \leq i \leq k} (r(x_i)) < \ell(w) < r(v) < r(y) \qquad (1)$$

and therefore y is adjacent to all x_i. See Fig. 2a for an illustration. As $\ell(y) < r(v) < r(y)$, $I(y)$ either right-intersects $I(v)$ or contains $I(v)$ and thus v and y are adjacent.   □

The rest of the proof consists of two claims. The first is that $D_s(v \rightarrow y)$ is dominating. The second is that $(v, y) \in M'$ for some minimum-cost matching M' between D_s and D_t.

Claim. $D_s(v \rightarrow y)$ is dominating.

Proof. Let $D' = D_s(v \rightarrow y)$. If y contains v, then $N[v] \subseteq N[y]$, therefore $D_s \subseteq D'_s$ and D'_s is dominating. Thus assume that y right-intersects v, which is the only remaining case as shown above.

Suppose $u \in N(v)$ is a vertex which is not dominated from D'_s. Note that u must not be adjacent to y and at the same time be adjacent to v, therefore u is to the left of y. Then u is to the left of all $w \in \sigma_M(v)$ as $\ell(y) < \ell(w)$, thus u is not dominated in $D_s(v \rightarrow w)$ and therefore $u = x_i$ for some i. This implies that u is not adjacent to y and this contradicts the choice of y.   □

Let $v' \in D_t$ such that $v' \neq v$ and $(v, v') \in M$. Similarly, let $y' \in D_s$ such that $y' \neq y$ and $(y', y) \in M$. We define the new matching M' as

$$M' = (M \setminus \{(v, v'), (y', y)\}) \cup \{(v, y), (y', v')\}.$$

Claim. M' is a minimum-cost matching between D_s and D_t.

Proof. We prove that $c(M') \leq c(M)$. Given that $d(v, y) = 1$ it suffices to show that

$$d(v, y) + d(v', y') \leq d(v, v') + d(y, y')$$
$$d(v', y') \leq d(v, v') + d(y, y') - 1.$$

Let $w_v \in \sigma(v, v')$ and $w_y \in \sigma(y', y)$.

Case 1. Vertices w_v and w_y are adjacent.

We can construct a walk W from v' to y' by concatenating shortest paths between each two consecutive vertices in (v', w_v, w_y, y'). It holds that $d(v', y')$ is at most the number of edges of W and therefore

$$\begin{aligned} d(v', y') &\leq d(v', w_v) + d(w_v, w_y) + d(w_y, y') \\ &= d(v', v) - 1 + 1 + d(y, y') - 1 \\ &= d(v, v') + d(y, y') - 1. \end{aligned}$$

Case 2. Vertices w_v and w_y are not adjacent and $I(w_v)$ is nested in $I(y)$.

Suppose that $v' = w_v$. Given that $I(w_v)$ is nested in $I(y)$, it follows that $N[v'] \subseteq N[y]$ and thus as $v', y \in D_t$ we have that $D_t \setminus \{v'\}$ is dominating. Therefore, $D_t(v' \rightarrow v)$ is dominating, a contradiction. See Fig. 2b for an illustration.

Assume further that $v' \neq w_v$ and therefore $d(v, v') \geq 2$. Let $w_v^2 \in \sigma(w_v, v')$ and note that w_v^2 must be adjacent to y as $N[w_v] \subseteq N[y]$. See Fig. 2c for an illustration. We can construct a walk between v' and y' by concatenating shortest paths between each two consecutive vertices in (y', y, w_v^2, v') of total length

$$d(y', y) + 1 + d(v, v') - 2 = d(v, v') + d(y, y') - 1$$

and therefore $d(v', y') \leq d(v, v') + d(y, y') - 1$.

Case 3. Vertices w_v and w_y are not adjacent and $I(w_v)$ is not nested in $I(y)$.

Recall that by Equation (1) it holds $\ell(y) < \ell(w_v)$. Furthermore, $r(y) < r(w_v)$, as otherwise $I(w_v)$ would be nested in $I(y)$.

Let us now consider the possible orderings of the right endpoints of $I(v)$, $I(y)$, $I(w_v)$, $I(w_y)$. The possibilities are restricted by the fact that by Equation (1) it holds $r(v) < r(y) < r(w_v)$, thus there remain 4 possible orderings. The case $r(v) < r(y) < r(w_v) < r(w_y)$ can be ruled out as it contradicts $I(y)$ and $I(w_y)$ intersecting and $I(w_v)$ and $I(w_y)$ not intersecting at the same time. Similarly

$r(v) < r(y) < r(w_y) < r(w_v)$ and $r(v) < r(w_y) < r(y) < r(w_v)$ is not possible as it would contradict $I(v)$ and $I(w_v)$ intersecting and at the same time $I(w_v)$ and $I(w_y)$ not intersecting.

Thus, the only remaining ordering is $r(w_y) < r(v) < r(y) < r(w_v)$. This by Lemma 4 implies that either $w_y = y'$ or $r(y') < r(y)$. In either case, it follows that $r(y') < r(v)$ which contradicts the choice of v.   □

We have shown that for any two dominating sets $D_s \neq D_t$ and a minimum-cost matching M between them, we can construct another minimum matching M' such that at least one of the following statements holds. There exists either $v \in D_s$ and $y \in D_t$ such that $(v, y) \in M'$ and $D_s(v \to y)$ is dominating or, by a symmetric proof with D_s and D_t swapped, there exists $v \in D_t, y \in D_s$ such that $(y, v) \in M'$ and $D_t(v \to y)$ is dominating. In either case, we have shown that v and y can be adjacent and thus $y \in \sigma(v, y)$. Furthermore, M' is constructed as described on line 7 and y can be found by testing all vertices in D_t. This concludes the proof.   □

Theorem 2. *Let G be an interval graph with n vertices and D_s, D_t its two dominating sets such that $|D_s| = |D_t|$. Then RECONFIG correctly computes a solution to SHORTEST RECONFIGURATION OF DOMINATING SETS UNDER TOKEN SLIDING in time $\mathcal{O}(n^3)$, where k is the size of the output.*

Proof. We first show that the resulting reconfiguration sequence has the shortest possible length.

Claim. The number of moves outputted by RECONFIG is $d_\mathcal{R}(D_s, D_t)$

Proof. We will show that $d_\mathcal{R}(D_s, D_t) = c^*(D_s, D_t)$ by induction over $c^*(D_s, D_t)$. If $c^*(D_s, D_t) = 0$, then $D_s = D_t$ and $d_\mathcal{R}(D_s, D_t) = 0$, which we can efficiently recognize.

Suppose that $c^*(D_s, D_t) > 0$. Let M be a minimum-cost matching between D_s and D_t. Without loss of generality, let $(u, v) \in M', u' \in \sigma(u, v)$ such that $D'_s = D_s(u \to u')$ is dominating. By Lemma 5, either such u, u', v already exist or we can recompute M' so that they exist.

Note that by Lemma 3, $c^*(D'_s, D_t) = c^*(D_s, D_t) - 1$ and thus by the induction hypothesis $d_\mathcal{R}(D'_s, D_t) = c^*(D'_s, D_t)$. Note that $d_\mathcal{R}(D_s, D_t) \leq d_\mathcal{R}(D'_s, D_t) + 1$ as D_s can be reached from D_s by a single token slide. At the same time, by Lemma 1, it holds $d_\mathcal{R}(D_s, D_t) \geq c^*(D_s, D_t) = c^*(D'_s, D_t) + 1 = d_\mathcal{R}(D'_s, D_t) + 1$. Thus $d_\mathcal{R}(D_s, D_t) = d_\mathcal{R}(D'_s, D_t)$ and with each output of a token slide, we decrease the distance in $d_\mathcal{R}$ by exactly one. Therefore, the resulting reconfiguration is shortest possible.   □

Claim. RECONFIG can be implemented to run in time $\mathcal{O}(n^3)$.

Proof. We initially compute a minimum-cost matching between D_s and D_t by reducing to minimum-cost matching in bipartite graphs, which can be solved in $\mathcal{O}(n^3)$ [12].

Now, we describe how to implement Algorithm 2 efficiently. If we want to find a suitable v in FIXMATCHING, we suppose that all greedy moves, i.e. moves along shortest paths to matches that result in a dominating set, have been done. This is not necessary, we can see that the assumption is invoked only on constantly many vertices for each call of FIXMATCHING. Checking if a greedy move can be performed requires only linear time and the total number of moves is at most $\mathcal{O}(n^2)$, thus the total running time is $\mathcal{O}(n^3)$. □

This concludes the proof of the theorem. □

5   Conclusion

Fig. 3. The minimum cost of matching between the light-gray and the dark-gray dominating sets is 2 but to reconfigure one into the other, we need at least 3 moves.

We have presented polynomial algorithms for finding a shortest reconfiguration sequence between dominating sets on trees and interval graphs, addressing the open question left by Bonamy et al. [4] regarding the complexity of the said problem on dually chordal graphs. While in case of trees and interval graphs, we can always match the lower bound of Lemma 1. That is not the case for dually chordal graph in general, see Fig. 3.

The general case of dually chordal graphs remains open. Additionally, the case of cographs is open and we conjecture that a polynomial-time solution is achievable.

Acknowledgments. J. M. Křišťan acknowledges the support of the Czech Science Foundation Grant No. 22-19557S. This work was supported by the Grant Agency of the Czech Technical University in Prague, grant No. SGS23/205/OHK3/3T/18. J. Svoboda acknowledges the support of the ERC CoG 863818 (ForM-SMArt) grant.

References

1. Bartier, V., Bousquet, N., Dallard, C., Lomer, K., Mouawad, A.E.: On girth and the parameterized complexity of token sliding and token jumping. Algorithmica **83**(9), 2914–2951 (2021). https://doi.org/10.1007/s00453-021-00848-1
2. Belmonte, R., Kim, E.J., Lampis, M., Mitsou, V., Otachi, Y., Sikora, F.: Token sliding on split graphs. Theory Comput. Syst. **65**(4), 662–686 (2020). https://doi.org/10.1007/s00224-020-09967-8

3. Bodlaender, H.L., Groenland, C., Swennenhuis, C.M.F.: Parameterized complexities of dominating and independent set reconfiguration. In: 16th International Symposium on Parameterized and Exact Computation, IPEC 2021. LIPIcs, vol. 214, pp. 9:1–9:16 (2021). https://doi.org/10.4230/LIPIcs.IPEC.2021.9
4. Bonamy, M., Dorbec, P., Ouvrard, P.: Dominating sets reconfiguration under token sliding. Discrete Appl. Math. **301**, 6–18 (2021). https://doi.org/10.1016/j.dam. 2021.05.014
5. Bousquet, N., Joffard, A.: TS-reconfiguration of dominating sets in circle and circular-arc graphs. In: Fundamentals of Computation Theory, FCT, pp. 114–134. Lecture Notes in Computer Science (2021). https://doi.org/10.1007/978-3-030-86593-1_8
6. Corneil, D.G., Olariu, S., Stewart, L.: The LBFS structure and recognition of interval graphs. SIAM J. Discrete Math. **23**(4), 1905–1953 (2010). https://doi.org/10.1137/s0895480100373455
7. Demaine, E.D., et al.: Linear-time algorithm for sliding tokens on trees. Theor. Comput. Sci. **600**, 132–142 (2015). https://doi.org/10.1016/j.tcs.2015.07.037
8. Gajjar, K., Jha, A.V., Kumar, M., Lahiri, A.: Reconfiguring shortest paths in graphs. Proc. AAAI Conf. Artif. Intelli. **36**(9), 9758–9766 (2022). https://doi.org/10.1609/aaai.v36i9.21211
9. Gopalan, P., Kolaitis, P.G., Maneva, E., Papadimitriou, C.H.: The connectivity of Boolean satisfiability: computational and structural dichotomies. SIAM J. Comput. **38**(6), 2330–2355 (2009). https://doi.org/10.1137/07070440X
10. Haddadan, A., et al.: The complexity of dominating set reconfiguration. Theor. Comput. Sci. **651**, 37–49 (2016). https://doi.org/10.1016/j.tcs.2016.08.016
11. Hoang, D.A., Khorramian, A., Uehara, R.: Shortest reconfiguration sequence for sliding tokens on spiders. In: Heggernes, P. (ed.) Algorithms and Complexity - 11th International Conference, CIAC, vol. 11485, pp. 262–273 (2019). https://doi.org/10.1007/978-3-030-17402-6_22
12. Jonker, R., Volgenant, T.: A shortest augmenting path algorithm for dense and sparse linear assignment problems. Computing **38**(4), 325–340 (1987). https://doi.org/10.1007/BF02278710
13. Kamiński, M., Medvedev, P., Milanič, M.: Shortest paths between shortest paths. Theor. Comput. Sci. **412**(39), 5205–5210 (2011). https://doi.org/10.1016/j.tcs. 2011.05.021
14. Lokshtanov, D., Mouawad, A.E.: The complexity of independent set reconfiguration on bipartite graphs. ACM Trans. Algorithms **15**(1), 1–19 (2019). https://doi.org/10.1145/3280825
15. Lokshtanov, D., Mouawad, A.E., Panolan, F., Ramanujan, M., Saurabh, S.: Reconfiguration on sparse graphs. J. Comput. Syst. Sci. **95**, 122–131 (2018). https://doi.org/10.1016/j.jcss.2018.02.004
16. Mouawad, A.E., Nishimura, N., Pathak, V., Raman, V.: Shortest reconfiguration paths in the solution space of Boolean formulas. SIAM J. Discrete Math. **31**(3), 2185–2200 (2017). https://doi.org/10.1137/16M1065288
17. Nishimura, N.: Introduction to reconfiguration. Algorithms **11**(4), 52 (2018). https://doi.org/10.3390/a11040052
18. Suzuki, A., Mouawad, A.E., Nishimura, N.: Reconfiguration of dominating sets. J. Comb. Optim. **32**(4), 1182–1195 (2015). https://doi.org/10.1007/s10878-015-9947-x
19. Yamada, T., Uehara, R.: Shortest reconfiguration of sliding tokens on subclasses of interval graphs. Theor. Comput. Sci. **863**, 53–68 (2021). https://doi.org/10.1016/j.tcs.2021.02.019

Computing Optimal Leaf Roots of Chordal Cographs in Linear Time

Van Bang Le and Christian Rosenke$^{(\boxtimes)}$

Institut für Informatik, Universität Rostock, Rostock, Germany
{van-bang.le,christian.rosenke}@uni-rostock.de

Abstract. A graph G is a *k-leaf power*, for an integer $k \geq 2$, if there is a tree T with leaf set $V(G)$ such that, for all vertices $x, y \in V(G)$, the edge xy exists in G if and only if the distance between x and y in T is at most k. Such a tree T is called a *k-leaf root* of G. The computational problem of constructing a k-leaf root for a given graph G and an integer k, if any, is motivated by the challenge from computational biology to reconstruct phylogenetic trees. For fixed k, Lafond [SODA 2022] recently solved this problem in polynomial time.

In this paper, we propose to study *optimal leaf roots* of graphs G, that is, the k-leaf roots of G with *minimum* k value. Thus, all k'-leaf roots of G satisfy $k \leq k'$. In terms of computational biology, seeking optimal leaf roots is more justified as they yield more probable phylogenetic trees. Lafond's result does not imply polynomial-time computability of optimal leaf roots, because, even for optimal k-leaf roots, k may (exponentially) depend on the size of G. This paper presents a linear-time construction of optimal leaf roots for chordal cographs (also known as trivially perfect graphs). Additionally, it highlights the importance of the parity of the parameter k and provides a deeper insight into the differences between optimal k-leaf roots of even versus odd k.

Keywords: k-leaf power · k-leaf root · optimal k-leaf root · trivially perfect leaf power · chordal cograph

1 Introduction

Leaf powers have been introduced by Nishimura, Ragde and Thilikos [12] to model the phylogeny reconstruction problem from computational biology: given a graph G that represents a set of species with vertices $V(G)$ and the interspecies similarity with edges $E(G)$, how can we reconstruct an evolutionary tree T with a given similarity threshold k? A *k-leaf root* of G, a tree T with species $V(G)$ as the leaf set and where species $x, y \in V(G)$ have distance at most k in T if and only if they are similar on account of $xy \in E(G)$, is considered a solution to this problem. In case T exists, the graph G is called a *k-leaf power*. The challenge of finding a k-leaf root for given G and k has, yet, been modelled as the *k-leaf power recognition problem*: given G and k, decide if G has a k-leaf root.

© The Author(s), under exclusive license to Springer Nature Switzerland AG 2023
H. Fernau and K. Jansen (Eds.): FCT 2023, LNCS 14292, pp. 348–362, 2023.
https://doi.org/10.1007/978-3-031-43587-4_25

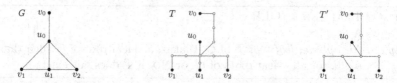

Fig. 1. A graph G (left), a 5-leaf root T of G (middle), a 4-leaf root T' of G (right).

For an example, see Fig. 1 with the graph G called *dart*. The similarities between the five species can be explained with similarity threshold $k = 5$ using the 5-leaf root T and with $k = 4$ by the 4-leaf root T', both depicted in Fig. 1.

For a deeper discourse into the heavily studied field of k-leaf powers, the reader is kindly referred to the survey [13]. Here, we just give a short overview.

Lately, Eppstein and Havvaei [8] showed that k-leaf power recognition for graphs G with n vertices can be solved in $\mathcal{O}(f(k, \omega) \cdot n)$ time with $f(k, \omega)$ exponential in k and ω, the clique number of G. Quite simply put, they reduce k-leaf power recognition to the decision of a certain monadic second order property in a graph derived from G having tree-width bounded by k and ω. Lafond's even more recent algorithm [10] solves k-leaf power recognition in $\mathcal{O}(n^{g(k)})$ time, where $g(k)$ grows superexponentially with k. It applies sophisticated dynamic programming on a tree decomposition of G and exploits structural redundancies in G. Observe that, for fixed k, the latter method runs in polynomial time.

Before these advances, k-leaf power recognition had only been solved for all fixed k between 2 and 6. The 2-leaf powers are exactly the graphs that have just cliques as their connected components, which makes the problem trivial. For $k = 3$ (see [12] and [3]), $k = 4$ (see [12] and [4]), $k = 5$ (see [5]) and $k = 6$ (see [7]) individual algorithms have been developed, all creating a certain (tree-) decomposition of the input graph G and then attempting to fit together candidate k-leaf roots for the components into one k-leaf root for G.

A general controversial aspect of modelling the reconstruction of a phylogenetic tree T with the k-leaf power recognition problem is that k is part of the input. In the biological context, the value of k describes an upper bound on the number of evolutionary events in T that lie between two similar species x, y, thus, species adjacent in the given graph G by an edge xy. Unlike the model suggests, biologists do not always have control over the parameter k. Instead, phylogenetic trees T with as few as possible evolutionary events between all pairs of similar species are preferred. That is because, in reality, a higher number of events between x and y makes a similarity between x and y less likely. Conversely, this means that a k-leaf root of G with a small parameter k models a more probable phylogenetic tree. This paper therefore proposes a subtle change in perspective towards considering the following optimization problem.

OPTIMAL LEAF ROOT (OLR)

Instance: A graph G.

Output: An *optimal leaf root* T of G, that is, a κ-leaf root of G such that $\kappa \leq k$ for all k-leaf roots of G, or NO, if T does not exist.

Subsequently, we use κ to indicate that the respective κ-leaf root is optimal. OLR is in a certain sense an optimization version of k-leaf power recognition. The answer NO states that the given graph G is not a k-leaf power for any k and, in particular, not for the given one. Getting an optimal κ-leaf root of G helps to decide if G is a k-leaf power in many cases. A difficulty is that, for all κ and k of different parity and with $2 \leq \kappa < k < 2\kappa - 2$, there are κ-leaf powers that are not k-leaf powers [14]. Then, checking $\kappa \leq k$ does not decide correctly.

As for k-leaf power recognition, there are no known general efficient solutions for OLR. If input was restricted to k-leaf powers with $k \leq K$ for some fixed K, we could repurpose Lafond's algorithm. Testing a given G with all $2 \leq k \leq K$ would finally reveal the minimum κ for which G is κ-leaf power. At that point, a κ-leaf root of G could also be extracted from the algorithm. But that classes of k-leaf powers have not been characterized well for any $k \geq 5$ makes restricting input in the proposed way difficult. Then again, it is unknown how to decide if a given graph G is a k-leaf power for any arbitrary k. And on top of that, the minimum value κ for which a given G is a κ-leaf power, if any, may exponentially depend on the size of G. This means that this brute force searching may take exponentially or even infinitely many runs of Lafond's algorithm.

It is known that, independent of k, all k-leaf powers are strongly chordal, but not vice versa. Ptolemaic graphs are strongly chordal and a *class of unbounded leaf powers*. That is, there is no bound β such that every Ptolemaic graph has a k-leaf root for some $k \leq \beta$. Nevertheless, every Ptolemaic graph on n vertices has a $2n$-leaf root [1,2]. Later, Theorem 2 shows that, often, this is not optimal. This paper considers a subclass of Ptolemaic graphs, the chordal cographs (also known as *trivially perfect graphs*), as input to OLR. By definition, they form the intersection of the well-known *chordal graphs* and the *cographs*. Accordingly, they are also characterized as the graphs without induced cycles on four vertices and without induced paths on four vertices [9,15,16]. As a side effect of Lemma 8, this paper proves that chordal cographs are still a class of unbounded leaf powers. This means that k-leaf power recognition on this class cannot be solved in polynomial time with the algorithm of Lafond or the one of Eppstein and Havvaei. Nevertheless, the following main result of our work states that OLR can be solved in linear time for chordal cographs.

Theorem 1. *Given a chordal cograph G on n vertices and m edges, a (compressed) κ-leaf root of G with minimum κ can be computed in $\mathcal{O}(n + m)$ time.*

To the best of our knowledge, chordal cographs are, thus, the first class of unbounded leaf powers with a polynomial-time solution for OLR. The word *compressed* in Theorem 1 means that the κ-leaf root T is returned in a denser representation, where long paths of degree two-vertices are compressed into single weighted edges. Otherwise, the size of T alone would be quadratic.

While, in general, an OLR-solution does not entirely work for k-leaf power recognition, as elaborated above, our OLR-approach can also be used for linear-time k-leaf power recognition on chordal cographs. The key to this is the ability of our method to solve OLR with a given parity, such that the computed κ-leaf root comes with the minimum κ of the given parity. Hence, if we choose the parity of the given k, we can tell that a given graph G is a k-leaf power if and only if the computed κ-leaf root with κ of the same parity as k satisfies $\kappa \leq k$.

As the desired parity of κ plays a certain role in our construction, we research this discrepancy here, and show, for certain chordal cographs, that the minimum κ can differ up to 25 percent depending on if it is wanted odd or even.

The next section presents basic notation, definitions, and facts on trees and k-leaf powers used in this paper. The optimal leaf root construction method for chordal cographs is introduced and proved correct in Sect. 3. Section 4 provides a respective linear-time implementation, thus, proving Theorem 1. A deepened evaluation of the difference between chordal cographs with κ-leaf roots of minimum odd versus even κ is carried out in the concluding Sect. 5.

Proofs are omitted and will be included in the full version of this paper[1].

2 Preliminaries

All considered graphs are finite and without multiple edges or loops. Let $G = (V, E)$ be a graph with vertex set $V(G) = V$ and edge set $E(G) = E$. A *universal vertex* in G is one that is adjacent to all other vertices. If all vertices of G are universal than G is *complete*. A vertex x that is adjacent to exactly one other vertex of G is called a *leaf* and the edge containing x is a *pendant edge*. Two adjacent vertices $x, y \in V(G)$ are *true twins* if $xz \in E(G)$, if and only if $yz \in E(G)$ for all $z \in V(G) \setminus \{x, y\}$.

A graph H is an *induced subgraph* of G if $V(H) \subseteq V(G)$ and $xy \in E(H)$ if and only if $xy \in E(G)$ for all $x, y \in V(H)$. All subgraphs considered in this paper are induced. For $X \subset V(G)$, $G - X$ denotes the induced subgraph H of G with $V(H) = V(G) \setminus X$. If X consists of one vertex x then we write $G - x$ for $G - \{x\}$. Complete subgraphs of G are called *cliques*.

As usual, an x, y-*path* in G is a sequence $v_1, \ldots, v_n$ of distinct vertices from $V(G)$ such that $x = v_1$, $y = v_n$ and $v_i v_{i+1} \in E(G)$ for all $i \in \{1, \ldots, n-1\}$. An x, y-path is called a *cycle* in G if $xy \in E(G)$. The length of the x, y-path, respective cycle, is the number of its edges, that is, $n - 1$ in the x, y-path and n in the cycle. If there is an x, y-path in G for all distinct $x, y \in V(G)$ then G is *connected*. Otherwise, G is *disconnected* and, therefore, composed of *connected components* $G_1, \ldots, G_n$, maximal induced subgraphs of G that are connected. A connected component is *non-trivial* if it has more than one vertex and, otherwise, it is called *isolated vertex*. We call $C \subseteq V(G)$ a *cut set* if $G - C$ has more connected components than G. If C is just a single vertex c then c is a *cutvertex*.

Graphs G and H are isomorphic if a bijection $\sigma : V(G) \to V(H)$ exists with $xy \in E(G)$ if and only if $\sigma(x)\sigma(y) \in E(H)$. If no induced subgraph of G is

---

[1] https://arxiv.org/abs/2308.10756.

isomorphic to a graph H then G is H-free. *Trees* are the connected cycle-free graphs. This means, a tree T contains exactly one x, y-path for all $x, y \in V(T)$.

In this paper, we learn that, dependent on the given parity, the construction of an optimal leaf root differs in several details. To avoid permanent case distinctions, we use $\pi(i)$ for the parity of an integer i, that is, $\pi(i) = i \bmod 2$.

2.1 Chordal Cographs and Cotrees

Chordal cographs, ccgs for short, are known as the graphs that are free of the path and the cycle on 4 vertices. See the top row of Fig. 2 for an example ccg. One particular ccg used in this paper is the *star (with t leaves)*, which consists of the vertices $u, v_1, \ldots, v_t$ for some $t \geq 2$ and the edges $uv_1, \ldots, uv_t$.

Like all cographs, ccgs can be represented with *cotrees* [6]. The second row of Fig. 2 shows the cotree of the example cograph in the first row. For every cograph G, the cotree T is a rooted tree with leaves $V(G)$ and where every internal node is labelled with ⓪ for *disjoint union* or ① for *full join*. In this way, the leaves define single vertex graphs and every internal node represents the cograph G combining the cographs $H_1, H_2, \ldots, H_n$ of its children with the respective graph operation. More precisely, $V(G) = V(H_1) \cup V(H_2) \cup \cdots \cup V(H_n)$ and $G = ⓪(H_1, H_2, \ldots, H_n)$ means the disconnected cograph on vertex set $V(G)$ and edge set $E(G) = E(H_1) \cup E(H_2) \cup \cdots \cup E(H_n)$ and $G = ①(H_1, H_2, \ldots, H_n)$ means the connected cograph with vertex set $V(G)$ and edge set $E(G) = E(H_1) \cup E(H_2) \cup \cdots \cup E(H_n) \cup \{xy \mid x \in V(H_i), y \in V(H_j), 1 \leq i < j \leq n\}$. The cotree T is unique, can be constructed in linear time, and has the following properties:

- Every internal node has at least two children.
- No two internal nodes with the same label, ⓪ or ①, are adjacent.
- The subtree T_X rooted at node X is the cotree of the subgraph G_X induced by the leaves of T_X. If X is labelled with ⓪ then G_X is the disjoint union of the cographs represented by the children of X and if it is labelled with ① than G_X is the full join of the children cographs.
- The cotree of an n-vertex cograph has at most $2n - 1$ nodes.

We mostly work with ccgs without true twins, like G in Fig. 2. These graphs have the following properties, as observed in the upper rows of the figure.

Proposition 1. *If G is a ccg without true twins and T is the cotree of G then every node of T labelled with ① has exactly two children, one leaf and one node labelled with ⓪.*

Proposition 2 (Wolk [15,16]). *Every connected ccg G without true twins has a unique universal vertex u and $G - u$ is disconnected (that is, u is a cutvertex).*

2.2 Diameter, Radius and Center in Trees

Let T be a tree. The following notions are throughout used in the paper:

- The *distance* between two vertices x and y in T, written $\text{dist}_T(x, y)$, is the length of the unique x, y-path in T.
- The *diameter* of T, denoted $\text{diam}(T)$, is the maximum distance between two vertices in T, that is, $\text{diam}(T) = \max\{\text{dist}_T(x, y) \mid x, y \in V(T)\}$.
- A *diametral path* in T is a path of length $\text{diam}(T)$.
- A vertex z is a *center vertex* of T if the maximum distance between z and any other vertex in T is minimum, that is, all $y \in V(T)$ satisfy $\max\{\text{dist}_T(x, z) \mid x \in V(T)\} \le \max\{\text{dist}_T(x, y) \mid x \in V(T)\}$.
- The *radius* of T, denoted $\text{rad}(T)$, is the maximum distance between a center z and other vertices of T, that is, $\text{rad}(T) = \max\{\text{dist}_T(v, z) \mid v \in V(T)\}$.

For convenience, we define $\pi(T) = \pi(\text{diam}(T))$, the parity of the diameter of T. It is well known for all trees T that $\text{diam}(T) = 2 \cdot \text{rad}(T) - \pi(T)$ and that there is a single center vertex if $\pi(T) = 0$ and two adjacent centers if $\pi(T) = 1$. Furthermore, it is obvious for any diametral path in T that the end-vertices x and y are leaves and the center coincides with the center of T. Thus, we have $\min\{\text{dist}_T(z, x), \text{dist}_T(z, y)\} = \text{rad}(T) - \pi(T)$ and $\max\{\text{dist}_T(z, x), \text{dist}_T(z, y)\} = \text{rad}(T)$ for any center vertex z of T.

We call a center vertex z a *min-max center* of T if, for all center vertices z' of T, $\min\{\text{dist}_T(z, v) \mid v \text{ is a leaf of } T\} \ge \min\{\text{dist}_T(z', v) \mid v \text{ is a leaf of } T\}$. Thus, a min-max center maximizes the distance to the closest leaf of T. For a min-max center z, the *leaf distance* is $d_T^{\min} = \min\{\text{dist}_T(z, v) \mid v \text{ is a leaf of } T\}$. The paper also needs the following technical lemma:

Lemma 1. *If* $T_1, \ldots, T_s$ *are* $s \ge 2$ *trees with* $\text{diam}(T_1) \ge \cdots \ge \text{diam}(T_s)$ *then*

  (i) $\text{rad}(T_1) \ge \text{rad}(T_2) \ge \cdots \ge \text{rad}(T_s)$,
 (ii) *for all* $1 \le i < j \le s$, *if* $\text{rad}(T_i) = \text{rad}(T_j)$ *then* $\pi(T_i) \le \pi(T_j)$, *and*
(iii) *for all* $1 \le i < j \le s$, $\text{rad}(T_i) - \pi(T_i) \ge \text{rad}(T_j) - \pi(T_j)$.

2.3   Leaf Powers, Leaf Roots and Their Basic Properties

Let $k \ge 2$ be an integer. A graph G is a k-*leaf power* if a k-*leaf root* T of G exists, a tree with leaves $V(G)$ such that xy is an edge in G if and only if $\text{dist}_T(x, y) \le k$. The example G in Fig. 2 therefore is an 11-leaf power because of the 11-leaf root T of G in the third row and also a 12-leaf power by the 12-leaf root T' in the bottom row. Note that Fig. 2 shows *compressed* illustrations of T and T' where some long paths of vertices with degree two are depicted by single weighted edges. It is well-known that

- a complete graph is a k-leaf power for all $k \ge 2$,
- a graph is a k-leaf power if and only if all of its connected components are k-leaf powers, and
- if x, y are true twins in G then G is a k-leaf power if and only if $G - x$ is a k-leaf power.

Note for the last fact that Lemma 7.3 and Corollary 7.4 in [11] imply the possibility to identify and remove all true twins from a graph in linear time. So, in the remainder of the paper, we smoothly focus on graphs without true twins.

Since the concept of k-leaf powers is slightly different for odd and even k, we formalize this discrepancy as follows: We say that a k-leaf root is of *even*, respectively *odd parity*, if k is even, respectively odd. A k-leaf root T of G is an *optimal even*, respectively *optimal odd* leaf root if k is even, respectively odd, and all k'-leaf roots of G with k' of the same parity as k satisfy $k \le k'$. Finally, a k-leaf root T of G is (just) *optimal* if $k \le k'$ for all k'-leaf roots of G (independent of parity). See the third row of Fig. 2 for an optimal odd leaf root T of the example graph G in the same figure and see the bottom row for an optimal even leaf root T' of G. Since T is an 11-leaf root and T' a 12-leaf root, it follows that T is an optimal leaf root of G.

We conclude this section by establishing a few properties for leaf roots as considered in this paper. The first one concerns a bound on the distance between center and leaves in T in case G is connected.

Lemma 2. *Every k-leaf root T of a connected graph satisfies $d_T^{\min} \le \frac{k}{2}$.*

See Fig. 2 with the 11-leaf root T of G having $d_T^{\min} = \text{dist}_T(u_0, z_0) = 2 \le \frac{11}{2}$ and the 12-leaf root T' with $d_{T'}^{\min} = \text{dist}_T(u_0, z_0) = 2 \le \frac{12}{2}$. Secondly, if G has a universal vertex u then the distance between u and the center in T cannot exceed the difference between k and the radius of T.

Lemma 3. *If G is a non-complete graph with a universal vertex u and T is a k-leaf root of G then $\text{dist}_T(u, z) \le k - \text{rad}(T) + \pi(T)$ for all center vertices z of T. If $z_1 \ne z_2$ are the center vertices of T, then $\text{dist}_T(u, z_1) \le k - \text{rad}(T)$ or $\text{dist}_T(u, z_2) \le k - \text{rad}(T)$.*

For an illustration, see Fig. 2, where the distance of u_0 and the farthest center vertex z_0 of T satisfies $\text{dist}_T(u_0, z_0) = 2 \le 11 - 10 + 1 = k - \text{rad}(T) + \pi(T)$ and, in T', $\text{dist}_{T'}(u_0, z_0) = 2 \le 12 - 11 + 1 = k' - \text{rad}(T') + \pi(T')$. Lemma 3 implies upper bounds on radius and diameter of T.

Corollary 1. *If G is a graph with a universal vertex and T is a k-leaf root of G then $\text{rad}(T) \le k - 1$ and, in particular, $\text{diam}(T) \le 2k - 2$.*

As a matter of fact, k-leaf roots tend to contain long paths of vertices with degree two. It is reasonable to *compress* such a path $P = v_0, \dots, v_n$ into a single *weighted* edge $v_0(n)v_n$. Clearly, weighted edges $v_0(n)v_n$ add their weight n to distances in T and, so, $\text{dist}_T(v_0, v_n) = n$.

3   Optimal Leaf Root Construction for CCGs

Aim of this section is the development of an optimal leaf-root construction approach for ccgs G. In very simple terms, we describe a divide and conquer method that splits G into smaller ccgs $G_1, G_2, \dots$, recursively obtains their optimal leaf roots $T_1, T_2, \dots$, and then extends them into an optimal leaf root for G.

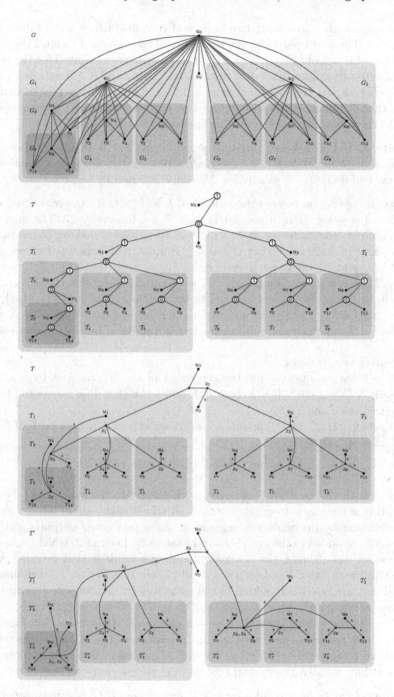

Fig. 2. A ccg G (top), the cotree T of G (2nd row), an optimal (odd) 11-leaf root T of G (3rd row) as computed by Algorithm 1 (with input $p = 1$), and an optimal even 12-leaf root T' of G (bottom) as computed by Algorithm 1 with input $p = 0$.

We start with introducing two basic leaf root operations and analyze their properties. The first operation, the *extension* of trees, is used to level the recursively found leaf roots $T_1, T_2, \ldots$ on the same k, which is essential for the subsequent composition into one k-leaf root. If T is a tree and $\delta \geq 0$ an integer then $T' = \eta(T, \delta)$ is the tree obtained from T by subdividing every pendant edge δ times, that is, replacing the edge with a new path of length $\delta + 1$ (hence, of $\delta + 1$ edges). The following property of this operation is well-known:

Lemma 4. *If T is a k-leaf root of a graph G and $\delta \geq 0$ an integer then $T' = \eta(T, \delta)$ is a $(k + 2\delta)$-leaf root of G with same center, same min-max center vertices, and $\mathrm{diam}(T') = \mathrm{diam}(T) + 2\delta$, $\mathrm{rad}(T') = \mathrm{rad}(T) + \delta$, $d_{T'}^{\min} = d_T^{\min} + \delta$.*

The second operation merges the individual k-leaf roots for the connected components of a graph G into one k-leaf root T for the entire G. The goal here is to minimize the diameter of T, which, in turn, allows making optimizations to the value of k. Assume that G has $s \geq 0$ non-trivial connected components $G_1, \ldots, G_s$ and $t \geq 0$ isolated vertices $v_1, \ldots, v_t$ such that $s + t \geq 2$ and let $T_1, \ldots, T_s$ be k-leaf roots for $G_1, \ldots, G_s$ with min-max center vertices $z_1, \ldots, z_s$. If $s > 0$, we define the *critical index* m as the smallest element of $\{1, \ldots, s\}$ with $d_{T_m}^{\min} = \min\{d_{T_i}^{\min} \mid 1 \leq i \leq s\}$ and call T_m the *critical root*. Then, the *merging* $\mu(k, T_1, \ldots, T_s, v_1, \ldots, v_t)$ results in the tree T produced by the following steps:

1. Create a new vertex c.
2. If $s > 0$ then connect c and the center z_m of the critical root by a path of length $\frac{k + \pi(k)}{2} - d_{T_m}^{\min}$. If $\pi(k) = 0$ and $d_{T_m}^{\min} = \frac{1}{2}k$ then this means to identify the vertices c and z_m.
3. For all $i \in \{1, \ldots, m-1, m+1, \ldots, s\}$, connect c and z_i by a path of length $\frac{k - \pi(k)}{2} + 1 - d_{T_i}^{\min}$.
4. For all $j \in \{1, \ldots, t\}$, connect c and v_j by a path of length $\frac{k - \pi(k)}{2} + 1$.

Notice that the μ-operation is sensitive with respect to the parity $\pi(k)$. For one thing, this is necessary to guarantee that all added paths are of integer length, which is done by in- or decreasing odd k. As a side note, we point out that the lengths of added paths are also non-negative by Lemma 2, which makes the μ-operation well-defined. On the other hand, the result is that merging works slightly different for odd and even k. For odd k, all trees $T_1, \ldots, T_s$, including the critical one, are essentially added in the same way by our construction method. This is because, for odd k, $\frac{k + \pi(k)}{2} - d_{T_m}^{\min} = \frac{k - \pi(k)}{2} + 1 - d_{T_m}^{\min}$. The special treatment of the critical root, thus, has an effect only if k is even. Specifically in that case, we can sometimes save one in the diameter of T, if we put the critical root closer to the center of T than the rest. The reason that this optimization works is that, usually, the critical root has the largest diameter.

Lemma 5. *Let G be a graph and $k \geq 2$ an integer. If G is disconnected with $s \geq 0$ non-trivial connected components $G_1, \ldots, G_s$ and $t \geq 0$ isolated vertices $v_1, \ldots, v_t$ such that $s + t \geq 2$ and if $T_1, \ldots, T_s$ are k-leaf roots for $G_1, \ldots, G_s$ then $T = \mu(k, T_1, \ldots, T_s, v_1, \ldots, v_t)$ is a k-leaf root of G.*

The two operations above simplify the description of the following leaf root construction algorithm for ccgs since they hide away many of the technical details. Foundation of the proposed recursive approach is that (i) induced subgraphs of ccgs are ccgs and (ii) every connected ccg without true twins has a unique universal cut vertex (see Proposition 2). Also, recall from Sect. 2.3 that true twins in graphs can be removed in linear time and, thus, be safely ignored. Therefore, we define for all ccgs G without true twins and a given parity $p \in \{0,1\}$ the result of the *root operation* $\rho(G,p)$ as the tree T and the number k produced by the following (recursive) procedure:

i. **If G is a star then** let u be the central vertex and $v_1, \ldots, v_t$ the leaves of G (with $t \geq 2$ because G does not have true twins)
   1. If $p = 1$ (for odd) then let $T' = \eta(G,1)$, obtain T by attaching a new leaf to u in T', and return $(T,3)$.
   2. If $p = 0$ (for even) and $t = 2$ then return $(T,4)$ with T obtained from a single vertex v by attaching the leaves u, v_1 and v_2 to v with paths of lengths one, two and three, respectively.
   3. If $p = 0$ and $t > 2$ then let $T' = \eta(G,2)$, obtain T by attaching a new leaf to u in T', and return $(T,4)$.
ii. **else if G is a connected graph then** let u be the universal cut vertex of G (by Proposition 2) and let $G_1, \ldots, G_s$ be the $s \geq 1$ non-trivial connected components and $v_1, \ldots, v_t$ the $t \geq 0$ isolated vertices of $G - u$.
   1. Recursively find $(T_1, k_1) = \rho(G_1, p), \ldots, (T_s, k_s) = \rho(G_s, p)$.
   2. If $s = 1$ then let $k = k_1 + 2(1 - \pi(T_1))$. Otherwise, let

$$k_a = \max\{k_1, \ldots, k_s\},$$
$$k_b = \max\{k_i \mid 1 \leq i \leq s, i \neq a\} \text{ and, if } s > 2 \text{ let}$$
$$k_c = \max\{k_i \mid 1 \leq i \leq s, i \neq a, i \neq b\}.$$

   If $p = 1$ (for odd) then let $k = k_a + k_b - 1 - 2 \cdot \pi(T_a) \cdot \pi(T_b)$ and, otherwise,

$$k = \begin{cases} k_a + k_b - 2 \cdot (\pi(T_a) + \pi(T_b) - \pi(T_a) \cdot \pi(T_b)), & \text{if } s = 2 \text{ or } k_a > k_c \\ k_a + k_b - 2 \cdot \pi(T_a) \cdot \pi(T_b), & \text{otherwise.} \end{cases}$$

   3. Get the extended leaf root $T_i' = \eta\left(T_i, \frac{k-k_i}{2}\right)$ for all $i \in \{1, \ldots, s\}$ and let $T' = \mu(k, T_1', \ldots, T_s', v_1, \ldots, v_t)$.
   4. Return (T,k) with T obtained from T' by attaching the leaf u to a center vertex of T'.
iii. **else G is a disconnected graph and then** let $G_1, \ldots, G_s$ be the $s \geq 0$ non-trivial connected components and $v_1, \ldots, v_t$ the $t \geq 0$ isolated vertices of G.
   1. Recursively find $(T_1, k_1) = \rho(G_1, p), \ldots, (T_s, k_s) = \rho(G_s, p)$.
   2. Let $k = \max\{k_1, \ldots, k_s, p + 2\}$ and let $T_i' = \eta\left(T_i, \frac{k-k_i}{2}\right)$ for all $i \in \{1, \ldots, s\}$.
   3. Return (T,k) with $T = \mu(k, T_1', \ldots, T_s', v_1, \ldots, v_t)$.

Hence, if the input graph G is not a star then the approach is to firstly divide G into smaller connected subgraphs $G_1, \ldots, G_s$ (and isolated vertices $v_1, \ldots, v_t$), secondly find corresponding k-leaf roots $T_1, \ldots, T_s$ by recursion and the η-operation, and, last, conquer by merging them into a single leaf root of G with the μ-operation. The divide-step is simple for disconnected G and, otherwise, is carried out by removing the unique universal (cutvertex) of G.

The ρ-operation is sensitive to the given parity p for using μ as a subroutine. Observe that p also decides how the resulting k is determined. There are four cases when G is connected and not a star. In the first one, when $s = 1$, the construction is the same for odd and even p and consists of adding $u, v_1, \ldots, v_t$ at the correct distance to the center of T_1 and computing k from k_1. Secondly, if $s > 1$ and $p = 1$, the μ-operation has only one way of merging the recursively found leaf roots $T_1, \ldots, T_s$ to minimize the diameter of the result T. Then, k widely depends on the two largest values of $k_1, \ldots, k_s$. But if $p = 0$, there is one situation that, on the one hand, allows μ to use a smaller diameter for T by prioritizing the critical leaf root and, on the other hand, lets ρ return a slightly better value for k. This happens only when $s = 2$, or whenever the k_c-leaf root, with k_c the third-largest value among $k_1, \ldots, k_s$, properly fits into the diametral space of T that is already required for the k_a-leaf root and the k_b-leaf root.

The third and bottom row of Fig. 2 illustrate the results $(T, 11)$ of $\rho(G, 1)$ and $(T', 12)$ of $\rho(G, 0)$ on the example G. By recursion, both are produced bottom-up, and it is difficult to follow their assembly at the deeper recursion levels. The highest recursion level of $\rho(G, 1)$, however, has received a 7-leaf root with odd diameter for subgraph G_1 and a 5-leaf root with even diameter for G_2 in Step (ii.1.). In Step (ii.2.), the ρ-procedure determines $k = k_1 + k_2 - 1 = 11$. The extension of the trees in Step (ii.3.) produces the 11-leaf roots T_1 and T_2 for G_1 and G_2, respectively, as shown in Fig. 2. Their following merging and the attachment of u_0 in Step (ii.4.) produces the shown 11-leaf root T of G. Similarly, $\rho(G, 0)$ receives an odd-diameter 8-leaf root of G_1 and an even-diameter 6-leaf root of G_2. Since $s = 2$, the critical root can be treated in the special way and, thus, $\rho(G, 0)$ determines $k' = k'_1 + k'_2 - 2 = 12$. After the extension, we get the 12-leaf roots T'_1 for G_1 and T'_2 for G_2 as in Fig. 2. Their merging and the attachment of u_0 yields the 12-leaf root T' of G as also illustrated there.

The following statement regards the correctness of our procedure.

Theorem 2. *Let G be a ccg on n vertices and without true twins and let $p \in \{0, 1\}$. Then $(T, k) = \rho(G, p)$ provides a k-leaf root T of G that is optimal for parity p (hence, $\pi(k) = p$) and with $k \leq n + 1$. If G is connected then*

(T1) $\operatorname{rad}(T) = k - 1$,
(T2) $d_T^{\min} = 1 + \pi(T)$, *and*
(T3) $\operatorname{diam}(T') \geq \operatorname{diam}(T) + k' - k$ *for all k'-leaf roots T' of G with $\pi(k') = p$.*

Note that, with respect to the optimality of the result, the theorem above makes a slightly stronger statement than our main result in Theorem 1. In fact, the ρ-operation can find a κ-leaf root with minimum κ for every ccg G simply by choosing the best from $(T, k) = \rho(G, 1)$ and $(T', k') = \rho(G, 0)$. To prove Theorem 1, the next section shows how to implement the ρ-operation in linear time.

4   Linear Time Leaf Root Construction for CCGs

The algorithm in this section is an implementation of the ρ-operation from Sect. 3. Here, the recursive subdivision of the input ccg G is replaced with a post-order iteration of the cotree of G. But before we go into the details, we analyze the used submodules and show that the operations η and μ run efficiently.

Lemma 6. *Let T be a compressed tree with n leaves and with explicitly given min-max center z, center Z, diameter $\mathrm{diam}(T)$, and leaf-distance $d_T^{\min}$. For all integers $\delta \geq 0$, the compressed tree $T' = \eta(T, \delta)$ with min-max center z', center Z', diameter $\mathrm{diam}(T')$, and leaf distance $d_{T'}^{\min}$ can be computed in $\mathcal{O}(n)$ time.*

Lemma 7. *Let $s \geq 0$ be an integer and, for all $i \in \{1, \ldots, s\}$, let T_i be a given, compressed tree with explicitly given min-max center z_i, center Z_i, diameter $\mathrm{diam}(T_i)$, and leaf-distance $d_{T_i}^{\min}$. For all integers $k \geq 2$ and vertices $v_1, \ldots, v_t$, the merged compressed tree $T' = \mu(k, T_1, \ldots, T_s, v_1, \ldots, v_t)$ with center Z' and diameter $\mathrm{diam}(T')$ can be computed in $\mathcal{O}(s + t)$ time.*

The recursive definition of $\rho(G, p)$ is implemented as an iterative traversal of the cotree of G. We observe that connected and disconnected graphs G can easily be distinguished with the cotree of G; the former have a root labelled by ① and the latter by ⓪. Likewise, we detect small input graphs, stars, with the cotree by checking if the root is labelled with ① and if the only child that is labelled with ⓪ has just leaf-children.

Recall that, for connected graphs G of sufficient size, the ρ-operation divides G at the unique universal vertex u, to recurse into the non-trivial connected components $G_1, \ldots, G_s$ of $G - u$, and to conquer by merging the according leaf roots $T_1, \ldots, T_s$ into a parity-optimal solution for G. This divide-and-conquer procedure is translated into a traversal of the cotree $\mathcal{T}$ as follows. Since input consists of ccgs without true twins, we rely on Proposition 1. That means that nodes with the label ①, like the root X of $\mathcal{T}$, always have exactly one leaf-child, say u, and one child with label ⓪, say Y. The leaf u marks the unique universal vertex in G and Y has children $Z_1, \ldots, Z_s$ with label ① and leaf-children $v_1, \ldots v_t$ that represent the non-trivial connected components $G_1, \ldots, G_s$ and the isolated vertices $v_1, \ldots, v_t$ of $G - u$. The chosen post-order traversal of $\mathcal{T}$ makes sure that, before processing X (and Y), the nodes $Z_1, \ldots, Z_s$ have been visited and finished. Because we use a stack to pass interim results upwards, we always find leaf roots $T_1, \ldots, T_s$ for $G_1, \ldots, G_s$ on the stack (in reverse order), when we need to compute a leaf root T for the subgraph that corresponds to $\mathcal{T}_X$.

We present the details of our construction in Algorithm 1: `OptimalLeafRoot` and summarize our results in the following theorem.

Theorem 3. *Given a chordal cograph G on n vertices and m edges and $p \in \{0, 1\}$, a (compressed) κ-leaf root of G with minimum integer κ of parity p can be computed in $\mathcal{O}(n + m)$ time.*

Algorithm 1: `OptimalLeafRoot`

Input: A ccg $G = (V, E)$ without true twins and a parity $p \in \{0, 1\}$
Output: A pair (T, k) of a k-leaf root T of G with smallest p-parity integer k.

1 initialize empty stack $\mathcal{S}$ and compute the cotree $\mathcal{T}$ of G
2 **foreach** *node X visited traversing $\mathcal{T}$ in post-order* **do**
3   **if** X *is labelled with* ① **then**
4   let Y be the ⓪-child and u the leaf-child of X
5   let s be the number of ①-children and $v_1, \ldots, v_t$ the leaf-children of Y
6   **if** $s = 0$ **then**   // Case i., base case, input is a star
7   build T like Case i., Section 3 for star on edges $uv_1, \ldots, uv_t$
8   push $(T, 4 - p)$ onto $\mathcal{S}$
9   **else**   // Case ii., input is a connected graph
10   **foreach** $i \in \{s, s-1, \ldots, 1\}$ **do** pop (T_i, k_i) from $\mathcal{S}$ **if** $s = 1$ **then**
            $k \leftarrow k_1 + 2(1 - \pi(T_1))$ **else**
11             $k_a \leftarrow \max\{k_1, \ldots, k_s\}$
12             $k_b \leftarrow \max\{k_i \mid 1 \leq i \leq s, i \neq a\}$
13   **if** $p = 1$ **then** $k \leftarrow k_a + k_b - 1 - 2 \cdot \pi(T_a) \cdot \pi(T_b)$ **else**
14   **if** $s > 2$ *and* $k_a > \max\{k_i \mid 1 \leq i \leq s, i \neq a, i \neq b\}$ **then**
15                   $k \leftarrow k_a + k_b - 2 \cdot \pi(T_a) \cdot \pi(T_b)$
16   **else** $k \leftarrow k_a + k_b - 2 \cdot (\pi(T_a) + \pi(T_b) - \pi(T_a) \cdot \pi(T_b))$
17   **end**
18   **end**
19   **foreach** $i \in \{1, \ldots, s\}$ **do** $T_i' \leftarrow \eta(T_i, (k - k_i)/2)$
            $T \leftarrow \mu(k, T_1', \ldots, T_s', v_1, \ldots, v_t)$ with $\leftarrow$ a center of T
20   attach u as a leaf to a center of T
21   push (T, k) onto $\mathcal{S}$
22   **end**
23   **end**
24   **if** X *is* ⓪-*node without parent* **then**   // Case iii., disconnected input
25   let s be the number of ①-children and $v_1, \ldots, v_t$ the leaf-children of X
26   **foreach** $i \in \{1, \ldots, s\}$ **do** pop (T_i, k_i) from $\mathcal{S}$ $k \leftarrow \max\{k_1, \ldots, k_s, p + 2\}$
27   **foreach** $i \in \{s, s-1, \ldots, 1\}$ **do** $T_i' \leftarrow \eta(T_i, (k - k_i)/2)$
         $T \leftarrow \mu(k, T_1', \ldots, T_s', v_1, \ldots, v_t)$
28   push (T, k) onto $\mathcal{S}$
29   **end**
30   pop (T, k) from $\mathcal{S}$
31   **return** (T, k)
32 **end**

---

5 Conclusion

With Theorem 3, we have shown that the OLR problem is linear-time solvable for chordal cographs. Our work also provides a linear-time solution for the k-leaf power recognition problem on chordal cographs. Specifically, for a given ccg G and an integer k, it is sufficient to compute $(T, \kappa) = \rho(G, \pi(k))$ (in linear time with Algorithm 1) to see by $\kappa \leq k$ if G is a k-leaf power.

We conclude the paper by exploring the differences in the construction of odd and even leaf-roots. As we have seen, merging three or more even leaf roots sometimes requires a slightly stronger increase in k than for odd leaf roots. This can accumulate to an arbitrary big gap between k and k' of an optimal odd k-leaf root and an optimal even k'-leaf root of a given ccg. For example, consider the (infinite) series $F_1, F_2, F_3, \ldots, F_i, \ldots$ of ccgs defined as follows. Let F_0 be the path on three vertices and for all integers $i > 0$ define

$$F_i = t_i \; ① \; ((x_i \; ① \; (F_{i-1} \; ⓪ \; u_i)) \; ⓪ \; (y_i \; ① \; (F_{i-1} \; ⓪ \; v_i)) \; ⓪ \; (z_i \; ① \; (F_{i-1} \; ⓪ \; w_i)))$$

with $g \in \{t_i, u_i, v_i, w_i, x_i, y_i, z_i\}$ denoting a graph with the single vertex g. By Sect. 2, $F_1, F_2, \ldots$ are a family of ccgs and, apparently, all these graphs are connected and without true twins.

Lemma 8. *For all integers $i \geq 1$, the graph F_i is a (odd) k_i-leaf power for $k_i = 2^{i+2} - 1$ but not a $(k_i - 2)$-leaf power and a (even) k_i'-leaf power for $k_i' = k_i + 2^i - 1$ but not a $(k_i' - 2)$-leaf power.*

This means that, although odd and even leaf root construction follows the same approach, there are k-leaf powers of odd k among the chordal cographs that have optimal even k'-leaf roots with k' roughly $1.25k$.

References

1. Brandstädt, A., Hundt, C.: Ptolemaic graphs and interval graphs are leaf powers. In: LATIN 2008, pp. 479–491 (2008). https://doi.org/10.1007/978-3-540-78773-0_42
2. Brandstädt, A., Hundt, C., Mancini, F., Wagner, P.: Rooted directed path graphs are leaf powers. Discret. Math. **310**(4), 897–910 (2010). https://doi.org/10.1016/j.disc.2009.10.006
3. Brandstädt, A., Le, V.B.: Structure and linear time recognition of 3-leaf powers. Inf. Process. Lett. **98**(4), 133–138 (2006). https://doi.org/10.1016/j.ipl.2006.01.004
4. Brandstädt, A., Le, V.B., Sritharan, R.: Structure and linear-time recognition of 4-leaf powers. ACM Trans. Algorithms **5**(1), 11:1–11:22 (2008). https://doi.org/10.1145/1435375.1435386
5. Chang, M., Ko, M.: The 3-steiner root problem. In: WG 2007, pp. 109–120 (2007). https://doi.org/10.1007/978-3-540-74839-7_11
6. Corneil, D.G., Lerchs, H., Burlingham, L.S.: Complement reducible graphs. Discrete Appl. Math. **3**(3), 163–174 (1981). https://doi.org/10.1016/0166-218X(81)90013-5
7. Ducoffe, G.: The 4-steiner root problem. In: WG 2019, pp. 14–26 (2019). https://doi.org/10.1007/978-3-030-30786-8_2
8. Eppstein, D., Havvaei, E.: Parameterized leaf power recognition via embedding into graph products. Algorithmica **82**(8), 2337–2359 (2020). https://doi.org/10.1007/s00453-020-00720-8
9. Golumbic, M.C.: Trivially perfect graphs. Discrete Math. **24**(1), 105–107 (1978). https://doi.org/10.1016/0012-365X(78)90178-4
10. Lafond, M.: Recognizing k-leaf powers in polynomial time, for constant k. In: SODA 2022, pp. 1384–1410. SIAM (2022). https://doi.org/10.1137/1.9781611977073.58
11. McConnell, R.M.: Linear-time recognition of circular-arc graphs. Algorithmica **37**(2), 93–147 (2003). https://doi.org/10.1007/s00453-003-1032-7
12. Nishimura, N., Ragde, P., Thilikos, D.M.: On graph powers for leaf-labeled trees. J. Algorithms **42**(1), 69–108 (2002). https://doi.org/10.1006/jagm.2001.1195
13. Rosenke, C., Le, V.B., Brandstädt, A.: Leaf powers, pp. 168–188. Encyclopedia of Mathematics and its Applications, Cambridge University Press (2021). https://doi.org/10.1017/9781108592376.011
14. Wagner, P., Brandstädt, A.: The complete inclusion structure of leaf power classes. Theor. Comput. Sci. **410**(52), 5505–5514 (2009). https://doi.org/10.1016/j.tcs.2009.06.031

15. Wolk, E.: The comparability graph of a tree. Proc. Am. Math. Soc. **13**, 789–795 (1962). https://doi.org/10.1090/S0002-9939-1962-0172273-0
16. Wolk, E.: A note on the comparability graph of a tree. Proc. Am. Math. Soc. **16**, 17–20 (1965). https://doi.org/10.1090/S0002-9939-1962-0172273-0

Verified Exact Real Computation with Nondeterministic Functions and Limits

Sewon Park[(✉)] [iD]

Kyoto University, Kyoto, Japan
sewon@kurims.kyoto-u.ac.jp

Abstract. The problems of computing limit points nondeterministically from sequences of nondeterministic real numbers appear ubiquitously in exact real computation along with root-finding of real and complex functions. To provide a rigorous foundation of verified computations with nondeterministic limits, we introduce a simple imperative language for exact real computation with a nondeterministic limit operator as its primitive. The operator's formal semantics is defined. To make nontrivial sequences of nondeterministic real numbers be defined within the language, we further extend the language with lambda expressions for constructing higher-order nondeterministic functions without side effects and countable nondeterministic choices. We devise proof rules for the new operations and prove their soundness. As an example, to demonstrate the strength of the proof rules, we verify the correctness of a program, a computational counterpart of a constructive Intermediate Value Theorem, computing nondeterministically a root of a continuous locally nonconstant real function whose signs at each endpoint of the unit interval differ.

Keywords: Exact real computation · Nondeterministic limits · Unbounded nondeterminism · Formal verification

1 Introduction

In exact real computation, real numbers are expressed and manipulated exactly without introducing rounding errors; e.g., see [10]. Instead of approximating real numbers to finite-precision floating-point numbers, which inevitably generate and accumulate round-off errors, real numbers in exact real computation are internally represented by infinite sequences and processed exactly via stream computations. Despite an inevitable inefficiency compared to using hardware-supported floating-point numbers, as the results of exact real computations are guaranteed to be free from numerical errors, it is suitable where arbitrarily high precision is required. Moreover, by abstracting representations away, the real numbers and their operations form closely the classical structure of real numbers

---

The author is a JSPS International Research Fellow supported by JSPS KAKENHI (Grant-in-Aid for JSPS Fellows) JP22F22071.

making it intuitive to reason about their computational behaviours. Implementations of exact real computation optimizing their performances suggest their usefulness in mission-critical application domains; see [3,15,18] for examples.

Nondeterminism is essential in exact real computation [16].[1] Instead of testing order comparisons naively by iterating through the infinite representations, which diverges (failing to terminate) when two real numbers that are compared coincide [23, Theorem 4.1.16], the following total nondeterministic variant, so-called soft comparison [5,16], is often used:

$$x <_k y := \equiv \{\texttt{true} \mid x < y + 2^{-k}\} \cup \{\texttt{false} \mid y < x + 2^{-k}\}. \tag{1}$$

Here, k is an integer and the set-valued function value denotes the set of possible nondeterministic results. It says when x and y are far apart, the order gets computed exactly and when they are close, either $\texttt{true}$ or $\texttt{false}$ gets returned nondeterministically, relative to 2^{-k}. It is a useful nondeterministic approximation of the order $x < y$, realized by evaluating the two possibly diverging tests $x < y + 2^{-k}$ and $y < x + 2^{-k}$ in parallel.

Constructing limits is a unique feature of exact real computation that makes it more expressive than algebraic or symbolic computation [19]. An interesting and inevitable case where nondeterminism gets engaged in exact real number computation is when it is combined with limits. (Note that such cases appear ubiquitously along with root-finding of real and complex functions.) The importance of constructing nondeterministic limits from their nondeterministic approximations is noticed in [17] and later it is implemented in **iRRAM**, a C++ library for exact real computation, with an example of complex square root function [18, Section 8]. In [13], a formal specification of nondeterministic limits is provided, as a theorem in a setting of constructive dependent type theory [14].

Programs in imperative programming languages can be specified naturally by their precondition and postcondition. Furthermore, Hoare-style proof rules can be designed for conveniently verifying the specifications; e.g., see [1]. A recent work [20] has formalized an imperative programming language designed to provide a framework of formal verification of first-order exact real computations with ordinary deterministic limits. There, the soundness of the verification calculus is proved with regards to its formal semantics. As a demonstrating example, the correctness of a (deterministic) root-finding program is proved. Dealing with nondeterministic limits and higher-order computations are left as future works.

In this paper, we bring the concept of nondeterministic limits into the framework of imperative programming and formal verification, to provide a rigorous foundation of verified computations with nondeterministic limits. Based on an imperative programming language for first-order exact real computation, we propose introducing a nondeterministic limit operator. We define the formal semantics of the limit operator by translating the type-theoretic specification in [13] to a set-theoretic function. To define nontrivial sequences, we extend the language with lambda expressions for constructing nondeterministic (higher-order)

---

[1] It is also often called multivaluedness or non-extensionality; see [6].

functions without side effects and countable nondeterminism. We further devise sound Hoare-style proof rules for the suggested operations.

Here, we do not specify the base language, though we have [20] in our mind since the specific design choices made in the base language are irrelevant to the contribution of this paper. We demand the base language to provide data types for (lazy) Boolean, integers, and real numbers; to admit an intuitive type system; and to offer fully specified exact first-order operations over them. It can be considered as a language modelling a core first-order fragment of **iRRAM**. Under this aspect, our work is extending the modelling language and its specification logic with higher-order computation that C++ already provides and nondeterministic limits. The latter includes modifying the nondeterministic limit operator in **iRRAM** and giving it formal semantics for formal verification.

In order to deal with our nondeterministic functions, in logical assertions, we adopt the program specification method of [12] with a minor modification and equip our assertion language with the proposition

$$\{\varphi\}\ f \bullet (e_1, \cdots, e_n) \searrow y\ \{\psi\}$$

saying the nondeterministic function f on its inputs $e_1, \cdots, e_n$ satisfying φ yields y satisfying ψ. Based on the specification language, we devise proof rules for the nondeterministic limit operator and the countable nondeterminism operator.

This paper is organized as follows. In Sect. 2, we specify how we model nondeterministic functions, define the semantics of a nondeterministic limit operator, and show that loops are well-defined. In Sect. 3, we define the formal syntax, type system, and denotational semantics of the new operations. In Sect. 4, we define the assertion language and program specification. Then, proof rules are introduced and proved sound in Sect. 5. They are used in Sect. 6 when we verify the correctness of a nondeterministic root-finding program.

2 Preliminaries

2.1 Nondeterministic Functions

We write $f : X \rightrightarrows Y$ for a partial nondeterministic function from X to Y.[2] It is denoted by a (potentially) set-valued function $f : X \rightarrow \mathsf{P}_\star(Y)$ where

$$\mathsf{P}_\star(Z) :\equiv \{S \subseteq Z \mid S \text{ is nonempty}\} \cup \{\bot\} \quad \text{for any set } Z.$$

We say $f(x)$ is defined when $f(x) \neq \bot$. In this case, the nonempty set $f(x) \subseteq Y$ denotes the set of all possible outcomes of the nondeterministic computation that $f(x)$ represents, guaranteeing that there is no failing nondeterministic branch.

---

[2] Often, a different notation $f :\subseteq X \rightrightarrows Y$ is used for a partial nondeterministic function and $f : X \rightrightarrows Y$ denotes a total nondeterministic function. However, we do not make a syntactic distinction between total and partial nondeterministic functions, and assume that all nondeterministic functions are possibly partial. Hence, in this paper, we use $X \rightrightarrows Y$ for the set of partial nondeterministic functions. For example, $(X \rightrightarrows Y) \rightrightarrows Z$ denotes the set of partial nondeterministic functions to Z from partial nondeterministic functions from X to Y.

Otherwise, when $f(x) = \perp$, we say $f(x)$ is undefined. It denotes that there exists a failing nondeterministic branch in the computation. Note that we consider failures contagious in the sense that we mark $\perp$ whenever there is a branch that fails. We may use the powerset of Y instead of $\mathsf{P}_\star(Y)$ and identify $\perp$ with $\emptyset$; however, then we need to modify the set union operator to deal with $\emptyset$ differently from other subsets when merging two nondeterministic results.

When $f : Y \rightrightarrows Z$ and $g : X \rightrightarrows Y$, their composition is expressed by $f^\dagger \circ g : X \rightrightarrows Z$ where $f^\dagger : \mathsf{P}_\star(Y) \to \mathsf{P}_\star(Z)$ is defined by

$$f^\dagger(S) = \begin{cases} \perp & \text{if } S = \perp \vee \exists y \in S.\ f(y) = \perp, \\ \bigcup_{y \in S} f(y) & \text{otherwise.} \end{cases}$$

A product domain can be extended by considering all possible combinations: $s_{X_1,\cdots,X_n} : (\mathsf{P}_\star(X_1) \times \cdots \times \mathsf{P}_\star(X_n)) \to \mathsf{P}_\star(X_1 \times \cdots \times X_n)$ defined by

$$s_{X_1,\cdots,X_n}(S_1,\cdots,S_n) :\equiv \begin{cases} \perp & \text{if } S_i = \perp \text{ for some } i, \\ S_1 \times \cdots \times S_n & \text{otherwise.} \end{cases}$$

Similarly, for any X, $s_{\mathbb{Z} \to X} : (\mathbb{Z} \to \mathsf{P}_\star(X)) \to \mathsf{P}_\star(\mathbb{Z} \to X)$ is defined by

$$s_{\mathbb{Z} \to X}(f) :\equiv \begin{cases} \perp & \text{if } f(z) = \perp \text{ for some } z \in \mathbb{Z}, \\ \{g \mid \forall z \in \mathbb{Z}.\ g(z) \in f(z)\} & \text{otherwise.} \end{cases}$$

For example, suppose we want to add nondeterministic real numbers. The addition $+ : \mathbb{R}^2 \to \mathbb{R}$ first gets lifted $((x,y) \mapsto \{x + y\}) : \mathbb{R}^2 \to \mathsf{P}_\star(\mathbb{R})$ to be a nondeterministic function. To get applied to another nondeterministic result, we extend its domain $((x,y) \mapsto \{x+y\})^\dagger \circ s_{\mathbb{R},\mathbb{R}} : \mathsf{P}_\star(\mathbb{R}) \times \mathsf{P}_\star(\mathbb{R}) \to \mathsf{P}_\star(\mathbb{R})$. Though the notation is complicated, it does exactly what we expect it to do:

$$((x,y) \mapsto \{x+y\})^\dagger \circ s_{\mathbb{R},\mathbb{R}}(S,T) = \begin{cases} \perp & \text{if } S = \perp \vee T = \perp, \\ \bigcup_{s \in S, t \in T}\{s + t\} & \text{if } S \neq \perp \wedge T \neq \perp. \end{cases}$$

By abuse of notation, for a (partial) ordinary or nondeterministic function f, we write $f^\dagger$ to refer to the partial nondeterministic function that is lifted and domain extended. For example, we simply write $S +^\dagger T$ for the above expression. It should be clear from the context which operations are implicitly applied.

Lazy Boolean denotes the set $\mathbb{L} = \{\texttt{true}, \texttt{false}, \texttt{div}\}$ of truth values where $\texttt{div}$ denotes the delayed divergence.[3] Based on lazy Boolean, a countable nondeterminism operator, similarly to [22] but on function arguments, is defined:

$$\text{choose}(f : \mathbb{Z} \to \mathbb{L}) :\equiv \{z \in \mathbb{Z} \mid f(z) = \texttt{true}\}_\star \tag{2}$$

where $S_\star = S$ if and only if $S \neq \emptyset$ and $S_\star = \perp$ if and only if $S = \emptyset$. Given indexed lazy Boolean objects f, $\text{choose}(f)$ nondeterministically selects an index

---

[3] In exact real computation, it is also called Kleenean; see [6].

for which f evaluates to **true**. Even when there are some that diverge, evaluating to div, the choose operator safely gets rid of them as long as there is an index it can choose. Note that $i \in \mathrm{choose}^\dagger(f : \mathbb{Z} \rightrightarrows \mathbb{L})$ if and only if **true** $\in f(i)$, $f(j) = \{$**true**$\}$ for some j, and $f(z) \neq \bot$ for all z. Otherwise, $\mathrm{choose}^\dagger(f : \mathbb{Z} \rightrightarrows \mathbb{L}) = \bot$.

The operation $\mathrm{choose}(f)$ can be realized by a scheduler that evaluates all $\{f(i) \mid i \in \mathbb{Z}\}$ in parallel until it finds an index i such that $f(i)$ terminates and evaluates to **true**.

2.2 Nondeterministic Limits

Suppose an ordinary limit operator as the following partial function:

$$\lim(f : \mathbb{Z} \to \mathbb{R}) = \begin{cases} y & \text{if } \forall p \in \mathbb{Z}.\ |y - f(p)| \leq 2^{-p}, \\ \bot & \text{otherwise.} \end{cases}$$

It is defined on $f : \mathbb{Z} \to \mathbb{R}$ which approximates a real number rigorously in the sense that when it receives a precision argument p, heading to ∞, it returns a 2^{-p} approximation to the real number. In other words, f is a (rapidly converging) Cauchy sequence as $p \to \infty$; see [4,8].

Note that, using the construction in Sect. 2.1, the ordinary limit operator can already be extended to accept sequences of nondeterministic real numbers:

$$\lim{}^\dagger(f : \mathbb{Z} \rightrightarrows \mathbb{R}) = \{y \mid \forall p \in \mathbb{Z}.\ \forall x \in f(p).\ |y - x| \leq 2^{-p}\}_\star.$$

Though it is a natural extension, the problem is that it only accepts sequences of nondeterministic real numbers that converge to deterministic points; observe that the value $\lim{}^\dagger(f)$ can have at most one real number.

Instead, we recall the nondeterministic limit theorem in [13, Theorem 2]. The theorem states indirectly on how to construct limits nondeterministically. Here, we translate the type-theoretic statement into our setting taking some minor changes. Consider any set X. For a closed subset $L \subseteq \mathbb{R}$ and indexed binary relations $(I_q)_{q \in \mathbb{Z}} \subseteq \mathbb{R} \times X$, a nondeterministic function $f : \mathbb{Z} \times ((\mathbb{R} \times \mathbb{Z}) \times X) \rightrightarrows \mathbb{R} \times X$ is called a *nondeterministic refinement procedure* for the *nondeterministic limit* L w.r.t. the *hint invariant* I if the following holds. For each inputs $p \in \mathbb{Z}, x \in \mathbb{R}, q \in \mathbb{Z}, h \in X$ such that $q < p, \exists y \in L.\ |x - y| \leq 2^{-q}$, and $(x, h) \in I_q$, it holds that

1. $f(p, ((x, q), h))$ is defined (i.e., $f(p, ((x, q), h)) \neq \bot$),
2. and for each $(x', h') \in f(p, ((x, q), h))$, the followings hold:

$$\exists y' \in L.\ |x' - y'| \leq 2^{-p}, \quad |x - x'| \leq 2^{-q} - 2^{-p}, \quad \text{and} \quad (x', h') \in I_p.$$

In words, f, given a 2^{-q} approximation x to some $y \in L$ and a hint h that satisfies I_q, nondeterministically refines it to a 2^{-p} approximation x' to some (possibly different) $y' \in L$, near the original x, and produces another hint h' that together with x' satisfies I_p. Let us write $\mathrm{A}_{L,I}(f)$ for this condition on f.

Furthermore, (x_0, q_0) and h_0 are called an *initial approximation and hint* of L w.r.t. I when there exists $y \in L$ such that $|y - x_0| \leq 2^{-q_0}$ and $(x_0, h_0) \in I_{q_0}$. Let us write $B_{L,I}(x_0, q_0, h_0)$ for this condition on (x_0, q_0) and h_0.

The theorem states that L gets *constructed* nondeterministically for some I when f and (x_0, q_0, h_0) are given such that $A_{L,I}(f) \wedge B_{L,I}(x_0, q_0, h_0)$. For such f and (x_0, q_0, h_0), what is happening operationally is intuitive. For some infinite increasing sequence of integers $\mathbf{p} = p_0, p_1, \cdots$ where $p_0 = q_0$, representing an implementation-specific protocol on how we increase precision, it creates an infinite sequence using this recursion-like procedure:

$$\text{pick one } (x_{n+1}, h_{n+1}) \in f(p_{n+1}, ((x_n, p_n), h_n)) \,.$$

Due to the conditions, this indefinite procedure nondeterministically constructs one deterministic sequence $(x_n)_{n \in \mathbb{N}}$ that converges with ratio $|x_n - x_m| \leq 2^{-p_n} + 2^{-p_m}$. Also, for each n, there is $y_n \in L$ that is 2^{-p_n} close to x_n. As L is sequentially closed, the limit point of $(x_n)_{n \in \mathbb{N}}$ which is the limit point of $(y_n)_{n \in \mathbb{N}}$ is in L. Therefore, applying an ordinary limit operation to $(x_n)_{n \in \mathbb{N}}$, a point in L gets computed. That means, depending on which deterministic sequence $(x_n)_{n \in \mathbb{N}}$ is constructed, the procedure constructs a limit point nondeterministically in L.

However, thus far discussion is not sufficient in giving a definition to an explicit nondeterministic limit operator of type

$$\lim_X^{\rightrightarrows} : ((\mathbb{Z} \times ((\mathbb{R} \times \mathbb{Z}) \times X)) \rightrightarrows (\mathbb{R} \times X)) \times ((\mathbb{R} \times \mathbb{Z}) \times X) \rightrightarrows \mathbb{R} \qquad (3)$$

parameterized by a set X. The reason is that for the arguments f and $((x_0, q_0), h_0)$, the set L satisfying the above conditions is not unique. In order to assign a specific value of $\lim_X^{\rightrightarrows}(f, ((x_0, q_0), h_0))$, let us consider the set of nondeterministic limits that the input *constructs*:

$$C_{f, x_0, q_0, h_0} \equiv \{L \mid \exists I. \, A_{L,I}(f) \wedge B_{L,I}(x_0, q_0, h_0)\}.$$

We take the most precise one:

Lemma 1. *When* C_{f, x_0, q_0, h_0} *is not empty,* $\bigcap_{L \in C_{f, x_0, q_0, h_0}} L \in C_{f, x_0, q_0, h_0}$.

The lemma says, the common limit points in all possible nondeterministic limits again form one nondeterministic limit. Hence, we define the limit operator as follows:

$$\lim_X^{\rightrightarrows}(f, ((x_0, q_0), h_0)) :\equiv \bigcap_{L \in C_{f, x_0, q_0, h_0}} L \qquad (4)$$

defined precisely at $\exists I. \, A_{L,I}(f) \wedge B_{L,I}(x_0, q_0, h_0)$.

2.3 While Loops

The denotation of a program is a partial nondeterministic function $f : X \rightrightarrows Y$ for some sets X and Y. Sequential compositions are done by the liftings defined earlier: when $f : X \rightrightarrows Y$ and $g : Y \rightrightarrows Z$ are the denotations of two programs, the composition of the two programs denotes $g^\dagger \circ f : X \rightrightarrows Z$.

For subsets $B \in \mathsf{P}_\star(\mathbb{L})$, $S, T \in \mathsf{P}_\star(X)$ for some X, define the following operator modeling the branching operator on a lazy Boolean condition:

$$\mathrm{ite}(B, S, T) :\equiv \begin{cases} \bot & \begin{aligned} &\text{if } B = \bot \text{ or } \mathbf{div} \in B \\ &\text{or } (\mathbf{true} \in B \wedge S = \bot) \\ &\text{or } (\mathbf{false} \in B \wedge T = \bot), \end{aligned} \\[2em] \cup \begin{cases} S & \text{if } \mathbf{true} \in B, \\ T & \text{if } \mathbf{false} \in B, \end{cases} & \text{otherwise.} \end{cases}$$

Note that the statement is defined to fail when the condition is $\mathbf{div}$ when evaluating the condition diverges.

For a set of states X, a loop condition $b : X \rightrightarrows \mathbb{L}$, and a loop body $c : X \rightrightarrows X$, the loop is a function $\mathrm{while}_{b,c} : X \rightrightarrows X$ that satisfies the recurrence equation:

$$\mathrm{while}_{b,c}(x) \doteq \mathrm{ite}(b(x), \mathrm{while}_{b,c}^\dagger(c(x)), \{x\}) \,.$$

Therefore, we define it to be a fixed-point of the operator:

$$\mathrm{W}_{b,c} : f \mapsto \left(x \mapsto \mathrm{ite}(b(x), \mathrm{while}_{b,c}^\dagger(c(x)), \{x\}) \right).$$

Lemma 2. *The operator* $\mathrm{W}_{b,c}$ *is monotone w.r.t. the point-wise ordering of functions to the flat domain* $\mathsf{P}_\star(X)$. *In addition, it admits a fixed-point.*

There are various ways to formalize denotations of loops in a setting of unbounded nondeterminism and all of them involve complications; see [2,9] for examples. Instead of specifying one construction, deferring it to a future work, in this paper we define $\mathrm{while}_{b,c}$ to be a fixed-point of the operator that is known to exist.

3   The Programming Language

3.1   Syntax and Typing Rules

We suppose there are the four base types: R for real numbers, Z for integers, IB for the lazy Boolean, and U for a singleton unit. Based on the base types, there are product types $\tau \times \sigma$ and function types $\tau_1 \times \cdots \times \tau_n \rightrightarrows \sigma$. In addition, for each type τ, there is $\mathrm{ro}(\tau)$ the read-only type of τ.

A context Γ is a finite partial mapping from variables to their data types and we write $\Gamma \vdash e : \tau$ for our typing rule judging an *expression* e to have type τ under a context Γ. We write $\Gamma \vdash c \triangleright \Delta$ for our typing rule judging a *command* c to be well-formed yielding a new context Δ under a context Γ.

The expression language includes integer, real, and lazy Boolean constants and operations. There are both a naive coercion $\mathbb{Z} \ni x \mapsto x \in \mathbb{R}$ and an exponential coercion $\mathbb{Z} \ni x \mapsto 2^x \in \mathbb{R}$ from integers to reals. Also reading read-only types is done implicitly in the sense that $\Gamma \vdash e : \tau$ holds when $\Gamma \vdash e : \mathrm{ro}(\tau)$. Of course, real number comparison $x < y$ is defined to be lazy Boolean typed that is meant to be div when x and y evaluate to the same real number.

Besides the operations on base types, there are lambda expressions $\lambda(x_1 : \tau_1, \cdots, x_n : \tau_n). c; e$ and function applications with the typing rules:

$$\frac{\text{ro}(\Gamma), v_1 : \tau_1, \cdots, v_n : \tau_n \vdash c \triangleright \Delta \quad \Delta \vdash e : \sigma}{\Gamma \vdash \lambda(v_1 : \tau_1, \cdots, v_n : \tau_n). c; e : \vec{\tau} \Rrightarrow \sigma} \qquad \frac{\Gamma \vdash f : \tau_1 \times \cdots \times \tau_n \Rrightarrow \sigma \quad \Gamma \vdash e_i : \tau_i \, (i = 1, \cdots, n)}{\Gamma \vdash f(e_1, \cdots, e_n) : \sigma}$$

Here, $\vec{\tau}$ is $\tau_1 \times \cdots \times \tau_n$ and $\text{ro}(\Gamma)$ is the context where each data type assigned in Γ is casted by $\text{ro}(\cdot)$. The lambda expression mimics the lambda expression in C++ and **iRRAM** $[=](\tau_1 \, v_1, \cdots, \tau_n \, v_n)\{c; \mathbf{return}(e); \}$ that can read-only the surrounding environment.

Countable choices and nondeterministic limits admit the following rules:

$$\frac{\Gamma \vdash f : \mathsf{Z} \Rrightarrow \mathsf{IB}}{\Gamma \vdash \mathtt{choose}(f) : \mathsf{Z}} \qquad \frac{\Gamma \vdash f : \mathsf{Z} \times ((\mathsf{R} \times \mathsf{Z}) \times \tau) \Rrightarrow \mathsf{R} \times \tau \quad \Gamma \vdash e : (\mathsf{R} \times \mathsf{Z}) \times \tau}{\Gamma \vdash \mathtt{lim}_\tau^{\Rrightarrow}(f, e) : \mathsf{R}}$$

The limit operator is polymorphic in the sense that $\mathtt{lim}_\tau^{\Rrightarrow}(f, e)$ for each data type τ has type R. Compare the rule with the type of the limit operator in Eq. (3).

3.2   Denotational Semantics

The denotations of data types are defined recursively as $[\![\mathsf{R}]\!] :\equiv \mathbb{R}$, $[\![\mathsf{Z}]\!] :\equiv \mathbb{Z}$, $[\![\mathsf{IB}]\!] :\equiv \mathbb{L}$, $[\![\tau \times \sigma]\!] :\equiv [\![\tau]\!] \times [\![\sigma]\!]$, $[\![\tau_1 \times \cdots \times \tau_n \Rrightarrow \sigma]\!] :\equiv [\![\tau_1]\!] \times \cdots \times [\![\tau_n]\!] \Rrightarrow [\![\sigma]\!]$, and $[\![\text{ro}(\tau)]\!] :\equiv [\![\tau]\!]$. A context Γ denotes the set of variable assignments. A well-typed expression $\Gamma \vdash e : \tau$ denotes a partial nondeterministic function $[\![\Gamma \vdash e : \tau]\!] : [\![\Gamma]\!] \Rrightarrow [\![\tau]\!]$ where $[\![\Gamma \vdash e : \tau]\!](\gamma) \subseteq [\![\tau]\!]$ denotes the set of nondeterministic results of evaluating e under a state $\gamma \in [\![\Gamma]\!]$. The evaluation failing is captured by $\bot$. A well-formed command $\Gamma \vdash c \triangleright \Delta$ denotes a partial nondeterministic function $[\![\Gamma \vdash c \triangleright \Delta]\!] : [\![\Gamma]\!] \Rrightarrow [\![\Delta]\!]$ where $[\![\Gamma \vdash c \triangleright \Delta]\!](\gamma)$, when it is not $\bot$, is the set of resulting states executing c from a state γ.

For the new features, we let choose denote the countable choice choose from Eq. (2), the nondeterministic limit operator $\mathtt{lim}_\tau^{\Rrightarrow}$ denote the nondeterministic limit $\lim_{[\![\tau]\!]}^{\Rrightarrow}$ from Eq. (4), and a while loop denotes the fixed point from Lemma 2. A lambda expression is defined to denote a partial nondeterministic function.

4   Specifications

4.1   Assertion Language for Nondeterministic Functions

We consider a many-sorted logic whose sorts are the data types as the assertion language. That means the language provides terms of type $\vec{\tau} \Rrightarrow \sigma$ which denote partial nondeterministic functions from $\vec{\tau}$ to σ. Instead of introducing power-sets and regarding nondeterministic functions be ordinary functions to them, we take an axiomatic treatment. For any $f : \vec{\tau} \rightrightarrows \sigma$ and $(e_1, \cdots, e_n) : \vec{\tau}$, instead of having $f(e_1, \cdots, e_n)$ in the term language of some type, as in [12] we extend the language of formulae with triples of the form:

$$\{\varphi\} \, f \bullet (e_1, \cdots, e_n) \searrow y \, \{\psi\}$$

It says that the partial nondeterministic f on arguments $(e_1, \cdots, e_n)$ satisfying φ is well-defined and the resulting values represented by variable y satisfy ψ.

For example, suppose we want to make a predicate on $f : \mathsf{R} \times \mathsf{Z} \times \mathsf{R} \rightrightarrows \mathsf{IB}$ saying that f is the soft comparison in Eq. (1). Then, instead of writing $f(x, k, y)$ which is not a well-typed term in our logical language, we use $\forall x, y : \mathsf{R}, k : \mathsf{Z}$.

$$\{\mathsf{True}\} \, f \bullet (x, k, y) \searrow b \, \{b \downarrow \wedge (b \Rightarrow x < y + 2^{-k}) \wedge (\neg b \Rightarrow y < x + 2^{-k})\}.$$

Here, for a term $b : \mathsf{IB}$, b used as a formula is an abbreviation for $b = \mathtt{true}$, $\neg b$ is an abbreviation for $b = \mathtt{false}$, and $b \downarrow$ is an abbreviation for $b \vee \neg b$.

We extend the inference rules with some valid proof rules including:

$$\frac{\{\varphi_1\} \, f \bullet \vec{e} \searrow y \, \{\psi_1\} \quad \{\varphi_2\} \, f \bullet \vec{e} \searrow y \, \{\psi_2\}}{\{\varphi_1 \wedge \varphi_2\} \, f \bullet \vec{e} \searrow y \, \{\psi_1 \wedge \psi_2\}} \tag{5}$$

Furthermore, we axiomatize that for each $f : \vec{\tau} \rightrightarrows \sigma$, there is a binary relation $\bar{f}$ between $\vec{\tau}$ and σ such that

$$\forall \vec{x}. \, \{\exists y. \, \vec{x} \bar{f} y\} \, f \bullet \vec{x} \searrow y \, \{\vec{x} \bar{f} y\}$$

which is optimal in the sense that for any $\forall \vec{x}. \, \{\varphi\} \, f \bullet \vec{x} \searrow y \, \{\psi\}$, it holds that $\forall \vec{x}. \, \varphi \Rightarrow \exists y. \, \vec{x} \bar{f} y$ and $\forall \vec{x}, y. \, \vec{x} \bar{f} y \Rightarrow \varphi \wedge \psi$.

Though nondeterministic function becomes a primitive notion, we often restrict to ordinary functions and reason about programs based on a classical theory of ordinary functions in mathematical analysis. For the purpose, our logical language provides also the ordinary function sorts $\vec{\tau} \rightarrow \sigma$ which now admit the ordinary function applications. We assume that the language is expressive enough to do enough classical analysis. For a nondeterministic function $f : \vec{\tau} \rightrightarrows \sigma$ and an ordinary function $g : \vec{\tau} \rightarrow \sigma$ define the proposition:

$$\mathsf{fun}(g, f) := \equiv \forall \vec{x} : \vec{\tau}. \, \{\mathsf{True}\} \, f \bullet \vec{x} \searrow y \, \{y = g(\vec{x})\}$$

saying that f turns out to be an ordinary function which is g.

When we have a nondeterministic function $f : \vec{\tau} \rightrightarrows \sigma$, we pose an assumption that f actually is a totally defined ordinary function by writing $f \searrow := \exists g : \vec{\tau} \rightarrow \sigma. \, \mathsf{fun}(g, f)$. Note that by the rule at Eq. (5), we can prove that if there exists such g, it is unique. For a formula P on $f : \vec{\tau} \rightrightarrows \sigma$ which regards f as an ordinary function of type $\vec{\tau} \rightarrow \sigma$, we write $[P]_f$ for

$$\forall g : \vec{\tau} \rightarrow \sigma. \, \mathsf{fun}(g, f) \Rightarrow P[g/f]$$

where g does not appear free in f and P.

Remark 1. For a partial nondeterministic $f : \vec{\tau} \rightrightarrows \sigma$, the well-formed formula $[P]_f$ is derivable if $P[g/f]$ is derivable for a variable $g : \vec{\tau} \rightarrow \sigma$ that does not appear free in P.

In other words, our language is expressive enough to restrict f to be an ordinary function, do classical reasoning on it, and apply it back on f.

As an example, suppose cont is a predicate on ordinary real functions for their continuity. For a nondeterministic $f : \mathsf{R} \rightrightarrows \mathsf{R}$, we write $f \searrow \wedge[\mathsf{cont}(f) \wedge f(0) \cdot f(1) < 0]_f$ to say that f is actually an ordinary function which is continuous and admits a sign change in $(0, 1)$. As our language proves the Intermediate Value Theorem, according to Remark 1, we can derive $f \searrow \wedge[\exists x.\, 0 < x < 1 \wedge f(x) = 0]_f$ which gives us the proposition on the nondeterministic f:

$$\exists x.\, 0 < x < 1 \wedge \{\mathsf{True}\}\, f \bullet x \searrow y\, \{y = 0\}.$$

4.2 Total Correctness Specifications

We adopt the specification method using *anchors* in [11, 12]. For a well-typed expression $\Gamma \vdash e : \tau$, we write $\{\varphi\}\, e :_y \{\psi\}$ for some context of auxiliary variables Υ to say that for any program state $\gamma \in \llbracket \Gamma \rrbracket$ and some free variables $\xi \in \llbracket \Upsilon \rrbracket$ satisfying $\gamma, \xi \models \varphi$, $\llbracket \Gamma \vdash e : \tau \rrbracket(\gamma)$ is well-defined and for each $x \in \llbracket \Gamma \vdash e : \tau \rrbracket(\gamma)$, it holds that $\gamma, y \mapsto x, \xi \models \psi$. In words, for any state γ and ξ satisfying the *precondition* φ, the evaluation of e does not fail and any possible value represented by the *anchor* y satisfies the *postcondition* ψ.

Similarly, we specify a well-typed command $\Gamma \vdash c \triangleright \Delta$ with a triple $\{\varphi\}\, c\, \{\psi\}$ for some context of auxiliary variables Υ to say that for any $\gamma \in \llbracket \Gamma \rrbracket$ and $\xi \in \llbracket \Upsilon \rrbracket$ validating the *precondition* φ, $\llbracket \Gamma \vdash c \triangleright \Delta \rrbracket(\gamma)$ is not $\perp$ and for each $\delta \in \llbracket \Gamma \vdash c \triangleright \Delta \rrbracket(\gamma)$, δ, ξ validates *postcondition* ψ.

5 Proof Rules

The rules for lambda expressions and function applications are minor modifications from [12]:

$$\frac{\{\varphi \wedge \psi \wedge \vec{v} = \vec{v}'\}\, c\, \{\psi'\} \qquad \{\psi'\}\, e :_y \{\theta[\vec{v}'/\vec{v}]\}}{\{\varphi\}\, \lambda(v_1 : \tau_1, \cdots, v_n : \tau_n).\, c; e :_f \{\forall \vec{v}.\, \{\psi\}\, f \bullet (\vec{v}) \searrow y\, \{\theta\}\}}$$

$$\frac{\{\varphi\}\, f :_g \{\theta\} \qquad \{\varphi\}\, e_i :_{v_i} \{\psi_i\} \quad (i = 1, \cdots, n)}{\{\varphi\}\, f(e_1, \cdots, e_n) :_y \{\psi\}}$$
$$\frac{\forall g, \vec{v}.\, \varphi \wedge \theta \Rightarrow \{\psi_1 \wedge \cdots \wedge \psi_n\}\, g \bullet (v_1, \cdots, v_n) \searrow y\, \{\psi\}}{\{\varphi\}\, f(e_1, \cdots, e_n) :_y \{\psi\}}$$

In the rule for lambda expressions, $\vec{v}'$ are fresh variables that capture the initial values of the input variables $\vec{v}$. Note that $\vec{v}$ in θ refer to the initial values not the values that $\vec{v}$ store at the end of c.

The rules for countable choices and limits are new. The rule for countable choices is as follows:

$$\frac{\{\varphi\}\ f :_g \{\mathsf{choosable}_g \wedge \forall n : \mathsf{Z}.\ \{\neg\psi\}\ g \bullet n \searrow y\ \{y \neq \mathbf{true}\}\}}{\{\varphi\}\ \mathsf{choose}(f) :_n \{\psi\}}$$

where $\mathsf{choosable}_g :\equiv (\forall n.\ \{\mathsf{True}\}\ g \bullet n \searrow y\ \{\mathsf{True}\}) \wedge (\exists n.\ \{\mathsf{True}\}\ g \bullet n \searrow y\ \{y\}).$

Suppose a well-typed expression $\mathsf{choose}(f)$. For any state φ as a precondition, first we need to ensure that the expression f evaluates well to a function g such that g satisfies (i) the abbreviated $\mathsf{choosable}_g$ and (ii) that for any input n not satisfying ψ, all return values y of g at n satisfy $y \neq \mathbf{true}$; i.e., g at such n can only be either $\mathbf{false}$ or div. Hence, ψ is a condition that any input value n of g must satisfy if $\mathbf{true}$ is one of the possible return values. Therefore, any n that choose chooses from g satisfies ψ.

The other condition, the abbreviated $\mathsf{choosable}_g$, ensures that the nondeterministic function g that f evaluates to has to be total and that there exists at least one input n' where the only possible return value of g on n' is $\mathbf{true}$. In other words, the rule ensures that the argument function is well-defined for all indices and there exists at least one index that choose can choose.

The rule for nondeterministic limits is as follows:

$$\frac{\{\varphi\}\ f :_g \{\mathsf{refine}(g)\} \qquad \{\varphi\}\ e :_{((x,q),h)} \{\psi \wedge x \approx_q \theta\}}{\{\varphi\}\ \mathtt{lim}^{\rightrightarrows}(f,e) :_y \{\theta\}}\ \varphi \Rightarrow \mathsf{closed}(\theta)$$

Here, the followings are abbreviations:

$$\mathsf{closed}(\theta) :\equiv \forall y.\ \neg\theta \Rightarrow \exists k \in \mathbb{Z}.\ \forall z.\ |y - z| \leq 2^{-k} \Rightarrow \neg\theta[z/y]$$

$$\mathsf{refine}(g) :\equiv \forall p, x, q, h.\ \{\psi \wedge x \approx_q \theta \wedge q < p\}$$
$$g \bullet (p, ((x, q), h)) \searrow (x', h')$$
$$\{\psi[x'/x, h'/h] \wedge x' \approx_p \theta \wedge |x - x'| \leq 2^{-q} - 2^{-p}\}$$

$$x \approx_q \theta :\equiv \exists y.\ \theta \wedge |x - y| \leq 2^{-q}$$

In order to apply the rule for limits and obtain $\{\varphi\}\ \mathtt{lim}^{\rightrightarrows}(f, e) :_y \{\theta\}$, we first need to prove the side-condition $\varphi \Rightarrow \mathsf{closed}(\theta)$ saying that under the assumption φ, θ is a closed, hence also a sequentially closed, set on its free variable y. Then, we need to prove that the evaluation of f leads to a nondeterministic function g which is a nondeterministic refinement procedure with a hint invariant ψ: for any $(p, ((x, q), h))$ such that $q < p$ satisfying ψ and $x \approx_q \theta$, which means that x approximates a real number that θ represents by 2^{-q}, each output (x', h') of g on $(p, ((x, q), h))$ satisfies (1) $\psi[x'/x, h'/h]$, (2) $x' \approx_p \theta$, and (3) $|x - x'| \leq 2^{-q} - 2^{-p}$. In words, (1) ensures that the refinements that f produces respect the invariant ψ, (2) ensures that f actually refines approximations, and (3) ensures that the sequences that f creates are consistent.

Furthermore, the second premise requires that e evaluates to $((x, q), h)$ that meets the hint invariant ψ and such that x is a 2^{-q} approximation to a real number represented by θ. Hence, they are an initial approximation and a hint.

Thus far discussion constitutes a proof for the new rules.

Lemma 3. *The proof rules for nondeterministic choices and limits are sound.*

6   Examples

6.1   Two Dimensional Searching

Consider any enumeration function $\hat{d} : \mathsf{Z} \to \mathsf{Z} \times \mathsf{Z}$ in our assertion language. Without specifying a lambda expression, we suppose we have an expression $\bar{d}$ admitting the specification:

$$\{\mathsf{True}\}\ \bar{d} :_d \{d \searrow \wedge \forall v.\ [d(v) = \hat{d}(v)]_d\}.$$

Using it, we can implement 2-dimensional searching:

$$\mathsf{choose}^2 :\equiv \lambda(f : \mathsf{Z} \times \mathsf{Z} \rightrightarrows \mathsf{IB}).\,\mathsf{skip}; \bar{d}\big(\mathsf{choose}(\lambda(z : \mathsf{Z}).\,\mathsf{skip}; f(\bar{d}(z))))\big).$$

Then, we can obtain the derivation using the proof rules:

$$\frac{\{\varphi\}\ f :_g \{\mathsf{choosable}_g^2 \wedge \forall i,j : \mathsf{Z}.\ \{\neg\psi\}\ g \bullet (i,j) \searrow y\ \{y \neq \mathsf{true}\}\}}{\{\varphi\}\ \mathsf{choose}^2(f) :_{(i,j)} \{\psi\}} \tag{6}$$

where $\mathsf{choosable}_g^2 :\equiv (\forall(i,j) : \mathsf{Z} \times \mathsf{Z}.\{\mathsf{True}\}\ g \bullet (i,j) \searrow y\ \{\mathsf{True}\}) \wedge (\exists(i,j) : \mathsf{Z} \times \mathsf{Z}.\{\mathsf{True}\}\ g \bullet (i,j) \searrow y\ \{y\})$.

The specification in Eq. (6) says that whenever a programming expression f turns out to be a nondeterministic function g satisfying that (1) $g(i,j)$ is not $\bot$ for all i,j, (2) $g(i,j) = \{\mathsf{true}\}$ for some i,j, and (3) for all i,j satisfying $\neg\psi$, $\mathsf{true} \notin g(i,j)$ holds, the expression $\mathsf{choose}^2(f)$ safely evaluates to (i,j) such that ψ holds. The condition (1) ensures that evaluating f never fails, (2) ensures that there exists an index (i,j) that ensures the termination of the searching procedure, and (3) ensures that for any i,j such that $\mathsf{true} \in g(i,j)$, ψ holds.

6.2   Intermediate Value Theorem

Consider a continuous real function f and $a < b$ such that $f(a) \cdot f(b) < 0$. Though the classical existence of a root in (a, b) is guaranteed by the Intermediate Value Theorem, to compute it, we need to make more assumptions. Among many variants, we assume f is continuous and locally non-constant [7]. Let us write C_f for a formula stating the conditions. Moreover, let us abbreviate $\mathsf{S}_{f,a,b}$ for $a < b \wedge f(a) \cdot f(b) < 0$ and let the initial search space be $(0, 1)$; i.e., $\mathsf{S}_{f,0,1}$ holds.

Suppose we have to refine an interval (a, b) where $|a - b| = 2^{-h+1}$ for an integer h into a sub-interval of width less than or equal to 2^{-p}. For each d such that $p \leq h + d$, we split the interval (a, b) into 2^{d+1} equally spaced intervals and consider the n'th and $n+1$'th adjacent intervals glued: $(c_{a,b,d}^{(n)}, c_{a,b,d}^{(n+2)})$ where $c_{a,b,d}^{(i)}$ abbreviates $a + i \cdot (b - a) \cdot 2^{-d-1}$ for each i. The condition $p \leq h + d$

ensures that the new glued intervals have length less than or equal to 2^{-p}. We nondeterministically choose (d, n) such that the glued interval that the pair of integers represents contains a root by passing the following lambda expression J into the two-dimensional searching:

$$J := \lambda((n, d) : \mathsf{Z} \times \mathsf{Z}).\, \mathbf{var}\, t : \mathsf{IB} = \mathtt{false};$$

$$\mathbf{if}\, D_{n,d,h,p}\, \mathbf{then}\, t := f(c_{a,b,d}^{(n)}) \cdot f(c_{a,b,d}^{(n+2)}) < 0\, \mathbf{else}\, \mathbf{skip}\, \mathbf{end};\, t$$

Here, $D_{n,d,h,p}$ abbreviates $p \leq h + d \wedge 0 < d \wedge 0 \leq n \leq 2^{d+1} - 2$ where $0 < d \wedge 0 \leq n \leq 2^{d+1} - 2$ is needed to make d a counter and n an index. Using the proof rules, we can prove $\{f \searrow \wedge [C_f \wedge S_{f,a,b}]_f\}\, J :_j \{j \searrow \wedge [j(n, d) \Leftrightarrow D_{n,d,h,p} \wedge S_{f,c_{a,b,d}^{(n)},c_{a,b,d}^{(n+2)}}]_{fj}\}$. By a classical reasoning on f inside $[\cdot]_{fj}$, according to Remark 1, based on the fact that f is locally non-constant, by postcondition weakening, we can add $[\exists n, d.\, j(n, d) = \mathtt{true}]_{fj}$ in the postcondition.

By the rule of function applications and Eq. (6), we get the triple:

$$\{f \searrow \wedge [C_f \wedge S_{f,a,b}]_f \wedge |a - b| = 2^{-h+1}\}\, \mathtt{choose}^2(J) :_{(n,d)} \{[S_{f,c_{a,b,d}^{(n)},c_{a,b,d}^{(n+2)}}]_f \wedge D_{n,d,h,p}\}.$$

It says indeed the new interval refines the original interval. Therefore, we can prove that the following lambda expression is a refinement

$$R := \lambda(p : \mathsf{Z}, ((x, q), h) : (\mathsf{R} \times \mathsf{Z}) \times \mathsf{Z}).\, \mathbf{var}\, a : \mathsf{R} = y - 2^{-h};\, \mathbf{var}\, b : \mathsf{R} = y + 2^{-h};$$

$$\mathbf{var}\, (n, d) : \mathsf{Z} \times \mathsf{Z} := \mathtt{choose}^2(J);\, (c_{a,b,d}^{(n+1)}, h + d)$$

for the limit points $\theta(y) := 0 < y < 1 \wedge [f(y) = 0]_f$ and the hint invariant $\psi := [S_{f,x-2^{-h},x+2^{-h}}]_f$. The initial approximation is $(2^{-1}, 1)$ and the initial hint is 1. The limit points $\theta(y)$ can be proven closed by the condition that f is continuous. Hence, applying the rule for limits, we get

$$\{f \searrow \wedge [C_f \wedge S_{f,0,1}]_f\}\, \mathtt{lim}_{\mathsf{Z}}^{\rightrightarrows}(R, ((2^{-1}, 1), 1)) :_y \{0 < y < 1 \wedge [f(y) = 0]_f\}$$

a formally proved specification for a program computing a root nondeterministically from a nondeterministic function f which happens to be a continuous and locally non-constant ordinary function whose sign changes at the unit interval.

7   Conclusion and Future Work

We presented an imperative programming language with countable nondeterminism, nondeterministic functions, and nondeterministic limit operations for exact real number computation. We also provided sound proof rules for proving total correctness specifications and demonstrated its practicality by verifying a constructive and nondeterministic Intermediate Value Theorem program.

Our future work includes extending the language with reference types, functions with side effects, and higher-order general recursions. Along with that, we plan to apply the verification methods to more advanced applications such as differential equation solving in exact real computation [21].

Acknowledgements. The author would like to thank Holger Thies and the anonymous reviewers for constructive comments on the manuscript.

References

1. Apt, K.R., Olderog, E.R.: Fifty years of Hoare's logic. Formal Aspects Comput. **31**, 751–807 (2019)
2. Apt, K.R., Plotkin, G.D.: Countable nondeterminism and random assignment. J. ACM (JACM) **33**(4), 724–767 (1986)
3. Balluchi, A., Casagrande, A., Collins, P., Ferrari, A., Villa, T., Sangiovanni-Vincentelli, A.: Ariadne: a framework for reachability analysis of hybrid automata. In: Proceedings of 17th International Symposium on Mathematical Theory of Networks and Systems, Kyoto (2006)
4. Bishop, E.A.: Foundations of constructive analysis (1967)
5. Brattka, V., Hertling, P.: Feasible real random access machines. J. Complex. **14**(4), 490–526 (1998). https://doi.org/10.1006/jcom.1998.0488
6. Brauße, F., Collins, P., Ziegler, M.: Computer science for continuous data. In: Boulier, F., England, M., Sadykov, T.M., Vorozhtsov, E.V. (eds.) Computer Algebra in Scientific Computing, pp. 62–82. Springer, Cham (2022)
7. Bridges, D.S.: A general constructive intermediate value theorem. Math. Log. Q. **35**(5), 433–435 (1989)
8. Bridges, D.S.: Constructive mathematics: a foundation for computable analysis. Theoret. Comput. Sci. **219**(1), 95–109 (1999). https://doi.org/10.1016/S0304-3975(98)00285-0
9. Di Gianantonio, P., Honsell, F., Plotkin, G.: Uncountable limits and the lambda calculus. Nordic J. Comput. (1995)
10. Geuvers, H., Niqui, M., Spitters, B., Wiedijk, F.: Constructive analysis, types and exact real numbers. Math. Struct. Comput. Sci. **17**(1), 3–36 (2007)
11. Honda, K., Yoshida, N.: A compositional logic for polymorphic higher-order functions. In: Proceedings of the 6th ACM SIGPLAN International Conference on Principles and Practice of Declarative Programming, pp. 191–202 (2004)
12. Honda, K., Yoshida, N., Berger, M.: An observationally complete program logic for imperative higher-order functions. In: 20th Annual IEEE Symposium on Logic in Computer Science (LICS 2005), pp. 270–279. IEEE (2005)
13. Konečný, M., Park, S., Thies, H.: Certified computation of nondeterministic limits. In: Deshmukh, J.V., Havelund, K., Perez, I. (eds.) NASA Formal Methods, pp. 771–789. Springer, Cham (2022)
14. Konečný, M., Park, S., Thies, H.: Axiomatic reals and certified efficient exact real computation. In: Silva, A., Wassermann, R., de Queiroz, R. (eds.) WoLLIC 2021. LNCS, vol. 13038, pp. 252–268. Springer, Cham (2021). https://doi.org/10.1007/978-3-030-88853-4_16
15. Konečný, M.: aern2-real: a Haskell library for exact real number computation. https://hackage.haskell.org/package/aern2-real (2021)
16. Luckhardt, H.: A fundamental effect in computations on real numbers. Theoret. Comput. Sci. **5**(3), 321–324 (1977). https://doi.org/10.1016/0304-3975(77)90048-2
17. Müller, N.T.: Implementing limits in an interactive realram. In: 3rd Conference on Real Numbers and Computers, 1998, Paris, vol. 13, p. 26 (1998)
18. Müller, N.T.: The iRRAM: exact arithmetic in C++. In: Blanck, J., Brattka, V., Hertling, P. (eds.) CCA 2000. LNCS, vol. 2064, pp. 222–252. Springer, Heidelberg (2001). https://doi.org/10.1007/3-540-45335-0_14
19. Neumann, E., Pauly, A.: A topological view on algebraic computation models. J. Complex. **44**, 1–22 (2018)

20. Park, S., et al.: Foundation of computer (algebra) ANALYSIS systems: Semantics, logic, programming, verification. arXiv preprint arXiv:1608.05787 (2016)
21. Selivanova, S., Steinberg, F., Thies, H., Ziegler, M.: Exact real computation of solution operators for linear analytic systems of partial differential equations. In: Boulier, F., England, M., Sadykov, T.M., Vorozhtsov, E.V. (eds.) CASC 2021. LNCS, vol. 12865, pp. 370–390. Springer, Cham (2021). https://doi.org/10.1007/978-3-030-85165-1_21
22. Tucker, J.V., Zucker, J.I.: Abstract versus concrete computation on metric partial algebras. ACM Trans. Comput. Logic (TOCL) 5(4), 611–668 (2004)
23. Weihrauch, K.: Computable Analysis. Springer, Berlin (2000)

Exact and Parameterized Algorithms for the Independent Cutset Problem

Johannes Rauch[1], Dieter Rautenbach[1], and Uéverton S. Souza[2]($\boxtimes$)

[1] Institute of Optimization and Operations Research, Ulm University, Ulm, Germany
{johannes.rauch,dieter.rautenbach}@uni-ulm.de
[2] Instituto de Computação, Universidade Federal Fluminense, Niterói, Brazil
ueverton@ic.uff.br

Abstract. The INDEPENDENT CUTSET problem asks whether there is a set of vertices in a given graph that is both independent and a cutset. Such a problem is NP-complete even when the input graph is planar and has maximum degree five. In this paper, we first present a $\mathcal{O}^*(1.4423^n)$-time algorithm to compute a minimum independent cutset (if any). Since the property of having an independent cutset is MSO_1-expressible, our main results are concerned with structural parameterizations for the problem considering parameters incomparable with clique-width. We present FPT-time algorithms for the problem considering the following parameters: the dual of the maximum degree, the dual of the solution size, the size of a dominating set (where a dominating set is given as an additional input), the size of an odd cycle transversal, the distance to chordal graphs, and the distance to P_5-free graphs. We close by introducing the notion of α-domination, which allows us to identify more fixed-parameter tractable and polynomial-time solvable cases.

Keywords: exact algorithms · parameterized algorithms · independent cutset

1 Introduction

The MATCHING CUTSET problem is a well-studied problem in the literature, both from a structural and from an algorithmic point of view. It asks whether a graph G admits a set of edges M such that $G - M$ is disconnected and no two distinct edges of M are incident. A natural variation of this problem is obtained by replacing the word "edges" by "vertices" and the word "incident" by "adjacent" in the previous problem definition. Doing this yields the INDEPENDENT CUTSET problem, which is also known as the STABLE CUTSET problem. Tucker [29] studied INDEPENDENT CUTSET in the context of perfect graphs and graph colorings in 1983. In 1993, Corneil and Fonlupt [8] explicitly asked for the complexity of INDEPENDENT CUTSET. They studied the problem in the context of perfect graphs, too.

© The Author(s), under exclusive license to Springer Nature Switzerland AG 2023
H. Fernau and K. Jansen (Eds.): FCT 2023, LNCS 14292, pp. 378–391, 2023.
https://doi.org/10.1007/978-3-031-43587-4_27

It is not hard to see that a graph G with minimum degree at least two has a matching cutset if and only if the line graph of G has an independent cutset. Therefore, the first NP-completeness proof of INDEPENDENT CUTSET is due to Chvátal [7], who presented in 1984 the first NP-completeness proof for the MATCHING CUTSET problem. Brandstädt, Dragan, Le and Szymczak [4] showed in 2000 that INDEPENDENT CUTSET stays NP-complete, even when restricted to K_4-free graphs. Note that the problem is trivial on K_3-free graphs, since the neighborhood of any vertex constitutes an independent cutset. They also concluded that it is NP-complete on perfect graphs. Le and Randerath [18] proved in 2003 that INDEPENDENT CUTSET is NP-complete on 5-regular line graphs of bipartite graphs [18]. In 2008, Le, Mosca and Müller [16] showed that the problem is NP-complete on planar graphs with maximum degree five.

On the other hand, several polynomial time solvable cases have been identified. In 2002, Chen and Yu [6] answered a question by Caro in the affirmative by showing that all graphs with n vertices and at most $2n - 4$ edges admit an independent cutset. Their proof can be used for a polynomial time algorithm that finds such a set. Le and Pfender [17] characterized in 2013 the extremal graphs having $2n - 3$ edges but no independent cutset. In particular, they showed that INDEPENDENT CUTSET can be decided for graphs with n vertices and $2n - 3$ edges in polynomial time. Le, Mosca and Müller [16] also showed that the problem can be decided in polynomial time for claw-free graphs of maximum degree 4, {claw, K_4}-free graphs, and claw-free planar graphs. The general case for maximum degree four is still open. Some more polynomial time solvable cases can be found in [4].

Regarding the parameterized complexity of INDEPENDENT CUTSET, Marx, O'Sullivan and Razgon [21] showed in 2013 that the problem of finding a minimum independent cutset in a graph can be solved in $\mathcal{O}(f(k) \cdot n)$ time, where n is the number of vertices of the input graph and k is an upper bound on the solution size. They transform the input graph to a graph of treewidth bounded by $g(k)$, where g is some function only depending on k. The transformed graph retains all cutsets of size at most k of the input graph. An application of Courcelle's Theorem [10] then shows the fixed-parameter tractability. Beside that, to the best of our knowledge, there is no other work about parameterized algorithms for INDEPENDENT CUTSET in the literature.

Moreover, although a $\mathcal{O}^*(2^n)$ time algorithm for INDEPENDENT CUTSET is trivial, there is also no discussion on more efficient exact exponential-time algorithms. Motivated by this, we present an $\mathcal{O}^*(3^{n/3})$ time algorithm for INDEPENDENT CUTSET in Sect. 2. It is based on iterating through all maximal independent sets of a graph. Note that $3^{1/3} < 1.4423$. In the same section we show how to adapt this algorithm to compute a minimum independent cutset in the same time (if there is one).

From the parameterized complexity point of view, an algorithmic meta-theorem of Courcelle, Makowsky and Rotics [9] states that any problem expressible in monadic second-order logic (MSO_1) can be solved in $\mathcal{O}(f(cw) \cdot n)$ time, where cw is the clique-width of the input graph. Originally this required a clique-width expression as part of the input. This restriction was removed when Oum and Seymour [23] gave an FPT algorithm, parameterized by the clique-width of the input graph, that finds a $2^{\mathcal{O}(cw)}$-approximation of an optimal clique-width expression. Since the property of having an independent cutset can be expressed in MSO_1, it holds that INDEPENDENT CUTSET is in FPT concerning clique-width parameterization. Therefore, our focus in this work is on structural parameters that measure the distance from the input instance to some "trivial class" that is relevant to the problem and does not have bounded clique-width.

We use the $\mathcal{O}^*$-notation to suppress polynomial factors in the $\mathcal{O}$-notation.

2   An Exact Exponential Algorithm

In this section, we present a single-exponential time algorithm for finding a minimum independent cutset (if any). It is structured as follows. First, we present a preliminary structural result in Lemma 1. Corollary 1 is an immediate consequence of Lemma 1. By combining these results with some known results from the literature, we are able to solve INDEPENDENT CUTSET in $\mathcal{O}^*(3^{n/3})$ time, which is given in Corollary 2. Furthermore, we get another polynomial-time solvable case, which is stated in Corollary 3. We conclude this section with Lemma 2 and Corollary 4, where we show how to adapt the algorithm to find a minimum independent cutset (if there is one).

We start with an easy but important structural observation.

Lemma 1. *If a connected graph G has an independent cutset S, then every independent set $S' \supseteq S$ is also a cutset of G.*

Proof. Let G be a connected graph, and let S be an independent cutset of G. Let $S' \supset S$ be another independent set of G. Since G is connected and S is a cutset, any component of $G - S$ has a vertex with a neighbor in S. This implies that any component of $G - S$ has a vertex not in S'. Thus, S' is also a cutset of G. □

Lemma 1 implies that it suffices to consider maximal independent sets to decide INDEPENDENT CUTSET. This is fact is used in Corollary 1.

Corollary 1. *Let G be a connected graph with n vertices. If there is an algorithm that enumerates all maximal independent sets of G in $\mathcal{O}^*(f(n))$ time, then INDEPENDENT CUTSET with G as input is solvable in $\mathcal{O}^*(f(n))$ time.*

Proof. This follows from Lemma 1 and the fact that checking if a given set S is a cutset of G can be done in polynomial time. □

By a result of Moon and Moser [22], a graph with n vertices has $\mathcal{O}(3^{n/3})$ maximal independent sets. Johnson, Yannakakis and Papadimitrou [15] showed that all maximal independent sets can be enumerated with $\mathcal{O}(n^3)$ delay. This, together with Corollary 1, gives a fast exponential algorithm for the problem, and a graph class for which the problem is efficiently solvable.

Corollary 2. INDEPENDENT CUTSET *can be solved in* $\mathcal{O}^*(3^{n/3})$ *time.*

Corollary 3. INDEPENDENT CUTSET *on* $2K_2$-*free graphs can be solved in polynomial time.*

Proof. $2K_2$-free graphs with n vertices only have $\mathcal{O}(n^2)$ maximal independent sets [13]. □

We showed in the proof of Lemma 1 that, in order to decide whether a graph admits an independent cutset, it suffices to consider all maximal independent sets. In the following lemma we show how to obtain small independent cutsets from maximal ones.

Lemma 2. *Given a connected graph G and an independent cutset S' of G, one can compute in polynomial time the smallest independent cutset S of G contained in S'.*

Proof. Let G and S' be as in the statement. We construct a hypergraph H. The vertices of H are the vertices of S' and the components of $G - S'$. The hyperedges of H are

$$\{v\} \cup \{K \mid K \text{ is a component of } G - S' \text{ adjacent to } v\} \text{ for every } v \in S'.$$

It is easy to see that a minimum edge cut of H corresponds to a smallest independent cutset S being a subset of S'. Since a minimum edge cut can be computed in polynomial time on hypergraphs, see for example [26], we can find such a set S' in polynomial time. □

The next corollary follows from Lemma 1 and Lemma 2.

Corollary 4. *Given a connected graph G that admits an independent cutset, a minimum independent cutset of G can be found in $\mathcal{O}^*(3^{n/3})$ time.*

3 Parameterized Algorithms

Intuitively, one could expect the following. If a graph has few edges, then there always exists an independent cutset, and it is easy to find one. This was made precise by Caro's Conjecture, and it was proved by Chen and Yu [6]. If a graph has many edges, then independent cutsets (if there are any) must have a specific structure, or there is no independent cutset at all. This can be seen by the given parameterizations presented in this paper, under which INDEPENDENT CUTSET is fixed-parameter tractable. Indeed, all our parameterizations have in common that they are small if the input graph (or at least a part of it) is "dense".

The section is structured as follows. In Subsect. 3.1, we consider the dual of the maximum degree and the dual of the solution size as a parameter. What follows in Subsect. 3.2 is our most important result, where we consider INDEPENDENT CUTSET with a dominating set as an additional input, and the size of the dominating set is the parameter. Then we consider the distance by vertex removals from the input graph to three different graph classes and take such distances as parameters in Subsects. 3.3, 3.4 and 3.5. The graph classes under consideration are bipartite graphs, chordal graphs, and P_5-free graphs. Finally, we generalize the distance to P_5-free graphs results in Subsect. 3.6.

3.1   Dual Parameterizations

We consider the dual of the maximum degree as a parameter in Theorem 1. Actually, it also follows from Theorem 3 that the problem is fixed-parameter tractable with respect to this parameter, because the size of a minimum dominating set of a graph G with n vertices and maximum degree Δ is at most $n - \Delta$. Nevertheless, we give a direct proof of this fact, since it comprises a faster algorithm.

Theorem 1. *Let G be the connected input graph with n vertices and maximum degree Δ, and let $\mathcal{O}^*(f(n))$ be the running time of an algorithm enumerating all maximal independent sets of G. It holds that* INDEPENDENT CUTSET *can be solved in $\mathcal{O}^*(2^k \cdot f(k))$ time, where $k = n - \Delta$.*

Proof. Let G be a connected graph with n vertices, let v be a vertex of G having maximum degree, and let $R = V(G) \setminus N_G[v]$.

Any independent cutset of G containing v is contained in $R \cup \{v\}$, since $|R| = k$, we can enumerate them in $\mathcal{O}(2^k)$ time. So, we may assume from here that every independent cutset of G (if any) does not contain v.

Let S^* be a minimal independent cutset of G that does not contain v.

Observe that $N_G[v] \setminus S^*$ is nonempty and belongs to one component of $G - S^*$. Thus, some subset $\emptyset \neq R' \subseteq R \setminus S^*$ belongs to the other components. Such a set R' can be guessed in $\mathcal{O}(2^k)$ time. Assuming that we are dealing with the correctly guessed set R', the cutset S^* can be seen as a minimal independent cutset separating v from R'. Let $I = N_G(v) \cap N_G(R')$. The set I must be a subset of S^*; in particular, the set I must be independent in G.

Note that S^* cannot contain a vertex $w \in N_G(v) \setminus I$; otherwise, as S^* is a minimal cutset, there is a path P from v to R' such that $V(P) \cap S^* = \{w\}$ and $V(P) \cap (R \setminus R') \neq \emptyset$. This implies that a vertex of $R \setminus R'$ is in neither S^* nor in the same component as v after the removal of S^*, contradicting the fact that R' is the correctly guessed set.

Thus, after guessing R', we can contract I into a single vertex x_I and remove $N_G[v] \setminus I$. At this point, the reduced graph has size k, and we can enumerate its maximal independent sets in $\mathcal{O}(f(k))$ time, one of them must contain $(S^* \setminus I) \cup \{x_I\}$. Let S' be such a set. By Lemma 1, the set $(S' \setminus \{x_I\}) \cup I$ is also a independent cutset of G (recall that $S' \supseteq S^*$). This concludes the proof.   □

Recall that $\mathcal{O}^*(2^k \cdot f(k))$ is faster than the running time $\mathcal{O}^*(3^k)$ of Theorem 3, since $f(k) = \mathcal{O}(3^{k/3})$ by Corollary 2.

We finish this subsection by considering a lower bound on the dual of the solution size as a parameter.

Theorem 2. *Let G be the connected input graph with n vertices, and let $\mathcal{O}^*(f(k))$ be the running time of an algorithm enumerating all vertex covers of size at most k of G. Then there is an algorithm that correctly decides in $\mathcal{O}^*(f(k))$ time if G has an independent cutset of size at least $n - k$.*

Proof. Let G, n and k be as in the statement. Assume that there is an independent cutset S^* of G such that $|S^*| \geq n - k$. The set $V(G) \setminus S^*$ is a vertex cover of size at most k of G. Therefore it suffices to iterate over all vertex covers X of size at most k of G, and check whether $V(G) \setminus X$ is an independent cutset of G. By assumption, this can be done in $\mathcal{O}^*(f(k))$ time.   □

We remark that all vertex covers of size at most k can be enumerated in $\mathcal{O}^*(2^k)$ time.

3.2   Dominating Set

In this subsection, we prove a central theorem of our paper. We consider INDEPENDENT CUTSET with a dominating set X of G as an additional input. We show that this variant is fixed-parameter tractable when parameterized by $|X|$. We split the proof of this fact over three lemmata. Among other things, we distinguish the cases when X is split by an independent cutset or not. In Lemma 3 we settle the case when X is not split by an independent cutset.

Lemma 3. *Let G be a connected graph, and let X be a dominating set of G. Assume that there is an independent cutset S^* of G such that $X \setminus S^*$ is contained in at most one component of $G - S^*$. Then there is an algorithm that returns an independent cutset of G in $\mathcal{O}^*(2^k)$ time, where $k = |X|$.*

Proof. Let G, X, S^* and k be as in the statement. In $\mathcal{O}^*(2^k)$ time, we can check whether any subset of X is an independent cutset of G. Therefore, we may assume from here that S^* is not a subset of X. Since X dominates G, the set X is not a proper subset of S^*. In particular, $S^* \neq X$.

To find an independent cutset of G, we iterate over all disjoint partitions (A, X') of X such that A is nonempty and X' is independent in G. In one iteration, we guess $X \cap S^*$ as X'. Let $F = N_G(X') \setminus A$ and $I = N_G(A) \setminus (X' \cup F)$. Note that (A, X', F, I) is a partition of $V(G)$, because X is a dominating set of G. A vertex of I is either in S^* or in the same component as a vertex of A. Let B be the set of vertices that are not in a component with a vertex of A in $G - (X' \cup I)$. Note that $B \subseteq F$, and since S^* is an independent cutset of G and $X \setminus S^*$ is contained in at most one component of $G - S^*$, it holds that $B \neq \emptyset$. This implies that there is a component K of $G[B]$ with $N_G(K) \cap I \subseteq S^*$. For such a component K, the set $X' \cup (N_G(K) \cap I)$ is an independent cutset of G. Since all this can be done in $\mathcal{O}^*(2^k)$ time, this finishes the proof.   □

For the case when the dominating set is split by an independent cutset, we need to consider the following situation. Let G' be a connected graph whose vertex set is the disjoint union of four sets A, B, N_A and N_B such that

(i) A and B are nonempty,
(ii) $A \cup B$ is independent in G', and
(iii) $N_{G'}(A) = N_A$ and $N_{G'}(B) = N_B$. (Note that $N_{G'}(A) \cap N_{G'}(B) = \emptyset$.)

We show how to decide efficiently if G' has an independent cutset $S' \subseteq N_A \cup N_B$ that separates A and B in G by testing the satisfiability of a 2-SAT formula. Let H be the bipartite graph induced by the set of edges between N_A and N_B, and let $K_1, \ldots, K_r$ be the vertex sets of the components of H. Let $N_{A,i} = K_i \cap N_A$ and $N_{B,i} = K_i \cap N_B$ for all $i \in [r]$. We construct a 2-SAT formula $f_{G'}$ over the Boolean variables x_i, $i \in [r]$, with the following clauses:

- For all $i \in [r]$, if $G'[N_{A,i}]$ contains an edge, then we add the clause (x_i), and if $G'[N_{B,i}]$ contains an edge, then we add the clause $(\bar{x}_i)$.
- For every two distinct $i, j \in [r]$, if there is an edge between $N_{A,i}$ and $N_{A,j}$ in G', then we add the clause $(x_i \vee x_j)$, and if there is an edge between $N_{B,i}$ and $N_{B,j}$ in G', then we add the clause $(\bar{x}_i \vee \bar{x}_j)$.

Lemma 4. *In the above setting, there is an independent cutset $S' \subseteq N_A \cup N_B$ separating A and B in G' if and only if $f_{G'}$ is satisfiable.*

Proof. Let G', $f_{G'}$ and all corresponding sets be as in the statement.

For one direction, assume that there is an independent cutset $S' \subseteq N_A \cup N_B$ separating A and B in G'. We claim that either $K_i \cap S' = N_{A,i}$ or $K_i \cap S' = N_{B,i}$ is true for every $i \in [r]$. Assume, for a contradiction, that the opposite is true for some $i \in [r]$. Let $N'_{A,i} = N_{A,i} \cap S'$ and $N'_{B,i} = N_{B,i} \cap S'$. Then both sets are nonempty. Recall that $N_G(A) = N_A$, $N_G(B) = N_B$, $H[K_i]$ is connected and S' is a cutset separating A and B in G'. Consider $d = \text{dist}_H(N'_{A,i}, N'_{B,i})$. Since $H[K_i]$ is connected, $d < \infty$, since H is bipartite, d is odd, and since S' is independent, $d \geq 2$. Altogether, we have $d \geq 3$. Since every vertex of $N'_{A,i}$ has a neighbor in A, and every vertex of $N'_{B,i}$ has a neighbor in B, the set S' does not separate A and B, a contradiction. Using the claim, the assignment

$$\alpha(x_i) = \begin{cases} \texttt{false}, & \text{if } K_i \cap S' = N_{A,i} \\ \texttt{true}, & \text{if } K_i \cap S' = N_{B,i} \end{cases} \quad \text{for } i \in [r]$$

is well-defined, and it satisfies $f_{G'}$.

For the opposite direction, let α be a satisfying assignment for $f_{G'}$. Then

$$S' = \left(\bigcup_{i \in [r]: \alpha(x_i) = \texttt{false}} N_{A,i} \right) \cup \left(\bigcup_{i \in [r]: \alpha(x_i) = \texttt{true}} N_{B,i} \right)$$

is a subset of $N_A \cup N_B$, and it is easy to verify that it is an independent cutset of G'. This completes the proof. $\square$

Lemma 5. *Let G be a connected graph, and let X be a dominating set of G. If G has an independent cutset S^* such that the vertices of $X \setminus S^*$ are in at least two different components of $G - S^*$, then an independent cutset of G can be found in $\mathcal{O}^*(3^k)$ time, where $k = |X|$.*

Proof. Let G, X and k be as in the statement. If G has an independent cutset S^* such that the vertices of $X \setminus S^*$ are in at least two different components of

$G - S^*$, then there is a partition (A^*, B^*, X^*) of X with the following properities. We have $X^* = X \cap S^*$, the set A is a maximal subset of vertices of X contained in one component of $G - S^*$, and $B = X \setminus (X^* \cup A)$.

In $\mathcal{O}^*(3^k)$ time one can enumerate all partitions of X into three sets A, B and X' such that A and B are nonempty, there is no edge between A and B, and X' is an independent set of G. In other words, X' is an independent cutset of $G[X]$ separating the nonempty sets A and B.

For each enumerated partition (A, B, X'), let

- $N = N_G(A) \cap N_G(B)$,
- $N_A = N_G(A) \setminus N$,
- $N_B = N_G(B) \setminus N$, and
- let H be the bipartite subgraph of G induced by the edges between N_A and N_B.

For the partition $(A, B, X') = (A^*, B^*, X^*)$ it holds that $(X' \cup N) \subseteq S^*$. Thus, $I = (X' \cup N)$ must be an independent set of G and $N_G(X' \cup N) \cap S^* = \emptyset$. Initialize $F = N_G(X' \cup N)$. (F is the set of vertices forbidden to be in S^*). Clearly, H has no edge with both endpoints in F; otherwise, we are not dealing with the partition $(A, B, X') = (A^*, B^*, X^*)$. Also, for any edge $uv \in E(H)$ such that $u \in F$ it holds that $v \in S^*$; otherwise S^* does not separate A and B. So, one can add v into I. Note that I is a subset of the vertices that must be in S^*. So, reversely, any edge of H with one endpoint in I must have the other endpoint forbidden to be in S^*, thus we can add it in F. This describes the rule to construct a bipartition (I_H, F_H) of the vertices of H that are connected to some vertex of $N_G(X' \cup N)$ in H. In particular, $I_H \subseteq I$, $F_H \subseteq F$, and I must be an independent set of G.

Let $F' = F \setminus (N_A \cup N_B)$. Note that the vertices of F' have neighbors in X' but no neighbor in $A \cup B$, also, there may exist paths of G from A to B passing through F'. For any path from A to B passing through F' having only one vertex of N_A, say a, and only one vertex of N_B, say b, it holds that either a or b must be in S^*. The same holds with edges $ab \in E(G)$ such that $a \in N_A$ and $b \in N_B$. Also, if there is a path of G from A to B passing through F', then there is a path from A to B passing through F' having only one vertex of N_A and only one vertex of N_B.

At this point, let G' be the graph obtained from G by

- contracting all components of A and B into a single vertex,
- inserting all possible edges between $N_G(K) \cap N_A$ and $N_G(K) \cap N_B$, where K is a component of F', and
- deleting all vertices of $I \cup F$.

Now, according to Lemma 4, we can use a 2-SAT formula to decide in polynomial time (cf. [1]) which vertices of N_A and N_B should be in S^*. These vertices, together with the vertices previously fixed in I, form an independent cutset of G separating A and B.

Therefore, by checking all partitions (A, B, X') of X, we can in $\mathcal{O}^*(3^k)$ time either find the required cutset, or conclude that G does not admit such a cutset. This completes the proof. □

Now we are in a position to formulate Theorem 3, which is a direct consequence of Lemmata 3 and 5.

Theorem 3. *Let G be the connected input graph, and let X be a dominating set of G. Then* INDEPENDENT CUTSET *with X as an additional input can be solved in time $\mathcal{O}^*(3^k)$, where $k = |X|$.*

As a corollary of Theorem 3, we obtain an FPT-algorithm for INDEPENDENT CUTSET parameterized by the independence number that has single-exponential dependence on the parameter. The fixed-parameter tractability of INDEPENDENT CUTSET parameterized by the independence number also follows from the result of Marx, O'Sullivan and Razgon [21]; however, their resulting dependence on the parameter is larger.

Corollary 5. INDEPENDENT CUTSET *can be solved in $\mathcal{O}^*(3^k)$ time, where k is the independence number of the input graph.*

Proof. This follows from Theorem 3, because any maximal independent set is a dominating set. $\qquad\square$

3.3 Distance to Bipartite Graphs

Let G be a graph. An *odd cycle transversal* of G is a set of vertices X such that $G - X$ is bipartite. The minimum cardinality of such a set can be seen as a distance measure of how far away a graph is from being bipartite. Given G and a nonnegative integer k, it is NP-hard to decide whether G admits an odd cycle transversal of size at most k [14]. Nevertheless, Reed, Smith and Vetta [28] proved that it is fixed-parameter tractable with respect to the solution size. Lokshtanov, Narayanaswamy, Raman, Ramanujan and Saket [19] showed that it can be decided in time $\mathcal{O}^*(2.3146^k)$ if G admits an odd cycle transversal of size at most k. If the answer is affirmative, an odd cycle transversal of size at most k can be determined as a byproduct. With this, we are able to formulate Corollary 6.

Corollary 6. INDEPENDENT CUTSET *can be solved in $\mathcal{O}^*(3^k)$ time, where k is the odd cycle transversal number of G.*

Proof. If a vertex is not part of a triangle in G, then we return its neighborhood as an independent cutset of G. Therefore we may assume every vertex is part of a triangle in G. We compute an odd cycle transversal X of size at most k of G. As stated, this is fixed-parameter tractable with respect to k as a parameter, and can be done faster than $\mathcal{O}^*(3^k)$ time. Since every vertex is part of a triangle in G, the set X is a dominating set of G. Therefore the statement follows from Theorem 3. $\qquad\square$

A *triangle-hitting set* of G is a set of vertices whose removal makes the resulting graph triangle-free. It is easy to see that the previous proof works even if X is a minimum triangle-hitting set instead of a minimum odd cycle transversal. Since a triangle-hitting set of G having size at most k can be found in $\mathcal{O}^*(3^k)$ time, the following corollary also holds.

Corollary 7. INDEPENDENT CUTSET *can be solved in time* $\mathcal{O}^*(3^k)$, *where* k *is the size of a minimum triangle-hitting set of* G.

3.4   Distance to Chordal Graphs

We begin by proving a more general theorem.

Dallard, Milanič and Štorgel [12] introduced a special kind of tree decomposition. Given a nonnegative integer ℓ, an ℓ-*refined tree decomposition* of a graph G is a pair $\hat{\mathcal{T}} = (T, \{(X_t, U_t) : t \in V(T)\})$ such that $\mathcal{T} = (T, \{X_t : t \in V(T)\})$ is a tree decomposition, and for every $t \in V(T)$ we have $U_t \subseteq X_t$ and $|U_t| \leq \ell$. We say $\mathcal{T}$ is the *underlying tree decomposition* of $\hat{\mathcal{T}}$. We extend any concept defined for tree decompositions to ℓ-refined tree decompositions by considering them on the underlying tree decomposition. The residual independence number of $\hat{\mathcal{T}}$ is defined as $\max_{t \in V(T)} \alpha(G[X_t \setminus U_t])$ and denoted by $\hat{\alpha}(\hat{\mathcal{T}})$.

Theorem 4. *Let* G *be the connected input graph, and let* $\hat{\mathcal{T}} = (T, \{(X_t, U_t) : t \in V(T)\})$ *be a nice* ℓ-*refined tree decomposition of* G *with residual independence number at most* k. INDEPENDENT CUTSET *with* $\hat{\mathcal{T}}$ *as an additional input can be decided in* $\mathcal{O}^*(3^\ell (2n)^k)$ *time.*

Proof. Let G and $\hat{\mathcal{T}}$ be as in the statement, and let r be the root of T. Let V_t denote the union of all sets $X_{t'}$ such that $t' \in V(T)$ is a descendant of t or $t' = t$. We call a partition (S, A, B) of X_t such that S is independent and there is no edge crossing from A to B in G a *potential* t-*partition*.

We outline a dynamic programming algorithm that operates on the nice tree decomposition $\hat{\mathcal{T}}$ in a bottom-up manner. At each node $t \in V(T)$, and for all potential t-partitions, we compute a Boolean variable $cut[t, S, A, B]$ that is **true** if and only if there is an independent cutset $S^* \supseteq S$ of G such that $S^* \cap X_t = S$, and A and B are in distinct components of $G[V_t] - S^*$. After computing these variables, we return $cut[r, \emptyset, \emptyset, \emptyset]$, which gives the correct answer. For the sake of exposition, we refer the reader on how to compute these variables to our arxiv article [27].

The key property of the algorithm is the following: Given that $|U_t| \leq \ell$, there are at most 3^ℓ ways to partition U_t into three sets. Since $\alpha(G[X_t \setminus U_t]) \leq k$, it holds that for any set $S \subseteq X_t \setminus U_t$, the number of components of $G[X_t \setminus (U_t \cup S)]$ is at most k. Thus, there are at most $\mathcal{O}(n^k)$ independent subsets S, and there are at most 2^k ways to partition the components of $G[X_t \setminus (U_t \cup S)]$. The computational effort to compute one variable $cut[t, S, A, B]$ is polynomial. Therefore the overall running time is $\mathcal{O}^*(3^l (2n)^k)$. At this point, it is not hard to see how to perform such a dynamic programming using a nice tree decomposition within the claimed running time. This completes the proof of Theorem 4.   □

Let G be a graph. A *chordal (vertex-)deletion* of G is a set of vertices X such that $G - X$ is chordal. The problem of deciding whether a graph admits a chordal deletion of a fixed size ℓ is NP-complete [14]. Marx [20] showed that this problem is fixed-parameter tractable when parameterized by ℓ.

Corollary 8. *Let G be the connected input graph, and let ℓ be the size of a chordal deletion of G. If $\mathcal{O}^*(f(\ell)$ is the running time of an algorithm computing a chordal deletion of size ℓ of a graph, then* INDEPENDENT CUTSET *can be solved in $\mathcal{O}^*(f(\ell) + 3^\ell)$ time.*

Proof. First, we compute a chordal deletion X of size ℓ of G. Then, since $G - X$ is chordal, we compute a clique tree T of $G - X$, which is possible in polynomial time [3]. To obtain a tree decomposition of G, we add the vertices of X to every bag of T. We transform this tree decomposition into a nice tree decomposition $\hat{T} = (T, \{X_t : t \in V(T)\})$ in polynomial time. Note that this can be done in a way such that for every node $t \in V(T)$, the set $X_t \setminus X$ still induces a clique in G. By letting $U_t = X \cap X_t$ for every node $t \in V(T)$, we augment $\hat{T}$ to a nice ℓ-refined tree decomposition with residual independence number 1. Now the statement follows from Theorem 4.   □

3.5   Distance to P_5-Free Graphs

A P_5-*hitting set* of G is a set of vertices X such that $G - X$ is P_5-free. In this subsection, we consider INDEPENDENT CUTSET parameterized by the size of a P_5-hitting set. Bacsó and Tuza [2] showed that any connected P_5-free graph has a dominating clique or a dominating P_3. Camby and Schaudt [5] generalized this result and showed that such a dominating set can be computed in polynomial time. We use the statement of the following corollary to prove a stronger statement in Theorem 5.

Corollary 9. INDEPENDENT CUTSET *can be solved in polynomial time for connected P_5-free graphs.*

Proof. Let G be a connected P_5-free graph. We compute a dominating set X as in [5] in polynomial time, that is, the set X induces either a clique or a P_3 in G. We invoke the algorithm of Theorem 3 with G and X as input. Since X induces a clique or a P_3 in G, there are at most $\mathcal{O}(n)$ partitions (A, X') or (A, B, X') of X such that there is no edge between A and B in G, and X' is independent in G. Given that the relevant partitions of X can be enumerated in polynomial time, the remaining steps of the algorithm of Theorem 3 can solve INDEPENDENT CUTSET in polynomial time for P_5-free graphs.   □

For Theorem 5 we need the following fact.

Proposition 1. *A P_5-hitting set of G with size k (if any) can be found in* FPT-*time with respect to k.*

Proof. A simple bounded search tree algorithm has $\mathcal{O}^*(5^k)$ running time.   □

We will now prove the main theorem of this subsection.

Theorem 5. *If a P_5-hitting set with at most k vertices of a connected graph G can be computed in $\mathcal{O}^*(f(k))$ time, then* INDEPENDENT CUTSET *can be solved in $\mathcal{O}^*(f(k) + 3^k)$ time.*

Proof. Let G and k be as in the statement. First, we compute a P_5-hitting set X of size at most k of G. Let $K_1, \ldots, K_\ell$ be the components of $G - X$. Then we compute a dominating set X_i of $G[K_i]$ as in [5], that is, the set X_i induces either a clique or a P_3 in $G[K_i]$. Note that $X_i \cup X$ dominates $G[K_i \cup X]$ for every $i \in [\ell]$. Assume that G admits an independent cutset S^*. There are two cases: Either $X \setminus S^*$ is contained in at most one component of $G - S^*$, or not. We explain for both cases how to compute an independent cutset of G in time $\mathcal{O}^*(3^k)$.

Case 1: $X \setminus S^*$ is contained in at most one component of $G - S^*$. This implies that there exists some $i \in [\ell]$ such that $S^* \cap (K_i \cup X)$ is an independent cutset of $G[K_i \cup X]$. We iterate over all disjoint partitions (A^*, X^*) of X such that X^* is independent in G. In one iteration, we guess $S^* \cap X$ as X^*. Now we invoke a modified version of the algorithm of Theorem 3 with $G[K_i \cup X]$ and $X_i \cup X$ as input. This version only considers partitions (A, X') or (A, B, X') of X with $A^* \subseteq A$ and $X^* \subseteq X'$. As in Corollary 9, this takes only polynomial time since the number of relevant partitions is polynomial. By Theorem 3, the modified algorithm returns an independent cutset of $G[K_i \cup X]$, which is also an independent cutset of G. The described procedure can be implemented to run $\mathcal{O}^*(2^k)$ time.

Case 2: $X \setminus S^*$ is contained in at least two components of $G - S^*$. We iterate over all disjoint partitions (A^*, B^*, X^*) of X such that X^* is an independent cutset of $G[X]$ that separates $A^* \neq \emptyset$ and $B^* \neq \emptyset$. In one iteration, we guess A^* and B^* such that they are in different components of $G - S^*$, and we guess $S^* \cap X$ as X^*. This time, we invoke a modified version of the algorithm of Lemma 5 with $G[K_i \cup X]$ and $X_i \cup X$ as input. This version only considers partitions (A, B, X') with $A^* \subseteq A$, $B^* \subseteq B$ and $X^* \subseteq X'$. As in Corollary 9, this takes only polynomial time. The modified algorithm returns, for every $i \in [\ell]$, an independent cutset S_i of $G[X_i \cup X]$ with the following properties: the set S_i separates A^* and B^* in $G[X_i \cup X]$, and $X^* \subseteq S_i$. Now, $\bigcup_{i \in \ell} S_i$ is an independent cutset of G. The described procedure for this can be implemented to run in $\mathcal{O}^*(3^k)$ time.

The overall running time is $\mathcal{O}^*(f(k) + 3^k)$, which completes the proof. $\square$

3.6 Generalizing Distance to P_5-Free Graphs

In this section, we generalize Corollary 9 and Theorem 5. Let G be a graph, and let $\mathcal{G}$ be a class of graphs. We say that a set of vertices X is a α_k-*dominating set* of G if X is a dominating set of G and $\alpha(G[X]) \leq k$. If G admits a α_k-dominating set, we say that G is α_k-*dominated*. In addition, if such a set can be computed in polynomial time, we say that G is *efficiently* α_k-*dominated*. The *(efficient)* α-*domination number* of G is the minimum k such that G is (efficient) α_k-dominated. We say that a graph class $\mathcal{G}$ is (efficiently) α_k-dominated if every graph $G \in \mathcal{G}$ is (efficiently) α_k-dominated. For example, Bacsó and Tuza [2] as well as Cozzens and Kelleher [11] independently proved that $\{P_5, C_5\}$-free graphs are α_1-dominated, that is, they contain a dominating clique. Another example is the class of P_5-free graphs, which is efficiently α_2-dominated.

We start by generalizing the polynomial time result on P_5-free graphs. The proof is similar to the one of Corollary 9.

Lemma 6. INDEPENDENT CUTSET *on* $\mathcal{G}$ *can be solved in polynomial time, whenever* $\mathcal{G}$ *is an efficiently* α_c-*dominated graph class for some constant* c.

In [24], Penrice presented some families of α_k-dominated graph classes. For example, Penrice shows that connected $\{P_6, H_{t+1}\}$-free graphs are α_t-dominated, where H_{t+1} denotes the graph obtained by subdividing each edge of a $K_{1,t+1}$. Penrice also shows that tK_2-free graphs without isolated vertices are α_{2t-2}-dominated. From a thorough reading in [24], it can be seen that tK_2-free graphs without isolated vertices are efficiently α_{2t-2}-dominated. Thus, from Lemma 6 and the results in [24] the following holds.

Corollary 10. INDEPENDENT CUTSET *can be solved in polynomial time for* tK_2-*free graphs for any constant* $t \geq 1$.

It has been remarked that Prisner proved in [25] that, for any integer t, any tK_2-free graph has polynomially many maximal independent sets. We were not able to verify this, because his article was inaccessible to us. With this, the statement of Corollary 10 already follows from Corollary 1.

Theorem 6 is a generalization of Theorem 5 and Lemma 6, and its proof is similar to the proof of Theorem 5.

Theorem 6. *Let* c *be a fixed constant and* G *be the input graph. If a set* X *of* k *vertices such that* $G - X$ *is efficiently* α_c-*dominated can be found in* $\mathcal{O}^*(f(k))$ *time, then* INDEPENDENT CUTSET *can be solved in* $\mathcal{O}^*(f(k) + 3^k)$ *time.*

References

1. Aspvall, B., Plass, M.F., Tarjan, R.E.: A linear-time algorithm for testing the truth of certain quantified boolean formulas. Inf. Process. Lett. **8**(3), 121–123 (1979). https://doi.org/10.1016/0020-0190(79)90002-4
2. Bacsó, G., Tuza, Z.: Dominating cliques in P_5-free graphs. Period. Math. Hung. **21**(4), 303–308 (1990)
3. Blair, J.R.S., Peyton, B.: An introduction to chordal graphs and clique trees. In: Graph Theory and Sparse Matrix Computation, pp. 1–29. Springer (1993)
4. Brandstädt, A., Dragan, F.F., Le, V.B., Szymczak, T.: On stable cutsets in graphs. Discret. Appl. Math. **105**(1–3), 39–50 (2000). https://doi.org/10.1016/S0166-218X(00)00197-9
5. Camby, E., Schaudt, O.: A new characterization of P_k-free graphs. Algorithmica **75**(1), 205–217 (2016)
6. Chen, G., Yu, X.: A note on fragile graphs. Discret. Math. **249**(1–3), 41–43 (2002). https://doi.org/10.1016/S0012-365X(01)00226-6
7. Chvátal, V.: Recognizing decomposable graphs. J. Graph Theory **8**(1), 51–53 (1984)
8. Corneil, D.G., Fonlupt, J.: Stable set bonding in perfect graphs and parity graphs. J. Combinatorial Theory, Series B **59**(1), 1–14 (1993)

9. Courcelle, B., Makowsky, J.A., Rotics, U.: Linear time solvable optimization problems on graphs of bounded clique-width. Theory Comput. Syst. **33**(2), 125–150 (2000)

10. Courcelle, B.: Graph rewriting: an algebraic and logic approach. In: van Leeuwen, J. (ed.) Handbook of Theoretical Computer Science, Volume B: Formal Models and Semantics, pp. 193–242. Elsevier and MIT Press (1990). https://doi.org/10.1016/b978-0-444-88074-1.50010-x

11. Cozzens, M.B., Kelleher, L.L.: Dominating cliques in graphs. Discret. Math. **86**(1–3), 101–116 (1990). https://doi.org/10.1016/0012-365X(90)90353-J

12. Dallard, C., Milanič, M., Štorgel, K.: Treewidth versus clique number. II, Tree-independence number (2022)

13. Farber, M.: On diameters and radii of bridged graphs. Discret. Math. **73**(3), 249–260 (1989). https://doi.org/10.1016/0012-365X(89)90268-9

14. Garey, M.R., Johnson, D.S.: Computers and Intractability; A Guide to the Theory of NP-Completeness. W. H. Freeman & Co., USA (1990)

15. Johnson, D.S., Yannakakis, M., Papadimitriou, C.H.: On generating all maximal independent sets. Inf. Process. Lett. **27**(3), 119–123 (1988)

16. Le, V.B., Mosca, R., Müller, H.: On stable cutsets in claw-free graphs and planar graphs. J. Discrete Algorithms **6**(2), 256–276 (2008). https://doi.org/10.1016/j.jda.2007.04.001

17. Le, V.B., Pfender, F.: Extremal graphs having no stable cutset. Electron. J. Combinatorics **20**(1), 35 (2013). https://doi.org/10.37236/2513

18. Le, V.B., Randerath, B.: On stable cutsets in line graphs. Theoret. Comput. Sci. **301**(1–3), 463–475 (2003). https://doi.org/10.1016/S0304-3975(03)00048-3

19. Lokshtanov, D., Narayanaswamy, N.S., Raman, V., Ramanujan, M.S., Saurabh, S.: Faster parameterized algorithms using linear programming. ACM Trans. Algorithms **11**(2), 15:1–15:31 (2014). https://doi.org/10.1145/2566616

20. Marx, D.: Chordal deletion is fixed-parameter tractable. Algorithmica **57**(4), 747–768 (2010). https://doi.org/10.1007/s00453-008-9233-8

21. Marx, D., O'Sullivan, B., Razgon, I.: Finding small separators in linear time via treewidth reduction. ACM Trans. Algorithms **9**(4), 30:1–30:35 (2013). https://doi.org/10.1145/2500119

22. Moon, J.W., Moser, L.: On cliques in graphs. Israel J. Math. **3**, 23–28 (1965)

23. Oum, S.I., Seymour, P.: Approximating clique-width and branch-width. Journal of Comb. Theory Ser. B **96**(4), 514–528 (2006)

24. Penrice, S.G.: Clique-like dominating sets (1995)

25. Prisner, E.: Graphs with few cliques. Graph Theory, Combinatorics, and Algorithms **1**, 2 (1995)

26. Queyranne, M.: A combinatorial algorithm for minimizing symmetric submodular functions. In: Clarkson, K.L. (ed.) Proceedings of the Sixth Annual ACM-SIAM Symposium on Discrete Algorithms, 22–24 January 1995. San Francisco, California, USA, pp. 98–101. ACM/SIAM (1995)

27. Rauch, J., Rautenbach, D., Souza, U.S.: Exact and parameterized algorithms for the independent cutset problem (2023)

28. Reed, B.A., Smith, K., Vetta, A.: Finding odd cycle transversals. Oper. Res. Lett. **32**(4), 299–301 (2004). https://doi.org/10.1016/j.orl.2003.10.009

29. Tucker, A.: Coloring graphs with stable cutsets. J. Comb. Theory Ser. B **34**(3), 258–267 (1983)

Kernelization for Finding Lineal Topologies (Depth-First Spanning Trees) with Many or Few Leaves

Emmanuel Sam[1]([✉])(iD), Benjamin Bergougnoux[2](iD), Petr A. Golovach[1](iD), and Nello Blaser[1](iD)

[1] Department of Informatics, University of Bergen, Bergen, Norway
{emmanuel.sam,petr.golovach,nello.blaser}@uib.no
[2] Institute of Informatics, University of Warsaw, Warsaw, Poland
benjamin.bergougnoux@mimuw.edu.pl

Abstract. For a given graph G, a depth-first search (DFS) tree T of G is an r-rooted spanning tree such that every edge of G is either an edge of T or is between a *descendant* and an *ancestor* in T. A graph G together with a DFS tree is called a *lineal topology* $\mathcal{T} = (G, r, T)$. Sam et al. (2023) initiated study of the parameterized complexity of the MIN-LLT and MAX-LLT problems which ask, given a graph G and an integer $k \geq 0$, whether G has a DFS tree with at most k and at least k leaves, respectively. Particularly, they showed that for the dual parameterization, where the tasks are to find DFS trees with at least $n - k$ and at most $n - k$ leaves, respectively, these problems are fixed-parameter tractable when parameterized by k. However, the proofs were based on Courcelle's theorem, thereby making the running times a tower of exponentials. We prove that both problems admit polynomial kernels with $\mathcal{O}(k^3)$ vertices. In particular, this implies FPT algorithms running in $k^{\mathcal{O}(k)} \cdot n^{\mathcal{O}(1)}$ time. We achieve these results by making use of a $\mathcal{O}(k)$-sized vertex cover structure associated with each problem. This also allows us to demonstrate polynomial kernels for MIN-LLT and MAX-LLT for the structural parameterization by the vertex cover number.

Keywords: DFS Tree · Spanning Tree · Kernelization · Parameterized Complexity

1 Introduction

Depth-first search (DFS) is a well-known fundamental technique for visiting the vertices and exploring the edges of a graph [6,30]. For a given connected undirected graph with vertex set $V(G)$ and edge set $E(G)$, DFS explores $E(G)$ by always choosing an edge incident to the most recently discovered vertex that

The research leading to these results has received funding from the Research Council of Norway via the projects (PCPC) (grant no. 274526) and BWCA (grant no. 314528).

H. Fernau and K. Jansen (Eds.): FCT 2023, LNCS 14292, pp. 392–405, 2023.
https://doi.org/10.1007/978-3-031-43587-4_28

still has unexplored edges. A selected edge, either leads to a new vertex or a vertex already discovered by the search. The set of edges that lead to a new vertex during the DFS define an r-rooted spanning tree T of G, called a *depth-first spanning* (DFS) tree, where r is the vertex from which the search started. This tree T has the property that each edge that is not in T connects an ancestor and a descendant of T. All rooted spanning trees of a finite graph with this property, irrespective of how they are computed, such as a *Hamiltonian path*, are generalized as *trémaux trees* [10]. Given a graph G and a DFS tree T rooted at a vertex $r \in V(G)$, it is easy to see that the family $\mathcal{T}$ of subsets of $E(G)$ induced by the vertices in all subtrees of T with the same root r as T constitute a topology on $E(G)$. For this reason, the triple (G, T, r) has been referred to as the *lineal topology* (LT) of G in [29]. Many existing applications of DFS and DFS trees — such as planarity testing and embedding [9,20], finding connected and biconnected components of undirected graphs [19], bipartite matching [21], and graph layout [1] — only require one to find an arbitrary DFS tree of the given graph, which can be done in time $O(n + m)$, where n and m are the number of vertices and edges of the graph.

An application of a DFS tree, noted by Fellows et al. [14], that calls for a DFS tree with minimum height is the use of DFS trees to structure the search space of backtracking algorithms for solving *constraint satisfaction problems* [17]. This motivated the authors to study the complexity of finding DFS trees of a graph G that optimize or near-optimize the maximum length or minimum length of the root-to-leaf paths in the DFS trees of G. They showed that the related decision problems are NP-complete and do not admit a polynomial-time absolute approximation algorithm unless P = NP.

In this paper, we look at the MINIMUM LEAFY LT (MIN-LLT) and MAXIMUM LEAFY LT (MAX-LLT) problems introduced by Sam et al. [29]. Given a graph G and an integer $k \geq 0$, MIN-LLT and MAX-LLT ask whether G has a DFS tree with at most k and at least k leaves, respectively. These two problems are related to the well-known NP-complete MINIMUM LEAF SPANNING TREE (MIN-LST) and MAXIMUM LEAF SPANNING TREE (MAX-LST) [18,27].

Sam et al. [29] proved that MIN-LLT and MAX-LLT are NP-hard. Moreover, they proved that when parameterized by k, MIN-LLT is para-NP-hard and MAX-LLT is W[1]-hard. They also considered the "dual" parameterizations, namely, DUAL MIN-LLT and DUAL MAX-LLT, where the tasks are to find DFS trees with at least $n - k$ and at most $n - k$ leaves, respectively. They proved that DUAL MIN-LLT and DUAL MAX-LLT are both FPT parameterized by k. These FPT algorithms are, however, based on Courcelle's theorem [7], which relates the expressibility of a graph property in *monadic second order* (MSO) logic to the existence of an algorithm that solves the problem in FPT-time with respect to *treewidth* [25]. As a by-product, their running times have a high exponential dependence on the treewidth and the length of the MSO formula expressing the property.

1.1 Our Results

We prove that MIN-LLT and MAX-LLT admit polynomial kernels when parameterized by the *vertex cover number* of the given graph. Formally, we prove the following theorem.

Theorem 1. MIN-LLT *and* MAX-LLT *admit kernels with* $\mathcal{O}(\tau^3)$ *vertices when parameterized by the vertex cover number* τ *of the input graph.*

Based on these kernels, we show that DUAL MIN-LLT, and DUAL MAX-LLT admit polynomial kernels parameterized by k.

Theorem 2. DUAL MIN-LLT *and* DUAL MAX-LLT *admit kernels with* $\mathcal{O}(k^3)$ *vertices.*

This last result follows from a win-win situation as either (1) the input graph has a large *vertex cover* in terms of k and, consequently, both problems are trivially solvable or (2) the input graph has a small vertex cover, and we can use Theorem 1. Finally, we use our polynomial kernels to prove that DUAL MIN-LLT, and DUAL MAX-LLT admit FPT algorithms parameterized by k with low exponential dependency.

Theorem 3. DUAL MIN-LLT *and* DUAL MAX-LLT *can be solved in* $k^{\mathcal{O}(k)} \cdot n^{\mathcal{O}(1)}$ *time.*

As the previously known FPT algorithm for each of these problems was based on Courcelle's theorem, our algorithms are the first FPT-algorithms constructed explicitly.

1.2 Related Results

Lu and Ravi [23] proved that the MIN-LST, problem has no constant factor approximation unless $P = NP$. From a parameterization point of view, Prieto et al. [26] showed that this problem is $W[P]$-hard parameterized by the solution size k. The MAX-LST problem is, however, FPT parameterized by k and has been studied extensively [2,3,13,15,24].

DUAL MIN-LLT is related to the well-studied k-INTERNAL SPANNING TREE problem [16,26], which asks to decide whether a given graph admits a spanning tree with at most $n - k$ leaves (or at least k internal vertices). Prieto et al. [26] were the first to show that the natural parameterized version of k-INTERNAL SPANNING TREE has a $\mathcal{O}^*(2^{k \log k})$-time FPT algorithm and a $\mathcal{O}(k^3)$-vertex kernel. Later, the kernel was improved to $\mathcal{O}(k^2)$, $\mathcal{O}(3k)$, and $\mathcal{O}(2k)$ by Prieto et al., Fomin et al. [16], and Li et al. [22] respectively. The latter authors also gave what is now the fastest FPT algorithm for k-INTERNAL SPANNING TREE, which runs in $\mathcal{O}^*(4^k)$ time.

An *independency tree* (IT) is a variant of a spanning tree whose leaves correspond to an independent set in the given graph. Given a connected graph on $n \geq 3$, G has no IT if it has no DFS tree in which the leaves and the root are pairwise nonadjacent in G [4]. From a parameterization point of view, the MIN

LEAF IT (INTERNAL) and MAX LEAF IT (INTERNAL) problems [5], which ask, given a graph G and an integer $k \geq 0$, whether G has an IT with at least k and at most k internal vertices, respectively, are related to DUAL MIN-LLT and DUAL MAX-LLT, respectively. Casel et al. [5] showed that, when parameterized by k, MIN LEAF IT (INTERNAL) has an $\mathcal{O}^*(4^k)$-time algorithm and a $2k$ vertex kernel. They also proved that MAX LEAF IT (INTERNAL) parameterized by k has a $\mathcal{O}^*(18^k)$-time algorithm and a $\mathcal{O}(k2^k)$-vertex kernel, but no polynomial kernel unless the polynomial hierarchy collapses to the third level. Their techniques, however, do not consider the properties of a DFS tree and, therefore, do not work for our problems.

1.3 Organization of the Paper

Section 2 contains basic terminologies relevant to graphs, DFS trees, and parameterized complexity necessary to understand the paper. In Sect. 3, we first prove a lemma about how, given a graph G and a vertex cover of G, the internal vertices of any spanning tree of G relate to the given vertex cover. We then use this lemma to demonstrate a polynomial kernel for MIN-LLT and MAX-LLT for the structural parameterization by the vertex cover number of the graph. This is followed by the kernelization algorithms for DUAL MIN-LLT and DUAL MAX-LLT parameterized by k. In Sect. 4, we devise FPT algorithms for DUAL MIN-LLT and DUAL MAX-LLT based on their polynomial kernels. Finally, we conclude the paper in Sect. 5 with remarks concerning future studies.

2 Preliminaries

We consider only simple finite graphs. We use $V(G)$ and $E(G)$ to denote the sets of vertices and edges, respectively, of a graph G. For a graph G, we denote the number of vertices $|V(G)|$ and the number of edges $|E(G)|$ of G by n and m, respectively, if this does not create confusion. For any vertex $v \in V(G)$, the set $N_G(v)$ denotes the neighbors of v in G and $N_G[v]$ denotes its *closed neighborhood* $N_G(v) \cup \{v\}$ in G. For a set of vertices $X \subseteq V$, $N_G(X) = \left(\bigcup_{v \in X} N_G(v)\right) \setminus X$. We omit the G in the subscript if the graph is clear from the context. For a vertex v, its *degree* is $d_G(v) = |N_G(v)|$. Given any two graphs $G_1 = (V_1, E_1)$ and $G_2 = (V_2, E_2)$, if $V_1 \subseteq V_2$ and $E_1 \subseteq E_2$ then G_1 is a *subgraph* of G_2, denoted by $G_1 \subseteq G_2$. If G_1 contains all the edges $uv \in E_2$ with $u, v \in V_1$, then we say G_1 is an *induced subgraph* of G_2, or V_1 induces G_1 in G_2, denoted by $G[V_1]$. If G_1 is such that it contains every vertex of G_2, i.e., if $V_1 = V_2$ then G_1 is a *spanning subgraph* of G_2. Given a set of vertices $X \subseteq V(G)$, we express the induced subgraph $G[V(G) \setminus X]$ as $G - X$. If $X = \{x\}$, we write $V(G) \setminus x$ instead of $V(G) \setminus \{x\}$ and $G - x$ instead of $G - \{x\}$. Given a graph G, a set of vertices $S \subseteq V(G)$ is a *vertex cover* of G if, for every edge $uv \in E(G)$, either $u \in S$ or $v \in S$; the *vertex cover number* of G, denoted by $\tau(G)$, is the minimum size of a vertex cover. A set $Y \subseteq V(G)$ is called an *independent set*, if for every vertex pair $u, v \in Y$, $uv \notin E(G)$. A *matching* M in a given graph G is a set of edges, no two of which share common vertices. A *pendant vertex* is a vertex with degree one.

For definitions of basic tree terminologies including root, child, parent, ancestor, and descendant, we refer the reader to [11]. Given a graph G, we denote a spanning tree of G rooted at a vertex $r \in V(G)$ by (T, r). When there is no ambiguity, we simply use T instead of (T, r). For a rooted tree T, a vertex v is a *leaf* if it has no descendants and v is an *internal* vertex if otherwise. A spanning tree T with a root r is a *DFS tree rooted in r* if for very every edge $uv \in E(G)$, either $uv \in E(T)$, or v is a descendant of u in T, or u is a descendant of v in T. Equivalently, T is a DFS tree if it can be produced by the classical depth-first search (DFS) algorithm [6]. We say that a path P in a rooted tree T is a *root-to-leaf* path if one of its end-vertices is the root and the other is a leaf of T.

Now we review some important concepts of Parameterized complexity (PC) relevant to the work reported herein. For more details about PC, we refer the reader to [8,12].

Definition 4. (Parameterized Problem). *Let Σ be a fixed finite alphabet. A parameterized problem is a language $P \subseteq \Sigma^* \times \mathbb{N}$. Given an instance $(x, k) \in \Sigma^* \times \mathbb{N}$ of a parameterized problem, $k \in \mathbb{N}$ is called the parameter, and the task is to determine whether (x, k) belongs to P. A parameterized problem P is classified as fixed-parameter tractable (FPT) if there exists an algorithm that answers the question "$(x, k) \in P$?" in time $f(k) \cdot poly(|x|)$, where $f : \mathbb{N} \to \mathbb{N}$ is a computable function.*

Definition 5. *A kernelization algorithm, or simply a kernel, for a parameterized problem P is a function ϕ that maps an instance (x, k) of P to an instance (x', k') of P such that the following properties are satisfied:*

1. *$(x, k) \in P$ if and only if $(x', k') \in P$,*
2. *$k' + |x'| \le g(k)$ for some computable function $g : \mathbb{N} \to \mathbb{N}$, and*
3. *ϕ is computable in time polynomial in $|x|$ and k.*

If the upper-bound $g(\cdot)$ of the kernel (Property 2) is polynomial (linear) in terms of the parameter k, then we say that P admits a polynomial (linear) kernel. It is common to write a kernelization algorithm as a series of reduction rules. A *reduction rule* is a polynomial-time algorithm that transform an instance (x, k) to an equivalent instance (x', k') such that Property 1 is fulfilled. Property 1 is referred to as the *safeness* or *correctness* of the rule.

3 Kernelization

In this section, we demonstrate polynomial kernels for DUAL MIN-LLT and DUAL MAX-LLT. But first, we show that MIN-LLT and MAX-LLT admit polynomial kernels when parameterized by the vertex cover number of the input graph. The following simple lemma is crucial for our kernelization algorithms.

Lemma 6. *Let G be a connected graph and let S be a vertex cover of G. Then every rooted spanning tree T of G has at most $2|S|$ internal vertices and at most $|S|$ internal vertices are not in S.*

Proof. Let T be a rooted spanning tree tree of G with a set of internal vertices X. For every vertex v of T, we denote by $\mathsf{child}(v)$ the set of its childred in T. For each internal vertex v of T, we have $\mathsf{child}(v) \neq \emptyset$ and if $v \notin S$, then $\mathsf{child}(v) \subseteq S$ because S is a vertex cover of G. Moreover, for any distinct internal vertices u and v of T, $\mathsf{child}(u) \cap \mathsf{child}(v) = \emptyset$. Given $X \setminus S = \{v_1, \ldots, v_t\}$, we deduce that $\mathsf{child}(v_1), \ldots, \mathsf{child}(v_t)$ are pairwise disjoint and non-empty subsets of S. We conclude that $|X \setminus S| \leq |S|$ and $|X| \leq 2|S|$.   □

We also use the following folklore observation.

Observation 7. *The set of internal vertices of any DFS tree T of a connected graph G is a vertex cover of G.*

Proof. To see the claim, it is sufficient to observe that any leaf of a DFS tree T is adjacent in G only to its ancestors, that is, to internal vertices.   □

We use Lemma 6 to show that, given a vertex cover, we can reduce the size of the input graph for both MIN-LLT and MAX-LLT.

Lemma 8. *There is a polynomial-time algorithm that, given a connected graph G together with a vertex cover S of size s, outputs a graph G' with at most $s^2(s-1) + 3s$ vertices such that for every integer $t \geq 0$, G has a DFS tree with exactly t internal vertices if and only if G' has a DFS tree with exactly t internal vertices.*

Proof. Let G be a connected graph and let S be a vertex cover of G of size s. As the lemma is trivial if $s = 0$, we assume that $s \geq 1$. Denote $I = V(G) \setminus S$; note that I is an independent set. We apply the following two reduction rules to reduce the size of G.

The first rule reduces the number of pendant vertices. To describe the rule, denote by $\mathsf{pendant}(v)$ for $v \in S$ the set of pendant vertices of I adjacent to v.

Rule 1

---
foreach $v \in S$ **do**
| **if** $|\mathsf{pendant}(v)| > 2$ **then** delete all but two vertices in $\mathsf{pendant}(v)$ from G
end

To see that Rule 1 is safe, denote by G' the graph obtained from G by the application of the rule. Notice that for every $v \in S$, at most one vertex of $\mathsf{pendant}(v)$ is the root and the other vertices are leaves that are children of v in any rooted spanning tree T of G.

Let T be a DFS tree of G rooted in r with t internal vertices. Because for every $v \in S$, the vertices of $\mathsf{pendant}(v)$ have the same neighborhood in G and Rule 1 does not delete all the vertices of $\mathsf{pendant}(v)$, we can assume without loss of generality that $r \in V(G')$. Let $T' = T[V(G')]$. Because the deleted vertices are leaves of T, we have that T' is a tree and, moreover, T' is a DFS tree of G' rooted in r. Clearly, each internal vertex of T' is an internal vertex of T. Let $v \in S$ be a vertex such that $|\mathsf{pendant}(v)| > 2$. Then v has a pendant neighbor $u \neq r$ in G' and u should be a child of v in T'. Thus, v is an internal vertex of T'. This implies that every leaf v of T' is not adjacent to any vertex of $V(G) \setminus V(G')$ in G. Hence, v is a leaf of T. Because the deleted vertices are leaves of T, we obtain that a vertex $v \in V(G)$ is an internal vertex of T if and only if v is an internal vertex of T'. Then T and T' have the same number of internal vertices.

For the opposite direction, let T' be a DFS tree of G' rooted in r with t internal vertices. We construct the tree T from T' by adding each deleted vertex u as a leaf to T': if $u \in V(G) \setminus V(G')$, then $u \in \mathsf{pendant}(v)$ for some $v \in S$ and we add u as a leaf child of v. Because the deleted vertices are pendants, we have that T is a DFS tree of G. Observe that each internal vertex of T' remains internal in T. In the same way as above, we observe that a vertex $v \in S$ with $|\mathsf{pendant}(v)| > 2$ cannot be a leaf of T', because v has a pendant neighbor in G' distinct from r that should be a child of v. Hence, every leaf v of T' is not adjacent to any vertex of $V(G) \setminus V(G')$ in G and, therefore, is a leaf of T. Since the deleted vertices are leaves of T, we obtain that a vertex $v \in V(G)$ is an internal vertex of T if and only if v is an internal vertex of T'. Thus, T and T' have the same number of internal vertices. This concludes the safeness proof.

The next rule is used to reduce the number of nonpendant vertices of I. For each pair of vertices $u, v \in S$, we use *common neighbor* of u and v to refer to a vertex $w \in I$ that is adjacent to both u and v and denote by W_{uv} the set of common neighbors of u and v. Rule 2 is based on the observation that if the size of W_{uv} for any vertex pair $u, v \in S$ is at least $2s + 1$, then it follows from Lemma 6 that every spanning tree T contains at most s internal vertices and at least $s + 1$ leaves from W_{uv}. We prove that it is enough to keep at most $2s$ vertices from W_{uv} for each $u, v \in S$.

Rule 2

forall the *pairs* $\{u, v\}$ *of distinct vertices of* S **do**
| Label $\max\{|W_{uv}|, 2s\}$ vertices in W_{uv};
end
Delete the unlabeled vertices of I with at least two neighbors in S from G.

To show that Rule 2 is safe, let $x \in I$ be a vertex with at least two neighbors in S which is not labeled by Rule 2. Let $G' = G - x$. We claim that G has a DFS tree with exactly t internal vertices if and only if G' has a DFS tree with exactly t internal vertices.

We use the following auxiliary claim, the proof of which can be found in the full version of this paper [28].

Claim 8.1. (i) For any DFS tree T of G, the vertices of $N_G(x)$ are vertices of a root-to-leaf path of T.

(ii) For any DFS tree T' of G', the vertices of $N_G(x)$ are vertices of a root-to-leaf path of T'.

(iii) For any DFS tree T' of G', every vertex of $N_G(x)$ is an internal vertex of T'.

We use Claim 8.1 to show the following property.

Claim 8.2. If G has a DFS tree with t internal vertices, then G has a DFS tree T with t internal vertices such that x is a leaf of T.

Proof of Claim. 8.2 Let T be a DFS tree of G with a root r that has exactly t internal vertices. We prove that if x is an internal vertex of T, then T can be modified in such a way that x would become a leaf. Observe that by Claim 8.1 (i), x has a unique child v in T. We have two cases depending on whether $x = r$ or has a parent u.

Suppose first that $x = r$. By Claim 8.1, the neighbors of x in G are vertices of some root-to-leaf path of T. Let u be the neighbor of x at maximum distance from r in T. Because $d_G(x) \geq 2$, $u \neq v$. Since x is not labeled by Rule 2, $|W_{uv}| > 2s$. By Lemma 6, there are at least $s + 1$ vertices W_{uv} that are leaves of T. These leaves have their parents in S which has size s. By the pigeonhole principle, there are distinct leaves $w, w' \in W_{uv}$ with the same parent. We rearrange T by making w a root with the unique child v and making x a leaf with the parent u. Denote by T' the obtained tree.

Because x is adjacent to u and some of its ancestors in T and w is adjacent only to some of its ancestors in T, we conclude that T' is a feasible DFS tree. Notice that w which was a leaf of T became an internal vertex of T' and x that was an internal vertex is now a leaf. Because x is a leaf of T', we have that $T'' = T' - x$ is a DFS tree of G' rooted in w. By Claim 8.1 (iii), u is an internal vertex of T''. This implies that u is an internal vertex of both T and T'. Since the parent of w in T has $w' \neq w$ as a child, we also have that w is an internal vertex of both T and T'. Therefore, T and T' have the same number of internal vertices. This proves that G has a DFS tree T' with t internal vertices such that x is a leaf of T'.

Assume now that x has a parent u in T. By Claim 8.1, the neighbors of x in G are vertices of some root-to-leaf path of T. Denote by v' be the neighbor of x at maximum distance from r in T; it may happen that $v' = v$. As x is not labeled by Rule 2, $|W_{uv}| > 2s$. Then by Lemma 6, there are at least $s + 1$ vertices W_{uv} that are leaves of T. These leaves have their parents in S which has size s. By the pigeonhole principle, there are distinct leaves $w, w' \in W_{uv}$ with the same parent. We rearrange T by making w a child of u and a parent of v and making x a leaf with the parent v'. Denote by T' the obtained tree.

Because x is adjacent to v' and some of its ancestors in T and w is adjacent only to some of its ancestors in T, including u and v, we have that T' is a feasible DFS tree. Notice that w was a leaf of T and is now an internal vertex of T', while x was an internal vertex in T and is now a leaf in T'. Because x is a leaf of T', we have that $T'' = T' - x$ is a DFS tree of G' rooted in w. By Claim 8.1 (iii), v' is an internal vertex of T''. Therefore, v' is an internal vertex of both T and T'. Since the parent of w in T has $w' \neq w$ as a child, we also have that w is an internal vertex of both T and T'. Thus, T and T' have the same number of internal vertices. We obtain that G has a DFS tree T' with t vertices such that x is a leaf of T'. This concludes the proof.   □

Now we are ready to proceed with the proof that G has a DFS tree with exactly t internal vertices if and only if G' has a DFS tree with exactly t internal vertices.

For the forward direction, let T be a DFS tree of G with t internal vertices. By Claim 8.2, we can assume that x is a leaf of T. Let $T' = T - x$. Because x is a leaf of T, T' is a DFS tree of G'. Let u be the parent of x in T. Because u is adjacent to x in G, we have that u is an internal vertex of T' by Claim 8.1 (iii). This means that the number of internal vertices of T and T' is the same, that is, G' has a DFS tree with t vertices.

For the opposite direction, let T' be a DFS tree of G' with t internal vertices with a root r. By Claim 8.1 (ii), the neighbors of x in G are vertices of some root-to-leaf path in T'. Let v be the neighbor of x at maximum distance from r in T'. We construct T by making x a leaf with the parent v. Because x is adjacent in G only to v and some of its ancestors in T', T is a DFS tree. By Claim 8.1(iii), v is an internal vertex of T'. Therefore, T' and T have the same set of internal vertices. We obtain that G has a DFS tree with t vertices. This concludes the proof of our claim.

Recall that G' was obtained from G by deleting a single unlabeled vertex $x \in I$ of degree at least two. Applying the claim that G has a DFS tree with exactly t internal vertices if and only if $G' = G - x$ has a DFS tree with exactly t internal vertices inductively for unlabeled vertices of I of degree at least two, we obtain that Rule 2 is safe.

Denote now by G' the graph obtained from G by the application of Rules 1 and 2. Because both rules are safe, for any integer $t \geq 0$, G has a DFS tree with exactly t internal vertices if and only if G' has a DFS tree with exactly t internal vertices. Because of Rule 1, $G' - S$ has at most $2s$ pendant vertices. Rule 2 guarantees that $G' - S$ has at most $2s\binom{s}{2} = s^2(s-1)$ vertices of degree at least two. Then the total number of vertices of G' is at most $s^2(s-1) + 2s + s = s^2(s-1) + 3s$.

It is straightforward to see that Rule 1 can be applied in $\mathcal{O}(sn)$ time and Rule 2 can be applied in $\mathcal{O}(s^2n)$ time. Therefore, the algorithm is polynomial. This concludes the proof.   □

As a direct consequence of Lemma 8 we obtain that MIN-LLT and MAX-LLT admit polynomial kernels when parameterized by the vertex cover number of the input graph.

We are ready to prove our kernels parameterized by vertex cover.

Proof of Theorem. 1 We show the theorem for MIN-LLT; the arguments for MAX-LLT are almost identical. Recall that the task of MIN-LLT is to decide, given a graph G and an integer $k \geq 0$, whether G has a DFS tree with at most k leaves. Equivalently, we can ask whether G has a DFS tree with at least $|V(G)| - k$ internal vertices. Let (G, k) be an instance of MIN-LLT. We assume that G is connected as, otherwise, (G, k) is a no-instance and we can return a trivial no-instance of MIN-LLT of constant size.

First, we find a vertex cover S of G. For this, we apply a folklore approximation algorithm (see, e.g., [8]) that greedily finds an inclusion-maximal matching M in G and takes the set S of endpoints of the edges of M. It is well-known that $|S| \leq 2\tau$. Then we apply the algorithm from Lemma 8. Let G' be the output graph. By Lemma 8, G' has $\mathcal{O}(\tau^3)$ vertices. We set $k' = k - |V(G)| + |V(G')|$ and return the instance (G', k') of MIN-LLT.

Suppose that G has a DFS tree with at most k leaves. Then G has a DFS tree with $t \geq |V(G)| - k$ internal vertices. By Lemma 8, G' also has a DFS tree with t internal vertices. Then G' has a DFS tree with $|V(G')| - t \leq |V(G')| - (|V(G)| - k) = k'$ leaves. For the opposite direction, assume that G' has a DFS tree with at most k' leaves. Then G' has a DFS tree with $t \geq |V(G')| - k' = |V(G)| - k$ internal vertices. By Lemma 8, G has a DFS tree with t internal vertices and, therefore, G has a DFS tree with at most k leaves.

Because S can be constructed in linear time and the algorithm from Lemma 8 is polynomial, the overall running time is polynomial. This concludes the proof. □

Now we demonstrate a polynomial kernel for DUAL MIN-LLT.

Theorem 9. DUAL MIN-LLT *admits a kernel with* $\mathcal{O}(k^3)$ *vertices.*

Proof. Recall that the task of DUAL MIN-DLL is to verify, given a graph G and an integer $k \geq 0$, whether G has a DFS tree with at most $n - k$ leaves. Equivalently, the task is to check whether G has a DFS tree with at least k internal vertices. Let (G, k) be an instance of DUAL MIN-LLT. If G is disconnected, then (G, k) is a no-instance and we return a trivial no-instance of DUAL MIN-DLL of constant size. From now, we assume that G is connected.

We select an arbitrary vertex r of G and run the DFS algorithm from this vertex. The algorithm produces a DFS tree T. Let S be the set of internal vertices of T. If $|S| \geq k$, then we conclude that (G, k) is a yes-instance. Then the kernelization algorithm returns a trivial yes-instance of DUAL MIN-LLT of constant size and stops. Assume that this is not the case and $|S| \leq k - 1$.

By Observation 7, we have that S is a vertex cover of G of size $s \leq k - 1$. We use S to call the algorithm from Lemma 8. Let G' be a graph produced by the algorithm. By Lemma 8, G' has $\mathcal{O}(k^3)$ vertices. Our kernelization algorithm returns (G', k) and stops.

To see correctness, it is sufficient to observe that by Lemma 8, for any integer $t \geq k$, G has a DFS tree with t internal vertices if and only if G' has a DFS tree with t internal vertices. Because the DFS algorithm runs in linear time (see, e.g., [6]) and the algorithm from Lemma 8 is polynomial, the overall running time is polynomial. This completes the proof. □

We use similar arguments to prove the following theorem. See the full version of this paper for the proof [28].

Theorem 10. DUAL MAX-LLT *admits a kernel with $\mathcal{O}(k^3)$ vertices.*

Theorems 9 and 10 implies Theorem 2.

4   FPT Algorithms

In this section, we give algorithms that solve DUAL MIN-LLT and DUAL MAX-LLT in FPT time using the kernels given in the previous section. Our algorithms are brute force algorithms which guess internal vertices.

Recall that the standard DFS algorithm [6] outputs a labeled spanning tree. More formally, given an n-vertex graph and a root vertex r, the algorithm outputs a DFS tree T rooted in r and assigns to the vertices of G distinct labels $d[v]$ from $\{1, \ldots, n\}$ giving the order in which the vertices were discovered by the algorithm. Thus, the algorithm outputs a linear ordering of vertices. Given an ordering $v_1, \ldots, v_n$ of $V(G)$, we say that a DFS tree T *respects* the ordering if T is produced by the DFS algorithm in such a way that $d[v_i] = i$ for every $i \in \{1, \ldots, n\}$. Observe that for an ordering of the vertices of G, there is a unique way to run the DFS algorithm to obtain T respecting the ordering. This gives us the following observation.

Observation 11. *It can be decided in linear time, given an ordering $v_1, \ldots, v_n$ of the vertices of a graph G, whether G has a DFS tree respecting the ordering. Furthermore, if such a tree T exists, it is unique and can be constructed in linear time.*

Let G be a graph and let $r \in V(G)$. For a tree $T \subseteq G$ with $r \in V(T)$, we say that T is *extendable* to a DFS tree rooted in r, if there is a DFS tree T' of G rooted in r such that T is a subtree of T'. We call T' an *extension* of T. The definition of a DFS tree immediately gives us the following necessary and sufficient conditions for the extendability of T.

Observation 12. *Let G be a graph with $r \in V(G)$ and let $T \subseteq G$ be a tree containing r. Then T is extendable to a DFS tree rooted in r if and only if*

(i) T is a DFS tree rooted in r of $G[V(T)]$,
(ii) for every connected component C of $G - V(T)$, the vertices of $N_G(V(C))$ are vertices of a root-to-leaf path of T.

Note that (i) and (ii) can be verified in polynomial (in fact, linear) time. We need the following variants of Observation 12 for special extensions in our algorithms.

Observation 13. *Let G be a graph with $r \in V(G)$ and let $T \subseteq G$ be a tree containing r. Then T is extendable to a DFS tree rooted in r with an extension T' such that the vertices of $V(T)$ are internal vertices of T' if and only if*

(i) T is a DFS tree rooted in r of $G[V(T)]$,
(ii) for every connected component C of $G - V(T)$, the vertices of $N_G(V(C))$ are vertices of a root-to-leaf path of T,
(iii) for every leaf v of T, there is $u \in V(G) \setminus V(T)$ that is adjacent to v.

Observation 14. *Let G be a graph with $r \in V(G)$ and let $T \subseteq G$ be a tree containing r. Then T is extendable to a DFS tree rooted in r with an extension T' such that the vertices of $L = V(G) \setminus V(T)$ are leaves of T' if and only if*

(i) T is a DFS tree rooted in r of $G[V(T)]$,
(ii) L is an independent set,
(iii) for every $v \in L$, the vertices of $N_G(v)$ are vertices of a root-to-leaf path of T.

Now, we are ready to describe our algorithms. For the proof of Lemma 15, see the full version of this paper [28].

Lemma 15. DUAL MIN-LLT *and* DUAL MAX-LLT *can be solved in $n^{\mathcal{O}(k)}$ time.*

Combining Lemma 15 and Theorem 2 implies Theorem 3 by providing $k^{\mathcal{O}(k)} \cdot n^{\mathcal{O}(1)}$ time algorithms for the dual problems.

5   Conclusion

We have shown that DUAL MIN-LLT and DUAL MAX-LLT admit kernels with $\mathcal{O}(k^3)$ vertices and can be solved in $k^{\mathcal{O}(k)} \cdot n^{\mathcal{O}(1)}$ time. A natural question is whether the problems have linear kernels, such as for k-INTERNAL SPANNING TREE [22]. Another question is whether the problems can be solved by single-exponential FPT algorithms.

As a byproduct of our kernelization algorithms for DUAL MIN-LLT and DUAL MAX-LLT, we also proved that MIN-LLT and MAX-LLT admit polynomial kernels for the structural parameterization by the vertex cover number. It is natural to wonder whether polynomial kernels exist for other structural parameterizations. In particular, it could be interesting to consider the parameterization by the *feedback vertex* number, i.e., by the minimum size of a vertex set X such that $G - X$ is a forest.

Acknowledgements. We acknowledge support from the Research Council of Norway grant "Parameterized Complexity for Practical Computing (PCPC)" (NFR, no. 274526) and "Beyond Worst-Case Analysis in Algorithms (BWCA)" (NFR, no. 314528).

References

1. Biedl, T.: The DFS-heuristic for orthogonal graph drawing. Comput. Geom. **18**(3), 167–188 (2001). https://doi.org/10.1016/S0925-7721(01)00006-2
2. Bonsma, P., Zickfeld, F.: Spanning trees with many leaves in graphs without diamonds and blossoms. In: Laber, E.S., Bornstein, C., Nogueira, L.T., Faria, L. (eds.) LATIN 2008: Theoretical Informatics, pp. 531–543. Springer, Berlin, Heidelberg (2008). https://doi.org/10.1007/978-3-540-78773-0_46

3. Bonsma, P.S., Brueggemann, T., Woeginger, G.J.: A faster FPT algorithm for finding spanning trees with many leaves. In: Rovan, B., Vojtáš, P. (eds.) MFCS 2003. LNCS, vol. 2747, pp. 259–268. Springer, Heidelberg (2003). https://doi.org/10.1007/978-3-540-45138-9_20

4. Böhme, T., Broersma, H., Göbel, F., Kostochka, A., Stiebitz, M.: Spanning trees with pairwise nonadjacent endvertices. Discrete Math. **170**(1), 219–222 (1997). https://doi.org/10.1016/S0012-365X(96)00306-8

5. Casel, K., et al.: Complexity of independency and Cliquy trees. Discrete Appl. Math. 272, 2–15 (2020). https://doi.org/10.1016/j.dam.2018.08.011, 15th Cologne-Twente Workshop on Graphs and Combinatorial Optimization (CTW 2017)

6. Cormen, T.H., Leiserson, C.E., Rivest, R.L., Stein, C.: Introduction to Algorithms, Third Edition. The MIT Press, 3rd edn. (2009)

7. Courcelle, B.: The monadic second-order logic of graphs. i. recognizable sets of finite graphs. Inf. Comput. **85**(1), 12–75 (1990). https://doi.org/10.1016/0890-5401(90)90043-H

8. Cygan, M., et al.: Parameterized Algorithms, 1st edn. Springer Publishing Company, Incorporated (2015). https://doi.org/10.1007/978-3-319-21275-3

9. De Fraysseix, H.: Trémaux trees and planarity. Electron. Notes Discrete Math. **31**, 169–180 (2008)

10. de Fraysseix, H., Ossona de Mendez, P.: Trémaux trees and planarity. Eur. J. Comb. **33**(3), 279–293 (2012). https://doi.org/10.1016/j.ejc.2011.09.012, topological and Geometric Graph Theory

11. Diestel, R.: Graph Theory, 5th edn. Springer Publishing Company, Incorporated (2017). https://doi.org/10.1007/978-3-662-53622-3

12. Downey, R.G., Fellows, M.R.: Fundamentals of Parameterized Complexity. Springer Publishing Company, Incorporated (2013). https://doi.org/10.1007/978-1-4471-5559-1

13. Estivill-Castro, V., Fellows, M.R., Langston, M.A., Rosamond, F.A.: FPT is p-time extremal structure I. In: Broersma, H., Johnson, M., Szeider, S. (eds.) Algorithms and Complexity in Durham 2005 - Proceedings of the First ACiD Workshop, 8–10 July 2005, Durham, UK. Texts in Algorithmics, vol. 4, pp. 1–41. King's College, London (2005)

14. Fellows, M.R., Friesen, D.K., Langston, M.A.: On finding optimal and near-optimal lineal spanning trees. Algorithmica **3**(1–4), 549–560 (1988)

15. Fellows, M.R., Langston, M.A.: On well-partial-order theory and its application to combinatorial problems of VLSI design. SIAM J. Discrete Math. **5**(1), 117–126 (1992)

16. Fomin, F.V., Gaspers, S., Saurabh, S., Thomassé, S.: A linear vertex kernel for MAXIMUM INTERNAL SPANNING TREE. In: Dong, Y., Du, D.-Z., Ibarra, O. (eds.) ISAAC 2009. LNCS, vol. 5878, pp. 275–282. Springer, Heidelberg (2009). https://doi.org/10.1007/978-3-642-10631-6_29

17. Freuder, E.C., Quinn, M.J.: Taking advantage of stable sets of variables in constraint satisfaction problems. In: Proceedings of the 9th International Joint Conference on Artificial Intelligence - Vol. 2. pp. 1076–1078. IJCAI'85, Morgan Kaufmann Publishers Inc., San Francisco, CA, USA (1985)

18. Garey, M.R., Johnson, D.S.: Computers and Intractability; A Guide to the Theory of NP-Completeness. W. H. Freeman & Co., USA (1990)

19. Hopcroft, J., Tarjan, R.: Algorithm 447: efficient algorithms for graph manipulation. Commun. ACM **16**(6), 372–378 (1973)

20. Hopcroft, J., Tarjan, R.: Efficient planarity testing. J. ACM (JACM) **21**(4), 549–568 (1974)

21. Hopcroft, J.E., Karp, R.M.: An $n^{5/2}$ algorithm for maximum matchings in bipartite graphs. SIAM J. Comput. **2**(4), 225–231 (1973). https://doi.org/10.1137/0202019
22. Li, W., Cao, Y., Chen, J., Wang, J.: Deeper local search for parameterized and approximation algorithms for maximum internal spanning tree. Inf. Comput. **252**, 187–200 (2017)
23. Lu, H.I., Ravi, R.: The power of local optimization: approximation algorithms for maximum-leaf spanning tree. In: Proceedings, Thirtieth Annual Allerton Conference on Communication, Control and Computing, pp. 533–542 (1996)
24. Michael, R.F., McCartin, C., Frances, A.R., Stege, U.: Coordinatized kernels and catalytic reductions: an improved FPT algorithm for max leaf spanning tree and other problems. In: Kapoor, S., Prasad, S. (eds.) FSTTCS 2000. LNCS, vol. 1974, pp. 240–251. Springer, Heidelberg (2000). https://doi.org/10.1007/3-540-44450-5_19
25. Nešetřil, J., de Mendez, P.O.: Bounded height trees and tree-depth. In: Sparsity. AC, vol. 28, pp. 115–144. Springer, Heidelberg (2012). https://doi.org/10.1007/978-3-642-27875-4_6
26. Prieto, E., Sloper, C.: Either/Or: using VERTEX COVER structure in designing FPT-algorithms — the case of k-INTERNAL SPANNING TREE. In: Dehne, F., Sack, J.-R., Smid, M. (eds.) WADS 2003. LNCS, vol. 2748, pp. 474–483. Springer, Heidelberg (2003). https://doi.org/10.1007/978-3-540-45078-8_41
27. Rahman, M.S., Kaykobad, M.: Complexities of some interesting problems on spanning trees. Inf. Process. Lett. **94**(2), 93–97 (2005). https://doi.org/10.1016/j.ipl.2004.12.016
28. Sam, E., Bergougnoux, B., Golovach, P.A., Blaser, N.: Kernelization for finding lineal topologies (depth-first spanning trees) with many or few leaves (2023). https://doi.org/10.48550/arXiv.2307.00362
29. Sam, E., Fellows, M., Rosamond, F., Golovach, P.A.: On the parameterized complexity of the structure of lineal topologies (depth-first spanning trees) of finite graphs: The number of leaves. In: Mavronicolas, M. (ed.) Algorithms and Complexity, pp. 353–367. Springer International Publishing, Cham (2023). https://doi.org/10.1007/978-3-031-30448-4_25
30. Tarjan, R.: Depth-first search and linear graph algorithms. In: 12th Annual Symposium on Switching and Automata Theory (swat 1971), pp. 114–121 (1971). https://doi.org/10.1109/SWAT.1971.10

Two UNO Decks Efficiently Perform Zero-Knowledge Proof for Sudoku

Kodai Tanaka[1]([✉]) [iD] and Takaaki Mizuki[2] [iD]

[1] Graduate School of Information Sciences, Tohoku University, Sendai, Japan
kodai.tanaka.r2@dc.tohoku.ac.jp
[2] Cyberscience Center, Tohoku University, Sendai, Japan

Abstract. Assume that there is a challenging Sudoku puzzle such that a prover knows a solution while a verifier does not know any solution. A zero-knowledge proof protocol allows the prover to convince the verifier that the prover knows the solution without revealing any information about it. In 2007, Gradwohl et al. constructed the first physical zero-knowledge proof protocol for Sudoku using a physical deck of playing cards; its drawback would be to have a soundness error. In 2018, Sasaki et al. improved upon the previous protocol by developing soundness-error-free protocols; their possible drawback would be to require many standard decks of playing cards, namely nine (or more) decks. In 2021, Ruangwises designed a novel protocol using only two standard decks of playing cards although it requires 322 shuffles, making it difficult to use in practical applications. In this paper, to reduce both the numbers of required decks and shuffles, we consider the use of UNO decks, which are commercially available: we propose a zero-knowledge proof protocol for Sudoku that requires only two UNO decks and 16 shuffles. Thus, the proposed protocol uses reasonable numbers of decks and shuffles, and we believe that it is efficient enough for humans to execute practically.

Keywords: Card-based cryptography · Sudoku · Zero-knowledge proof

1   Introduction

Sudoku is one of the most popular pencil puzzles in the world. A standard Sudoku puzzle consists of a 9×9 grid divided into 9 sub-grids (whose sizes are 3×3); we call such a sub-grid a *block* throughout this paper. Some of the cells on the grid are already filled with numbers from 1 through 9, and the goal of the puzzle is to place a number on each empty cell so that exactly one number from 1 to 9 appears in each row, each column, and each block. Figure 1 shows an example of a puzzle instance and its solution.

Sudoku and UNO are trademarks or registered trademarks of Nikoli Co., Ltd. and Mattel, Inc., respectively.

© The Author(s), under exclusive license to Springer Nature Switzerland AG 2023
H. Fernau and K. Jansen (Eds.): FCT 2023, LNCS 14292, pp. 406–420, 2023.
https://doi.org/10.1007/978-3-031-43587-4_29

1.1 Zero-Knowledge Proof for Sudoku

Assume that a player P had created a challenging Sudoku puzzle and P showed the puzzle to another player V. The player V has spent much time and effort finding a solution, but V cannot find any solution. The player V gets skeptical about whether the puzzle really has a solution. As easily imaged, this is the right opportunity to make use of the *zero-knowledge proof* [3]. Thus, this paper deals with zero-knowledge proof protocols for Sudoku.

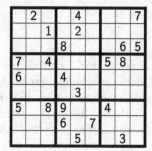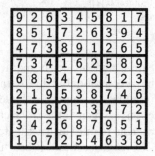

Fig. 1. Example of a standard Sudoku puzzle and its solution

In addition, we solicit *physical* zero-knowledge proof protocols that do not use any electronic devices such as computers, but use everyday objects such as a physical deck of playing cards. Physical protocols tend to be simple and easy-to-understand so that lay-people can easily execute them.

1.2 The Existing Protocols

In 2007, for the first time, Gradwohl et al. [4,5] developed several physical zero-knowledge proof protocols for Sudoku. Among them, the most easy-to-implement protocol would be the one which uses a physical deck of playing cards. This protocol is the first *card-based* zero-knowledge proof protocol for Sudoku; roughly speaking, after a prover P places face-down cards to commit to the solution P has, the protocol applies a series of actions, such as shuffling and turning over cards, to convince a verifier V that P surely knows the solution. Since the protocol has a soundness error, it must be repeated many times to make the soundness error probability negligibly small.

Later in 2018, Sasaki et al. [28] improved upon the previous protocol by devising a new technique (called the "uniqueness verification protocol" as will be seen in Sect. 2.3) to eliminate the soundness error. Specifically, they proposed three zero-knowledge proof protocols for Sudoku having no soundness error [27]; we name them *Sasaki's Protocols A, B,* and *C,* whose performances are shown as the first three protocols listed in Table 1. For instance, Sasaki's Protocol A uses 9 sets of 9 numbered cards 1 2 3 4 5 6 7 8 9 as well as arbitrary

Table 1. Comparison of the proposed protocol with the existing protocols

Protocols	Number of cards	Accommodating decks	Number of shuffles	Interactive?
Sasaki's Protocol A [27]	90	9 standard decks / 5 UNO decks	45	No
Sasaki's Protocol B [27]	171	18 standard decks / 9 UNO decks	36	No
Sasaki's Protocol C [27]	243	27 standard decks / 14 UNO decks	28	Yes
Ruangwises's Protocol A [22]	120	3 standard decks	108	Yes
Ruangwises's Protocol B [22]	108	2 standard decks	322	Yes
Ours	117	2 UNO decks	16	No

distinct 9 cards, and hence, it requires 90 cards in total and all the required cards can be accommodated by 9 standard decks of playing cards. As for the number of shuffles, it applies a "pile-scramble shuffle" (which is a kind of a shuffling action as seen later in Sect. 2.2) 45 times. As known from Table 1, Sasaki's protocols require at least 9 standard decks and at least 28 shuffles. Note that Sasaki's Protocols A and B are "non-interactive" whereas Sasaki's Protocol C is "interactive."[i] A *non-interactive* card-based zero-knowledge proof protocol uses no prover's knowledge after a prover commits to a solution with face-down cards (and hence, the protocol can be executed by either a prover or a verifier, or even a third party) [12]. By contrary, an *interactive* card-based zero-knowledge proof protocol requires prover's knowledge during the execution of the protocol.

In 2021, to reduce the number of required standard decks, Ruangwises [21,22] proposed two novel interactive protocols, which we name *Ruangwises's Protocols A* and *B*, using two or three standard decks of playing cards. For instance, Ruangwises's Protocol B requires two sets of $13 \times 4 = 52$ cards

along with two jokers (i.e., two cards of different patterns); therefore, two standard decks can accommodate all the required cards. Thus, Ruangwises's Protocol B is very efficient in terms of the number of required decks. However, the number of required shuffles is 322, which is too many for humans to execute. Overall, as known from Table 1, Ruangwises's protocols use only two or three decks while requiring at least 108 shuffles.

1.3 Contribution of This Paper

In this paper, instead of standard decks, we consider the use of UNO decks, which are also commercially available over the world. Specifically, we construct a zero-knowledge proof protocol for Sudoku that works on two UNO decks using only 16 shuffles. Our proposed protocol is non-interactive (and has no soundness error); see the last line in Table 1.

---

[i] The usage of these terms is valid only for card-based zero-knowledge proof protocols.

The specific numbers of required cards are as follows. We use four sets of yellow numbered cards

(a) Yellow, red, and blue numbered cards

(b) 27 distinct cards

Fig. 2. UNO cards to use

three sets of red numbered cards

and three sets of blue numbered cards

(as partially illustrated in Fig. 2(a)) along with 27 arbitrary distinct cards (as illustrated in Fig. 2(b)). One can easily confirm that two UNO decks can accommodate all the necessary cards mentioned above.

Making use of "color information" of cards, as will be seen, we can aggregate necessary shuffles, resulting in only the 16 shuffles. The shuffling operation our protocol uses is also the pile-scramble shuffle; our protocol uses such a shuffle for 27 cards whereas the existing ones use it for 9 cards or less. Note that a pile-scramble shuffle for 27 cards is also easy-to-implement.

Needless to say, zero-knowledge proofs play an important role in providing security and privacy. As Hanaoka [6] mentioned, for technology diffusion, we require easy-to-understand structures of cryptographic tools whereby potential users can easily understand the essential mechanisms. Since Sudoku is one of the most famous puzzles and our protocol is quite simple, our protocol might motivate potential users to try to use zero-knowledge proof technology.

1.4 Related Work

Aside from Sudoku, many physical zero-knowledge proof protocols have been constructed for other pencil puzzles. Examples are as follows: Akari [1], Hashiwokakero (Bridges) [25], Heyawake [19], Kakuro [1,14], Makaro [2,26], Masyu [10], Numberlink [23,24], Nurikabe [18,19], Slitherlink [10,11], Suguru [17,20], and Takuzu [1,13].

2 Preliminaries

In this section, we formally describe the cards and actions on them, and explain the "pile-scramble shuffle." Then, we introduce the "uniqueness verification protocol," which we will use as a sub-protocol in our protocol.

2.1 Cards and Actions

First, as mentioned in Sect. 1.3, we use numbered cards of three colors (yellow, red, and blue); the face of every card has a number from 1 through 9 and the patterns of the backs of all cards are identical, illustrated as

face up: R1 Y6 B9 ,   face down: ? ? ? .

Specifically, as mentioned before, we use four sets of yellow numbered cards Y1 Y2 $\cdots$ Y9 , three sets of red numbered cards R1 R2 $\cdots$ R9, and three sets of blue numbered cards B1 B2 $\cdots$ B9. We call these cards *encoding cards*, which we will use to represent a solution to a Sudoku puzzle.

In addition to the encoding cards, 27 distinct cards h_1 h_1 $\cdots$ h_{27} with the identical backs ? ? $\cdots$? are used; these are called *helping cards*. We choose helping cards from UNO decks as already illustrated in Fig. 2(b).

Our proposed protocol follows the standard computational model of card-based cryptographic protocols [8,9,15,16], in which a protocol is formally defined as an abstract machine. Here, we introduce three main actions, which are applied to a sequence of cards; below, we assume a sequence of m cards.

Permute.

This is denoted by (perm, π), where π is a permutation applied to the sequence of cards as follows:

$$\overset{1\quad2\qquad\quad m}{\boxed{?}\;\boxed{?}\cdots\boxed{?}} \xrightarrow{(\mathsf{perm},\pi)} \overset{\pi^{-1}(1)\;\pi^{-1}(2)\qquad\;\pi^{-1}(m)}{\boxed{?}\;\;\boxed{?}\;\cdots\;\boxed{?}} \;.$$

Turn.

This is denoted by (turn, T), where T is a set of indices, indicating that for every $t \in T$, the t-th card is turned over as follows:

$$\begin{array}{cccc} \overset{1}{\fbox{?}}\overset{2}{\fbox{?}}\cdots\overset{t\in T}{\fbox{?}}\cdots\overset{m}{\fbox{?}} \end{array} \xrightarrow{(\mathsf{turn},T)} \begin{array}{cccc} \overset{1}{\fbox{?}}\overset{2}{\fbox{?}}\cdots\overset{t\in T}{\fbox{R1}}\cdots\overset{m}{\fbox{?}} \end{array}.$$

Shuffle.

This is denoted by (shuf, Π), where Π is a set of permutations, indicating that $\pi \in \Pi$ is drawn uniformly and is applied to the sequence of cards as follows:

$$\begin{array}{ccc} \overset{1}{\fbox{?}}\overset{2}{\fbox{?}}\cdots\overset{m}{\fbox{?}} \end{array} \xrightarrow{(\mathsf{shuf},\Pi)} \begin{array}{ccc} \overset{\pi^{-1}(1)}{\fbox{?}}\overset{\pi^{-1}(2)}{\fbox{?}}\cdots\overset{\pi^{-1}(m)}{\fbox{?}} \end{array}.$$

Although a protocol is supposed to be defined with a combination of the actions above (or an abstract machine), in the sequel, we use a natural language to describe a protocol for simplicity.

2.2 Pile-Scramble Shuffle

In our protocol, we use the *pile-scramble shuffle*, which was devised in [7].

Assume that there are k piles of cards, denoted by p_i for every i, $1 \le i \le k$, such that all the piles have the same number of cards:

$$\underset{\mathsf{p}_1}{\fbox{?}}\ \underset{\mathsf{p}_2}{\fbox{?}}\ \cdots\ \underset{\mathsf{p}_k}{\fbox{?}}\ .$$

A pile-scramble shuffle is a shuffling action that applies a random permutation $r \in S_k$ to the sequence of piles, denoted by $[\cdot | \cdots | \cdot]$:

$$\left[\underset{\mathsf{p}_1}{\fbox{?}}\ \underset{\mathsf{p}_2}{\fbox{?}}\ \Big|\ \cdots\ \Big|\ \underset{\mathsf{p}_k}{\fbox{?}} \right] \rightarrow \underset{\mathsf{p}_{r^{-1}(1)}}{\fbox{?}}\ \underset{\mathsf{p}_{r^{-1}(2)}}{\fbox{?}}\ \cdots\ \underset{\mathsf{p}_{r^{-1}(k)}}{\fbox{?}}\ ,$$

where S_k denotes the symmetric group of degree k.

A pile-scramble shuffle can be implemented in practice by placing each pile of cards in a sleeve or envelope and stirring the envelopes (until the original sequence is no longer discernible).

2.3 Uniqueness Verification Protocol

This subsection explains the *uniqueness verification protocol* using the pile-scramble shuffle described in Sect. 2.2. This protocol has been proposed by Sasaki et al. [27,28].

Assume that there is a sequence of k face-down cards $\boxed{?}\,\boxed{?}\cdots\boxed{?}$ such that it consists of k distinct cards $\boxed{a_1}\,\boxed{a_2}\cdots\boxed{a_k}$ arranged in an unknown order. The uniqueness verification protocol enables us to confirm that the sequence surely consists of these k distinct cards without revealing the unknown order while keeping the original sequence unchanged. The protocol uses k helping cards $\boxed{h_1}\,\boxed{h_2}\cdots\boxed{h_k}$ and proceeds as follows.

1. Place the k helping cards below the unknown sequence, turn them face down, and apply a pile-scramble shuffle as follows:

$$\begin{matrix}\boxed{?}\,\boxed{?}\cdots\boxed{?}\\ \boxed{h_1}\,\boxed{h_2}\cdots\boxed{h_k}\end{matrix} \to \left[\begin{matrix}\boxed{?}\\ \boxed{?}\end{matrix}\ \begin{matrix}\boxed{?}\\ \boxed{?}\end{matrix}\ \cdots\ \begin{matrix}\boxed{?}\\ \boxed{?}\end{matrix}\right] \to \begin{matrix}\boxed{?}\,\boxed{?}\cdots\boxed{?}\\ \boxed{?}\,\boxed{?}\cdots\boxed{?}\end{matrix}.$$

2. Turn over all the cards of the top sequence to verify that it consists of $\{a_1, a_2, \ldots, a_k\}$:

$$\begin{matrix}\boxed{?}\,\boxed{?}\cdots\boxed{?}\\ \boxed{?}\,\boxed{?}\cdots\boxed{?}\end{matrix} \to \begin{matrix}\boxed{a_7}\,\boxed{a_k}\cdots\boxed{a_3}\\ \boxed{?}\,\boxed{?}\cdots\boxed{?}\end{matrix}.$$

3. After turning over all the cards of the top sequence, apply a pile-scramble shuffle:

$$\left[\begin{matrix}\boxed{?}\\ \boxed{?}\end{matrix}\ \begin{matrix}\boxed{?}\\ \boxed{?}\end{matrix}\ \cdots\ \begin{matrix}\boxed{?}\\ \boxed{?}\end{matrix}\right] \to \begin{matrix}\boxed{?}\,\boxed{?}\cdots\boxed{?}\\ \boxed{?}\,\boxed{?}\cdots\boxed{?}\end{matrix}.$$

Fig. 3. The names of the 9 blocks

4. Turn over all the cards of the bottom sequence and sort the piles so that the bottom sequence becomes $h_1, h_2, \ldots, h_k$ in this order:

$$
\begin{array}{|c|c|c|c|}
\hline
? & ? & \cdots & ? \\
\hline
h_2 & h_5 & \cdots & h_3 \\
\hline
\end{array}
\rightarrow
\begin{array}{|c|c|c|c|}
\hline
? & ? & \cdots & ? \\
\hline
h_1 & h_2 & \cdots & h_k \\
\hline
\end{array}.
$$

This restores the original sequence.

As mentioned above, we utilize this uniqueness verification protocol as a sub-protocol when constructing our protocol.

3 Building Blocks

Our proposed protocol consists of several sub-protocols: the color verification, row verification, color change, and column verification sub-protocols. In this section, we present these sub-protocols as building blocks for our proposed protocol (which will be shown in the next section). Hereinafter, we call the 9 blocks of a Sudoku puzzle Blocks $A, B, C, \ldots, H, I$ as shown in Fig. 3.

We first explain how a prover P commits to a solution with face-down encoding cards in Sect. 3.1, and then the sub-protocols will be presented in the succeeding subsections.

3.1 Commitment to a Solution

Assume that a prover knows a solution of a given 9×9 Sudoku puzzle. According to the solution, the prover P secretly arranges face-down encoding cards as follows.

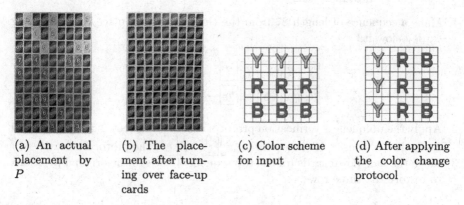

(a) An actual placement by P

(b) The placement after turning over face-up cards

(c) Color scheme for input

(d) After applying the color change protocol

Fig. 4. Commitment to a solution by a prover and color schemes for input blocks and those after color change

1. Prepare three sets of encoding cards Y1 Y2 ⋯ Y9 R1 R2 ⋯ R9 B1 B2 ⋯ B9.
2. P and V (or someone) publicly place an encoding card on each cell that already has a number written on it such that the number of the card is equal to the written number (on the cell) and its color matches the color scheme for blocks shown in Fig. 4(c) (i.e., yellow cards are placed in Blocks A, B, and C, red cards in Blocks D, E, and F, and blue cards in Blocks G, H, and I).
3. According to the solution, P secretly places face-down encoding cards on the remaining empty cells (without V's seeing the faces), based on the same color and number schemes as in the previous step.

In this way, the prover P commits to the solution with face-down encoding cards; Fig. 4(a) illustrates an actual placement by P, and Fig. 4(b) shows the placement after turning over all the face-up cards, which we sometimes call a *commitment* to the solution.

3.2 Color Verification Sub-protocol

Given a commitment to a solution (as illustrated in Fig. 4(b)), we first want to check that the face-down cards satisfy the color scheme properly. That is, given three rows of length 9:

?	?	?	?	?	?	?	?	?
?	?	?	?	?	?	?	?	?
?	?	?	?	?	?	?	?	?

we want to verify that all cards have the same color, say yellow. More precisely, the following *color verification sub-protocol* can confirm that the 27 cards consist of three sets of Y1 Y2 ⋯ Y9 using 27 helping cards.

1. Make a sequence of length 27 from the three rows, and place the 27 helping cards below it:

?	?	?	?	?	⋯	?	?	?
h_1	h_2	h_3	h_4	h_5	⋯	h_{25}	h_{26}	h_{27}

2. Apply the uniqueness verification protocol described in Sect. 2.3. If the opened cards do not consist of three sets of Y1 Y2 ⋯ Y9, the protocol aborts.
3. Move the face-down cards (of the top sequence) back to the original positions to restore the three rows.

3.3 3-Row Verification Sub-protocol

Suppose that all the cards in Blocks A, B, and C are yellow, all the cards in Blocks D, E, and F are red, and all the cards in Blocks G, H, and I are blue. The 3-*row verification sub-protocol* can verify that each of the three rows (of different colors) consists of numbers 1 through 9. Take Rows $1, 4$, and 7 as an example for row verification:

Row 1: | ? | ? | ? | ? | ? | ? | ? | ? | ? |
Row 4: | ? | ? | ? | ? | ? | ? | ? | ? | ? |.
Row 7: | ? | ? | ? | ? | ? | ? | ? | ? | ? |

1. Make a sequence of 27 cards from the three row and place 27 helping cards below it:

| ? | ? | ? | ? | ? | $\cdots$ | ? | ? | ? |
|h_1|h_2|h_3|h_4|h_5| $\cdots$ |h_{25}|h_{26}|h_{27}|

2. Apply the uniqueness verification protocol described in Sect. 2.3. If the opened cards do not consist of

$$\boxed{Y1}\boxed{Y2}\cdots\boxed{Y9}\boxed{R1}\boxed{R2}\cdots\boxed{R9}\boxed{B1}\boxed{B2}\cdots\boxed{B9},$$

   the protocol aborts.
3. Move the face-down cards (of the top sequence) back to the original positions to restore the three rows.

   The same can be applied to the other three rows (of different colors).

3.4   Color Change Sub-protocol

In this subsection, we present the *color change sub-protocol*, which changes only the colors of the cards in blocks without changing their numbers as well as provides block verification.

Assume that there are three blocks, the first one consisting of yellow cards, the second one consisting of red cards, and the third one consisting of blue cards. The following procedure exchanges the colors of the first and second blocks as well as changes the color of the third block into yellow[ii].

1. Make a sequence of 27 cards from the three blocks and place 27 helping cards below it:

| ? | ? | ? | ? | ? | $\cdots$ | ? | ? | ? |
|h_1|h_2|h_3|h_4|h_5| $\cdots$ |h_{25}|h_{26}|h_{27}|

[ii] The sub-protocol works for other combinations of colors.

2. Apply a pile-scramble shuffle to them. Turn the encoding cards face up, making sure that the yellow, red, and blue cards from 1 to 9 appear without omission (otherwise it aborts):

$$\begin{bmatrix} \fbox{?} & \fbox{?} & \cdots & \fbox{?} \\ \fbox{?} & \fbox{?} & & \fbox{?} \end{bmatrix} \rightarrow \begin{matrix} \boxed{Y3} & \boxed{R3} & \boxed{B6} & \cdots & \boxed{Y5} \\ \fbox{?} & \fbox{?} & \fbox{?} & \cdots & \fbox{?} \end{matrix}.$$

3. For yellow and red cards, swap the cards with the same number (for example, swap the positions of the yellow 3 and the red 3). For blue cards, preparing 9 free yellow cards $\boxed{Y1}\boxed{Y2}\cdots\boxed{Y9}$,[iii] swap the cards with the same number:

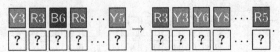

$$\begin{matrix} \boxed{Y3} & \boxed{R3} & \boxed{B6} & \boxed{R8} & \cdots & \boxed{Y5} \\ \fbox{?} & \fbox{?} & \fbox{?} & \fbox{?} & \cdots & \fbox{?} \end{matrix} \rightarrow \begin{matrix} \boxed{R3} & \boxed{Y3} & \boxed{Y6} & \boxed{Y8} & \cdots & \boxed{R5} \\ \fbox{?} & \fbox{?} & \fbox{?} & \fbox{?} & \cdots & \fbox{?} \end{matrix}.$$

4. Turn the encoding cards face down and apply a pile-scramble shuffle:

$$\begin{bmatrix} \fbox{?} & \fbox{?} & \cdots & \fbox{?} \\ \fbox{?} & \fbox{?} & & \fbox{?} \end{bmatrix} \rightarrow \begin{matrix} \fbox{?} & \fbox{?} & \cdots & \fbox{?} \\ \fbox{?} & \fbox{?} & & \fbox{?} \end{matrix}.$$

5. Turn over all the cards of the bottom sequence and sort the piles so that the bottom sequence become $h_1, h_2, \ldots, h_{27}$ in this order. Return the encoding cards to their original positions:

$$\begin{matrix} \fbox{?} & \fbox{?} & \cdots & \fbox{?} \\ \boxed{h_2} & \boxed{h_5} & \cdots & \boxed{h_3} \end{matrix} \rightarrow \begin{matrix} \fbox{?} & \fbox{?} & \cdots & \fbox{?} \\ \boxed{h_1} & \boxed{h_2} & \cdots & \boxed{h_{27}} \end{matrix}.$$

Note that this sub-protocol also checks that each of the three blocks consists of numbers from 1 to 9.

3.5   3-Column Verification Sub-Protocol

Suppose that all the cards in Blocks $A, D,$ and G are yellow, all the cards in Blocks $B, E,$ and H are red, and all the cards in Blocks $C, F,$ and I are blue. The 3-*column verification sub-protocol* can verify that each of the three columns (of different colors) consists of numbers 1 through 9 while it does not revert the blocks contrary to the 3-row verification sub-protocol presented in Sect. 3.3. Take Columns 1, 4, and 7 as an example for column verification:

Column 1: $\fbox{?}\fbox{?}\fbox{?}\fbox{?}\fbox{?}\fbox{?}\fbox{?}\fbox{?}\fbox{?}$

Column 4: $\fbox{?}\fbox{?}\fbox{?}\fbox{?}\fbox{?}\fbox{?}\fbox{?}\fbox{?}\fbox{?}$.

Column 7: $\fbox{?}\fbox{?}\fbox{?}\fbox{?}\fbox{?}\fbox{?}\fbox{?}\fbox{?}\fbox{?}$

[iii] This is why our protocols needs four sets of yellow cards.

1. Make a sequence of 27 cards from the three columns and apply a pile-scramble shuffle (i.e., a normal shuffle):

$$\boxed{?}\,\boxed{?}\cdots\boxed{?} \;\rightarrow\; \left[\,\boxed{?}\,\boxed{?}\,\middle|\cdots\middle|\,\boxed{?}\,\right] \;\rightarrow\; \boxed{?}\,\boxed{?}\cdots\boxed{?}\,.$$

2. Turn over all the 27 cards. If the opened cards do not consist of

$$\mathrm{Y1}\,\mathrm{Y2}\cdots\mathrm{Y9}\,\mathrm{R1}\,\mathrm{R2}\cdots\mathrm{R9}\,\mathrm{B1}\,\mathrm{B2}\cdots\mathrm{B9},$$

the protocol aborts.

Since this sub-protocol will be used at the last step of our proposed protocol, there is no need to restore the three retrieved columns. Therefore, the uniqueness verification protocol described in Sect. 2.3 is not used. The same can be applied to other three columns (of different colors).

4   Our Protocol for 9 × 9 Sudoku

In this section, we construct a zero-knowledge proof protocol for a standard Sudoku puzzle.

A prover P who knows a solution places a commitment to the solution as described in Sect. 3.1, and this is the input to the protocol. After the input is given, the protocol does not require the knowledge of the prover P and can be executed by anyone thereafter (because our protocol is non-interactive).

Given the placement by P along with 27 helping cards and free 9 cards $\mathrm{Y1}\,\mathrm{Y2}\cdots\mathrm{Y9}$, our protocol proceeds as follows.

1. Apply the color verification sub-protocol given in Sect. 3.2 to each of Blocks A, B, C, Blocks D, E, F, and Blocks G, H, I. If it does not abort, Blocks A, B, C consist of three sets of $\mathrm{Y1}\,\mathrm{Y2}\cdots\mathrm{Y9}$, Blocks D, E, F consist of three sets of $\mathrm{R1}\,\mathrm{R2}\cdots\mathrm{R9}$, and Blocks G, H, I consist of three sets of $\mathrm{B1}\,\mathrm{B2}\cdots\mathrm{B9}$.
2. Apply the 3-row verification sub-protocol given in Sect. 3.3 to Rows 1, 4, and 7. Apply also the 3-row verification sub-protocol to Rows 2, 5, and 8. If it does not abort, Rows 1, 2, 4, 5, 7, and 8 consist of numbers 1 through 9. In addition, this result together with the result of the color verification in the previous step automatically implies that the remaining Rows 3, 6, and 9 must consist of numbers 1 through 9.
3. In order to be able to verify the columns, we want to change the color of Blocks A, D, and G to yellow, Blocks B, E, and H to red, and Blocks C, F, and I to blue, as shown in Fig. 4(d). First, apply the color change sub-protocol described in Sect. 3.4 to Blocks B, D, and G with colors yellow, red, and blue in this order. This swaps the colors of Blocks B and D, and changes the color

of Block G to yellow. (This produces 9 free cards B1 B2 $\cdots$ B9 .) Second, apply the color change sub-protocol to Blocks H, F, and C with colors blue, red, and yellow in this order. This swaps the colors of Blocks H and F, and changes the color of Block C to blue. Through these color transformations, Blocks B, C, D, F, G, and H are verified to consist of numbers 1 through 9. In addition, based on the results of Step 1, it is automatically guaranteed that the remaining Blocks A, E, and I consist of numbers 1 through 9.

4. Apply the 3-column verification sub-protocol described in Sect. 3.5 to Columns 1, 4, and 7. The same procedure is applied to Columns 2, 5, and 8. The remaining Columns 3, 6, and 9 are automatically verified as in the row case.

This is the proposed protocol for convincing a verifier that the numbers 1 to 9 appear exactly once in each row, each column, and each block.

The number of required cards is 117 and the number of shuffles is 16 although we omit the breakdowns due to the space limitation. We also omit the security proof in this extended abstract.

5   Conclusion

In this paper, we proposed a new card-based zero-knowledge proof protocol for Sudoku. The main idea behind our proposed protocol is to make use of UNO decks, which are commercially available over the world, and hence, they are easy to prepare. The property that there are several UNO cards having the same pair of a color and a number on their faces leads to reducing the number of required shuffles. Specifically, two UNO decks and 16 shuffles are sufficient for constructing a zero-knowledge proof protocol for a standard Sudoku puzzle. We believe that this is efficient enough for humans to execute practically. The proposed protocol can also be implemented with standard decks of playing cards although it requires four decks along with two jokers of different designs.

Our techniques can be applied to a general $n \times n$ Sudoku although we omit the details in this extended abstract.

Acknowledgements. We thank the anonymous referees, whose comments have helped us to improve the presentation of the paper. This work was supported in part by JSPS KAKENHI Grant Numbers JP21K11881 and JP23H00479.

References

1. Bultel, X., Dreier, J., Dumas, J.G., Lafourcade, P.: Physical zero-knowledge proofs for Akari, Takuzu, Kakuro and KenKen. In: Fun with Algorithms. LIPIcs, vol. 49, pp. 8:1–8:20. Schloss Dagstuhl, Dagstuhl, Germany (2016)
2. Bultel, X., et al.: Physical zero-knowledge proof for Makaro. In: Izumi, T., Kuznetsov, P. (eds.) SSS 2018. LNCS, vol. 11201, pp. 111–125. Springer, Cham (2018). https://doi.org/10.1007/978-3-030-03232-6_8

3. Goldwasser, S., Micali, S., Rackoff, C.: The knowledge complexity of interactive proof-systems. In: Annual ACM Symposium on Theory of Computing, STOC 1985, pp. 291–304. ACM, New York (1985)
4. Gradwohl, R., Naor, M., Pinkas, B., Rothblum, G.N.: Cryptographic and physical zero-knowledge proof systems for solutions of Sudoku puzzles. In: Crescenzi, P., Prencipe, G., Pucci, G. (eds.) FUN 2007. LNCS, vol. 4475, pp. 166–182. Springer, Heidelberg (2007). https://doi.org/10.1007/978-3-540-72914-3_16
5. Gradwohl, R., Naor, M., Pinkas, B., Rothblum, G.N.: Cryptographic and physical zero-knowledge proof systems for solutions of Sudoku puzzles. Theory Comput. Syst. **44**(2), 245–268 (2009)
6. Hanaoka, G.: Towards user-friendly cryptography. In: Paradigms in Cryptology-Mycrypt 2016. Malicious and Exploratory Cryptology. LNCS, vol. 10311, pp. 481–484. Springer, Cham (2017)
7. Ishikawa, R., Chida, E., Mizuki, T.: Efficient card-based protocols for generating a hidden random permutation without fixed points. In: Calude, C.S., Dinneen, M.J. (eds.) UCNC 2015. LNCS, vol. 9252, pp. 215–226. Springer, Cham (2015). https://doi.org/10.1007/978-3-319-21819-9_16
8. Kastner, J., Koch, A., Walzer, S., Miyahara, D., Hayashi, Y., Mizuki, T., Sone, H.: The minimum number of cards in practical card-based protocols. In: Takagi, T., Peyrin, T. (eds.) ASIACRYPT 2017. LNCS, vol. 10626, pp. 126–155. Springer, Cham (2017). https://doi.org/10.1007/978-3-319-70700-6_5
9. Koch, A., Walzer, S., Härtel, K.: Card-based cryptographic protocols using a minimal number of cards. In: ASIACRYPT 2015. LNCS, vol. 9452, pp. 783–807. Springer, Heidelberg (2015)
10. Lafourcade, P., Miyahara, D., Mizuki, T., Robert, L., Sasaki, T., Sone, H.: How to construct physical zero-knowledge proofs for puzzles with a "single loop" condition. Theor. Comput. Sci. **888**, 41–55 (2021)
11. Lafourcade, P., Miyahara, D., Mizuki, T., Sasaki, T., Sone, H.: A physical ZKP for Slitherlink: How to perform physical topology-preserving computation. In: Heng, S.-H., Lopez, J. (eds.) ISPEC 2019. LNCS, vol. 11879, pp. 135–151. Springer, Cham (2019). https://doi.org/10.1007/978-3-030-34339-2_8
12. Miyahara, D., Haneda, H., Mizuki, T.: Card-based zero-knowledge proof protocols for graph problems and their computational model. In: Huang, Q., Yu, Yu. (eds.) ProvSec 2021. LNCS, vol. 13059, pp. 136–152. Springer, Cham (2021). https://doi.org/10.1007/978-3-030-90402-9_8
13. Miyahara, D., et al.: Card-based ZKP protocols for Takuzu and Juosan. In: Fun with Algorithms. LIPIcs, vol. 157, pp. 20:1–20:21. Schloss Dagstuhl, Dagstuhl, Germany (2020)
14. Miyahara, D., Sasaki, T., Mizuki, T., Sone, H.: Card-based physical zero-knowledge proof for Kakuro. IEICE Trans. Fundam. **102**(9), 1072–1078 (2019)
15. Mizuki, T., Shizuya, H.: A formalization of card-based cryptographic protocols via abstract machine. Int. J. Inf. Secur. **13**(1), 15–23 (2014)
16. Mizuki, T., Shizuya, H.: Computational model of card-based cryptographic protocols and its applications. IEICE Trans. Fundam. **E100.A**(1), 3–11 (2017)
17. Robert, L., Miyahara, D., Lafourcade, P., Mizuki, T.: Physical zero-knowledge proof for Suguru puzzle. In: Devismes, S., Mittal, N. (eds.) SSS 2020. LNCS, vol. 12514, pp. 235–247. Springer, Cham (2020). https://doi.org/10.1007/978-3-030-64348-5_19
18. Robert, L., Miyahara, D., Lafourcade, P., Mizuki, T.: Interactive physical ZKP for connectivity: Applications to Nurikabe and Hitori. In: Connecting with Computability. LNCS, vol. 12813, pp. 373–384. Springer, Cham (2021)

19. Robert, L., Miyahara, D., Lafourcade, P., Mizuki, T.: Card-based ZKP for connectivity: applications to Nurikabe, Hitori, and Heyawake. New Gener. Comput. **40**, 149–171 (2022)
20. Robert, L., Miyahara, D., Lafourcade, P., Libralesso, L., Mizuki, T.: Physical zero-knowledge proof and NP-completeness proof of Suguru puzzle. Inf. Comput. **285**, 1–14 (2022)
21. Ruangwises, S.: Two standard decks of playing cards are sufficient for a ZKP for Sudoku. In: Chen, C.-Y., Hon, W.-K., Hung, L.-J., Lee, C.-W. (eds.) COCOON 2021. LNCS, vol. 13025, pp. 631–642. Springer, Cham (2021). https://doi.org/10.1007/978-3-030-89543-3_52
22. Ruangwises, S.: Two standard decks of playing cards are sufficient for a ZKP for Sudoku. New Gener. Comput. **40**, 49–65 (2022)
23. Ruangwises, S., Itoh, T.: Physical zero-knowledge proof for Numberlink. In: Fun with Algorithms. LIPIcs, vol. 157, pp. 22:1–22:11. Schloss Dagstuhl, Dagstuhl, Germany (2020)
24. Ruangwises, S., Itoh, T.: Physical zero-knowledge proof for Numberlink puzzle and k vertex-disjoint paths problem. New Gener. Comput. **39**(1), 3–17 (2021)
25. Ruangwises, S., Itoh, T.: Physical ZKP for connected spanning subgraph: applications to Bridges puzzle and other problems. In: Unconventional Computation and Natural Computation, pp. 149–163. Springer, Cham (2021)
26. Ruangwises, S., Itoh, T.: Physical ZKP for Makaro using a standard deck of cards. In: Theory and Applications of Models of Computation. LNCS, vol. 13571, pp.43–54. Springer, Cham (2022, to appear)
27. Sasaki, T., Miyahara, D., Mizuki, T., Sone, H.: Efficient card-based zero-knowledge proof for Sudoku. Theor. Comput. Sci. **839**, 135–142 (2020)
28. Sasaki, T., Mizuki, T., Sone, H.: Card-based zero-knowledge proof for Sudoku. In: Fun with Algorithms. LIPIcs, vol. 100, pp. 29:1–29:10. Schloss Dagstuhl, Dagstuhl, Germany (2018)

Power of Counting by Nonuniform Families of Polynomial-Size Finite Automata

Tomoyuki Yamakami[✉]

Faculty of Engineering, University of Fukui, 3-9-1 Bunkyo, Fukui 910-8507, Japan
tomoyukiyamakami@gmail.com

Abstract. Lately, there have been intensive studies on strengths and limitations of nonuniform families of promise decision problems solvable by various types of polynomial-size finite automata families, where a "polynomial-size" finite automata family has polynomially-bounded state complexity. In this line of study, we further expand the scope of these studies to partial counting functions and their relevant (promise) decision problems, defined in terms of nonuniform families of polynomial-size nondeterministic finite automata. With no unproven hardness assumption, we show numerous separations and collapses of complexity classes of those counting function families and their relevant decision problem families. We also investigate their relationships to pushdown automata families of polynomial stack-state complexity.

Keywords: nonuniform polynomial state complexity · counting functions · counting complexity classes · closure properties

1 Background and Quick Overview

1.1 Counting in Computational Complexity Theory

Counting is one of the most intensively studied subjects in computational complexity theory. The basic notion of *counting functions* was formally introduced by Valiant [16,17] to count the total number of accepting computation paths of each run of underlying polynomial-time nondeterministic Turing machines (NTMs). These counting functions form a function class known as #P. Typical #P-functions can compute the permanent of a given positive-integer matrix and the total number of perfect matchings of a given graph. By contrast, a gap function computes the difference between the number of accepting computation paths and the number of rejecting computation paths of a polynomial-time NTM. Such gap functions form the function class GapP [4]. The power of counting is demonstrated by an inclusion of the polynomial(-time) hierarchy (PH) within $P^{\#P}$ [15]. Those two function classes, #P and GapP, naturally induce so-called "counting complexity classes" of decision problems, such as UP, $C_=P$,

H. Fernau and K. Jansen (Eds.): FCT 2023, LNCS 14292, pp. 421–435, 2023.
https://doi.org/10.1007/978-3-031-43587-4_30

$\oplus$P, PP, and SPP (Stoic PP) [4], whose structural properties have been vigorously investigated over decades in the mainstream computational complexity theory. Unfortunately, none of them is known to be equal to or separated from each other.

Outside of the polynomial-time setting, *logarithmic-space* counting complexity classes, such as UL, $C_{=}L$, $\oplus$L, and PL, have been studied by Allender and Ogihara [1], Àlvarez and Jenner [2], and Ruzzo, Simon, and Tompa [12]. Using more restrictive models of *one-tape linear-time Turing machines*, counting complexity classes, such as 1-$C_{=}$LIN, 1-$\oplus$LIN, and 1-PLIN, and their advised variants, 1-$C_{=}$LIN/lin, 1-$\oplus$LIN/lin, and 1-PLIN/lin, were studied in [14,20], where "advice" is an external information source providing extra helpful information to underlying Turing machines. Advice has been also supplemented to enhance the computational power of finite automata in, e.g., [14,19–22]. Furthermore, counting aspects of auxiliary pushdown automata were explored in [9,18] and counting pushdown automata were discussed in [11].

Along the line of the literature, we wish to look into lower complexity levels in the study of counting functions and their induced decision problems. More specifically, we wish to expand the scope of them by taking a new, distinctive approach with nonuniform families of nondeterministic finite automata of polynomially-bounded state complexity. In the subsequent subsection, we quickly review the background of these nonuniform families.

1.2 Nonuniform Families of Polynomial-Size Finite Automata

Finite automata have long served as a simple memoryless machine model to help us understand the *essence* of computation. Associated with such automata, the number of inner states of a finite automaton has been considered as an important resource to measure the *descriptional complexity* of memoryless computation produced by the automaton. Such a number therefore plays the role of a complexity measure, known as the *state complexity* of the automaton. Berman and Lingas [3] and Sakoda and Sipser [13] studied in the late 19701970ss the "nonuniform" families $\{M_n\}_{n\in\mathbb{N}}$ of two-way finite automata M_n whose state complexities are upper-bounded by fixed polynomials in n.

In computational complexity theory, "nonuniformity" has played a distinctive role. Nonuniform families of polynomial-size circuits, for instance, have been treated as a nonuniform version of P and it has been discussed in comparison with NP. As a clear difference from circuit families, nonetheless, automata can take inputs of arbitrary lengths simply by moving their tape heads along semi-infinite input tapes until the end of the inputs. Nonuniformity can be also implemented in terms of "advice".

To ease our further discussion, we now introduce two notations 1D and 1N to denote respectively the collections of families of promise decision problems over fixed alphabets[1] solvable by nonuniform families of one-way deterministic

---

[1] In this exposition, we always fix an alphabet for all promise problems in the same family. This situation slightly differs from [5–8].

finite automata (or 1dfa's, for short) and by nonuniform families of one-way non-deterministic finite automata (or 1nfa's) using polynomially many inner states. By replacing 1dfa's and 1nfa's with their two-way head models called 2dfa's and 2nfa's, we obtain 2D and 2N.

Although 1nfa's are known to be equivalent in computational power to 1dfa's, their state complexities are "exponentially" apart. This exponential state complexity gap instantly leads to 1D $\neq$ 1N. After an early study of [3,13], Kapoutsis [5,6] revitalized a study of nonuniform polynomial state complexity by expanding the scope of underlying machines to probabilistic and alternating finite automata. Nonuniform polynomial state complexity has lately become a subject of intensive studies in close connection to logarithmic-space computation. Kapoutsis [7] and Kapoutsis and Pighizzini [8] exhibited a close relationship between the two-way nonuniform polynomial state complexity classes 2D and 2N and the logarithmic space-bounded complexity classes L and NL equipped with polynomial-size advice. Yamakami further expanded those subjects and covered the computational model of quantum finite automata [26], various restrictions of 1nfa's and 2nfa's [27], width-bounded 2nfa's in direct connection to the *linear space hypothesis* (LSH) [23], and relativization of nonuniform polynomial sate complexity classes [24]. As a natural extension of finite automata families, nonuniform families of pushdown automata were also studied lately [25].

Unlike the two-way models of finite automata, we can determine the computational complexity of the one-way models of various finite automata with *no unproven hardness assumption*. For example, numerous nonuniform complexity classes located between 1D and 1N were investigated in [27] with respect to the accepting behaviors of underlying nondeterministic finite automata. It was shown in [27] that 1N and its unambiguous variant 1U are different whereas their two-way head models, 2N and 2U, are equal when all valid instances are polynomially long.

An expansion of the scope of the theory of nonuniform polynomial state complexity has been expected to promote our basic understandings of this intriguing theory. The importance of "counting" in theoretical computer science therefore makes us conduct an initial study in this exposition on the power of "counting" within the theory of nonuniform polynomial state complexity.

1.3   Quick Overview of Main Contributions

A series of recent studies along the line of Sakoda and Sipser [13] has made a steady progress on the theory of nonuniform polynomial state complexity. A main purpose of this exposition is therefore to enrich the field of nonuniform polynomial state complexity by promoting the understanding of descriptional complexity of nonuniform finite automata families.

As natural analogues of #P and GapP, we introduce two fundamental function classes 1# and 1Gap. Unlike #P and GapP, however, these classes are, in fact, collections of nonuniform families of "partial" functions whose values are "defined" only on predetermined domains. With the use of such partial function families in 1# and 1Gap, we can define 1U, 1$\oplus$, 1C$_=$, 1SP, and 1P, as

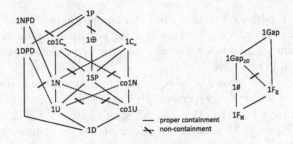

Fig. 1. Containments and separations of nonuniform polynomial state complexity classes shown in this exposition. The class separations 1D $\neq$ 1U $\neq$ 1N and 1U $\neq$ co-1U come from [27].

nonuniform-state-complexity analogues of UP, $\oplus$P, C$_=$P, SPP, and PP, respectively, although 1U and 1P were already studied in [5,6,26] in a slightly different context. See Sect. 3 for their precise definitions.

As the main contribution of this exposition, we study the containments/separations among the aforementioned nonuniform polynomial state complexity classes. A key is a discovery of a close connection between nonuniform families of finite automata and one-tape linear-time Turing machines with advice. We further compare the complexity classes 1U, 1N, and 1P with the complexity classes 1DPD and 1NPD induced respectively by one-way deterministic and nondeterministic pushdown automata families [25]. In summary, Fig. 1 depicts the relationships among those complexity classes.

All omitted proofs, due to the page limit, will be included in the forthcoming complete version of this exposition.

2 Preparation: Notions and Notation

2.1 Numbers, Strings, and Promise Decision Problems

Let us provide the basic notions and notation necessary to read through the rest of this exposition. The notation $\mathbb{Z}$ indicates the set of all integers. All nonnegative integers are called *natural numbers* and form a set denoted by $\mathbb{N}$. We also set $\mathbb{N}^+$ to express $\mathbb{N} - \{0\}$. For two integers m and n with $m \leq n$, $[m, n]_{\mathbb{Z}}$ indicates the *integer set* $\{m, m+1, m+2, \ldots, n\}$. When $n \in \mathbb{N}^+$, we abbreviate $[1, n]_{\mathbb{Z}}$ as $[n]$. Given a set S, the notation $\mathcal{P}(S)$ stands for the *power set* of S.

We assume the reader's familiarity with basic automata theory. Let Σ denote an *alphabet*, which is a finite nonempty set of "symbols" or "letters". A finite sequence of symbols in Σ is called a *string* over Σ. The *empty string* is always denoted by λ. A subset of Σ^* is a *language* over Σ. Given a number $n \in \mathbb{N}$, Σ^n (resp., $\Sigma^{\leq n}$) denotes the set of all strings of length exactly n (resp., at most n).

A *promise (decision) problem* over alphabet Σ is a pair (A, R) of sets (or languages) satisfying that $A \cup R \subseteq \Sigma^*$ and $A \cap R = \varnothing$, where instances in A are called *positive* and those in R are *negative*. All strings in $A \cup R$ are customary called *valid* or *promised*. A language can be seen as a special case of a promise problem. Given a string $w \in \Sigma^*$ and a symbol $\sigma \in \Sigma$, $\#_\sigma(w)$ denotes the total number of occurrences of σ in w.

For any positive integer k, the notation $bin(k)$ denotes the binary representation of k leading with 1. For convenience, we also set $bin(0)$ to be the empty string λ. We further translate integers k to binary strings $trans(k)$ as follows: 0 is translated into $trans(0) = \lambda$, a positive integer k is translated into $trans(k) = 1bin(k)$, and a negative integer $-k$ is interpreted as $trans(-k) = 0bin(k)$.

For any $k \in \mathbb{N}^+$ and any $i_1, i_2, \ldots, i_k \in \mathbb{N}^+$, we use the special notation $[i_1, i_2, \ldots, i_k]$ to denote the binary string of the form $1^{i_1}01^{i_2}0 \cdots 01^{i_k}0$. Given $n \in \mathbb{N}^+$, let A_n denote the collection of all such strings $[i_1, i_2, \ldots, i_k]$ for arbitrary $k \in \mathbb{N}^+$ and $i_1, i_2, \ldots, i_k \in [n]$. We write A_∞ for $\bigcup_{n \in \mathbb{N}} A_n$. For any $r = [i_1, i_2, \ldots, i_k] \in A_\infty$, we write $Set(r)$ to denote the set $\{i_1, i_2, \ldots, i_k\}$. In comparison, we use another notation $MSet(r)$ for the corresponding "multi" set of k elements. Obviously, it follows that $|Set(r)| \leq |MSet(r)| = k$. For any index $e \in [n]$, the notation $(r)_{(e)}$ denotes the e-th entry i_e in r. We further define $\Sigma_2^{(n)} = \{r_1 \# r_2 \mid r_1, r_2 \in A_n\}$. Moreover, given k numbers $i_1, i_2, \ldots, i_k \in [0, n]_\mathbb{Z}$ with $m \geq n$, we define $[[i_1, i_2, \ldots, i_k]]_m$ to be $1^{i_1}0^{m-i_1}01^{i_2}0^{m-i_2}0 \cdots 01^{i_k}0^{m-i_k}$, where we treat 1^0 and 0^0 as λ. Let $A_n(m)$ denote the set of all strings of the form $[[i_1, i_2, \ldots, i_k]]_m$ with $i_1, i_2, \ldots, i_k \in [0, n]_\mathbb{Z}$. It follows that $|[i_1, i_2, \ldots, i_k]| = \sum_{l=1}^{k} i_l + k$, whereas $|[[i_1, i_2, \ldots, i_k]]_m| = km + k - 1$. See [27] for more information.

2.2 Various Types of One-Way Finite Automata

In association with the theme of "counting", we use various models of finite automata. Following [26], all finite automata in this exposition are restricted so that they should move their tape heads only in one direction *without making any λ-move*, where a "λ-move" refers to a step in which the tape head scans no input symbol but it may change the automaton's inner states. Such machines are briefly called *one way* in this exposition. A *one-way nondeterministic finite automaton* (or a 1nfa, for short) is of the form $(Q, \Sigma, \{\triangleright, \triangleleft\}, \delta, q_0, Q_{acc}, Q_{rej})$, where Q is a finite set of inner states, Σ is an (input) alphabet, $\triangleright$ and $\triangleleft$ are two endmarkers, δ is a transition function from $(Q - Q_{halt}) \times \check{\Sigma}$ to $\mathcal{P}(Q)$, where $\check{\Sigma} = \Sigma \cup \{\triangleright, \triangleleft\}$, and Q_{acc} and Q_{rej} are respectively sets of accepting states and of rejecting states with $Q_{halt} = Q_{acc} \cup Q_{rej} \subseteq Q$ and $Q_{acc} \cap Q_{rej} = \varnothing$. Since we deal with promise problems, we need Q_{acc} and Q_{rej} to allow the circumstances where, after reading the right endmarker $\triangleleft$, M enters neither accepting states nor rejecting states. Since M makes no λ-move, it should halt within $|x| + 2$ steps for any input x. Given a computation path ζ, we say that M *halts properly on the path* ζ if M enters a halting state in Q_{halt} on ζ. When M reads off $\triangleleft$ without entering any halting state, on the contrary, M is said to *halt improperly on the path* ζ. For any input x, we say that M *halts properly on x* if there is a computation path ζ of M on x for which M halts properly on ζ. For a *one-way deterministic finite automaton* (or a 1dfa), its transition function δ maps $(Q - Q_{halt}) \times \check{\Sigma}$ to Q. We write $sc(M)$ for the *state complexity* $|Q|$ of M.

A finite automaton M is said to *solve* a promise problem $(L^{(+)}, L^{(-)})$ if M accepts all instances in $L^{(+)}$ and rejects all instances in $L^{(-)}$. However, we do not impose any condition on invalid instances.

Given an input string x, $M(x)$ stands for the "outcome" of M on x whenever M halts properly on x. To express the total number of accepting computation paths of M on input x, we use the notation $\#M(x)$; by contrast, $\#\overline{M}(x)$ expresses the total number of rejecting computation paths of M on x.

Given a 1nfa M, we can modify it into another "equivalent" 1nfa N so that (1) N makes exactly c nondeterministic choices at every step and (2) N produces exactly $c^{|\triangleright x \triangleleft|}$ computation paths on all inputs x, where c is an appropriately chosen constant in $\mathbb{N}^+$. For convenience, we call this specific form the *branching normal form*.

Lemma 1 (branching normal form). *Let M be any 1nfa solving a promise problem $(L^{(+)}, L^{(-)})$. There exists another 1nfa N such that N is in a branching normal form and $sc(N) \le 3sc(M) + 1$.*

When a finite automaton is further equipped with a *write-once*[2] output tape, we call it a *(finite) transducer* to distinguish it from the aforementioned finite automata. In this exposition, we consider only one-way (deterministic) finite transducers (or 1dft's, for short) equipped with transition functions δ mapping $Q \times \check{\Sigma}$ to $Q \times \Gamma^*$ for two alphabets Σ and Γ. We remark that, since a 1dft always moves its input-tape head, it is allowed to write multiple symbols at once onto its output tape by moving its tape head over multiple tape cells *in a single step*.

A *one-way nondeterministic pushdown automaton* (or a 1npda, for short) M is a tuple $(Q, \Sigma, \{\triangleright, \triangleleft\}, \Gamma, \delta, q_0, \bot, Q_{acc}, Q_{rej})$, additionally equipped with a stack alphabet Γ and the bottom marker $\bot$ for a stack, where δ maps $(Q - Q_{halt}) \times \check{\Sigma}_\lambda \times \Gamma$ to $\mathcal{P}(Q \times \Gamma^*)$, where $\check{\Sigma}_\lambda = \check{\Sigma} \cup \{\lambda\}$. We remark that, different from 1nfa's, M is allowed to make λ-moves. A deterministic variant of a 1npda is called a *one-way deterministic pushdown automaton* (or a 1dpda) if the extra condition $|\delta(q, \sigma, \gamma) \cup \delta(q, \lambda, \gamma)| \le 1$ is met for all triplets $(q, \sigma, \gamma) \in Q \times \check{\Sigma} \times \Gamma$. The value $ssc(M) = |Q||\Gamma^{\le e}|$ is called the *stack-state complexity* of M, where e is the *push size* of M defined by $e = \max_{p,q,\sigma,\gamma}\{|a| : (p, a) \in \delta(q, \sigma, \gamma)\}$ with $p, q \in Q$, $\sigma \in \check{\Sigma}_\lambda$, and $\gamma \in \Gamma$.

2.3 Nonuniform Families of Promise Problems and Finite Automata

Given a fixed alphabet Σ, we consider a family $\mathcal{L} = \{(L_n^{(+)}, L_n^{(-)})\}_{n \in \mathbb{N}}$ of promise problems, each of which $(L_n^{(+)}, L_n^{(-)})$ indexed by n is a promise problem over Σ. The set $L_n^{(+)} \cup L_n^{(-)}$ of all promised instances is succinctly denoted by $\Sigma^{(n)}$. Notice that instances in $\Sigma^{(n)}$ may not be limited to length-n strings. It is important to note that the underlying alphabet Σ is fixed for all promise problems $(L_n^{(+)}, L_n^{(-)})$ in $\mathcal{L}$ in accordance with the setting of [23–27].

---

[2] A *write-once tape* means that its tape head always moves to the next blank cell whenever the tape head writes a non-blank symbol.

The *complement* of $\mathcal{L}$ is $\{(L_n^{(-)}, L_n^{(+)})\}_{n \in \mathbb{N}}$ and is denoted co-$\mathcal{L}$. Given two families $\mathcal{L} = \{(L_n^{(+)}, L_n^{(-)})\}_{n \in \mathbb{N}}$ and $\mathcal{K} = \{(K_n^{(+)}, K_n^{(-)})\}_{n \in \mathbb{N}}$, we define the *intersection* $\mathcal{L} \cap \mathcal{K}$ to be $\{(L_n^{(+)} \cap K_n^{(+)}, L_n^{(-)} \cup K_n^{(-)})\}_{n \in \mathbb{N}}$ and the *union* $\mathcal{L} \cup \mathcal{K}$ to be $\{(L_n^{(+)} \cup K_n^{(+)}, L_n^{(-)} \cap K_n^{(-)})\}_{n \in \mathbb{N}}$.

To solve (or recognize) $\mathcal{L}$, we use a nonuniform family $\mathcal{M} = \{M_n\}_{n \in \mathbb{N}}$ of finite automata indexed by natural numbers. Formally, we say that $\mathcal{M}$ *solves* (or *recognizes*) $\mathcal{L}$ if, for any index $n \in \mathbb{N}$, M_n solves $(L_n^{(+)}, L_n^{(-)})$. We say that $\mathcal{M}$ is of *polynomial size* if there is a polynomial p satisfying $sc(M_n) \leq p(n)$ for all $n \in \mathbb{N}$. The complexity class 1D (resp., 1N) consists of all families of promise problems solvable by nonuniform families of polynomial-size 1dfa's (resp., 1nfa's). Similarly, we define the complexity classes 1DPD and 1NPD using 1dpda's and 1npda's, respectively [25].

We further consider nonuniform families of partial functions. Let f denote a partial function from Σ^* to Γ^* for two alphabets Σ and Γ, where "partial" means that f is treated as "defined" only on its domain D, which is a subset of Σ^*, and f is treated as "undefined" on $\Sigma^* - D$. To clarify the domain of f, we intend to write (f, D) instead of f. Given two alphabets Σ and Γ, we write $\mathcal{F} = \{(f_n, D_n)\}_{n \in \mathbb{N}}$ for a family of partial functions mapping Σ^* to Γ^*.

A family $\{M_n\}_{n \in \mathbb{N}}$ of 1dft's is said to *compute* $\mathcal{F}$ if, for any $n \in \mathbb{N}$ and for all $x \in D_n$, M_n begins with the input x and produces $f_n(x)$ on its write-once output tape. Remark that, even if $x \notin D_n$, M_n may possibly write a certain string on the output tape. The notation 1F denotes the collection of all families $\{(f_n, D_n)\}_{n \in \mathbb{N}}$ of partial functions computed by appropriate families $\{M_n\}_{n \in \mathbb{N}}$ of polynomial-size 1dft's.

3 Introduction of Counting Functions and Counting Complexity Classes

Let us explain numerous nonuniform state complexity classes associated with "counting" used in the rest of this exposition.

In a similar way of defining #P and GapP by polynomial-time nondeterministic Turing machines (NTMs), we introduce two counting-function classes 1# and 1Gap using nondeterministic finite automata families as follows. The former class 1# (pronounced "one sharp" or "one pound") is the collection of all families $\{(f_n, D_n)\}_{n \in \mathbb{N}}$ of partial functions such that there exists a nonuniform family $\{M_n\}_{n \in \mathbb{N}}$ of polynomial-size 1nfa's satisfying $f_n(x) = \#M_n(x)$ for all indices $n \in \mathbb{N}$ and all strings $x \in D_n$. The latter class 1Gap is composed of all families $\{(f_n, D_n)\}_{n \in \mathbb{N}}$ of partial functions of the form $f_n(x) = \#M_n(x) - \#\overline{M}_n(x)$ for all $x \in D_n$ for a certain nonuniform family $\{M_n\}_{n \in \mathbb{N}}$ of polynomial-size 1nfa's. Additionally, $1\mathrm{Gap}_{\geq 0}$ denotes the class of all partial functions in 1Gap that take only nonnegative integer values unless their outputs are undefined.

As a concrete example of counting function families, let us recall $\Sigma_2^{(n)}$ and define $f_n(x) = |\{e \in \mathbb{N}^+ \mid (r_1)_{(e)} \in Set(r_2)\}|$, where A_n is defined in Sect. 2.1. It is not difficult to show that the function family $\{(f_n, \Sigma_2^{(n)})\}_{n \in \mathbb{N}}$ belongs to 1#.

In Sect. 2.3, we have defined 1F as the class consisting of all families of partial functions, mapping Σ^* to Γ^* for arbitrary alphabets Σ and Γ, which are computed by nonuniform families of polynomial-size 1dft's. In accordance to 1# and 1Gap, using the translation between integers and binary strings described in Sect. 2.1, we treat some of these functions as functions mapping Σ^* to $\mathbb{Z}$.

Given a partial function $f : \Sigma^* \to \mathbb{Z}$, we define $f^{(trans)}$ by setting $f^{(trans)}(x) = trans(f(x))$ whenever $f(x)$ is defined. We write $1F_{\mathbb{Z}}$ (resp., $1F_{\mathbb{N}}$) to indicate the class composed of all families $\{(f_n, D_n)\}_{n\in\mathbb{N}}$ of partial functions mapping Σ^* to $\mathbb{Z}$ (resp., to $\mathbb{N}$) for arbitrary alphabets Σ satisfying that $\{(f^{(trans)}, D_n)\}_{n\in\mathbb{N}}$ belongs to 1F. Here, we claim basic relationships among the aforementioned function families.

Lemma 2. $1F_{\mathbb{N}} \subsetneqq 1\# \subsetneqq 1Gap_{\geq 0}$, $1F_{\mathbb{Z}} \not\subseteq 1Gap_{\geq 0}$, and $1F_{\mathbb{Z}} \subsetneqq 1Gap$.

Following [4], we define $\mathcal{U} - \mathcal{V}$ for any two function classes $\mathcal{U}$ and $\mathcal{V}$ as the collection of families $\{(f_n, D_n)\}_{n\in\mathbb{N}}$ such that there are two families $\mathcal{G} = \{(g_n, D_n)\}_{n\in\mathbb{N}} \in \mathcal{U}$ and $\mathcal{G}' = \{(g'_n, D'_n)\}_{n\in\mathbb{N}} \in \mathcal{V}$ for which $D_n = D'_n$ and $f_n = g_n - g'_n$ (i.e., $f_n(x) = g_n(x) - g'_n(x)$ for all $x \in D_n$) for any index $n \in \mathbb{N}$.

Lemma 3. $1Gap = 1\# - 1\# = 1\# - 1F_{\mathbb{N}} = 1F_{\mathbb{N}} - 1\#$.

In this exposition, we define counting classes 1U, $1\oplus$, $1C_=$, 1SP, and 1P in terms of function families in 1# and 1Gap although 1P is originally defined in terms of unbounded-error probabilistic finite automata in, e.g., [26]. We first define the parity class $1\oplus$ (pronounced "one parity") as the collection of all families $\{(L_n^{(+)}, L_n^{(-)})\}_{n\in\mathbb{N}}$ of promise problems such that there exists a function family $\{(f_n, D_n)\}_{n\in\mathbb{N}}$ in 1# satisfying that, for any index $n \in \mathbb{N}$, (1) $L_n^{(+)} \cup L_n^{(-)} \subseteq D_n$ and (2) $f_n(x)$ is odd for all $x \in L_n^{(+)}$, and $f_n(x)$ is even for all $x \in L_n^{(-)}$. In a similar way, the unambiguous class 1U is obtained by replacing (2) with the following condition : $f_n(x) = 1$ for all $x \in L_n^{(+)}$ and $f_n(x) = 0$ for all $x \in L_n^{(-)}$. We remark that the aforementioned definition of 1N can be rephrased in terms of function families in 1# by requiring the following condition: $f_n(x) > 0$ for all $x \in L_n^{(+)}$ and $f_n(x) = 0$ for all $x \in L_n^{(-)}$.

The exact counting class $1C_=$ (pronounced "one C equal") is defined as the collection of all families $\{(L_n^{(+)}, L_n^{(-)})\}_{n\in\mathbb{N}}$ of promise problems such that there exists a function family $\{(f_n, D_n)\}_{n\in\mathbb{N}}$ in 1Gap satisfying the following: for any index $n \in \mathbb{N}$, (1') $L_n^{(+)} \cup L_n^{(-)} \subseteq D_n$ and (2') $f_n(x) = 0$ for all $x \in L_n^{(+)}$, and $f_n(x) \neq 0$ for all $x \in L_n^{(-)}$. In contrast, the stoic probabilistic class 1SP refers to the collection of all families $\{(L_n^{(+)}, L_n^{(-)})\}_{n\in\mathbb{N}}$ defined by replacing (2') with the following condition: $f_n(x) = 1$ for all $x \in L_n^{(+)}$, and $f_n(x) = 0$ for all $x \in L_n^{(-)}$. Finally, the bounded-error probabilistic class 1P is defined in a similar way but with the following condition: $f_n(x) > 0$ for all $x \in L_n^{(+)}$ and $f_n(x) \leq 0$ for all $x \in L_n^{(-)}$.

The following property of 1# is useful.

Lemma 4. *For any 1nfa M with a set Q_M of inner states, there exists another 1nfa N with a set Q_N of inner states such that, for any x, if $\#M(x) = \#\overline{M}(x)$, then $\#N(x) = \#\overline{N}(x)$, and if $\#M(x) \neq \#\overline{M}(x)$, then $\#N(x) < \#\overline{N}(x)$. Moreover, $|Q_N| \leq 2|Q_M|$ holds.*

The complexity classes 1C$_=$ and 1SP are closely related to co-1N and 1U, respectively, because the latter two classes are obtained directly from the above definitions of the former ones by replacing 1Gap in their definitions with 1#.

4 Relationships Among Counting Complexity Classes

Let us discuss relationships among various counting complexity classes introduced in Sect. 3. In particular, we focus our attention on containments and separations of these classes, which are illustrated in Fig. 1.

In the polynomial-time setting, containment/separation relationships among UP, NP, SPP, ⊕P, C$_=$P, and PP are not yet known except for trivial ones. On the contrary, it is possible in our setting to completely determine the relationships among 1U, 1N, 1SP, 1⊕, 1C$_=$, and 1P.

4.1 Basic Closure Properties

We briefly discuss basic closure properties of 1⊕, 1SP, and 1C$_=$. We begin with the closure under complementation. It was shown that co-1P coincides with 1P [26]. A similar closure property holds for 1⊕ and 1SP.

Lemma 5. $1⊕ =$ co-$1⊕$ *and* $1SP =$ co-$1SP$.

It is known that 1N is closed under intersection and union but not under complementation [6,26]. A similar argument used for this fact shows the following.

Lemma 6. $1C_=$ *is closed under intersection and* co-$1C_=$ *is closed under union.*

Given two alphabets Σ and Γ, a homomorphism $h : \Sigma \to \Gamma^*$ is said to be *non-erasing* if $h(\sigma) \neq \lambda$ for all $\sigma \in \Sigma$, and h is called *prefix-free* if there is no pair $\sigma, \tau \in \Sigma$ such that $h(\sigma)$ is a proper prefix of $h(\tau)$. As usual, h is naturally expanded to a map from Σ^* to Γ^*. We write $h^{-1}(y)$ for the set $\{x \in \Sigma^* \mid h(x) = y\}$ for any string $y \in \Gamma^*$. These homomorphisms and also inverse homomorphisms can be applied to partial functions as well.

Lemma 7. *1# is closed under homomorphism and under inverse of non-erasing prefix-free homomorphism.*

4.2 Complexity Class 1U

We next discuss containments and separations concerning 1U. It was shown in [27] that $1U \neq$ co-$1U$ (equivalently, co-$1U \not\subseteq 1U$). We first remark that $1U \subseteq 1SP$ and co-$1U \subseteq 1SP$. It is possible to show that these inclusions are, in fact, proper. Let $T_n = \{x\#y \mid x, y \in \{0,1\}^{2^n}\}$ for each index $n \in \mathbb{N}$

Proposition 8. $1U \subsetneq 1SP$.

Proof Sketch. For the proof of $1U \subseteq 1SP$, let $\mathcal{L} = \{(L_n^{(+)}, L_n^{(-)})\}_{n\in\mathbb{N}}$ be any family in $1U$ and take a family $\mathcal{M} = \{M_n\}_{n\in\mathbb{N}}$ of polynomial-size 1nfa's solving $\mathcal{L}$, where each M_n is unambiguous at least on $\Sigma^{(n)}$ $(= L_n^{(+)} \cup L_n^{(-)})$. The following 1nfa N_n solves $\mathcal{L}$. Simulate M_n on x. If M_n enters an accepting state, then N_n does the same. If M_n enters a rejecting state, then N_n branches into two computation paths and accepts on one path and rejects on the other. For the proof of $1U \neq 1SP$, we define $L_n^{(+)} = \{x\#y \in T_n \mid \#_0(x) = \#_0(y) + 1\}$, $L_n^{(-)} = \{x\#y \in T_n \mid \#_0(x) = \#_0(y)\}$, and $\Sigma^{(n)} = L_n^{(+)} \cup L_n^{(-)}$. We then set $\mathcal{L}_{sp} = \{(L_n^{(+)}, L_n^{(-)})\}_{n\in\mathbb{N}}$. With the use of this family, it is possible to prove that $\mathcal{L}_{sp} \in 1SP$ and $\mathcal{L}_{sp} \notin 1U$. $\qquad\square$

We next extend the aforementioned class separation of co-$1U \not\subseteq 1U$ to the following class separation.

Proposition 9. co-$1U \not\subseteq 1N$.

4.3 Complexity Class $1C_=$

We continue discussing containment/separation relationships concerning $1C_=$. We remark a close connection between nonuniform polynomial state complexity classes in this exposition and one-tape linear-time Turing machines with linear-size advice discussed in [14,20]. This close connection helps us adopt two key lemmas of [20] into our setting and exploit them to prove the class separations between $1N$ and $1C_=$, as shown below.

Proposition 10. *1.* $1N \subsetneq$ co-$1C_=$ *and* co-$1N \subsetneq 1C_=$.
2. $1C_= \not\subseteq 1N$ *and* $1N \not\subseteq 1C_=$.

As an immediate consequence of this proposition, we obtain the following non-closure property of $1C_=$.

Corollary 11. $1C_=$ *is not closed under complementation.*

Proof. If $1C_= =$ co-$1C_=$, then $1N \subseteq 1C_=$ follows because $1N \subseteq$ co-$1C_=$. However, Proposition 10(2) shows that $1N \not\subseteq 1C_=$. $\qquad\square$

In what follows, we intend to prove Proposition 10. The desired proof requires an idea from [20]. More specifically, in order to prove that $1\text{-}C_=\text{LIN}/lin$ is not closed under complementation, a useful, characteristic property of $1\text{-}C_=\text{LIN}/lin$ was presented in [20, Lemma 4.3]. As noted in Sect. 1.2, there is a close connection between $1\text{-}C_=\text{LIN}/lin$ and $1C_=$. We can exploit this connection to adapt the above property of $1\text{-}C_=\text{LIN}/lin$ into the setting of $1C_=$ and to achieve the desired separation results.

Lemma 12. *Let $\{(L_n^{(+)}, L_n^{(-)})\}_{n \in \mathbb{N}}$ denote any family in $1C_=$ over alphabet Σ. There exists a polynomial p that satisfies the following statement. Let n and l be any two numbers in $\mathbb{N}$ with $l \leq n - 1$, let $z \in \Sigma^l$, and let $A_{n,l,z} = \{x \in \Sigma^{n-l} \mid xz \in L_n^{(+)}\}$. There exists a subset $S \subseteq A_{n,l,z}$ with $|S| \leq p(n)$ such that, for any $y \in \Sigma^l$, if $\{wy \mid w \in S\} \subseteq L_n^{(+)}$, then $\{xy \mid x \in A_{n,l,z}\} \subseteq L_n^{(+)}$.*

With the use of Lemma 12, we can prove Proposition 10.

Proof Sketch of Proposition 10. Here, we prove only the second statement of (2). An example family $\mathcal{L}_N = \{(L_n^{(+)}, L_n^{(-)})\}_{n \in \mathbb{N}}$ is defined as follows. We first write I_n for the set $\{[[i_1, i_2, \ldots, i_n]]_{n^2} \mid i_1, i_2, \ldots, i_n \in [0, n^2]_{\mathbb{Z}}\}$. For any string $z \in I_n$, $|z| = n(n^2 + 1)$ follows. The desired sets $L_n^{(+)}$ and $L_n^{(-)}$ are respectively defined by $L_n^{(+)} = \{u\#v \mid u, v \in I_n, Set(u) \neq Set(v)\}$ and $L_n^{(-)} = \{u\#v \mid u, v \in I_n, Set(u) = Set(v)\}$. It is not difficult to show that $\mathcal{L}_N \in 1N$.

Next, we wish to prove that $\mathcal{L}_N \notin 1C_=$. Assuming for contradiction that $\mathcal{L}_N \in 1C_=$, we intend to apply Lemma 12 to $\mathcal{L}_N$. Take a polynomial p that satisfies the lemma. Let $m = 2n(n^2) + 1$ and $l = n(n^2 + 1)$. Take $z = [[1, 2, \ldots, n]]_{n^2}$ in I_n. Let us consider the set $A_{m,l,z}$, which in fact equals $\{x\# \in \Sigma^{m-l} \mid x\#z \in L_n^{(+)}\}$. There exists a set $S \subseteq A_{m,l,z}$ with $|S| \leq p(m)$ satisfying the lemma. For convenience, we introduce $P_n = \{[[i_1, i_2, \ldots, i_n]]_{n^2} \mid (i_1, i_2, \ldots, i_n)$ is a permutation of $(1, 2, \ldots, n)\}$. Choose a string y in $I_n - \{u \mid u \in P_n$ or $\exists w\# \in S[Set(w) = Set(u)]\}$. Clearly, the string $y\#$ belongs to $A_{m,l,z}$. It also follows by the definition of y that $\{w\#y \mid w\# \in S\} \subseteq L_n^{(+)}$. The lemma then concludes that $\{x\#y \mid x\# \in A_{m,l,z}\} \subseteq L_n^{(+)}$. Since $y\# \in A_{m,l,z}$, we obtain $y\#y \in L_n^{(+)}$, a contradiction. As a consequence, $\mathcal{L}_N \notin 1C_=$ follows. $\square$

As another consequence of Proposition 10, we obtain the following.

Proposition 13. $1C_= \subsetneq 1P$.

Proposition 14. $1SP \subsetneq co\text{-}1C_=$ and $1SP \subsetneq 1C_=$.

4.4   Complexity Class 1P

We turn our attention to another complexity class 1P. In a demonstration of the power of CFL over 1-PLIN/*lin*, a useful property of 1-PLIN/*lin* was presented in [20, Lemma 4.7]. By exploiting a close connection between 1-PLIN/*lin* and 1P, we can show a similar property in our setting.

Lemma 15. *Let $\{(L_n^{(+)}, L_n^{(-)})\}_{n \in \mathbb{N}}$ be any family in 1P. There exists a polynomial p that satisfies the following statement. Let n and l be any numbers in $\mathbb{N}$ with $l \leq n - 1$. There exists a set $S = \{w_1, w_2, \ldots, w_{p(n)}\} \subseteq \Sigma^{n-l}$ with $|S| = p(n)$ for which the following implication holds: for any subset $R \subseteq \Sigma^l$, if $|\{a_1^{(y)} a_2^{(y)} \cdots a_{p(n)}^{(y)} \mid y \in R\}| \geq 2^{p(n)}$, where $a_i^{(y)} = 1$ if $w_i y \in L_n^{(+)}$, $a_i^{(y)} = 0$ if $w_i y \in L_n^{(-)}$, and $a_i^{(y)}$ is undefined otherwise, then it follows that, for any $x \in \Sigma^{n-l}$, there exists a pair $y, y' \in R$ satisfying that $xy \in L_n^{(+)}$ and $xy' \in L_n^{(-)}$.*

We use Lemma 15 to prove the following class separation.

Proposition 16. $1\oplus \nsubseteq 1P$.

Proof. For two binary strings x, y with $|x| = |y|$, the *bitwise inner product* $x \odot y$ of x and y is defined to be $\sum_{i=1}^{n} x_i y_i$, where $x = x_1 x_2 \cdots x_n$ and $y = y_1 y_2 \cdots y_n$.

Toward a contradiction, we assume that $1\oplus \subseteq 1P$. We then take an example family $\mathcal{L}_{\oplus} = \{(L_n^{(+)}, L_n^{(-)})\}_{n \in \mathbb{N}}$ defined as follows. Fix n arbitrarily. We first define J_n to be $\{u_1 \$ u_2 \$ \cdots \$ u_n \mid \forall i \in [n][u_i \in \{0,1\}^{\lfloor \log n \rfloor}]\}$, where $\$$ is a special separator. Using this set J_n, we define $L_n^{(+)} = \{u\#v \mid u, v \in J_n, \sum_{i=1}^{n} u_i \odot v_i \equiv 1 \pmod{2}\}$, where $u = u_1 \$ u_2 \$ \cdots \$ u_n$ and $v = v_1 \$ v_2 \$ \cdots \$ v_n$, and $L_n^{(-)} = \{u\#v \mid u, v \in J_n\} - L_n^{(+)}$. Consider the following 1nfa N_n. On input $u\#v$, nondeterministically choose a number $i \in [n]$, read u_i in u and remember it in the form of inner states. After passing $\#$, read v_i in v and calculate $a_i = u_i \odot v_i \pmod{2}$. If $a_i = 1$, then accept the input, and otherwise, reject the input. The definition of N_n places $\mathcal{L}_{\oplus}$ in $1\oplus$. Our assumption then implies $\mathcal{L}_{\oplus} \in 1P$.

Apply Lemma 15 and take a polynomial p provided by the lemma. By taking a sufficiently large $n \in \mathbb{N}$ satisfying $p(n) < 2^{n \log n}$, we set $\hat{n} = 2n\lfloor \log n \rfloor$ and $l = \hat{n}/2$. There is a subset $S = \{w_1, w_2, \ldots, w_{p(n)}\}$ of $\{0,1\}^{l}$ that satisfies the lemma. For each string $r \in \Sigma^{p(n)}$ of the form $r_1 r_2 \cdots r_{p(n)}$ with $r_i \in \{0,1\}$ for all $i \in [p(n)]$, we choose a string $y_r \in \Sigma^{n-l}$ such that, for any $i \in [p(n)]$, $w_i \odot y_r \equiv r_i \pmod{2}$ holds. We then define the set Y_n to be $\{y_r \mid r \in \{0,1\}^{p(n)}\}$. We also define $a_y = a_1 a_2 \cdots a_{p(n)}$ with $w_i \odot y \equiv a_i \pmod{2}$ for any $i \in [p(n)]$. It then follows that there exists a string $x \in \{0,1\}^{l}$ satisfying $x \odot y_r \equiv 0 \pmod{2}$ for all $r \in \{0,1\}^{p(n)}$. Notice that $|\{a_y \mid y \in Y_n\}| \geq 2^{p(n)}$. By Lemma 15, there is a pair $y, y' \in Y_n$ such that $xy \in L_n^{(+)}$ and $xy' \in L_n^{(-)}$. This consequence is in contradiction to the choice of x.   □

From Proposition 16, we obtain a class separation between 1SP and $1\oplus$.

Proposition 17. $1SP \subsetneq 1\oplus$.

5   Relations to Nonuniform Families of Pushdown Automata

The complexity classes 1DPD and 1NPD are induced naturally from nonuniform families of polynomial-size pushdown automata that run in polynomial time. In particular, their two-way head models, 2DPD and 2NPD, are closely related to LOGDCFL/poly and LOGCFL/poly [25]. For the one-way head models, in contrast, it is proven in [25] that $1N \nsubseteq 1DPD$ and $1DPD \nsubseteq 1N$. Nonetheless, we can strengthen the former separation to $1U \nsubseteq 1DPD$ and the latter one to $1DPD \nsubseteq 1P$.

Proposition 18. *(1)* $1U \nsubseteq 1DPD$. *(2)* $1DPD \nsubseteq 1P$

Proof Sketch. Here, we prove only (1). Since 1DPD = co-1DPD [25], 1U $\subseteq$ 1DPD is equivalent to co-1U $\subseteq$ 1DPD. Let us consider the family $\mathcal{L}_U = \{(L_n^{(+)}, L_n^{(-)})\}_{n\in\mathbb{N}}$ of promise problems with $L_n^{(+)} = \{u\#v \mid u,v \in A_n(n), \exists!e \in [n]((u)_{(e)} \neq (v)_{(e)})\}$ and $L_n^{(-)} = \{u\#v \mid u,v \in A_n(n), \forall e \in [n]((u)_{(e)} = (v)_{(e)})\}$. Notice that $|u\#v| = 2k(n+1)+1$ if $|MSet(u)| = |MSet(v)| = k$. Clearly, co-$\mathcal{L}_U$ falls into co-1U. We next want to verify that co-$\mathcal{L}_U \notin$ 1DPD. Toward a contradiction, we assume that co-$\mathcal{L}_U \in$ 1DPD. Take a nonuniform family $\{M_n\}_{n\in\mathbb{N}}$ of polynomial-size 1dpda's that solves co-$\mathcal{L}_U$. Let Q_n denote a set of inner states of M_n. It is possible to assume that M_n has only one accepting state, say, $q_{acc,n}$ and that M_n always empties its stack at the end of computation.

A *configuration* of M_n is a triplet $(q, \sigma w, \gamma)$, which indicates that M_n is in inner state q, a tape head is scanning σ, σw is a suffix of an input with the endmarkers, and γ is the current stack content. The notation $\vdash$ refers to a transition between two configurations in a single step and $\vdash^*$ is the transitive closure of $\vdash$. We define S_n to be the collection of all quintuples (x, q_1, y, q_2, z) satisfying that (i) $xyz \in L_n^{(+)}$ with $y \neq \lambda$ and (ii) $(q_0, \triangleright xyz\triangleleft, \bot) \vdash^* (q_1, yz\triangleleft, \gamma) \vdash^* (q_2, z\triangleleft, \gamma) \vdash^* (a_{acc,n}, \lambda, \bot)$, provided that M_n's stack height does not go below $|\gamma|$ while reading y. We then define $B_{q_1,q_2,\ell_1,\ell_2}$ to be the set of all such tuples (x, q_1, y, q_2, z) in S_n for which $|x| = \ell_1$ and $|y| = \ell_2$, where $q_1, q_2 \in Q_n$ and $\ell_1\ell_2 \in [0, 2n(n+1)+2]_{\mathbb{Z}}$.

By an argument similar to the proof of [25, Proposition 5.2], the following claim can be proven.

Claim. There exists a quadruple $(q_1, q_2, \ell_1, \ell_2)$ satisfying $|B_{q_1,q_2,\ell_1,\ell_2}| \geq 2$.

Take a quadruple $(q_1, q_2, \ell_1, \ell_2)$ that satisfies the claim. Choose two distinct elements (x, q_1, y_1, q_2, z_1) and $(x_2, q_1, y_2, q_2, z_2)$ from $B_{q_1,q_2,\ell_1,\ell_2}$. This implies that $x_1 y_1 z_1$ and $x_2 y_2 z_2$ are accepted by M_n. Note that $(q_1, y_1 z_1\triangleleft, \gamma_1) \vdash^* (q_2, z_1\triangleleft, \gamma_1)$ and $(q_1, y_2 z_2\triangleleft, \gamma_2) \vdash^* (q_2, z_2\triangleleft, \gamma_2)$ are interchangeable in the corresponding computations of M_n on $x_1 y_1 z_1$ and $x_2 y_2 z_2$. Hence, $x_2 y_1 z_2$ is also accepted by M_n. Since y_1 and y_2 are nonempty and distinct, $x_2 y_1 z_2$ and $x_2 y_1 z_2$ must satisfy the condition of $L_n^{(-)}$. This is absurd because of the definition of $L_n^{(-)}$. Therefore, we conclude that co-$\mathcal{L}_U \notin$ 1DPD. $\qquad\square$

Lemma 19. 1N $\subsetneq$ 1NPD.

Counting is an important research subject in theoretical computer science. We have initiated a study on "counting" within the framework of nonuniform models of polynomial-size finite automata families. We would like to see further progress on this study for the better understandings of the nature of counting in low complexity classes.

References

1. Allender, E., Ogihara, M.: Relationships among PL, #L, and the determinant. Informatique théorique et Applications **30**, 1–21 (1996)

2. Àlvarez, C., Jenner, B.: A very hard log-space counting class. Theoret. Comput. Sci. **107**, 3–30 (1993)

3. Berman, P., Lingas, A.: On complexity of regular languages in terms of finite automata. Report 304, Institute of Computer Science, Polish Academy of Science, Warsaw (1977)

4. Fenner, S., Fortnow, L., Kurtz, S.: Gap-definable counting classes. J. Comput. System Sci. **48**, 116–148 (1994)

5. Kapoutsis, C.A.: Size complexity of two-way finite automata. In: Diekert, V., Nowotka, D. (eds.) DLT 2009. LNCS, vol. 5583, pp. 47–66. Springer, Heidelberg (2009). https://doi.org/10.1007/978-3-642-02737-6_4

6. Kapoutsis, C.A.: Minicomplexity. J. Automat. Lang. Combin. **17**, 205–224 (2012)

7. Kapoutsis, C.A.: Two-way automata versus logarithmic space. Theory Comput. Syst. **55**, 421–447 (2014)

8. Kapoutsis, C.A., Pighizzini, G.: Two-way automata characterizations of L/poly versus NL. Theory Comput. Syst. **56**, 662–685 (2015)

9. Niedermeier, R., Rossmanith, P.: Unambiguous auxiliary pushdown automata and semi-unbounded fan-in circuits. Inf. Comput. **118**, 227–245 (1995)

10. Ogiwara, M., Hemachandra, L.: A complexity theory for feasible closure properties. J. Comput. System Sci. **46**, 295–325 (1993)

11. Reinhardt, K.: Counting and empty alternating pushdown automata. In: The Proceedings of the 7th IMYCS, pp. 198–207 (1992)

12. Ruzzo, W., Simon, J., Tompa, M.: Space-bounded hierarchies and probabilistic computation. J. Comput. System Sci. **28**, 216–230 (1984)

13. Sakoda, W.J., Sipser, M.: Nondeterminism and the size of two-way finite automata. In: The Proceedings of STOC 1978, pp. 275–286 (1978)

14. Tadaki, K., Yamakami, T., Lin, J.C.H.: Theory of one-tape linear-time Turing machines. Theor. Comput. Sci. **411**, 22–43 (2010)

15. Toda, S.: PP is as hard as the polynomial-time hierarchy. SIAM J. Comput. **20**, 865–877 (1991)

16. Valiant, L.G.: Relative complexity of checking and evaluating. Inform. Process. Lett. **5**, 20–23 (1975)

17. Valiant, L.G.: The complexity of enumeration and reliability problems. SIAM J. Comput. **8**, 410–421 (1979)

18. Vinay, V.: Counting auxiliary pushdown automata and semi-unbounded arithmetic circuits. In: Conference on Structure in Complexity Theory, pp. 270–285 (1991)

19. Yamakami, T.: Swapping lemmas for regular and context-free languages. Manuscript. (2008). arXiv:0808.4122

20. Yamakami, T.: The roles of advice to one-tape linear-time Turing machines and finite automata. Int. J. Found. Comput. Sci. **21**, 941–962 (2010)

21. Yamakami, T.: Immunity and pseudorandomness of context-free languages. Theor. Comput. Sci. **412**, 6432–6450 (2011)

22. Yamakami, T.: Pseudorandom generators against advised context-free languages. Theor. Comput. Sci. **613**, 1–27 (2016)

23. Yamakami, T.: State complexity characterizations of parameterized degree-bounded graph connectivity, sub-linear space computation, and the linear space hypothesis. Theor. Comput. Sci. **798**, 2–22 (2019)

24. Yamakami, T.: Relativizations of nonuniform quantum finite automata families. In: McQuillan, I., Seki, S. (eds.) UCNC 2019. LNCS, vol. 11493, pp. 257–271. Springer, Cham (2019). https://doi.org/10.1007/978-3-030-19311-9_20

25. Yamakami, T.: Parameterizations of logarithmic-space reductions, stack-state complexity of nonuniform families of pushdown automata, and a road to the LOGCFL⊆LOGDCFL/poly question (2021). arXiv:2108.12779
26. Yamakami, T.: Nonuniform families of polynomial-size quantum finite automata and quantum logarithmic-space computation with polynomial-size advice. Inform. Comput. **286**, article 104783 (2022)
27. Yamakami, T.: Unambiguity and fewness for nonuniform families of polynomial-size nondeterministic finite automata. In: The Proceedings of RP 2022. LNCS, vol. 13608, pp. 77–92. Springer, Cham (2022). https://doi.org/10.1007/978-3-031-19135-0_6

Correction to: Shortest Dominating Set Reconfiguration Under Token Sliding

Jan Matyáš Křišťan [iD] and Jakub Svoboda [iD]

Correction to:
Chapter "24" in: H. Fernau and K. Jansen (Eds.):
Fundamentals of Computation Theory, **LNCS 14292,**
https://doi.org/10.1007/978-3-031-43587-4_24

In the original version of this paper the funding information was missing. This was corrected.

The updated version of this chapter can be found at
https://doi.org/10.1007/978-3-031-43587-4_24

Author Index

Printed in the United States
by Baker & Taylor Publisher Services